THERMOPHYSICS OF ATMOSPHERIC ENTRY

Edited by
T. E. Horton
Department of Mechanical Engineering
The University of Mississippi, University, Mississippi

Volume 82
PROGRESS IN
ASTRONAUTICS AND AERONAUTICS

Martin Summerfield, Series Editor-in-Chief
Princeton Combustion Research Laboratories, Inc.
Princeton, New Jersey

Technical papers from the AIAA 19th Aerospace Sciences Meeting, January 1981, and the AIAA 16th Thermophysics Conference, June 1981, and subsequently revised for this volume.

Published by the American Institute of Aeronautics and Astronautics, Inc.
1290 Avenue of the Americas, New York, N.Y. 10104.

American Institute of Aeronautics and Astronautics, Inc.
New York, New York

Library of Congress Cataloging in Publication Data
Main entry under title:

Thermophysics of atmospheric entry.

(Progress in astronautics and aeronautics; v. 82)
Technical papers from the AIAA 19th Aerospace Sciences Meeting, January 1981, and the AIAA 16th Thermophysics Conference, June 1981.
Includes index.
1. Space vehicles—Atmospheric entry—Congresses. 2. Space vehicles—Thermodynamics—Congresses. I. Horton, T.E. (Thomas E.) II. American Institute of Aeronautics and Astronautics. III. AIAA Aerospace Sciences Meeting (19th: 1981: St. Louis, Mo.) IV. AIAA Thermophysics Conference (16th: 1981: Palo Alto, Calif.) V. Series.
TL507.P75 vol. 82 [TL1060] 629.ls 82-6686
ISBN 0-915928-66-3 [629.47'152] AACR2

Table of Contents

Preface

Thermophysics represents a harmonious blend of the classical engineering sciences of materials, thermofluids, heat transfer, and electromagnetic theory with the microsciences of solid state, physical optics, and atomic and molecular dynamics. The impetus for the formation of a thermophysics community during the predawn of the "space age" was the need for a science/technology base which could cope with the thermal management problems encountered in the early satellites and in ballistic re-entry. During the past two decades the thermophysics community has met ever-increasing mission requirements for more effective space systems, as well as the demands of transfer of these technologies to terrestrial energy problems. Today and in the near future we see a continuation of the challenges in the thermophysics field presented by entry systems, spacecraft thermal control, and laser technology.

This volume is devoted to the science and technology of atmosphere entry systems. From the perspective of current major project activity, this field can be divided into three areas. First is the area of strategic systems, which is concerned with the refinement and improvement of ballistic entry systems. Second is the area of manned re-entry systems, which is currently centered on the Space Shuttle Orbiter. The third project area is the Galileo Probe of the Jovian atmosphere. All of these areas can be viewed as depending upon the answers to a series of common questions—the questions being the basis of the science of thermophysics. Although the questions are common, the answers are not redundant as each area of application represents a different range of parameters and thus different dominant phenomena.

The volume presents a view of timely advances in atmospheric entry thermophysics, which was drawn from over 160 papers which were contributed to thermophysics sessions at the AIAA 19th Aerospace Sciences Meeting in St. Louis, Missouri in January 1981, and the AIAA 16th Thermophysics Conference in Palo Alto, California in June 1981. These papers have been revised, updated, and organized into five coherent chapters which treat thermophysical properties, aerothermodynamics, Space Shuttle studies, Galileo studies, and future planetary missions.

The first chapter deals with the characterization of the transport properties, both kinetic and radiative, of the high-temperature species found in the shock layer of ablative entry bodies. Although this material transcends specific applications, the questions addressed by these authors will be of immediate value in the design of ablating probes for use in entry missions to the outer planets. The kinetic transport properties of gaseous mixtures are determined by binary collision integrals, which are functions of the interaction potentials of the collision partners. In the first paper of this chapter, *Rainwater, Holland,* and *Biolsi* discuss revised integrals based upon spectroscopic data and Hulburt-Hirschfelder interaction potentials. These revisions have been incorporated into the computed transport properties of carbonaceous ablation products over an extensive range of temperatures by *Biolsi, Fenton,* and *Owenson.* Further tabulations of computed viscosity, conductivity, and diffusion coefficients for high-temperature species composed of nitrogen, carbon, and hydrogen are presented in the third paper by *Flori* and *Biolsi.* Calculations of transport properties of the complex high-temperature mixture typical of a shock layer requires extensive computational time and storage capacity—items which are in short supply when performing a flowfield determination. Thus, accurate procedures for approximating transport properties are a necessity. *Armaly* and *Sutton* had previously presented an effective approximation for the viscosity of a mixture, and in the fourth paper they present a companion approximation for the translational thermal conductivity of high-temperature ionized mixtures. In the final paper, *Shinn* reports on absorption spectroscopy experiments in a shock tube, confirming the oscillator strength of the uv absorption band of C_3, an ablation layer species which blocks shock-layer radiation in outer planetary entry.

Chapter II treats the assorted hypersonic gasdynamic problems which comprise the field of aerothermodynamics. The range of parameters addressed in this chapter correspond to those encountered in ballistic re-entry. In the first paper, *Goulard* explores nondimensional parameters, which may prove to be of value in correlating both radiation cooling of the shock layer and radiation blockage of the ablative layer which can serve as the basis of simple engineering models for estimating stagnation-point radiative heating for the severe environment of outer planetary entry missions. Correlations of convective surface parameters for spherically blunted cones are presented by *Griffith, Majors,* and *Adams.* These correlations for zero angle of attack resulted from computer experiments in which cone geometry, freestream Mach number, and ratios of wall-to-stagnation temperature are varied. In the third paper, *Nestler* and *McCauley* report a correlation for predicting the tripping of a boundary layer by positioning an array of three-dimensional roughness elements on spherically blunted cones. Axisymmetric boundary-layer transition can induce shape changes which induce further flow asymmetries. The result is a significant loss in vehicle targeting accuracy. The next two papers present data pertinent to this shape change problem. *Kobayashi* and *Saperstein* report on a series of wind-tunnel tests which simulate re-entry trajectories, using several low-temperature ablator (camphor) model configurations. *Morrison, Yanta,* and *Voisinet* present a comprehensive compilation of flowfield data for the severely indented body, which is indicative of the shape to which some vehicles evolve during re-entry. The final paper by *Park* and *Bowen* represents an intriguing terrestrial mission—the projection of an ablative body to escape velocity by a ground-based mass driver. For such a mission, extreme shock-layer temperatures and pressures are encountered at low altitudes.

The successful completion of the first re-entry flight of the Space Shuttle Orbiter marks a significant achievement in thermal protection design. Chapter III is devoted to thermal performance studies associated with this system. In the first paper, *Zoby* presents a relatively simple technique for computing convective heating rates on large angle-of-attack bodies. The reliability of the technique is demonstrated by comparison with more rigorous treatments and data from model studies. The approach extends previous work by the author by representing variations along a plane of symmetry using the ''equivalent axisymmetric body'' concept. The next two papers deal with the question of catalytic efficiency of the thermal protection tiles. Because of the uncertainties in the high-temperature catalytic efficiency, designs have not taken full advantage of the reduced heating expected with a noncatalytic glassy surface. *Stewart* and *Rakich* describe both ground test and flight experiments in which side-by-side measurements of the noncatalytic and catalytic overcoated surfaces can be compared. In the third paper, *Scott* reports on calculations of heating rates based upon temperature-dependent surface recombination coefficients.

The extreme heating anticipated with the Galileo Probe of the atmosphere of Jupiter represents a severe test of thermal protection assessment and design. The heat shield for this probe is a blunt configuration (40 deg half angle) which will enter at a relative velocity of 48 km/s at a shallow entry angle. Chapter IV addresses the issues which make this the most demanding entry mission as yet encountered. The first paper, authored by *Howe, Pitts,* and *Lundell,* presents a comprehensive review of the computational and experimental studies which have been conducted in support of this probe. *Green* and *Davy* present an analysis of the transient response of the ablative heat-shield material during entry and heat-shield size data. Proper turbulence modeling is an area in which current uncertainties are reflected in a significant addition in heat-shield weight. *Moss* and *Kumar* have studied the significance of turbulence modeling upon radiative heating and ablative mass-loss predictions. In the final paper of this chapter, *Holden* reports on an experimental study conducted in a gas that approximates the Jovian atmosphere, in which injectants are used to simulate ablation gases. The paper provides data demonstrating the influence of massive blowing on the extent of turbulence in the flowfield and on shock stability.

Chapter V considers the problems associated with maneuverable vehicle concepts proposed for future missions. The concepts involve aerobraking or aerocapture—aeromaneuver. The latter concept uses high lift-to-drag ratio designs to achieve aerodynamic deceleration for transfer of orbits. In the first paper, *French* reviews possible future missions utilizing these concepts. The second paper by *Tiwari, Chow,* and *Moss* is an assessment of the convective and radiative heating for aerocapture by Titan to achieve a transfer to Saturn orbit. They examined the role of atmospheric composition by using several models that have been suggested. *Andrews, Caluori*, and *Bloetscher* describe the use of the ballote in an aerobraking orbital transfer vehicle used to return from a geosynchronous orbit. In the final paper, *Florence* presents a configuration which can be used for a variety of proposed aerocapture missions and which, with slight additions, can be used as an aeromaneuver vehicle. For this "generic" configuration and the proposed missions, an integrated design study is summarized.

As editor I gratefully acknowledge the assistance of the Editorial Committee, Ms. Ruth F. Bryans, former Associate Series Editor, Ms. Brendo Hio, Managing Editor of the Series, and Dr. Martin Summerfield, Editor-in-Chief of the AIAA *Progress in Astronautics and Aeronautics* series. I should also like to recognize the contribution of Mr. Len Stimpson who organized the thermophysics sessions at the AIAA 19th Aerospace Sciences Meeting, and the support of Dr. James Moss as Technical Program Chairman of the AIAA 16th Thermophysics Conference.

T. E. Horton
February 1982

Editorial Committee for Volume 82

Progress in Astronautics and Aeronautics

Martin Summerfield,
Series Editor-in-Chief
Princeton Combustion Research Laboratories, Inc.

Norma J. Brennan,
Director, Editorial Department
AIAA

Brenda J. Hio,
Series Managing Editor
AIAA

VOLUMES	EDITORS
*1. **Solid Propellant Rocket Research.** 1960	Martin Summerfield *Princeton University*
*2. **Liquid Rockets and Propellants.** 1960	Loren E. Bollinger *The Ohio State University* Martin Goldsmith *The Rand Corporation* Alexis W. Lemmon Jr. *Battelle Memorial Institute*
*3. **Energy Conversion for Space Power.** 1961	Nathan W. Snyder *Institute for Defense Analyses*
*4. **Space Power Systems.** 1961	Nathan W. Snyder *Institute for Defense Analyses*
*5. **Electrostatic Propulsion.** 1961	David B. Langmuir *Space Technology Laboratories, Inc.* Ernst Stuhlinger *NASA George C. Marshall Space Flight Center* J. M. Sellen Jr. *Space Technology Laboratories, Inc.*
*6. **Detonation and Two-Phase Flow.** 1962	S. S. Penner *California Institute of Technology* F. A. Williams *Harvard University*

*Now out of print.

*7.	**Hypersonic Flow Research.** 1962	Frederick R. Riddell *AVCO Corporation*
*8.	**Guidance and Control.** 1962	Robert E. Roberson *Consultant* James S. Farrior *Lockheed Missiles and Space Company*
*9.	**Electric Propulsion Development.** 1963	Ernst Stuhlinger *NASA George C. Marshall Space Flight Center*
*10.	**Technology of Lunar Exploration.** 1963	Clifford I. Cummings and Harold R. Lawrence *Jet Propulsion Laboratory*
*11.	**Power Systems for Space Flight.** 1963	Morris A. Zipkin and Russell N. Edwards *General Electric Company*
*12.	**Ionization in High-Temperature Gases.** 1963	Kurt E. Shuler, Editor *National Bureau of Standards* John B. Fenn, Associate Editor *Princeton University*
*13.	**Guidance and Control—II.** 1964	Robert C. Langford *General Precision Inc.* Charles J. Mundo *Institute of Naval Studies*
*14.	**Celestial Mechanics and Astrodynamics.** 1964	Victor G. Szebehely *Yale University Observatory*
*15.	**Heterogeneous Combustion.** 1964	Hans G. Wolfhard *Institute for Defense Analyses* Irvin Glassman *Princeton University* Leon Green Jr. *Air Force Systems Command*
*16.	**Space Power Systems Engineering.** 1966	George C. Szego *Institute for Defense Analyses* J. Edward Taylor *TRW Inc.*

***17. Methods in Astrodynamics and Celestial Mechanics.** 1966	Raynor L. Duncombe *U. S. Naval Observatory* Victor G. Szebehely *Yale University Observatory*
***18. Thermophysics and Temperature Control of Spacecraft and Entry Vehicles.** 1966	Gerhard B. Heller *NASA George C. Marshall Space Flight Center*
***19. Communication Satellite Systems Technology.** 1966	Richard B. Marsten *Radio Corporation of America*
***20. Thermophysics of Spacecraft and Planetary Bodies: Radiation Properties of Solids and the Electromagnetic Radiation Environment in Space.** 1967	Gerhard B. Heller *NASA George C. Marshall Space Flight Center*
***21. Thermal Design Principles of Spacecraft and Entry Bodies.** 1969	Jerry T. Bevans *TRW Systems*
***22. Stratospheric Circulation.** 1969	Willis L. Webb *Atmospheric Sciences Laboratory, White Sands, and University of Texas at El Paso*
***23. Thermophysics: Applications to Thermal Design of Spacecraft.** 1970	Jerry T. Bevans *TRW Systems*
24. Heat Transfer and Spacecraft Thermal Control. 1971	John W. Lucas *Jet Propulsion Laboratory*
25. Communications Satellites for the 70's: Technology. 1971	Nathaniel E. Feldman *The Rand Corporation* Charles M. Kelly *The Aerospace Corporation*
26. Communications Satellites for the 70's: Systems. 1971	Nathaniel E. Feldman *The Rand Corporation* Charles M. Kelly *The Aerospace Corporation*

27. **Thermospheric Circulation.** 1972 — Willis L. Webb, *Atmospheric Sciences Laboratory, White Sands, and University of Texas at El Paso*

28. **Thermal Characteristics of the Moon.** 1972 — John W. Lucas, *Jet Propulsion Laboratory*

29. **Fundamentals of Spacecraft Thermal Design.** 1972 — John W. Lucas, *Jet Propulsion Laboratory*

30. **Solar Activity Observations and Predictions.** 1972 — Patrick S. McIntosh and Murray Dryer, *Environmental Research Laboratories, National Oceanic and Atmospheric Administration*

31. **Thermal Control and Radiation.** 1973 — Chang-Lin Tien, *University of California, Berkeley*

32. **Communications Satellite Systems.** 1974 — P. L. Bargellini, *COMSAT Laboratories*

33. **Communications Satellite Technology.** 1974 — P. L. Bargellini, *COMSAT Laboratories*

34. **Instrumentation for Airbreathing Propulsion.** 1974 — Allen E. Fuhs, *Naval Postgraduate School*; Marshall Kingery, *Arnold Engineering Development Center*

35. **Thermophysics and Spacecraft Thermal Control.** 1974 — Robert G. Hering, *University of Iowa*

36. **Thermal Pollution Analysis.** 1975 — Joseph A. Schetz, *Virginia Polytechnic Institute*

37. **Aeroacoustics: Jet and Combustion Noise; Duct Acoustics.** 1975 — Henry T. Nagamatsu, Editor, *General Electric Research and Development Center*; Jack V. O'Keefe, Associate Editor, *The Boeing Company*; Ira R. Schwartz, Associate Editor, *NASA Ames Research Center*

38. Aeroacoustics: Fan, STOL, and Boundary Layer Noise; Sonic Boom; Aeroacoustics Instrumentation. 1975	Henry T. Nagamatsu, Editor *General Electric Research and Development Center* Jack V. O'Keefe, Associate Editor *The Boeing Company* Ira R. Schwartz, Associate Editor *NASA Ames Research Center*
39. Heat Transfer with Thermal Control Applications. 1975	M. Michael Yovanovich *University of Waterloo*
40. Aerodynamics of Base Combustion. 1976	S. N. B. Murthy, Editor *Purdue University* J. R. Osborn, Associate Editor *Purdue University* A. W. Barrows and J. R. Ward, Associate Editors *Ballistics Research Laboratories*
41. Communication Satellite Developments: Systems. 1976	Gilbert E. LaVean *Defense Communications Engineering Center* William G. Schmidt *CML Satellite Corporation*
42. Communication Satellite Developments: Technology. 1976	William G. Schmidt *CML Satellite Corporation* Gilbert E. LaVean *Defense Communications Engineering Center*
43. Aeroacoustics: Jet Noise, Combustion and Core Engine Noise. 1976	Ira R. Schwartz, Editor *NASA Ames Research Center* Henry T. Nagamatsu, Associate Editor *General Electric Research and Development Center* Warren C. Strahle, Associate Editor *Georgia Institute of Technology*
44. Aeroacoustics: Fan Noise and Control; Duct Acoustics; Rotor Noise. 1976	Ira R. Schwartz, Editor *NASA Ames Research Center* Henry T. Nagamatsu, Associate Editor *General Electric Research and Development Center* Warren C. Strahle, Associate Editor *Georgia Institute of Technology*

45. Aeroacoustics: STOL Noise; Airframe and Airfoil Noise. 1976	Ira R. Schwartz, Editor *NASA Ames Research Center* Henry T. Nagamatsu, Associate Editor *General Electric Research and Development Center* Warren C. Strahle, Associate Editor *Georgia Institute of Technology*
46. Aeroacoustics: Acoustic Wave Propagation; Aircraft Noise Prediction; Aeroacoustic Instrumentation 1976	Ira R. Schwartz, Editor *NASA Ames Research Center* Henry T. Nagamatsu, Associate Editor *General Electric Research and Development Center* Warren C. Strahle, Associate Editor *Georgia Institute of Technology*
47. Spacecraft Charging by Magnetospheric Plasmas. 1976	Alan Rosen *TRW Inc.*
48. Scientific Investigations on the Skylab Satellite. 1976	Marion I. Kent and Ernst Stuhlinger *NASA George C. Marshall Space Flight Center* Shi-Tsan Wu *The University of Alabama*
49. Radiative Transfer and Thermal Control. 1976	Allie M. Smith *ARO Inc.*
50. Exploration of the Outer Solar System. 1977	Eugene W. Greenstadt *TRW Inc.* Murray Dryer *National Oceanic and Atmospheric Administration* Devrie S. Intriligator *University of Southern California*
51. Rarefied Gas Dynamics, Parts I and II (two volumes). 1977	J. Leith Potter *ARO Inc.*
52. Materials Sciences in Space with Application to Space Processing. 1977	Leo Steg *General Electric Company*

53. Experimental Diagnostics in Gas Phase Combustion Systems. 1977	Ben T. Zinn, Editor *Georgia Institute of Technology* Craig T. Bowman, Associate Editor *Stanford University* Daniel L. Hartley, Associate Editor *Sandia Laboratories* Edward W. Price, Associate Editor *Georgia Institute of Technology* James G. Skifstad, Associate Editor *Purdue University*
54. Satellite Communications: Future Systems. 1977	David Jarett *TRW Inc.*
55. Satellite Communications: Advanced Technologies. 1977	David Jarett *TRW Inc.*
56. Thermophysics of Spacecraft and Outer Planet Entry Probes. 1977	Allie M. Smith *ARO Inc.*
57. Space-Based Manufacturing from Nonterrestrial Materials. 1977	Gerard K. O'Neill, Editor *Princeton University* Brian O'Leary, Assistant Editor *Princeton University*
58. Turbulent Combustion. 1978	Lawrence A. Kennedy *State University of New York at Buffalo*
59. Aerodynamic Heating and Thermal Protection Systems. 1978	Leroy S. Fletcher *University of Virginia*
60. Heat Transfer and Thermal Control Systems. 1978	Leroy S. Fletcher *University of Virginia*
61. Radiation Energy Conversion in Space. 1978	Kenneth W. Billman *NASA Ames Research Center*
62. Alternative Hydrocarbon Fuels: Combustion and Chemical Kinetics. 1978	Craig T. Bowman *Stanford University* Jϕrgen Birkeland *Department of Energy*

63. Experimental Diagnostics in Combustion of Solids. 1978	Thomas L. Boggs *Naval Weapons Center* Ben T. Zinn *Georgia Institute of Technology*
64. Outer Planet Entry Heating and Thermal Protection. 1979	Raymond Viskanta *Purdue University*
65. Thermophysics and Thermal Control. 1979	Raymond Viskanta *Purdue University*
66. Interior Ballistics of Guns. 1979	Herman Krier *University of Illinois at Urbana-Champaign* Martin Summerfield *New York University*
67. Remote Sensing of Earth from Space: Role of "Smart Sensors." 1979	Roger A. Breckenridge *NASA Langley Research Center*
68. Injection and Mixing in Turbulent Flow. 1980	Joseph A. Schetz *Virginia Polytechnic Institute and State University*
69. Entry Heating and Thermal Protection. 1980	Walter B. Olstad *NASA Headquarters*
70. Heat Transfer, Thermal Control, and Heat Pipes. 1980	Walter B. Olstad *NASA Headquarters*
71. Space Systems and Their Interactions with Earth's Space Environment. 1980	Henry B. Garrett and Charles P. Pike *Hanscom Air Force Base*
72. Viscous Flow Drag Reduction. 1980	Gary R. Hough *Vought Advanced Technology Center*
73. Combustion Experiments in a Zero-Gravity Laboratory. 1981	Thomas H. Cochran *NASA Lewis Research Center*
74. Rarefied Gas Dynamics, Parts I and II (two volumes). 1981	Sam S. Fisher *University of Virginia at Charlottesville*

75. Gasdynamics of Detonations and Explosions. 1981	J. R. Bowen *University of Wisconsin at Madison* N. Manson *Université de Poitiers* A. K. Oppenheim *University of California at Berkeley* R. I. Soloukhin *Institute of Heat and Mass Transfer, BSSR Academy of Sciences*
76. Combustion in Reactive Systems. 1981	J. R. Bowen *University of Wisconsin at Madison* N. Manson *Université de Poitiers* A. K. Oppenheim *University of California at Berkeley* R. I. Soloukhin *Institute of Heat and Mass Transfer, BSSR Academy of Sciences*
77. Aerothermodynamics and Planetary Entry. 1981	A. L. Crosbie *University of Missouri-Rolla*
78. Heat Transfer and Thermal Control. 1981	A. L. Crosbie *University of Missouri-Rolla*
79. Electric Propulsion and Its Applications to Space Missions. 1981	Robert C. Finke *NASA Lewis Research Center*
80. Aero-Optical Phenomena. 1982	Keith G. Gilbert and Leonard J. Otten *Air Force Weapons Laboratory*
81. Transonic Aerodynamics 1982	David Nixon *Nielsen Engineering & Research, Inc.*
82. Thermophysics of Atmospheric Entry. 1982	T.E. Horton *The University of Mississippi*

83. Spacecraft Radiative Transfer and Temperature Control. 1982 — T.E. Horton, *The University of Mississippi*

(Other volumes are planned.)

Chapter I Thermophysical Properties

Numerical Calculation of Gaseous Transport Properties from the Hulburt-Hirschfelder Potential with Applications to Planetary Entry Thermal Protection

James C. Rainwater*
National Bureau of Standards, Boulder, Colo.
and
Paul M. Holland† and Louis Biolsi‡
University of Colorado/NOAA, Boulder, Colo.

Abstract

Transport properties of dilute monatomic gases have been evaluated by means of the Hulburt-Hirschfelder (H-H) potential, which contains nonadjustable parameters determined completely from spectroscopic data. The H-H potential is shown to encompass five distinct patterns of binary collision dynamics, some of which include double orbiting. Since the collision dynamics are more complicated than those of commonly used potentials such as the Lennard-Jones, the numerical routines to evaluate Chapman-Enskog collision integrals have been extensively revised. For applications to thermal protection during entry into atmospheres of the outer planets, transport properties of monatomic carbon gas at $T \sim 10^4$ K have been evaluated.

Presented as Paper 81-1172 at the AIAA 16th Thermophysics Conference, Palo Alto, Calif., June 23-25, 1981.

*Physicist, Thermophysical Properties Division, National Engineering Laboratory.

†Research Assistant, Cooperative Institute for Research in Environmental Sciences (Chemist, The Procter & Gamble Co., Cincinnati, Ohio).

‡Visiting Fellow, Cooperative Institute for Research in Environmental Sciences (Professor of Chemistry, University of Missouri-Rolla, Rolla, Mo.).

Nomenclature

a	= parameter of reduced H-H potential
a_1, a_2, a_3	= parameters used to construct H-H potential
b	= impact parameter
B_e	= rotation constant
$C_{\ell s}$	= constant in collision integral
d	= ratio of r_e to σ
D	= diffusion coefficient
E	= relative total energy
E_1	= energy at minimum of $z(r)$
E_2	= energy at second maximum of $z(r)$
E_3	= energy bound of double-orbiting region
E_c	= energy upper bound of orbiting region
E_L	= energy of maximum in potential
k	= Boltzmann's constant
ℓ	= index of collision integral
r	= interatomic separation
r_e	= distance of potential minimum
r_L	= distance of potential maximum
r_m	= distance of closest approach for collision
s	= index of collision integral
T	= temperature
T^*	= reduced temperature
$z(r)$	= locus of effective potential extrema
α_e	= rotation-vibration coupling constant
β	= parameter of reduced H-H potential
γ	= parameter of reduced H-H potential
ε	= depth of potential well
ε'	= wavenumber corresponding to ε
η	= viscosity
λ_{tr}	= translational part of thermal conductivity
σ	= distance at which potential is zero
$\phi(r)$	= potential
$\phi_{eff}(r)$	= effective potential
χ	= scattering angle
ω_e	= fundamental vibration frequency
$\omega_e x_e$	= anharmonicity constant
$\Omega^{(\ell,s)*}$	= reduced collision integral

Introduction

The transport properties of monatomic gases may be evaluated according to the well-known Chapman-Enskog solution of the Boltzmann equation[1] if the intermolecular potential is known. Within the first Enskog approximation, the self-diffusion coefficient of a gas is inversely proportional to the $\Omega^{(1,1)*}$ integral, and the shear viscosity and thermal conductivity are inversely proportional to the $\Omega^{(2,2)*}$ integral. The general form of the reduced transport collision integral is

$$\Omega^{(\ell,s)*} = C_{\ell s} \int_0^\infty E^{s+1}\, e^{-E/kT}\, dE \int_0^\infty b db(1-\cos^\ell \chi) \qquad (1)$$

where $C_{\ell s}$ is a constant, $C_{11} = 1$ and $C_{22} = 1/2$, k is Boltzmann's constant, T the temperature, E and b the relative total energy and impact parameter of a binary collision, and χ the scattering angle for that collision which is given by

$$\chi = \pi - 2b \int_{r_m}^\infty \frac{dr}{r^2\left[1 - \frac{\phi_{eff}(r)}{E}\right]^{1/2}} \qquad (2)$$

where

$$\phi_{eff}(r) = \phi(r) + (Eb^2/r^2) \qquad (3)$$

Here $\phi(r)$ is the interparticle potential; $\phi_{eff}(r)$ is an effective interparticle potential including a centrifugal barrier term; and r_m, the distance of closest approach, is the largest value of r such that the expression in brackets in Eq. (2) is zero. We make use of reduced units in Eqs. (1-3), such that the unit of energy is ε, defined as min $[\phi(r)] = -\varepsilon$, and the unit of length is σ, defined to be the smallest root of the equation $\phi(\sigma) = 0$.

The Hulburt-Hirschfelder Potential

Calculation of gaseous transport properties requires the interparticle potential as input. Frequently, potentials are determined by adjusting the parameters of a particular mathematical form of the potential, e.g., the Lennard-Jones, to fit other bulk physical properties such as the second virial coefficient. However, situations occur in which bulk property data are not available, for example the layer of monatomic carbon gas at $T \sim 10^4$ K surrounding a carbon-phenolic heat shield during planetary entry. A preferred method of determining the potential in such cases is to use the Hulburt-Hirschfelder (H-H) potential[2] whose three parameters are determined entirely from widely available spectroscopic data and which therefore has no adjustable parameters.

The required spectroscopic parameters in the construction of the H-H potential are ε, the depth of the potential well; r_e, the interatomic separation corresponding to ε; ω_e, the fundamental vibrational frequency; $\omega_e X_e$, the anharmonicity constant; B_e, the rotation constant; and α_e, the rotation-vibration coupling constant. In a suitably reduced form the H-H potential is

$$\begin{aligned}\phi(r) &= \exp\,[-\,2a\,(r/d - 1)] \\ &\quad - 2\exp\,[-\,a\,(r/d - 1)] \\ &\quad + \beta\,(r/d - 1)^3\,[1 + \gamma\,(r/d - 1)] \\ &\quad \exp\,[-\,2a\,(r/d - 1)]\end{aligned} \tag{4}$$

where

$$a_0 = \omega_e^2/4B_e \tag{5}$$

$$a_1 = -1 - \alpha_e\omega_e/6B_e^2 \tag{6}$$

$$a_2 = \frac{5}{4}\,a_1^2 - 2\omega_e X_e/3B_e \tag{7}$$

$$a = \omega_e/2\,(B_e\varepsilon')^{1/2} \tag{8}$$

$$\beta = a^3\left[1 + a_1\,(\varepsilon'/a_0)^{1/2}\right] \tag{9}$$

$$\gamma = a\left[2 - a^3\beta^{-1}\left(\frac{7}{12} - \frac{\varepsilon' a_2}{a_0}\right)\right] \tag{10}$$

$$d = r_e/\sigma \tag{11}$$

and ε' is the wavenumber corresponding to energy ε. This potential contains the three parameters a, β, and γ which are independent although they are completely determined by spectroscopic properties. The variable d is a function of these three variables and is determined by the constraint, in reduced units, that $\phi(1) = 0$. In general d must be evaluated numerically.

The numerical integration program of O'Hara and Smith[3] has been the most efficient technique for calculating gaseous transport properties with potentials such as the Lennard-Jones. However, in certain cases the H-H potential

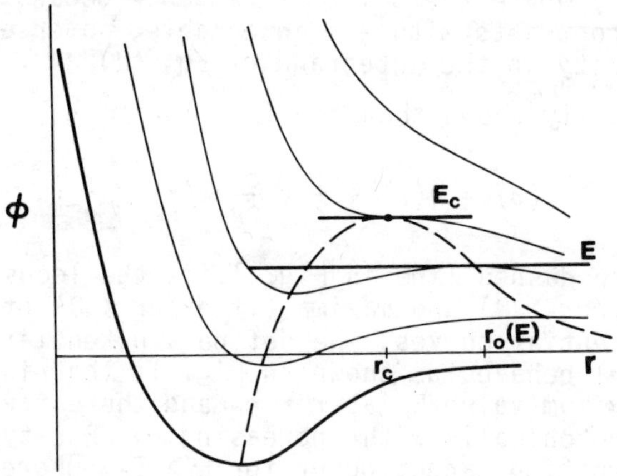

Fig. 1 Effective potential curves and locus of their extrema (dashed line) for a type 1 potential (bold line).

has a radically different pattern of binary collision dynamics than the Lennard-Jones or similar potentials, and the unmodified O'Hara-Smith collision integral program fails in such cases.

Classification of Potentials

The problem at hand is to analyze the binary collision dynamics for all possible cases of the H-H potential and then to revise the O'Hara-Smith collision integral program so that it can accommodate all such cases. The resulting program makes possible the calculation of the transport properties of a wide variety of gaseous systems, including, but not limited to, noble gases, alkali metal vapors, ablation products during planetary entry, and the Earth's upper atmosphere.

Intermolecular potentials of interest may be classified according to their binary collision dynamics, and in particular to the occurrance of orbiting collisions.[4] The phenomenon of orbiting is best described by means of Fig. 1, which shows the simplest case of the H-H potential and its associated family of effective potential curves. When E and b are chosen such that the line of constant E is tangent to the associated effective potential curve, orbiting occurs. There is a nonintegrable infinity in the integrand of Eq. (2), and the scattering angle becomes negatively infinite. Physically, this corresponds to the two colliding particles orbiting each other indefinitely at the distance $r_0(E)$ where the point of tangency occurs. This singularity contrasts with the integrable, inverse-square-root singularity in the integrand of Eq. (2) at $r = r_m$.

It is easily shown that the function

$$z(r) = \phi(r) + \frac{1}{2} r \frac{d}{dr} \phi(r) \tag{12}$$

plotted as the dashed line in Fig. 1, is the locus of minima (if $dz/dr > 0$) and maxima (if $dz/dr < 0$) of the effective potential curves. We define a potential to be type 1 if $z(r)$ behaves as shown in Fig. 1; that is, it reaches a maximum value E_c at $r = r_c$ and thereafter decreases monotonically with increasing r. For type 1 potentials orbiting cannot occur for $E > E_c$, whereas for each energy $E < E_c$ there exists a unique impact parameter $b_c(E)$ such that orbiting occurs. The b integrand of Eq. (1) thus undergoes oscillations of infinite frequency at $b = b_c(E)$.

Our analysis shows that the H-H potential, for all physically reasonable choices of parameters, may be classified into five separate types, each with qualitatively distinct binary collision dynamics. The potential type depends on the parameters a, β, and γ, and types of higher number are obtained by increasing β and/or γ.

Figure 2 shows the effective potential family for a type 2 potential. The potential itself is qualitatively similar to a type 1 potential; it has a repulsive core, a negative minimum, and thereafter with increasing r it increases monotonically to zero. However, the function $z(r)$, as shown, does not decrease monotonically from its maximum. Rather, $z(r)$ has a minimum at an energy E_1 and a second maximum at an energy E_2.

There exists an effective potential curve, not shown, with two equal maxima at an energy E_3 approximately midway between E_1 and E_2. For $E < E_3$ and $E_2 < E < E_c$ there exists a unique orbiting impact parameter $b_c(E)$. However, for $E_3 < E < E_2$ there are two separate orbiting impact parameters. In this case the b integrand of Eq. (1) undergoes oscillations of infinite frequency at two points, not only one.

Figure 3 shows the corresponding diagram for a type 3 potential, which has a negative maximum and, at a larger value of r, a second negative minimum. Note from Eq. (12)

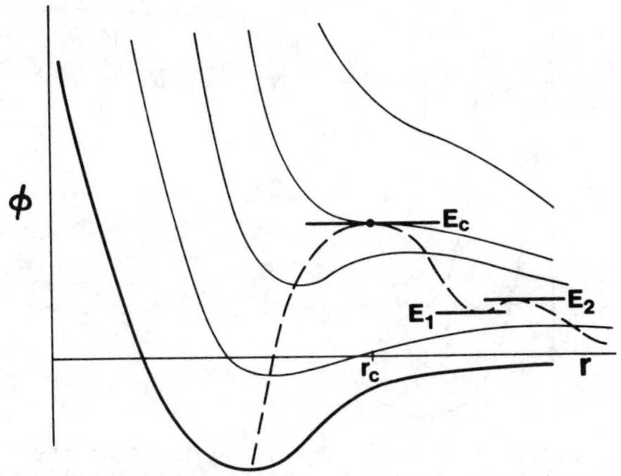

Fig. 2 Effective potential curves and locus of their extrema (dashed line) for a type 2 potential (bold line).

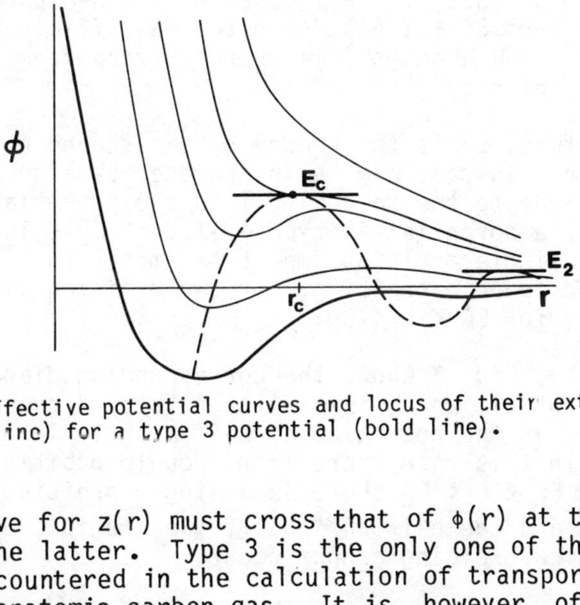

Fig. 3 Effective potential curves and locus of their extrema (dashed line) for a type 3 potential (bold line).

that the curve for $z(r)$ must cross that of $\phi(r)$ at the extrema of the latter. Type 3 is the only one of the five types not encountered in the calculation of transport properties for monatomic carbon gas. It is, however, of practical interest since the H-H potential for potassium vapor is type 3. Except that E_1 is negative and not physically relevant, the pattern of orbiting closely parallels that of type 2.

The corresponding diagram for a type 4 potential is shown in Fig. 4. This potential has a small positive maximum E_L at $r = r_L$ and, at larger distance, a second negative

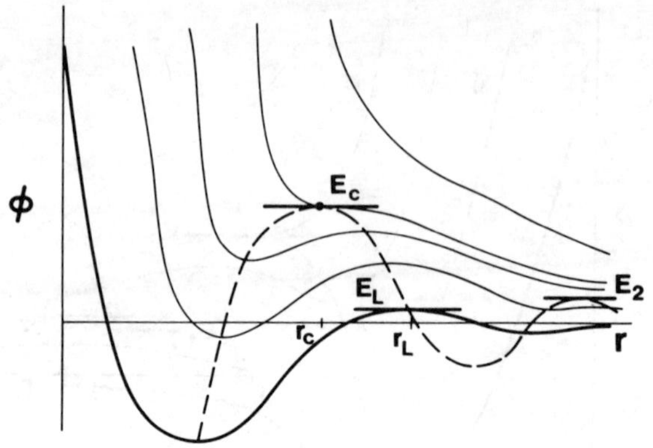

Fig. 4 Effective potential curves and locus of their extrema (dashed line) for a type 4 potential (bold line).

minimum. Note that, for all cases of the H-H potential, the second term of Eq. (4) dominates at sufficiently large r and thus all H-H potentials approach zero from below as r approaches infinity.

As before, E_2 is the energy of the second maximum in z(r) and in this case may be interpreted as a second critical energy due to the second well in the potential. By definition, a potential is type 4 if $E_2 > E_L$. In this case there is a single orbiting impact parameter for $E < E_L$ and $E_2 < E_c$ and "double orbiting," or two orbiting impact parameters, for $E_L < E < E_2$.

Finally, Fig. 5 shows the corresponding diagram for a type 5 potential, which is similar to type 4 except that the maximum in the potential is sufficiently large so $E_L > E_2$. In this case there is no double orbiting. For $E < E_2$ and $E_L < E < E_c$ there is a single-orbiting impact parameter, and there is an "energy gap" region, $E_2 < E < E_L$, for which orbiting <u>cannot</u> occur.

Numerical Integration Methods

Our starting point for numerical calculations has been a slightly modified version of the O'Hara-Smith[3] collision integral program developed at the National Bureau of Standards and coded by James F. Ely.[5] This program uses a Clenshaw-Curtis quadrature[3] for numerical integration. A particularly convenient feature is that the user can specify a desired accuracy. The program automatically uses

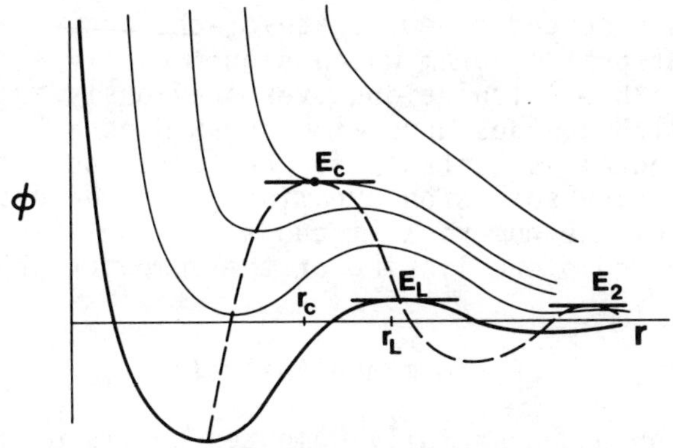

Fig. 5 Effective potential curves and locus of their extrema (dashed line) for a type 5 potential (bold line).

less computer time for larger accuracy inputs and more computer time for smaller accuracy inputs, as required.

The unmodified program requires the potential to be infinite at the origin and hence is not suitable for the H-H potential which, at the origin, is finite. However, for type 1 potentials only a minor modification is required to make the program applicable. Collision integrals have been computed for the (type 1) Morse potential, a special case of the H-H potential ($\beta = 0$), and agree with previously published results[6] to high accuracy.

Included in the program are subroutines to scan the potential, to locate the critical impact parameters and distances of closest approach by interpolation and root-finding techniques, and to perform the required numerical integrals. When orbiting occurs, the integrals must be split at the singularities due to the orbiting. Then a transformation of variable is required to introduce a weight factor which suppresses the singular end of the integration interval, and thereby transforms the integrand into a well-behaved function.

We have redesigned all appropriate subroutines to accommodate the most general H-H potential. The scanning subroutine has been revised to determine the potential type. Then, modified subroutines determine all possible orbiting collisions, split the integrals at the points of singularity, and, conforming to the techniques of O'Hara and Smith,[3] suppress the singularities with the appropriate transformation of variable. The features that enable the user to specify a desired accuracy are retained.

We have produced a new, state-of-the-art, transport collision integral program which significantly extends the range of possible calculations over previously available programs. Although designed with the H-H potential in mind, the program is suitable for any other potential with equivalent binary collision dynamics. A more detailed description of the numerical methods,[7] as well as a publication of the complete listing of the computer program,[8] are in progress.

Performance Analysis

The program successfully computes collision integrals for all of the five types of potential described above. Typically, our runs have calculated the $\Omega^{(\ell,s)*}$ integrals for $\ell \leq s$, $s \leq 3$ for 46 reduced temperatures ranging from a low of $T^* = 0.004$ ($T^* = kT/\varepsilon$) to a high of $T^* = 200$. The times for such a run are generally larger for the middle potential types, but the program is quite efficient for all cases. As examples, on a CDC 6400 computer,* a run of a type 1 potential took 34.3 s at 10^{-2} accuracy and 179.7 s at 10^{-4} accuracy, whereas a run of a type 3 potential took 56.2 and 363.2 s, respectively. For the sake of comparison, equivalent runs of the Lennard-Jones 12-6 potential with the unmodified program took 26.3 and 91.4 s, respectively.

The program has been tested for self-consistency by running with successive accuracies of 10^{-2}, 10^{-3}, 10^{-4}, and 10^{-5}, and monitoring the relative error of successive results. Of the 276 collision integrals of a given run, at no time did more than nine exceed the allowed errors of 10^{-2}-10^{-4}. Most of those in excess where at very low temperatures and the excesses were never large. However, at an accuracy of 10^{-5} many more discrepancies were found; this indicates that the program is at this point being pushed too far, although it might be successful if the maximum number of Clenshaw-Curtis abcissas were increased from 64 to 128.

Our first test of the method was to compute the transport properties of argon gas. Results were compared with the viscosity data of Kestin et al.[9] and Maitland and

*The use of a specific tradename is necessary in order to specify precisely the speed and efficiency of a computer program. Use of the tradename in no way implies approval, endorsement, or recommendation by the National Bureau of Standards.

Smith,[10] and the thermal conductivity data of Schaeffer and Reiter,[11] and others. Agreement is within 3% for all but the smallest temperatures.[12] Although state-of-the-art argon potentials, such as that of Barker, Fisher, and Watts,[13] give better experimental agreement, our method is a general recipe with no adjustable parameters and so we find such agreement to be most satisfactory. Deviations at low temperature are well understood in terms of the (incorrect) exponential tail of the H-H potential; such deviations are expected on theoretical grounds to be less severe for the higher potential types.

Application to Ablative Carbon Gas

Our first motivation for the method was to apply it to the problem of ablative carbon gas during planetary entry. During entry into the Jovian atmosphere, it is predicted[14] that the carbon-phenolic ablation heat shield will be surrounded by a layer of carbon gas at a temperature $T \sim$ 10,000 K. At this temperature, carbon gas is predominantly monatomic and the carbon atoms are predominately in their ground (^3P) atomic states.

Under such conditions, the carbon atoms undergo binary collisions according to one of 18 possible potential energy curves, which correspond to the 18 molecular states of the carbon dimer. To calculate the transport properties of gaseous carbon, the transport collision integrals [Eq. (1)] for each state must be evaluated and an appropriate weighted average over the states must then be computed.[1]

Five of the states $\left(^5\Sigma_u^-,\ ^5\Pi_u,\ ^5\Delta_g,\ ^3\Sigma_{u2}^+, \text{ and } ^5\Sigma_{g2}^+\right)$ possess purely repulsive potentials. For these states it suffices to fit the potential curves to the exponential repulsive potential and then obtain the collision integrals from tabulated values,[6] as described previously.[14,15]

The remaining 13 molecular carbon states have potential curves with at least one attractive well. The spectroscopic constants required to construct the H-H potential, Eqs. (4-11) and preceding discussion, are available. The molecular states, reduced H-H potential parameters, and type classifications are listed in Table 1. Note that the $^5\Sigma_g$, $^1\Pi_g$, $^3\Delta_u$, and $^1\Sigma_u^-$ states are predicted to have long-range maxima and second minima. Such features have been predicted previously from quantum-mechanical calculations[16] for the $^1\Pi_g$ and $^1\Sigma_u^-$ states.

Table 1 Reduced H-H parameters and type classification for partly attractive C_2 states

State	a	β	γ	d	Type
$^1\Sigma_g^+$	3.0531	3.5703	2.2026	1.3087	1
$^3\Pi_u$	2.8562	1.2326	3.2115	1.3228	1
$^3\Sigma_g^-$	2.8362	0.4243	5.9194	1.3218	1
$^1\Pi_u$	3.0536	3.0182	2.6921	1.3015	1
$^3\Sigma_u^+$	3.6798	13.4984	3.5801	1.2449	2
$^1\Delta_g$	3.3049	5.6709	3.2183	1.2739	1
$^1\Sigma_{g2}^+$	3.4481	11.2288	3.0636	1.2692	1
$^3\Pi_g$	3.8176	18.3271	2.4169	1.2544	1
$^5\Pi_g$	3.6002	13.0005	3.0991	1.2564	1
$^5\Sigma_g^+$	4.3308	32.4911	4.3260	1.2048	5
$^1\Pi_g$	5.2665	68.2301	5.4861	1.1628	5
$^3\Delta_u$	4.0971	24.1056	4.0672	1.2173	4
$^1\Sigma_u^-$	4.5978	46.2266	4.6065	1.1931	5

The transport collision integrals for all the above potentials are required for calculation of transport properties of both pure gaseous monatomic carbon and gaseous mixtures containing monatomic carbon. With the numerical methods described above, we have evaluated transport collision integrals for all potentials listed in Table 1. From these results, the viscosity η, translational part of the thermal conductivity λ_{tr}, and diffusion coefficient D for pure carbon gas have been calculated, and are listed in Table 12 of Ref. 17.

In earlier work by Biolsi,[14,15] the H-H potential was fitted to the nearest Morse potential and the transport properties of carbon gas were deduced from published tables of collision integrals for the Morse potential. It is, of course, more accurate to calculate the H-H collision integrals directly. Comparison of the two techniques shows that the earlier method of Biolsi significantly underestimates transport properties of carbon gas, cf. Table 5 of Ref. 14. Our new results should prove useful in the design of probes for atmospheric entry into the outer planets.

Analysis of various additional gaseous sytems is in progress. An initial calculation of transport properties of alkali metal vapors shows promising agreement with experiment, once again without adjustable parameters.

Acknowledgments

This research was supported in part by NASA grant NASA NSG 1369.

References

[1]Hirschfelder, J. O., Curtiss, C. F., and Bird, R. B., Molecular Theory of Gases and Liquids, John Wiley & Sons, New York, 1954.

[2]Hulburt, H. M. and Hirschfelder, J. O., "Potential Energy Functions for Diatomic Molecules," Journal of Chemical Physics, Vol. 9, Jan. 1941, pp. 61-69; "Erratum: Potential Energy Functions for Diatomic Molecules," Journal of Chemical Physics, Vol. 35, Nov. 1961, p. 1901.

[3]O'Hara, H. and Smith, F. J., "The Efficient Calculation of the Transport Properties of a Dilute Gas to a Prescribed Accuracy," Journal of Computational Physics, Vol. 5, April 1970, pp. 328-344.

[4]Stogryn, D. E. and Hirschfelder, J. O., "Contribution of Bound, Metastable, and Free Molecules to the Second Virial Coefficient and Some Properties of Double Molecules," Journal of Chemical Physics, Vol. 31, Dec. 1959, pp. 1531-1545.

[5]Ely, J. F. and Hanley, H. J. M., "The Statistical Mechanics of Non-Spherical Polyatomic Molecules: Application to the Properties of Carbon Dioxide," Molecular Physics, Vol. 30, Aug. 1975, pp. 565-578.

[6]Smith, F. J. and Munn, R. J., "Automatic Calculation of the Transport Collision Integrals with Tables for the Morse Potential," Journal of Chemical Physics, Vol. 41, Dec. 1, 1964, pp. 3560-3568.

[7]Rainwater, J. C., Holland, P. M., and Biolsi, L., "Binary Collision Dynamics and Numerical Evaluation of Dilute Gas Transport Properties for Potentials with Multiple Extrema," submitted to Journal of Chemical Physics.

[8]Rainwater, J. C., Ely, J. F., and Hanley, H. J. M., "A Computer Program to Evaluate Transport Collision Integrals for Potentials With Single and Double Minima," in preparation.

[9]Kestin, J., Ro, S. T., and Wakeham, W. A., "Viscosity of the Noble Gases in the Temperature Range 25-700 C," Journal of Chemical Physics, Vol. 56, April 15, 1972, pp. 4119-4124.

[10]Maitland, G. C. and Smith, E. B., "Critical Reassessment of Viscosities of Eleven Common Gases," Journal of Chemical Engineering Data, Vol. 17, April 1972, pp. 150-156.

[11]Schaeffer, K. L. and Reiter, F. W., "Measurement of the Absolute Thermal Conductivity of Gases Up to 1100 C," Naturwissenschaften, Vol. 43, May 1956, pp. 296-297.

[12]Holland, P. M., Biolsi, L., and Rainwater, J. C., "The Hulburt-Hirschfelder Potential: Calculation of the Transport Properties and Second Virial Coefficient of Argon Without Adjustable Parameters," to be published.

[13]Barker, J. A., Fisher, R. A., and Watts, R. O., "Liquid Argon: Monte Carlo and Molecular Dynamics Calculations," Molecular Physics, Vol. 21, Sept. 1971, pp. 657-673.

[14]Biolsi, L., "Transport Properties of Monatomic Carbon," Journal of Geophysical Research, Vol. 83, June 1, 1978, pp. 2476-2480.

[15]Biolsi, L. and Biolsi, K. J., "Transport Properties of Monatomic Carbon II: Contributions From Excited Electronic States," Journal of Geophysical Research, Vol. 84, Sept. 1, 1979, pp. 5311-5318.

[16]Fougere, P. F. and Nesbet, R. K., "Electronic Structure of C_2," Journal of Chemical Physics, Vol. 44, Jan. 1, 1966, pp. 285-298.

[17]Biolsi, L., Fenton, J. T., and Owenson, B., "Transport Properties for Mixtures of the Ablation Products C, C_2 and C_3," Thermophysics of Atmospheric Entry, Progress in Astronautics and Aeronautics, Vol. 82, edited by T. E. Horton, AIAA, New York, 1982, pp.17-36.

Transport Properties for a Mixture of the Ablation Products C, C_2, and C_3

Louis Biolsi,* Jeff Fenton,† and Brian Owenson‡
University of Missouri-Rolla, Rolla, Mo.

Abstract

The ablation of carbon-phenolic heat shields upon entry into the atmosphere of one of the outer planets leads to the injection of large amounts of C, C_2, and C_3 into the shock layer. These species must be included in the calculation of transport properties in the shock layer. The kinetic theory of gases has been used to obtain accurate results for the transport properties of monatomic carbon. The Hulburt-Hirschfelder potential, the most accurate general purpose atom-atom potential for states with an attractive minimum, was used to represent such states and repulsive states were represented by fitting quantum mechanical potential energy curves with the exponential repulsive potential. These results were orientation averaged according to the peripheral force model to obtain transport collision integrals for the C-C_2 and C_2-C_2 interaction. Results for C_3 were obtained by ignoring the presence of the central carbon atom. The thermal conductivity, viscosity, and diffusion coefficients for pure C, C_2, and C_3, and for mixtures of these gases were then calculated at 1000K-25,000K.

Presented as Paper 81-1173 at the AIAA 16th Thermophysics Conference, Palo Alto, Calif., June 23-25, 1981.

*Associate Professor, Chemistry Department.
†Graduate Student, Chemistry Department.
‡Undergraduate Student, Chemistry Department.

Introduction

Entry vehicle heat shields are usually made of carbonaceous materials such as carbon-phenolic ablator; typically 92% carbon, 6% oxygen, and 2% hydrogen by mass. Ablative injection of the carbon-phenolic material into the shock layer is an important mechanism for reducing the intense heating at the probe surface during entry into the atmosphere of one of the outer planets.[1-3]

The mole fractions of the various gaseous species, including ablation products, at the surface of an entry probe are given[4] in Table 1. It has been assumed that the composition of the atmosphere of the outer planet is X_{H_2} = 0.89 and X_{He} = 0.11, where X denotes mole fraction. This is the composition of the atmosphere of Jupiter. The species listed in Table 1 primarily determine the transport properties at the surface of the entry probe. The transport properties are needed in order to calculate heating and viscous flow rates at the surface of the entry vehicle due to conductive transport. The purpose of this paper is to evaluate the contribution to the transport properties from the species C, C_2, and C_3 with X = 0.357, i.e., almost 36% of the total number of moles of gas at the surface of the probe is due to these three species. Contributions to the transport properties from the species H and H_2 have been considered previously.[5]

The kinetic theory of gases will be used to calculate the transport properties. This requires knowledge of all the possible two-body interaction potentials in the gas mixture. These are listed in Table 2 for the C, C_2, C_3 system.

Monatomic Carbon

The transport properties of monatomic carbon have been considered in detail previously.[6,7] When two ground-state (^3P) carbon atoms interact, they can follow any of 18 potential energy curves, corresponding to 18 molecular states of C_2. Thirteen of these states have an attractive minimum in the potential energy and five states are purely repulsive. Previous results[6] for the transport properties were obtained by fitting the states with an attractive minimum with a Morse potential and the three repulsive states for which theoretical potential energy curves are available[8] were fitted with an exponential repulsive (ER) potential.

Table 1 Mole fractions of various species at the surface of an entry probe upon entry into the atmosphere of Jupiter[4]

Species	Mole fraction	Species	Mole fraction
C_3	0.303	C_3H	0.099
H	0.149	C_2	0.028
C_4H	0.138	C	0.026
C_2H	0.115	H_2	0.024
CO	0.108	C_2H_2	0.010
		Total	1.000

Table 2 Two-body interactions in a gaseous mixture of C, C_2, and C_3

Atom-atom	Atom-molecule	Molecule-molecule
C-C	C-C_2	C_2-C_2
	C-C_3	C_3-C_3
		C_2-C_3

Table 3 Spectroscopic constants[12] for the states of C_2 that dissociate into ^3P carbon atoms

State	ε,eV[8]	r_e,Å	ω_e,cm^{-1}	$\omega_e\chi_e$,cm^{-1}	B_e,cm^{-1}	α_e,cm^{-1}
$^1\Sigma_g^+$	6.36	1.2425	1854.71	13.34	1.81984	0.01765
$^3\Pi_u$	6.27	1.3119	1641.35	11.67	1.63246	0.01661
$^3\Sigma_g^-$	5.56	1.3693	1470.45	11.19	1.49852	0.01634
$^1\Pi_u$	5.32	1.3184	1608.35	12.078	1.61634	0.01686
$^3\Sigma_u^+$	4.71	1.23	1961.6	13.65	1.87	[0.018]
$^1\Delta_g$	4.45	[1.39]	[1510]	[11.1]	[1.454]	[0.015]
$^1\Sigma_{g2}^+$	4.03	[1.38]	[1510]	[10.0]	[1.475]	[0.013]
$^3\Pi_g$	3.88	1.266	1788.2	16.44	1.7527	0.01608
$^5\Pi_g$	3.54	[1.46]	[1400]	[9.9]	[1.324]	[0.012]
$^5\Sigma_g^+$	2.55	[1.35]	[1550]	[11.7]	[1.557]	[0.015]
$^1\Pi_g$	2.11	1.2552	1809.1	15.81	1.7334	0.018
$^3\Delta_u$	2.02	[1.51]	[1160]	[9.2]	[1.23]	[0.013]
$^1\Sigma_u^-$	1.98	[1.90]	[1025]	[5.8]	[0.778]	[0.005]

These calculations have been redone. The Hulburt-Hirschfelder (H-H) potential[9,10] has been used to represent the states with an attractive minimum. This is the most accurate general purpose potential for representing atom-atom interactions with an attractive minimum.[11] The potential has the form[9,10]

$$V(\text{H-H}) = \varepsilon[(1-e^{-x}) + cx^3e^{-2x}(1 + bx) - 1] \tag{1}$$

where

$$x = \frac{\omega_e}{2\sqrt{B_e\varepsilon}}\left(\frac{r-r_e}{r_e}\right) \qquad c = 1 + a_1\sqrt{\frac{\varepsilon}{a_o}}$$

$$b = 2 - \frac{7/12 - a_2/a_o}{c} \qquad a_o = \frac{\omega_e^2}{4B_e}$$

$$a_1 = -1 - \frac{\alpha_e\omega_e}{6B_e^2} \qquad a_2 = \frac{5}{4}a_1^2 - \frac{2\omega_e\chi_e}{3B_e}$$

Also, r is the internuclear separation, ε the depth of the potential well, r_e the interatomic separation corresponding to ε, ω_e the fundamental vibrational frequency, $\omega_e\chi_e$ the anharmonicity constant, B_e the rotational constant, and α_e the rotation-vibration coupling constant. An important feature of the H-H potential is that all parameters are determined experimentally from spectroscopic data, i.e., it contains no adjustable parameters. The required spectroscopic constants are available for many diatomic molecules.[12] Thus the H-H potential can be used to calculate reliable estimates of transport properties for systems that are difficult or impossible to investigate experimentally. The spectroscopic constants[12] used for C_2 are shown in Table 3.

These constants were used in Eq. (1) and a recently developed computer code was used to calculate the diffusion collision integral, $\sigma^2\Omega^{(1,1)*}$, and the viscosity collision integral, $\sigma^2\Omega^{(2,2)*}$, for each state.

Potential energy curves for the two repulsive states for which theoretical potential energy curves are not available (the $^3\Sigma^+_{u2}$ and $^5\Sigma^+_{g2}$ states) were estimated by using the perfect pairing method.[3] The results were curve fit with the ER potential, i.e.,

$$V(ER) = A\exp(-Br) \qquad (2)$$

where A and B are empirical parameters. The parameters for the five repulsive states of C_2 are shown in Table 4.

Transport collision integrals for these states were obtained from Monchick.[14] The collision integrals for all 18 states were then averaged according to their degeneracies.[15] Results are given in Table 5.

The C-C_2 and C-C_3 Interactions

Consider the C-C_2 interaction. Unfortunately there is no suitable general-purpose interaction potential available for atom-molecule interactions. However, Amdur and Mason and their co-workers have proposed[16] a relatively simple method for estimating such potentials: the peripheral force method. They have tested this method over a period of many years by comparing their derived potentials with potentials obtained directly from molecular beam scattering data. The results are generally quite good.

The peripheral force method is based on the assumption that atom-molecule and molecule-molecule potentials can be "built up" by an appropriate orientation averaging of the atom-atom potentials; in this case, the C-C potential which has already been considered in detail. For the C-C_2 interaction, the orientation averaged atom-molecule potential has the form[16,17]

$$V(r)_{av} = \frac{2}{4\Pi}\int_0^{\Pi} V(R)2\Pi\sin\theta d\theta \qquad (3)$$

where V(R) is the atom-atom potential, R the distance between the monatomic carbon atom and one of the carbon atoms in C_2, r the distance between the monatomic carbon atom and

Table 4 ER potential parameters for the states of C_2 that dissociate into ^3P carbon atoms

State	A,eV	B,$\mathring{A}^{-1}$	State	A,eV	B,$\mathring{A}^{-1}$
$^5\Sigma_u^-$	309.5	0.4060	$^3\Sigma_{u2}^+$	13.33	1.515
$^5\Pi_u$	400.0	0.3601	$^5\Sigma_{g2}^+$	186.5	2.270
$^5\Delta_g$	504.3	0.3931			

the center of geometry of C_2, and θ the angle between r and the C_2 bond axis.

When V(R) is the H-H potential, Eq. (3) gives the result

$$V(HH)_{av} = \frac{2\varepsilon}{\alpha(xr)^2}\, e^{xr_e}[e^{-xr(1+\alpha)}(xr[1+\alpha]+1)$$

$$- e^{-xr(1-\alpha)}(xr[1-\alpha]+1)] + \frac{\varepsilon}{\alpha}\, cx^2 e^{2xr_e} \cdot [B+E+F+G+H] \quad (4)$$

where

$$B = [(1+\alpha)^5 e^{-2xr(1+\alpha)} - (1-\alpha)^5 e^{-2xr(1-\alpha)}](-\tfrac{1}{2}bxr^3)$$

$$E = [(1+\alpha)^4 e^{-2xr(1+\alpha)}$$

$$- (1-\alpha)^4 e^{-2xr(1-\alpha)}]r^2(-\frac{1}{2} - \frac{5b}{4} + 2bxr_e)$$

$$F = [(1+\alpha)^3 e^{-2xr(1+\alpha)} - (1-\alpha)^3 e^{-2xr(1-\alpha)}]r$$

$$\cdot (-\frac{1}{x} + \frac{3}{2}r_e + 4br_e - \frac{5b}{2x} - 3bxr_e^2)$$

$$G = [(1+\alpha)^2 e^{-2xr(1+\alpha)} - (1-\alpha)^2 e^{-2xr(1-\alpha)}]$$

$$\cdot (-\frac{3}{2x^2} + \frac{9r_e}{4x} - \frac{3r_e^2}{2} + \frac{6br_e}{x} - \frac{9br_e^2}{2} + 2bxr_e^3 - \frac{15b}{4x^2})$$

and

$$H = [e^{-2xr(1+\alpha)}(2xr[1+\alpha]+1) - e^{-2xr(1-\alpha)}$$

$$\cdot (2xr[1-\alpha]+1)]\cdot(-\frac{1}{4cx^4r^2} - \frac{15b}{8x^4r^2} + \frac{3br_e}{x^3r^2}$$

$$- \frac{9br_e^2}{4x^2r^2} + \frac{br_e^3}{xr^2} - \frac{3}{4x^4r^2} + \frac{9r_e}{8x^3r^2} - \frac{3r_e^2}{4x^2r^2} + \frac{r_e^3}{4xr^2} - \frac{br_e^4}{4r^2})$$

Table 5 Collision integrals for monatomic carbon

$T, 10^{-3}$ K	$\sigma^2\Omega^{(1,1)*}$, Å^2	$\sigma^2\Omega^{(2,2)*}$, Å^2	$T, 10^{-3}$ K	$\sigma^2\Omega^{(1,1)*}$, Å^2	$\sigma^2\Omega^{(2,2)*}$, Å^2
1	8.5041	9.5400	14	4.0147	4.6428
2	7.1364	8.0090	15	3.8922	4.5207
3	6.4729	7.2222	16	3.7735	4.4041
4	6.0162	6.6956	17	3.6645	4.2938
5	5.6664	6.3116	18	3.5628	4.1888
6	5.5108	6.0069	19	3.4675	4.0910
7	5.2757	5.7629	20	3.3774	3.9958
8	4.9502	5.5491	21	3.2922	3.9055
9	4.7560	5.3625	22	3.2098	3.8175
10	4.5776	5.1913	23	3.1322	3.7329
11	4.4227	5.0396	24	3.0622	3.6517
12	4.2782	4.9022	25	2.9934	3.5765
13	4.1428	4.7697			

Also

$$\alpha = a/r$$

and a is half the C_2 bond length. When V(R) is the ER potential, Eq. (3) gives the result[18]

$$V(ER)_{av} = \frac{A}{\alpha(Br)^2}[e^{-Br(1-\alpha)}(Br[1-\alpha]+1) - e^{-Br(1+\alpha)}(Br[1+\alpha]+1)] \quad (5)$$

Using Eqs. (4) and (5), $V(r)_{av}$ was obtained for each of the 18 possible C-C interactions. The results for $V(r)_{av}$ obtained by using Eq. (4) were best fit with the Morse potential (MP), given by

$$V(MP) = \varepsilon[e^{-2A(r-r_e)} - 2e^{-A(r-r_e)}] \quad (6)$$

The best fit values of the parameters ε and r_e are given in Table 6. There is a third parameter, A, in Eq. (6) which can be used to optimize the fit of V(MP) to $V(H\text{-}H)_{av}$ for the calculation of transport collision integrals. The value of A which optimizes this fit near r_e was obtained. Then the collision integrals $\sigma_{MP}\Omega_{MP}^{(2,2)*}$ were found as a function of temperature, using tabulated results.[19] Hirschfelder and Eliason[20] have suggested that an improved curve fit for the calculation of transport properties can

Table 6 Parameters for the MP for the C-C_2 interactions with an attractive minimum

C-C state	ε,eV	r_e,Å	A,Å^{-1}	r range,Å
$^1\Sigma_g^+$	6.80	1.538	2.55	2.5-4.0
$^3\Pi_u$	7.18	1.618	2.20	2.75-4.25
$^3\Sigma_g^-$	6.46	1.691	2.075	2.80-4.5
$^1\Pi_u$	5.66	1.638	2.40	2.5-4.0
$^3\Sigma_u^+$	3.78	1.572	3.45	2.00-2.15
			3.65	2.15-2.35
			3.80	2.35-3.00
$^1\Delta_g$	4.27	1.750	2.55	2.400-2.625
			2.61	2.625-3.750
$^1\Sigma_{g2}^+$	3.53	1.738	2.79	2.20-2.35
			2.95	2.35-2.60
			3.10	2.60-3.50
$^3\Pi_g$	3.08	1.606	3.20	1.95-2.05
			3.35	2.05-2.25
			3.60	2.25-3.00
$^5\Pi_g$	2.95	1.853	2.680	2.200-2.350
			2.805	2.350-2.625
			3.050	2.625-3.500
$^5\Sigma_g^+$	1.53	1.770	3.80	1.80-2.15
			4.30	2.15-2.35
			5.10	2.35-2.80
$^1\Pi_g$	0.92	1.693	5.45	1.600-1.725
			5.00	1.725-1.925
			5.60	1.925-2.025
			6.50	2.025-2.350
$^3\Delta_u$	1.35	1.963	3.10	1.95-2.35
			3.30	2.35-2.55
			3.60	2.55-2.75
			4.10	2.75-3.10
$^1\Sigma_u^-$	1.03	2.512	3.00	2.400-2.975
			3.30	2.975-3.150
			3.90	3.150-3.600

be obtained by setting

$$\sigma_{ers}^2 = \sigma_{MP}^2 \Omega_{MP}^{(2,2)*}(T)$$

where σ_{ers}^2 is the "equivalent rigid sphere" cross section. The curve fitting is then redone by optimizing the fit between V(MP) and V(H-H)$_{av}$ at values of r near σ_{ers}. Since

σ_{ers} is a function of temperature, there is a range of r values to be used in optimizing the fit. The best fit value of A and the range of r values used in the fitting procedure are shown in Table 6. A similar procedure was used for the repulsive C-C_2 states; i.e., the results for $V(r)_{av}$ obtained by using Eq. (5) were fit with the ER potential, using the optimization procedure of Hirschfelder and Eliason.[20] Results are shown in Table 7.

Transport collision integrals are tabulated[14,19] for the MP and ER potentials. Thus collision integrals have been obtained for each of the 18 "states" corresponding to the C-C_2 interaction, using the peripheral force method. These have been averaged according to their degeneracies. The results are given in Table 8. The quantities A* and B* are ratios of collision integrals which are required to calculate the transport properties of gas mixtures, a mixture of C and C_2 in this case. These quantities are defined by

$$A^* = \frac{\Omega^{(2,2)*}}{\Omega^{(1,1)*}} \qquad B^* = \frac{5\Omega^{(1,2)*} - 4\Omega^{(1,3)*}}{\Omega^{(1,1)*}}$$

Now consider the C-C_3 interaction. The species C_3 is linear.[20] There is good evidence[17,21-23] that the central atom in a triatomic species and that nonperipheral atoms in polyatomic species can be ignored in applying the peripheral force method. Thus the C-C_3 collision integrals are the same as the C-C_2 collision integrals.

The C_2-C_2, C_2-C_3, and C_3-C_3 Interactions

Consider the C_2-C_2 interaction. Using the peripheral force method, the orientation averaged molecule-molecule potential for the C_2-C_2 interaction has the form[16]

$$V(r)_{av} = \frac{4}{(4\Pi)^2} \int_0^{\Pi}\int_0^{\Pi} V(R) 2\Pi \sin\theta_1 d\theta_1 2\Pi \sin\theta_2 d\theta_2 \qquad (7)$$

When the H-H potential is used in Eq. (7), a lengthy algebraic expression results. Substitution of the spectroscopic constants for the ground $^1\Sigma_g^+$ state of C_2 into this expression leads to a purely repulsive interaction. It is possible, but unlikely, that the bound state C-C inter-

Table 7 Parameters for the ER potential for the C-C_2 repulsive interactions

C-C state	A,eV	B,Å^{-1}	r range,Å
$^5\Sigma_u^-$	891.85	2.6996	2.20-3.50
$^5\Pi_u$	991.98	2.7330	2.20-3.50
$^5\Delta_g$	1238.3	2.5167	2.50-3.90
$^3\Sigma_{u2}^+$	26.272	1.4828	1.70-3.90
$^5\Sigma_{g2}^+$	429.06	2.2434	2.30-4.00

action leads to a repulsive C_2-C_2 interaction. It is more likely that the relatively simple orientation-averaging procedure used in the peripheral force method is being forced to average a potential that is algebraically too "complex" when the six-parameter H-H potential is used to represent atom-atom interactions for molecule-molecule potentials.

Since there are substantial uncertainties associated with the use of the peripheral force model for the C_2-C_2 interaction, it is reasonable to estimate the transport collision integrals for this interaction by using another method. Boushehri et al.[24] have proposed the use of a "universal" set of collision integrals. The primary justification for the assumption that these collision integrals can be applied to a variety of interactions is the observation[24] that the transport properties of the rare gases are relatively insensitive to the details of the interaction potential, i.e., the repulsive wall of the potential (which tends to vary only slightly from interaction to interaction) makes the primary contribution to the transport collision integrals at high temperatures. This set of collision integrals has been applied to a variety of atom-molecule and molecule-molecule interactions with good results.[24-29]

Table I in Boushehri et al.[24] contains the collision integrals $\Omega^{(1,1)*}$, $\Omega^{(2,2)*}$, and B* as a function of log (T*) where

$$T^* = kT/\varepsilon$$

This table has been used to calculate the transport collision integrals for the C_2-C_2 interaction. The parameters

Table 8 Collision integrals for the C-C_2 interactions

T, 10^{-3}K	$\sigma^2\Omega^{(1,1)*}$, $\mathring{A}^2$	$\sigma^2\Omega^{(2,2)*}$, $\mathring{A}^2$	A*	B*
1	10.9159	12.3063	1.1310	1.1197
2	9.5378	10.6120	1.1232	1.1357
3	8.7476	9.7077	1.1202	1.1468
4	8.1928	9.0751	1.1194	1.1579
5	7.7548	8.5858	1.1203	1.1708
6	7.3954	8.1976	1.1223	1.1834
7	7.1098	7.8987	1.1255	1.1972
8	6.8408	7.6229	1.1295	1.2093
9	6.6018	7.3817	1.1333	1.2187
10	6.3899	7.1739	1.1376	1.2297
11	6.1933	6.9806	1.1417	1.2370
12	6.0584	6.8512	1.1451	1.2476
13	5.8777	6.6728	1.1489	1.2516
14	5.7045	6.5007	1.1527	1.2548
15	5.5504	6.3467	1.1554	1.2571
16	5.4062	6.2019	1.1584	1.2586
17	5.2741	6.0681	1.1609	1.2594
18	5.1490	5.9403	1.1634	1.2606
19	5.0335	5.8212	1.1655	1.2605
20	4.9233	5.7072	1.1675	1.2604
21	4.8246	5.6064	1.1696	1.2616
22	4.7178	5.4938	1.1716	1.2606
23	4.6257	5.3959	1.1731	1.2595
24	4.5280	5.2918	1.1744	1.2585
25	4.4400	5.1968	1.1758	1.2569

ε and σ, the effective rigid sphere diameter, are those recommended by Esch et al.,[30] i.e.,

$$\varepsilon/k - 78.8\text{K} \qquad \sigma = 3.913\mathring{A}$$

Results are given in Table 9. Unfortunately, the correlations of Boushehri et al.[24] begin to "break down" when T* > 90. Thus results are only given to 8000K.

Exactly the same procedure has been used for the C_3-C_3 interaction, using the parameters[30]

$$\varepsilon/k = 128.0\text{K} \qquad \sigma = 4.450\mathring{A}$$

Results are given in Table 10.

Transport collision integrals for the C_2-C_3 interaction were also obtained by using the tabulations of Boushehri

Table 9 Collision integrals for the C_2-C_2 interaction

$T, 10^{-3}K$	$\sigma^2\Omega^{(1,1)*}, Å^2$	$\sigma^2\Omega^{(2,2)*}, Å^2$
1	10.81	12.23
2	9.511	10.92
3	8.795	10.17
4	8.302	9.637
5	7.933	9.229
6	7.639	8.899
7	7.390	8.620
8	7.174	8.379

Table 10 Collision integrals for the C_3-C_3 interaction

$T, 10^{-3}K$	$\sigma^2\Omega^{(1,1)*}, Å^2$	$\sigma^2\Omega^{(2,2)*}, Å^2$
1	15.30	17.09
2	13.46	15.30
3	12.49	14.31
4	11.82	13.62
5	11.32	13.09
6	10.91	12.65
7	10.58	12.29
8	10.29	11.97
9	10.05	11.70
10	9.828	11.45
11	9.630	11.23
12	9.450	11.02
13	9.312	10.84

Table 11 Collision integrals for the C_2-C_3 interaction

$T, 10^{-3}K$	$\sigma^2\Omega^{(1,1)*}, Å^2$	$\sigma^2\Omega^{(2,2)*}, Å^2$	A*	B*
1	12.90	14.51	1.125	1.111
2	11.36	12.98	1.143	1.117
3	10.52	12.11	1.151	1.122
4	9.947	11.51	1.157	1.126
5	9.512	11.04	1.161	1.127
6	9.167	10.66	1.163	1.128
7	8.880	10.34	1.164	1.127
8	8.633	10.06	1.165	1.127
9	8.415	9.822	1.167	1.126
10	8.220	9.605	1.168	1.126

et al.[24] The parameters ε and σ were obtained by using the empirical combining rules[30,31]

$$\frac{\varepsilon}{k} = \frac{\sqrt{\varepsilon_1 \varepsilon_2}}{k} = 100 \qquad\qquad \sigma = \frac{\sigma_1 + \sigma_2}{2} = 4.182$$

Results are given in Table 11.

Transport Properties

Using the kinetic theory of gases, the viscosity, η, and the translational contribution to the thermal conductivity, λ_{tr}, of a pure gas are given by[31]

$$\eta = 2.669 \times 10^{-5} \frac{\sqrt{MT}}{\sigma^2 \Omega^{(2,2)*}}, \quad \text{g/cm·s} \tag{8}$$

and

$$\lambda_{tr} = 0.1989 \frac{\sqrt{T/M}}{\sigma^2 \Omega^{(2,2)*}}, \quad \text{mcal/cm·s·K} \tag{9}$$

where M is the molecular weight. The binary diffusion coefficient, D_{ij}, is given by[31]

$$D_{ij} = \frac{2.628 \times 10^{-3} T^{3/2}}{p\sigma^2 \Omega^{(1,1)*}} \sqrt{\frac{M_i + M_j}{2M_i M_j}}, \ \text{cm}^2/\text{s} \tag{10}$$

where p is the pressure in atmospheres. To the extent that it is reasonable to assume that the transport of internal energy is due only to diffusion, the internal contribution to the thermal conductivity, λ^i_{int}, is given by[32]

$$\lambda^i_{int} = \lambda_{int} = \frac{pD}{RT}\left(C_p - \frac{5}{2} R\right), \ \text{mcal/cm·s·K} \tag{11}$$

where $D = D_{ii}$ is the self-diffusion coefficient, C_p the constant pressure heat capacity, and R the gas constant. If contributions from excited electronic states are ignored, λ_{int} for monatomic carbon is zero. The values of C_p used for C_2 and C_3 to 6000K came from the JANAF thermochemical tables.[21] The values of C_p for 6000K are used at higher temperatures. Results for these transport properties are shown in Tables 12-15.

Table 12 Transport properties of C

T, 10^{-3}K	η, 10^{4}g/cm·s	λ_{tr}, mcal/cm·s·K	D,[a] cm^2/s
1	3.07	0.190	2.82
2	5.17	0.321	9.50
3	7.02	0.435	19.3
4	8.74	0.542	31.9
5	10.4	0.643	47.3
6	11.9	0.740	64.0
7	13.4	0.833	84.2
8	14.9	0.925	110
9	16.4	1.02	136
10	17.8	1.11	166
11	19.3	1.19	198
12	20.7	1.28	233
13	22.1	1.37	271
14	23.6	1.46	313
15	25.1	1.56	358
16	26.6	1.65	407
17	28.1	1.74	459
18	29.6	1.84	514
19	31.2	1.93	573
20	32.7	2.03	635
21	34.3	2.13	701
22	35.9	2.23	771
23	37.6	2.33	845
24	39.2	2.44	921
25	40.9	2.56	1001

[a]At 1 atm pressure.

Consider mixtures. The viscosity, η^{mix}, is given by[31]

$$\eta^{mix} = - \frac{\begin{vmatrix} H_{ij} & X_i \\ X_j & 0 \end{vmatrix}}{\begin{vmatrix} H_{ij} \end{vmatrix}} \tag{12}$$

where

$$H_{ii} = \frac{X_i^2}{\eta_i} + \sum_{k \neq i} \frac{2X_i X_k}{\eta_{ik}} \frac{M_i M_k}{(M_i + M_k)^2} \cdot \left(\frac{5}{3A_{ik}^*} + \frac{M_k}{M_i}\right)$$

and

$$H_{ij}\ (j \neq i) = - \frac{2X_i X_j}{\eta_{ik}} \frac{M_i M_j}{(M_i + M_j)^2} \cdot \left(\frac{5}{3A_{ij}^*} - 1\right)$$

Table 13 Transport properties of C_2

T, 10^{-3}K	η, 10^4g/cm·s	λ_{tr}, mcal/cm·s·K	λ_{int}, mcal/cm·s·K	D, cm^2/s[a]
1	3.38	0.105	0.069	1.57
2	5.36	0.166	0.132	5.04
3	7.04	0.219	0.197	10.0
4	8.58	0.266	0.263	16.3
5	10.0	0.311	0.333	23.9
6	11.4	0.353	0.407	32.6
7	12.7	0.394	0.455	42.5
8	14.0	0.433	0.501	53.5

[a]At 1 atm pressure.

Table 14 Transport properties of C_3

T 10^{-3}K	η, 10^4g/cm·s	λ_{tr}, mcal/cm·s·K	λ_{int}, mcal/cm·s·K	D, cm^2/s[a]
1	2.96	0.061	0.063	0.905
2	4.68	0.097	0.128	2.91
3	6.13	0.127	0.185	5.76
4	7.44	0.154	0.238	9.37
5	8.65	0.179	0.287	13.7
6	9.81	0.203	0.335	18.7
7	10.9	0.226	0.373	24.2
8	12.0	0.248	0.410	30.4
9	13.0	0.269	0.448	37.2
10	14.0	0.289	0.479	44.6
11	15.0	0.309	0.513	52.5
12	15.9	0.329	0.546	60.9
13	16.9	0.349	0.577	69.7

[a]At 1 atm pressure.

The translational contribution to the thermal conductivity of a multicomponent gas mixture, λ_{tr}^{mix}, is given by[33]

$$\lambda_{tr}^{mix} = 4 \frac{\begin{vmatrix} L_{ij} & X_i \\ X_j & 0 \end{vmatrix}}{|L_{ij}|} \tag{13}$$

where

$$L_{ii} = -\frac{4X_i^2}{\lambda_i} - \sum_{k\neq i} \frac{2X_iX_k}{(M_i+M_k)^2 A_{ik}^* \lambda_{ik}} \left(\frac{15}{2} M_i^2 + \frac{25}{4} M_k^2 - 3B_{ik}^* M_k^2 + 4A_{ik}^* M_i M_k\right)$$

and

$$L_{ij}(j \neq i) = \frac{2X_i X_j}{(M_i+M_j)^2 A^*_{ij}\lambda_{ij}} \left(\frac{55}{4} - 3B^*_{ij} - 4A^*_{ij}\right)$$

The internal contribution to the thermal conductivity, λ^{mix}_{int}, is given by[32]

$$\lambda^{mix}_{int} = \sum_i \lambda^i_{int} \Big/ \sum_j \frac{X_j}{X_i} \frac{D_{ii}}{D_{ij}} \tag{14}$$

Clearly the mixture transport properties depend sensitively on mole fraction. Since many ablation products, in addition to C, C_2, and C_3, must be considered to do a complete transport calculation for an ablating heat shield, it is not prudent to do extensive calculations for the C, C_2, C_3 mixture. However, some results for λ, the most important transport property associated with planetary entry, are shown in Table 16. The mole fractions chosen have the same ratio as those given in Table 1. Clearly, the contribution from C_3, the most abundant species, dominates.

Discussion

The accuracy of these results is determined by the accuracy of the interaction potentials. The C-C, C-C_2,

Table 15 Binary diffusion coefficients, D,cm^2/s

T, 10^{-3}K	C-C_2	C-C_3	C_2-C_3	T, 10^{-3}K	C-C_2	C-C_3
1	1.90	1.79	1.20	14	191	180
2	6.16	5.81	3.85	15	217	205
3	12.3	11.6	7.65	16	246	232
4	20.3	19.1	12.5	17	276	260
5	29.9	28.2	18.2	18	308	290
6	41.3	38.9	24.8	19	342	322
7	54.1	51.0	32.3	20	377	356
8	68.7	64.8	40.6	21	414	391
9	84.9	80.1	49.7	22	454	428
10	103	96.9	59.6	23	495	467
11	122	115		24	539	508
12	142	134		25	584	551
13	166	156				

Table 16 Thermal conductivity, mcal/cm·s·K, of a mixture of C, C_2, and C_3

$X_C = 0.07$	$X_{C_2} = 0.08$	$X_{C_3} = 0.85$
$T, 10^{-3}K$	λ_{tr}^{mix}	λ_{int}^{mix}
1	0.069	0.061
2	0.176	0.231
3	0.287	0.402

Table 17 "Universal" collision integrals[24] for the C-C and C-C_2 interactions

$T, 10^{-3}K$	$\sigma^2\Omega^{(2,2)*}(C\text{-}C), Å^2$	$\sigma^2\Omega^{(2,2)*}(C\text{-}C_2), Å^2$
1	7.819	9.851
2	6.867	8.731
3		8.086

and C-C_3 interaction potentials should be accurate to within 5-10% since they have been accurately determined and/or based on well-tested procedures. Previous results[17] indicate that the method of Boushehri et al.[24] disagrees with the peripheral force method by only a few percent for interactions involving ablation products.

Collision integrals for the C-C and C-C_2 interactions, obtained by using the method of Boushehri et al.[24] with the parameters[30]

$$\varepsilon/k = 30.6K \qquad \sigma = 3.385Å$$

for the C-C interaction and

$$\varepsilon/k = 49.1K \qquad \sigma = 3.649Å$$

for the C-C_2 interaction, are shown in Table 17.

The "universal" collision integrals differ from those given in Tables 5 and 8 by about 20%. Thus it is reasonable to conclude that the collision integrals determined by using the method of Boushehri et al.[24] may be in error by as much as 50%. However, this does not mean that the error in the transport properties is this large. Roughly speaking, an error of a factor of two in the potential causes an error of 20-40% in the collision integrals.[34] Thus the error in the collision integrals is probably not greater than 20-25%.

Acknowledgment

This work was supported, in part, by NASA grant NASA NSG 1369. The authors would like to thank J. N. Moss for his help and advice.

References

[1]Moss, J. N., Anderson, E. C., and Bolz, C. W., Jr., "Aerothermal Environment for Jovian Entry Probes," AIAA Paper 76-469, July 1976.

[2]Moss, J. N., Anderson, E. C., and Bolz, C. W., Jr., "Viscous-Shock-Layer Solutions with Radiation and Ablation Injection for Jovian Entry," AIAA Paper 75-671, May 1975.

[3]Moss, J. N., Jones, J. J., and Simmonds, A. L., "Radiative Flux Penetration Through a Blown Shock Layer for Jupiter Entry Conditions," Outer Planet Entry Heating and Thermal Protection, Progress in Astronautics and Aeronautics, Vol. 64, edited by R. Viskanta, AIAA, New York, 1979, pp. 22-41.

[4]Moss, J. N., Private communication.

[5]Biolsi, L., "Transport Properties in the Jovian Atmosphere," Journal of Geophysical Research, Vol. 83, March 1, 1978, pp. 1125-1131.

[6]Biolsi, L., "Transport Properties of Monatomic Carbon," Journal of Geophysical Research, Vol. 83, June 1, 1978, pp. 2476-2480.

[7]Biolsi, L. and Biolsi, K. J., "Transport Properties of Monatomic Carbon, II: Contributions from Excited Electronic States," Journal of Geophysical Research, Vol. 84, Sept. 1, 1979, pp. 5311-5318.

[8]Fougere, P. F. and Nesbet, R. K., "Electronic Structure of C_2," The Journal of Chemical Physics, Vol. 44, Jan. 1, 1966, pp. 285-298.

[9]Hulburt, H. M. and Hirschfelder, J. O., "Potential Energy Functions for Diatomic Molecules," The Journal of Chemical Physics, Vol.9, Jan. 1941, pp. 61-69.

[10]Hulburt, H. M. and Hirschfelder, J. O., "Erratum: Potential Energy Functions for Diatomic Molecules," The Journal of Chemical Physics, Vol. 35, Nov. 1961, p. 1901.

[11]Steele, D., Lippincott, E. R., and Vanderslice, J. T., "Comparative Study of Empirical Internuclear Potential Functions," Reviews of Modern Physics, Vol. 34, April 1962, pp. 239-251.

[12]Huber, K. P. and Herzberg, G., Molecular Spectra and Molecular Structure, IV: Constants of Diatomic Molecules, Van Nostrand Reinhold, New York, 1979.

[13]Mason, E. A. and Monchick, L., "Methods for the Determination of Intermolecular Forces," Intermolecular Forces, edited by J. O. Hirschfelder, Interscience, New York, 1967, pp. 329-387.

[14]Monchick, L., "Collision Integrals for the Exponential Repulsive Potential," The Physics of Fluids, Nov. 1959, pp. 695-700.

[15]Vanderslice, J. T., Weissman, S., Mason, E. A., and Fallon, R. J., "High-Temperature Transport Properties of Dissociating Hydrogen," The Physics of Fluids, Vol. 5, Feb. 1962, pp. 155-164.

[16]Amdur, I., Mason, E. A., and Jordan, J. E., "Scattering of High-Velocity Neutral Particles, X: He-N_2; A-N_2. The N_2-N_2 Interaction," The Journal of Chemical Physics, Vol. 27, Aug. 1957, pp. 527-531.

[17]Biolsi, L., "Proposed Computational Method for Transport Properties of Ablation Products," AIAA Journal, Vol. 18, May 1980, pp. 596-597.

[18]Vanderslice, J. T., Mason, E. A., and Lippincott, E. R., "Interactions Between Ground-State Nitrogen Atoms and Molecules: The N-N, N-N_2, and N_2-N_2 Interactions," The Journal of Chemical Physics, Vol. 30, Jan. 1959, pp. 129-136.

[19]Smith, F. J. and Munn, R. J., "Automatic Calculation of the Transport Collision Integrals with Tables for the Morse Potential," The Journal of Chemical Physics, Vol. 41, Dec. 1961, pp. 3560-3568.

[20]Stull, D. R. and Prophet, H. (Eds.), JANAF Thermochemical Tables, NSRDS-NBS 37, June 1971.

[21]Amdur, I., Peters, W. A., Jordan, J. E., and Mason, E. A., "Scattering of High-Velocity Ar Atoms by CO_2, OCS, and CS_2," The Journal of Chemical Physics, Vol. 64, Feb. 15, 1976, pp. 1538-1540.

[22]Amdur, I., Longmire, M. S., and Mason, E. A., "Scattering of High-Velocity Neutral Particles. XII: He-CH_4; He-CF_4; CH_4-CH_4;

and CF_4-CF_4 Interactions," The Journal of Chemical Physics, Vol. 35, Sept. 1961, pp. 895-898.

[23]Amdur, I., Longmire, M. S., Jordan, J. E., and Mason, E. A., "Scattering of High-Velocity Neutral Particles, XVIII: He-(CH_3F, CH_2F_2, CHF_3)," The Journal of Chemical Physics, Vol. 63, Oct. 1, 1975, pp. 2926-2928.

[24]Boushehri, A., Viehland, L. A., and Mason, E. A., "On the Extended Principle of Corresponding States and the Pair Interaction Potential," Physica, Vol. 91A, May 1978, pp. 424-436.

[25]Kestin, J., Ro, S. T., and Wakeham, W. A., "Viscosity of the Noble Gases in the Temperature Range 25-700°C," The Journal of Chemical Physics, Vol. 56, April 15, 1972, pp. 4119-4124.

[26]Hellemans, J. M., Kestin, J., and Ro, S. T., "Viscosity of the Binary Gaseous Mixtures of Nitrogen with Argon and Krypton," The Journal of Chemical Physics, Vol. 57, Nov. 1, 1972, pp. 4038-4042.

[27]Hellemens, J. M., Kestin, J., and Ro, S. T., "The Viscosity of Oxygen and of Some of Its Mixtures with Other Gases," Physica, Vol. 65, April 15, 1973, pp. 362-375.

[28]Hellemens, J. M., Kestin, J., and Ro, S. T., "The Viscosity of CH_4, CF_4, and SF_6 Over a Range of Temperatures," Physica, Vol. 65, April 15, 1973, pp. 376-380.

[29]Kestin, J., Khalifa, H. E., Ro, S. T., and Wakeham, W. A., "The Viscosity and Binary Diffusion Coefficients of Eighteen Binary Gaseous Systems," Physica, Vol. 88A, Aug. 1977, pp. 242-260.

[30]Esch, D. D., Siripong, A., and Pike, R. W., "Thermodynamic Properties in Polynomial Form for Carbon, Hydrogen, Nitrogen, and Oxygen Systems from 300 to 15,000°K," NASA CR-111989, 1970.

[31]Hirschfelder, J. O., Curtiss, C. F., and Bird, R. B., Molecular Theory of Gases and Liquids, John Wiley & Sons, New York, 1954, Chaps. 7 and 8.

[32]Monchick, L., Yun, K. S., and Mason, E. A., "Formal Kinetic Theory of Transport Phenomena in Polyatomic Gas Mixtures," The Journal of Chemical Physics, Vol. 39, Aug. 1963, pp. 654-669.

[33]Muckenfuss, C. and Curtiss, C. F., "Thermal Conductivity of Multicomponent Gas Mixtures," The Journal of Chemical Physics, Vol. 29, Dec. 1958, pp. 1273-1277.

[34]Krupenie, P. H., Mason, E. A., and Vanderslice, J. T., "Interaction Energies and Transport Coefficients of Li + H and O + H Gas Mixtures at High Temperatures," The Journal of Chemical Physics, Vol. 39, Nov. 15, 1963, pp. 2399-2408.

Transport Properties Associated with Entry into the Atmosphere of Titan

Beverly Flori* and Louis Biolsi†
University of Missouri-Rolla, Rolla, Mo.

Abstract

In this paper the transport properties in the atmosphere of Titan and at the shock front and the probe surface for nonablative entry into Titan's atmosphere are calculated, using the kinetic theory of gases. The species N_2, H_2, CH_4, HCN, N, and H contribute to the transport properties under these conditions. The transport properties were calculated by using accurate potential energy curves for each of the 13 two-body interactions which occur between these species. Results are given for the binary diffusion coefficients, the nonreactive and reactive thermal conductivity, and the viscosity at 1000-8000 K. This information is needed for calculating flowfield properties and surface heating rates. The most important conclusions are that the nonreactive thermal conductivity is much larger behind the shock layer than it is in the ambient atmosphere and that the nonreactive thermal conductivity appears to be nearly constant in the shock layer.

Introduction

Titan, the largest moon of Saturn, has an atmosphere which is primarily composed of methane and molecular nitrogen.[1,2] The reddish-brown hue of Titan's atmosphere is attributed to the formation of large organic molecules, possibly including biological polymers, from the methane and nitrogen. This interesting atmosphere makes Titan a likely candidate for future unmanned exploration.

Presented as Paper 81-0278 at the AIAA 19th Aerospace Sciences Meeting, St. Louis, Mo., Jan. 12-15, 1981.

*Graduate Student, Chemistry Department.
†Associate Professor, Chemistry Department.

Preliminary calculations of the entry conditions into Titan's atmosphere have been completed.[2] In order to accurately determine the heating diffusion, and viscous flow rates at the surface of an entry vehicle, good estimates of the transport properties are needed. The purpose of this paper is to report reliable estimates of the transport properties of importance in determining these rates during entry into Titan's atmosphere. The kinetic theory of gases has been used to calculate the transport properties.

Two-Body Interaction Potentials

In order to use the kinetic theory of gases, it is necessary to know the two-body interaction potentials between the various species in the atmosphere. Assuming that Titan's atmosphere is 90% (mole) molecular nitrogen and 10% (mole) methane, Moss'[2] test run for nonablative entry gives the mole fractions shown in Table 1. The resulting two-body interactions are shown in Table 2.

Reliable, experimentally determined estimates of the interaction potential are available for some of the interactions listed in Table 2. The experimental data are usually well presented by the exponential repulsive (ER) potential, with the form

$$V(ER) = A\exp[-Br] \tag{1}$$

where A and B are empirical constants and r is the distance between the species. This has been done for the N-N, N-N_2, and N_2-N_2 interactions.[3,4]

Experimental data[5,6] for the H-N_2 and CH_4-CH_4 interactions were originally curve fit[5,6] with the repulsive inverse power (RIP) potential. However, we have curve fit the experimental results with the ER potential without any loss of accuracy. This has been done to minimize the number of different potentials used to calculate the transport properties.

The H-H($^1\Sigma$) interaction[7,8] was curve fit[9-11] with the attractive inverse power (AIP) potential, i.e.,

$$V(AIP) = -C/r^D \tag{2}$$

where C and D are empirical constants. The H-H($^3\Sigma$) interaction[7,12] was curve fit[9-11] with the ER potential.

Table 1 Mole fractions of species for nonablative entry into the atmosphere of Titan[2]

Species	Atmosphere	Shock front[a]	Probe surface[b]
N_2	0.90	0.22	0.78
CH_4	0.10		
N		0.51	
H		0.21	
H_2			0.13
HCN			0.06
Total	1.00	0.94	0.97

[a]The stagnation point temperature is 7352K and the pressure is 1.98 atm.

[b]The stagnation point temperature is 2778K and the pressure is 1.98 atm.

Table 2 Two-body interactions between species for nonablative entry into the atmosphere of Titan

Atmosphere	Shock front	Probe surface
N_2-N_2	N_2-N_2	N_2-N_2
CH_4-CH_4	N-N	H_2-H_2
N_2-CH_4	H-H	HCN-HCN
	N-H	N_2-H_2
	N-N_2	N_2-HCN
	H-N_2	H_2-HCN

Table 3 Two-body interaction potentials derived from experimental data

Interaction	Potential parameters[a]		Refs.
N-N	A = 86	B = 2.68	3,4
N-N_2	A = 285	B = 2.83	3,4
N_2-N_2	A = 450	B = 2.62	3,4
H-N_2	A = 1228	B = 4.26	5
CH_4-CH_4	A = 5×10^6	B = 5.63	6
H-H($^1\Sigma$)	C = 37.97	D = 6	7,8
	C = 12.43	D = 4.36	
H-H($^3\Sigma$)	A = 60.42	B = 3.013	7,8
H_2-H_2	α = 14.0	r_m = 37.3	7,13
		ε/k = 37.3	
	A = 116.5	B = 2.859	7,8

[a]The parameters are chosen so that V is in eV and r is in Å.

At low temperatures, the H_2-H_2 interaction potential parameters were obtained by making a best fit to viscosity and second virial coefficient data,[9,13] using the exponential-six (ES) potential, i.e.,

$$V(ES) = \frac{\varepsilon}{1-6/\alpha}\left[\frac{6}{\alpha}\, e^{\alpha(1-r/r_m)} - \left(\frac{r_m}{r}\right)^6\right] \tag{3}$$

where α, ε, and r_m are constants. At high temperatures, the ER potential was used to make a best fit[9,13] to a potential obtained[14] by using the perfect pairing method. This potential may be in error by as much as a factor of two.[14] It is the least reliable potential discussed so far.

The potentials discussed above, usually derived directly from experimental data, are summarized in Table 3. The other potentials needed to calculate the transport properties are discussed below. Most of them are obtained only indirectly from experimental data.

The N_2-CH_4 potential was obtained from the N_2-N_2 and CH_4-CH_4 potentials by using the empirical combining rule[15]

$$V(A\text{-}B) = \sqrt{V(A\text{-}A)V(B\text{-}B)} \tag{4}$$

The N-H potential was obtained by using the experimentally determined[5] N_2-H potential in conjunction with the peripheral force method, i.e.,[16,17]

$$V(N_2\text{-}H) = \frac{34}{r^{6.9}} = \frac{K}{r^s}\left[\frac{(1+d)^{s-2}-(1-d)^{s-2}}{2d(s-2)(1-d^2)^{s-2}}\right] \tag{5}$$

where K and s are the empirical constants of the RIP potential. Also

$$d = a/r$$

where a is half the N_2 bond length. Usually a, K, and s are known and V is calculated. However, in this case, V and a are known and K and s are to be calculated for the N-H interaction. The constant s was obtained by using data at two values of r to eliminate K from Eq. (5) and solving by trial and error. The K was found using Eq. (5). The results were best fit with the ER potential.

The N_2-H_2 potential was obtained by assuming that the H_2 molecule sees N_2 as a "point molecule" and averaging the experimentally determined H-N_2 potential listed in Table 3 over both H-N_2 interactions, using the peripheral force method.[16,17] For ER potentials, the averaged atom-molecule potential is[18]

$$V(r) = \frac{A}{2aB^2r} e^{-Br}[2(Br+1)\sinh(Ba) - (2Ba)\cosh(Ba)] \qquad (6)$$

where a is half the H_2 bond length. The results were best fit with the ER potential.

The procedure discussed above for finding the N_2-H_2 potential is not entirely consistent with the spirit of the peripheral force method.[16,17] An alternative approach is to use the N_2-N_2 and H_2-H_2 potentials listed in Table 3 and Eq. (4). However, the validity of the empirical combining laws is not well established for molecule-molecule interactions. Another alternative approach is to average the N-H potential over the four possible interactions according to the peripheral force method.[16-18] However, the N-H potential was itself obtained by using the peripheral force method. The authors believe that it is desirable to use experimentally determined potentials in as direct a way as possible as often as possible.

The N_2-HCN potential was obtained by assuming that HCN sees N_2 as a "point molecule" and averaging the H-N_2 and N-N_2 potentials listed in Table 3 by using Eq. (6) and taking a to be half the length of HCN. The results were best fit with the ER potential. The HCN molecule was assumed to be tumbling fast enough so that N_2 does not see the central carbon atom in HCN. This has been shown to be a good assumption for similar systems.[19]

The H_2-HCN potential was obtained by assuming that HCN sees H_2 as a "point molecule." This requires use of the H-H_2 and N-H_2 potentials, assuming that the C-H_2 interaction can be ignored. The H-H_2 potential is given by[14]

$$V(H\text{-}H_2) = 61.5\exp[-2.95r] \qquad (7)$$

However, this was obtained by using the perfect pairing method. As mentioned previously, it may be in error by a factor of two. The N-H_2 potential was obtained by using Eq. (6) to orientation average the N-H potential. The results were best fit with the ER potential, giving

$$V(N\text{-}H_2) = 2838\exp[-4.18r] \qquad (8)$$

Equation (8) was obtained by orientation averaging a potential that was obtained by orientation averaging. Clearly such a potential is less reliable than one which has been obtained from experimentally determined potentials in a more direct manner. The H_2-HCN potential was obtained by orientation averaging Eqs. (7) and (8), using Eq. (6) with a taken to be half the length of HCN. The results were best fit with the ER potential. This is the least reliable of the potentials.

The HCN-HCN potential was obtained by using the orientation averaged molecule-molecule potential for ER potentials, i.e.[18]

$$V(r) = \frac{A}{4B^3a^2r} e^{-Br}[2(Br+2)\{\cosh(2Ba)-1\} - (Ba)\sinh(2Ba)] \tag{9}$$

where a is half the length of HCN. It was assumed again that the carbon atom is not involved in the interaction. Thus the N-H, N-N, and H-H potentials were used in Eq. (9). However, the use of the H-H potential presents a problem. From Table 3, it is seen that there are different forms for the potential for the $^1\Sigma$ and $^3\Sigma$ states. One is tempted to obtain an average potential for the two states, but this is a poor procedure for calculating transport properties. A better procedure[9] is to calculate a different HCN-HCN potential for each H-H potential, then calculate the transport collision integrals for each interaction, and finally average the collision integrals. For the AIP potential representing the H-H($^1\Sigma$) state, the orientation averaged potential is

$$V(r) = - \frac{C}{r^D} \frac{(1+d)^{D-2}-(1-d)^{D-2}}{2d(D-2)(1-d^2)^{D-2}} \tag{10}$$

The HCN-HCN potentials were obtained by using the procedures described above. The results were best fit with the ER potential. The fit is not good for the interaction involving the H-H($^1\Sigma$) potential. This is to be expected since one of the terms contributing to the result involves an AIP potential. The fit is reasonable for the interaction involving the H-H($^3\Sigma$) potential but not as good as for the potentials discussed previously.

The potentials used to calculate the transport properties are summarized in Table 4.

Table 4 Potentials used to calculate the transport properties for nonablative entry into the atmosphere of Titan

Interactions	Potential parameters[a]	
Atmosphere		
N_2-N_2	$A = 450$	$B = 2.62$
CH_4-CH_4	$A = 5 \times 10^6$	$B = 5.63$
N_2-CH_4	$A = 4.75 \times 10^4$	$B = 4.13$
Shock front		
N-N	$A = 86$	$B = 2.68$
H-H($^1\Sigma$)	$C = 37.97$	$D = 6$
	$C = 12.43$	$D = 4.36$
H-H($^3\Sigma$)	$A = 60.42$	$B = 3.013$
N_2-N_2	$A = 450$	$B = 2.62$
N-H	$A = 1226$	$B = 4.26$
N-N_2	$A = 285$	$B = 2.83$
H-N_2	$A = 1228$	$B = 4.26$
Probe surface		
N_2-N_2	$A = 450$	$B = 2.62$
H_2-H_2	$\alpha = 14.0$	$r_m = 3.337$
		$\varepsilon/k = 37.3$
	$A = 116.5$	$B = 2.859$
HCN-HCN($^1\Sigma$)	$A = 23791$	$B = 3.80$
HCN-HCN($^3\Sigma$)	$A = 6086$	$B = 3.11$
N_2-H_2	$A = 2875$	$B = 4.19$
N_2-HCN	$A = 1700$	$B = 3.06$
H_2-HCN	$A = 446$	$B = 3.44$

[a]The parameters are chosen so that V is in eV volts and r in Å.

The N_2-N_2 transport collision integrals were actually obtained from the expressions[4]

$$\sigma^2\Omega^{(1,1)*} = \frac{60.4}{T^{0.274}} \qquad \sigma^2\Omega^{(2,2)*} = \frac{51.3}{T^{0.231}} \tag{11}$$

where $\sigma^2\Omega^{(1,1)*}$ is the diffusion collision integral, $\sigma^2\Omega^{(2,2)*}$ is the viscosity collision integral, and T is the temperature. These expressions were used in order to incorporate as much different experimental information as possible into the determination of the collision integrals.

Transport Properties

The collision integrals are available[9] for the H-H and H_2-H_2 interactions. The other interactions listed in Table 4 are represented by the ER potential. Monchick[20] tabulated the collision integrals for this potential as a function of the parameters A and B and temperature. The authors have calculated the collision integrals for each of the interactions listed in Table 4.

The binary diffusion coefficient, D_{ij}, is given by[15]

$$D_{ij} = \frac{2.6280 \times 10^{-3} T^{3/2}}{\sigma^2 \Omega^{(1,1)*} p} \sqrt{\frac{M_i + M_j}{2 M_i M_j}}, \ \mathrm{cm^2/s} \tag{12}$$

where p is the pressure and M_i is the molecular weight of species i. The viscosity, η, of a pure gas is given by[5]

$$\eta = 2.669 \times 10^{-5} \frac{\sqrt{MT}}{\sigma^2 \Omega^{(2,2)*}}, \ \mathrm{g/cm{\cdot}s} \tag{13}$$

and the translational contribution to the thermal conductivity, λ_{tr}, of a pure gas is given by[15]

$$\lambda_{tr} = 0.1989 \frac{\sqrt{T/M}}{\sigma^2 \Omega^{(2,2)*}}, \ \mathrm{mcal/cm{\cdot}s{\cdot}K} \tag{14}$$

To the extent that it is reasonable to assume that the transport of internal energy is due only to diffusion, the internal contribution to the thermal conductivity, λ^i_{int}, of a pure gas is given by[21]

$$\lambda^i_{int} = \lambda_{int} = \frac{pD_{ii}}{RT} (C_p - \tfrac{5}{2}R), \ \mathrm{mcal/cm{\cdot}s{\cdot}K} \tag{15}$$

where C_p is the constant pressure molar heat capacity and R is the gas constant. The values of C_p used in the calculations were obtained from the JANAF thermochemical tables.[22] Rotation and vibration were assumed to be fully excited above 6000K. Electronic excitation[23] of the species has been ignored. Thus λ_{int} is zero for the atoms. Results for D_{ij}, η, λ_{tr}, and λ_{int} are shown in Tables 5-8, respectively.

The viscosity of a multicomponent gas mixture, η^{mix}, is given by[15]

$$\eta^{mix} = - \frac{\begin{vmatrix} H_{ij} & X_i \\ X_j & 0 \end{vmatrix}}{|H_{ij}|} \tag{16}$$

Table 5 Binary diffusion coefficients, D_{ij} (cm^2/s), at 2 atm pressure

T, 10^{-3}K	N_2	CH_4	H_2	HCN	N	H	N-H
1	0.86	1.11	5.62	0.81	1.98	7.91	6.82
2	2.94	3.41	19.0	2.59	7.03	28.3	22.6
3	6.04	6.57	40.1	5.11	14.9	60.2	45.9
4	10.1	10.5	68.5	8.30	25.7	103	76.1
5	14.9	15.1	104	12.1	39.2	153	113
6	20.6	20.3	147	16.5	55.6	211	156
7	27.1	26.1	197	21.4	74.9	278	205
8	34.4	32.4	255	26.9	97.1	357	260

T, 10^{-3}K	N-N_2	H-N_2	N_2-CH_4	N_2-H_2	N_2-HCN	H_2-HCN
1	1.36	6.71	0.99	3.90	0.86	2.42
2	4.65	22.2	3.12	12.7	2.84	7.86
3	9.63	45.1	6.15	25.6	5.73	15.7
4	16.2	74.8	9.96	42.1	9.47	25.8
5	24.4	111	14.5	62.1	14.0	37.9
6	34.1	153	19.7	85.3	19.3	52.0
7	45.3	201	25.5	112	25.3	68.0
8	58.0	255	32.0	141	32.1	85.8

Table 6 Viscosity, η (10^4 g/cm·s), of the individual species

T, 10^{-3}K	N_2	CH_4	N	H	H_2	HCN
1	4.29	3.27	4.59	1.42	2.00	3.90
2	7.13	5.01	8.03	2.53	3.18	6.18
3	9.59	6.42	11.3	3.56	4.39	8.11
4	11.8	7.68	14.4	4.53	5.58	9.85
5	13.9	8.82	17.5	5.41	6.76	11.5
6	15.9	9.87	20.6	6.33	7.91	13.0
7	17.8	10.9	23.7	7.32	9.07	14.4
8	19.6	11.8	26.8	8.31	10.2	15.8

Table 7 Translational contribution to the thermal conductivity, λ_{tr} (10^1 mcal/cm·s·K), of the individual species

T, 10^{-3}K	N_2	CH_4	N	H	H_2	HCN
1	1.14	1.52	2.44	10.5	7.38	1.08
2	1.90	2.33	4.27	18.7	11.8	1.70
3	2.55	2.99	6.00	26.4	16.2	2.24
4	3.15	3.57	7.67	33.5	20.6	2.72
5	3.70	4.10	9.33	40.0	25.0	3.16
6	4.23	4.59	11.0	46.8	29.2	3.58
7	4.74	5.05	12.6	54.1	33.5	3.98
8	5.22	5.49	14.3	61.5	37.7	4.37

where

$$H_{ii} = \frac{X_i^2}{\eta_i} + \sum_{k\neq i} \frac{2X_iX_k}{\eta_{ik}} \frac{M_iM_k}{(M_i+M_k)^2} \left(\frac{5}{3A^*_{ik}} + \frac{M_k}{M_i}\right)$$

and

$$H_{ij}\ (j\neq i) = -\frac{2X_iX_j}{\eta_{ij}} \frac{M_iM_j}{(M_i+M_j)^2} \left(\frac{5}{3A^*_{ij}} - 1\right)$$

Also, X_i is the mole fraction of species i and A^*_{ij} is a ratio of collision integrals. The translational contribution to the thermal conductivity of a multicomponent gas mixture, λ_{tr}^{mix}, is given by[24]

$$\lambda_{tr}^{mix} = 4 \frac{\begin{vmatrix} L_{ij} & X_i \\ X_j & 0 \end{vmatrix}}{|L_{ij}|} \tag{17}$$

where

$$L_{ii} = -\frac{4X_i^2}{\lambda_i} - \sum_{k\neq i} \frac{2X_iX_k}{(M_i+M_k)^2 A^*_{ik}\lambda_{ik}} \left(\frac{15}{2} M_i^2 + \frac{25}{4} M_k^2 - 3B^*_{ik}M_k^2 + 4A^*_{ik}M_iM_k\right)$$

and

$$L_{ij}(j\neq i) = \frac{2X_iX_j}{(M_i+M_j)^2 A^*_{ij}\lambda_{ij}} \left(\frac{55}{4} - 3B^*_{ij} - 4A^*_{ij}\right)$$

Also, B^*_{ij} is a ratio of collision integrals. The internal contribution to the thermal conductivity of a multicomponent gas mixture, λ_{int}^{mix}, is given by[21]

$$\lambda_{int}^{mix} = \sum_i \lambda_{int}^i \Big/ \sum_j \frac{X_j}{X_i} \frac{D_{ii}}{D_{ij}} \tag{18}$$

Also

$$\lambda_{total}^{mix} = \lambda_f = \lambda_{tr}^{mix} + \lambda_{int}^{mix} \tag{19}$$

Table 8 Internal contribution to the thermal conductivity, λ_{int} (10^1 mcal/cm·s·K), of the individual species

T, 10^{-3}K	N_2	CH_4	H_2	HCN
1	0.597	3.29	3.08	0.696
2	1.30	7.30	7.45	1.41
3	1.91	10.3	12.7	2.00
4	2.46	12.7	18.3	2.52
5	2.99	14.9	24.3	2.99
6	3.52	16.8	30.7	3.41
7	3.97	18.9	35.3	3.80
8	4.40	20.6	39.9	4.17

Table 9 Mixture transport properties for the chemical composition in the atmosphere of Titan

T^a	$\eta^{mix\,b}$	$\lambda_{tr}^{mix\,c}$	$\lambda_{int}^{mix\,c}$
1	4.19	1.16	0.841
2	6.90	1.91	1.85
3	9.24	2.55	2.69
4	11.4	3.13	3.43
5	13.3	3.68	4.12
6	15.2	4.19	4.79
7	17.0	4.68	5.41
8	18.7	5.15	5.96

[a]T = 10^{-3}K.
[b]η^{mix} = 10^4g/cm·s
[c]λ^{mix} = 10^1mcal/cm·s·K.

Table 10 Mixture transport properties for the chemical composition at the shock front

T^a	$\eta^{mix\,b}$	$\lambda_{tr}^{mix\,c}$	$\lambda_{int}^{mix\,c}$	λ_r^c
1	4.46	0.867	0.230	
2	7.60	1.47	0.502	
3	10.5	2.03	0.739	
4	13.2	2.55	0.959	
5	15.8	3.05	1.17	
6	18.3	3.54	1.39	
7	20.8	4.02	1.57	112
8	23.3	4.50	1.76	

[a]T = 10^{-3}K.
[b]η^{mix} = 10^4g/cm·s.
[c]λ = 10^1mcal/cm·s·K.

where λ_f is called the "frozen" (i.e., nonreactive) thermal conductivity.

The contribution of chemical reactions to the thermal conductivity, λ_r, may greatly exceed the contribution of λ_f, where[9]

$$\lambda_{total} = \lambda_r + \lambda_f \tag{20}$$

The results shown in Table 1 indicate that the most important chemical reaction is the dissociation reaction

$$N_2 \rightleftarrows 2N$$

Also, this reaction is only important at the shock front. For a single dissociation reaction[25]

$$\lambda_r = \frac{pD_{ij}}{RT}\,\frac{(\Delta H)^2}{RT^2}\,\frac{X_i X_j}{(2X_i + X_j)^2} \tag{21}$$

where ΔH is the molar enthalpy of reaction, obtained from the JANAF thermochemical tables.[22] Results for the mixture transport properties in the atmosphere of Titan, at the shock front, and at the probe surface are given in Tables 9-11, respectively.

Discussion

The transport properties given in Tables 9-11 are needed as input in order to calculate heating and viscous

Table 11 Mixture transport properties for the Chemical Composition at the probe surface

T, 10^{-3}K	η^{mix}, 10^4 g/cm·s	λ_{tr}^{mix}, 10^1 mcal/cm·s·K	λ_{int}^{mix}, 10^1 mcal/cm·s·K
1	4.27	1.64	0.886
2	7.06	2.70	1.96
3	9.49	3.64	2.96
4	11.7	4.49	3.92
5	13.8	5.30	4.86
6	15.8	6.07	5.81
7	17.6	6.81	6.53
8	19.5	7.53	7.23

Table 12 Mixture transport properties in Titan's atmosphere, at the shock front, and at the probe surface for stagmation point peak heating during nonablative entry

	η^{mix}, 10^4 g/cm·s	λ_{tr}^{mix}	λ_{tr}^{mix}	λ_f	λ_r
Atmosphere (120K)[a]	0.918	0.265	0.065	0.330	
Shock front (7352K)	21.7	4.19	1.64	5.83	106
Probe Surface (2778K)	8.97	3.44	2.75	6.19	

[a]The temperature of Titan's ambient atmosphere is 120K.
λ is in mcal/cm·s·K.

flow rates. Almost all of the uncertainty in these calculations is due to the uncertainty in the two-body interaction potentials. This has been discussed above. Accurate, experimentally tested potentials have been used for as many interactions as possible. For the remaining interactions, the peripheral force method[16,17] has been used to obtain the potentials, with experimentally determined potentials as input. This method has been tested[19] for a variety of linear, nonlinear, symmetric, and unsymmetric molecules by comparison with molecular beam data and found to be a satisfactory method for estimating potentials for interactions between small atoms and/or molecules. Thus the potentials used for these calculations should be quite reliable.

The trends in the viscosity and thermal conductivity obtained as one moves from the pure atmosphere to the probe surface are shown in Table 12. These results indicate that the thermal conductivity and viscosity are much greater in the shock layer than in Titan's atmosphere. More important, the nonreactive thermal conductivity, λ_f, remains nearly constant across the shock layer; in fact, it increases slightly. In addition, the viscosity decreases across the shock layer. Of course, injection of gaseous products from ablation of the heat shield might drastically alter the chemical composition, changing the transport properties.[26]

The results for the total thermal conductivity, λ_{total}, including the reactive thermal conductivity, λ_r, shown in Table 12 indicate that there is a drastic reduction in λ_{total} across the shock layer. However, this result may be misleading. There are small amounts of other species at the shock front and at the probe surface which have not been included in these calculations. For instance, the species N^+, C, and C^+ are present at the shock front at very low concentration. Thus there are contributions to

λ_r from the reactions

$$N \rightleftarrows N^+ + e$$

and

$$C \rightleftarrows C^+ + e$$

Also, at the probe surface, there is an appreciable concentration of H and small amounts of CN. Thus the reactions

$$H_2 \rightleftarrows 2H$$

and

$$HCN \rightleftarrows H + CN$$

contribute to λ_r. Even though the mole fractions of the species mentioned above are small, λ_r dominates λ_f for this particular entry problem by such a large amount that contributions to λ_r should be considered even for species present at very low concentration. However, it is reasonable to also include the effects of ablation on the chemical composition before attempting such a detailed calculation.

Acknowledgment

This work was supported by NASA Grant NSG 1369. The authors would like to thank J. N. Moss for his help and advice.

References

[1]Gross, S. H., "The Atmosphere of Titan," Reviews of Geophysics and Space Physics, Vol. 12, Aug. 1974, pp. 435-446.

[2]Moss, J. N., Private communication.

[3]Jordan, J. E., Colgate, S. O., Amdur, I., and Mason, E. A., "Scattering of High-Velocity Neutral Particles, XVII: Ar-O_2, Ar-CO," The Journal of Chemical Physics, Vol. 52, Feb. 1, 1970, pp. 1143-1149.

[4]Cubley, S. J. and Mason, E. A., "Atom-Molecule and Molecule-Molecule Potentials and Transport Collision Integrals for High-Temperature Air Species," The Physics of Fluids, Vol. 18, Sept. 1975, pp. 1109-1111.

[5]Belyaev, Yu. N. and Leonas, V. B., "Features of Scattering of Fast Beams of H, N, and O Atoms in Molecular Gases (N_2, O_2)," JETP Letters, Vol. 4, Aug. 15, 1966, pp. 92-94.

[6]Amdur, I., Longmire, M. S., and Mason, E. A., "Scattering of High-Velocity Neutral Particles, XII: He-CH_4, He-CF_4, CH_4-CH_4, and CF_4-CF_4 Interactions," The Journal of Chemical Physics, Vol. 35, Sept. 1961, pp. 895-898.

[7]Vanderslice, J. T., Weissman, S., Mason, E. A., and Fallon, R. J., "High Temperature Transport Properties of Dissociating Hydrogen," The Physics of Fluids, Vol. 5, Feb. 1962, pp. 155-164.

[8]Biolsi, L., "Transport Properties in the Jovian Atmosphere," Journal of Geophysical Research, Vol. 83, March 1, 1978, pp. 1125-1131.

[9]Dalgarno, A. and Lynn, N., "Resonance Forces at Large Separations," Proceedings of the Physical Society (London), Sec. A., Vol. 69, Nov. 1956, pp. 821-829.

[10]Vanderslice, J. T., Mason, E. A., and Maisch, W. G., "Ground State of Hydrogen by the Rydberg-Klein-Rees Method," Journal of Molecular Spectroscopy, Vol. 3, Feb. 1959, pp. 17-29.

[11]James, H. H., Coolidge, A. S., and Present, R. D., "The Heitler-London Repulsive State of Hydrogen," The Journal of Chemical Physics, Vol. 4, March 1936, pp. 187-193.

[12]Fallon, R. J. Mason, E. A., and Vanderslice, J. T., "Energies of Various Interactions Between Hydrogen and Helium Atoms and Ions," Astrophysical Journal, Vol. 131, Jan. 1960, pp. 12-14.

[13]Mason, E. A. and Rice, W. E., "The Intermolecular Potentials of Helium and Hydrogen," The Journal of Chemical Physics, Vol. 22, March 1954, pp. 522-535.

[14]Vanderslice, J. T. and Mason, E. A., "Interaction Energies for the H-H_2 and H_2-H_2 Systems," The Journal of Chemical Physics, Vol. 33, Aug. 1960, pp. 492-494.

[15]Hirschfelder, J. O., Curtiss, C. F., and Bird, R. B., Molecular Theory of Gases and Liquids, John Wiley & Sons, New York, 1954, Chaps. 7 and 8.

[16]Amdur, I., Mason, E. A., and Jordan, J. E., "Scattering of High-Velocity Neutral Particles, X: He-N_2; A-N_2. The N_2-N_2 Interaction," The Journal of Chemical Physics, Vol. 27, Aug. 1957, pp. 527-531.

[17]Biolsi, L., "Proposed Computational Method for Transport Properties of Ablation Products," AIAA Journal, Vol. 18, May 1980, pp. 596-597.

[18]Vanderslice, J. T., Mason, E. A., and Lippincott, E. R., "Interactions Between Ground State Nitrogen Atoms and Molecules. The N-N, N-N_2, and N_2-N_2 Interactions," The Journal of Chemical Physics, Vol. 30, Jan. 1959, pp. 129-136.

[19]Amdur, I., Peters, W. A., Jordan, J. E., and Mason, E. A., "Scattering of High-Velocity Ar Atoms by CO_2, OCS, and CS_2," The Journal of Chemical Physics, Vol. 64, Feb. 15, 1976.

[20]Monchick, L., "Collision Integrals for the Exponential Repulsive Potential," The Physics of Fluids, Vol. 2, Nov. 1959, pp. 695-700.

[21]Monchick, L., Yun, K. S., and Mason, E. A., "Formal Kinetic Theory of Transport Phenomena in Polyatomic Gas Mixtures," The Journal of Chemical Physics, Vol. 39, Aug. 1963, pp. 654-669.

[22]Stull, D. R. and Prophet, H. (Eds.), JANAF Thermochemical Tables, NSRDS-NBS 37, June 1971.

[23]Biolsi, L. and Biolsi, K. J., "Transport Properties of Monatomic Carbon, II: Contributions from Excited Electronic States," Journal of Geophysical Research, Vol. 84, Sept. 1, 1979, pp. 5311-5318.

[24]Muckenfuss, C. and Curtiss, C. F., "Thermal Conductivity of Multicomponent Gas Mixtures," The Journal of Chemical Physics, Vol. 29, Dec. 1958, pp. 1273-1277.

[25]Butler, J. N. and Brokaw, R. S., "Thermal Conductivity of Gas Mixtures in Chemical Equilibrium," The Journal of Chemical Physics, Vol. 26, June 1957, pp. 1636-1643.

[26]Biolsi, L. and Wallace, L. R., "Some Effects of Ablation on Transport Properties in the Jovian Atmosphere," Outer Planet Entry Heating and Thermal Protection, Progress in Astronautics and Aeronautics, Vol. 64, edited by R. Viskanta, AIAA, New York, 1979, pp. 65-79.

Thermal Conductivity of Partially Ionized Gas Mixtures

Bassem F. Armaly*
University of Missouri-Rolla, Rolla, Mo.
and
Kenneth Sutton†
NASA Langley Research Center, Hampton, Va.

Abstract

A method is proposed for predicting the translational component of the thermal conductivity of partially ionized gas mixtures. It is approximate but simple in form and offers a significant improvement over commonly utilized approximations. It does not require large computer run times nor storage, and thus is suitable for use with complex flowfields and heat-transfer calculations. Results for gas mixtures which are representative of the atmosphere of Jupiter, Earth, and Venus are presented and they compare favorably with results from detailed kinetic theory analyses.

Introduction

The flowfield of and the heat transfer to bodies either re-entering the Earth's atmosphere or entering the atmospheres of other planets can be predicted by detailed and complex computer codes.[1-4] The high temperatures associated with these probes ionize and dissociate the surrounding gases, and this produces a large number of different molecular species in the boundary layer and in the viscous shock layer that surround the probes. The viscosity

---

Presented as Paper 81-1174 at the AIAA 16th Thermophysics Conference, Palo Alto, Calif., June 23-25, 1981.

*Professor, Thermal Radiative Transfer Group, Department of Mechanical and Aerospace Engineering.

†Research Leader, Aerothermodynamics Branch, Space Systems Division.

and the thermal conductivity of such a mixture are needed in order to predict the flowfield and the heat transferred to the probe. It has been shown[5] that a 25-40% difference in convective heating rates can result from using different models for predicting viscosity and thermal conductivity of gas mixtures.

Several investigators have examined the thermal conductivity and viscosity of partially ionized gas mixtures. Biolsi[6] has published a report on the H_2-He gas mixtures in the atmosphere of Jupiter, Yos[7] on the air in the Earth's atmosphere, and Freeman and Oliver[8] on the CO_2-N_2 gas mixtures in the atmosphere of Venus. They all utilized the Chapman-Enskog[9] kinetic theory formulas and provided appropriate collision integrals for the various interactions that occur in these mixtures at high temperatures. Because their detailed method requires extensive computer time and storage, it is not practical to include it directly in existing detailed and complex flowfield codes. It is, however, used as a reference for evaluating approximate methods which can be easily added to flowfield and heat-transfer codes.

Several approximate methods for evaluating the viscosity[10-14] and thermal conductivity[15-20] of gas mixtures have been proposed in the literature. The most widely used was suggested by Wilke[11] for viscosity and by Mason and Saxena[16] for thermal conductivity. Recently, Armaly and Sutton[21] presented a new approximation for the mixture's viscosity which resulted in a significant improvement over Wilke's approximation. The purpose of this paper is to present a similar approximation for the translational component of the thermal conductivity which will offer an improvement over the Mason and Saxena approximation.

Development

The thermal conductivity of a gas mixture, λ_{mix}, can be expressed in terms of three components as follows:

$$\lambda_{mix} = \lambda_{tr}^{mix} + \lambda_{int}^{mix} + \lambda_{r}^{mix} \tag{1}$$

where λ_{tr}^{mix} is the translational component, λ_{int}^{mix} the internal component, and λ_{r}^{mix} the reactive component. At high temperatures when molecules dissociate and ionize the internal component becomes negligible and in the absence of chemical reaction the translational component becomes the dominant part. An approximation to that component will be developed in this paper.

According to the rigorous kinetic theory, the translational component is given by[16]

$$\lambda_{tr}^{mix} = 4 \begin{vmatrix} L_{11} & & L_{10} & X_1 \\ \cdot & & \cdot & \cdot \\ \cdot & & \cdot & \cdot \\ \cdot & & \cdot & \cdot \\ L_{n1} & \cdots & L_{nn} & X_n \\ X_1 & \cdots & X_n & 0 \end{vmatrix} \Bigg/ \begin{vmatrix} L_{11} & & L_{1n} \\ \cdot & & \cdot \\ \cdot & & \cdot \\ \cdot & & \cdot \\ L_{n1} & \cdots & L_{nn} \end{vmatrix} \tag{2}$$

in which X_i is the mole fraction of the ith component, and the elements L_{ij} are given by

$$L_{ii} = -4\, X_i^2/\lambda_i - \sum_{\substack{k=1 \\ k\neq i}}^{n} 2X_iX_k[(15/2)M_i^2 + (5/2)M_k^2 + 4\, M_iM_kA_{ik}^*]/((M_i + M_k)^2A_{ik}^*\lambda_{ik}) \tag{3}$$

and

$$L_{ij} = 2\, X_iX_jM_iM_j(10 - 4A_{ij}^*)/((M_i + M_j)^2\lambda_{ij}A_{ij}^*) \tag{4}$$

in which λ_i is the thermal conductivity of ith pure component, λ_{ij} the binary thermal conductivity for components i and j, M the molecular weights, and A_{ij}^* the dimensionless ratios of the collision integrals $\Omega_{ij}^{(2,2)*}/\Omega_{ij}^{(1,1)*}$. The thermal conductivity of a pure gas, λ_i, is given by

$$\lambda_i = C\,\sqrt{T/M_i}/(\sigma_i^2\Omega_{ii}^{(2,2)*}) \tag{5}$$

and the thermal conductivity of a binary mixture is given by

$$\lambda_{ij} = C\,\sqrt{T(M_i + M_j)/2M_iM_j}/(\sigma_{ij}^2\Omega_{ij}^{(2,2)*}) \tag{6}$$

in which C is a constant that depends on the units used in the evaluation, T the absolute temperature, and σ_{ij} the collision diameter associated with the interaction ij.

The complexities and the large number of interaction potentials, which are required for calculating the collision integrals $\Omega_{ij}^{(2,2)*}$, $\Omega_{ij}^{(1,1)*}$, make the above detailed formulation impractical for use in a complex flowfield calculation. The proposed approximation for predicting the translational component of the thermal conductivity is similar to the one developed by the authors[21] for predicting the viscosity of a gas mixture. Equation (2) can be replaced by an infinite series equivalent to

$$\lambda_{tr}^{mix} = -4 \sum_{i=1}^{n} \frac{X_i^2}{L_{ii}} + 4 \sum_{i=1}^{n} \sum_{\substack{j=1 \\ j \neq i}}^{n} \frac{X_i X_j L_{ij}}{L_{ii} L_{jj}} + \ldots \qquad (7)$$

and the off-diagonal elements L_{ij} have been shown[6] to be much smaller than the diagonal elements L_{ii}, thus permitting the following approximation

$$\lambda_{tr}^{mix} = -4 \sum_{i=1}^{n} X_i^2/L_{ii} \qquad (8)$$

To utilize Eq. (8), values for σ_i, A_{ij}^*, and $\sigma_{ij}^2\Omega_{ij}^{(2,2)*}$ are needed. These values are not normally available in the literature. Biolsi[6,22] reported the behavior of these functions for the case of an H_2-He gas mixture in the temperature range of 1,000-25,000K and accounted for both dissociation and ionization by examining the 21 different types of interactions that occur in this mixture. His study and the one by Yun and Mason[23] are the only detailed reports that give data of this type and deal with interactions of a specific mixture. Others have examined and published collision integrals for various atom-atom interactions.[24-29] The following observations can be made:

1) The parameter A_{ij}^* can be approximated as being independent of temperature, but its magnitude differs for different types of interactions and it is different than the value of 5/2 assumed in all previous approximations.

2) The parameter $\sigma_{ij}^2\Omega_{ij}^{(2,2)*}$ is a strong function of temperature and its magnitude differs for different types of interactions.

The last feature is undesirable because it requires a different functional relation for each interaction. To remedy this behavior, it is proposed to express σ_{ij}^2 in terms of the pure-gas collision diameter σ_i, σ_j, similar to the approximation used by the authors[21] in developing the expression for the viscosity. With this approximation it is possible to replace Eq. (6) by the following expression:

$$\lambda_{ij} = \lambda_i \sqrt{8(M_i + M_j)/M_j}/[F_{ij} + \left(\frac{\lambda_i}{\lambda_j}\right)^{1/2} \left(\frac{M_i}{M_j}\right)^{1/4} B_{ij}]^2 \qquad (9)$$

where

$$F_{ij} = \left\{ \frac{\Omega_{ij}^{(2,2)*}}{\Omega_{ii}^{(2,2)*}} \right\}^{1/2}, \quad B_{ij} = \left\{ \frac{\Omega_{ij}^{(2,2)*}}{\Omega_{jj}^{(2,2)*}} \right\}^{1/2} \qquad (10)$$

The advantages of Eq. (9) over Eq. (6) result from the fact that λ_{ij} is now expressed in terms of individual pure-gas properties, which are more readily available than the ones associated with interactions between two different species, and the parameters F_{ij}, B_{ij} are not a strong function of temperature. For the same general type of interaction, i.e., atom-atom and atom-ion, etc., F_{ij} and B_{ij} can be considered as constants independent of the species involved in these interactions.

The Mason and Saxena[16] approximation can be deduced from the above development if the following restrictions are imposed:

1) $A_{ij}^* = 5/2$

2) $F_{ij} = 1$

3) $B_{ij} = 1$

4) $(3 M_i + M_j)/(2(M_i + M_j)) = 1$ or $M_i = M_j$

These restrictions are not realistic and will result in significant errors when ionizations take place in the gas mixture as will be shown in the following sections. The use of these restrictions will result in the following simplified expression for the translational component of the thermal conductivity of a gas mixture:

$$\lambda_{tr}^{mix} = \sum_{i=1}^{n} \lambda_i/(1 + \sum_{\substack{k=1 \\ k\neq i}}^{n} (X_k/X_i)G_{ik}) \quad (11)$$

where

$$G_{ik} = [1 + (\lambda_i/\lambda_k)^{1/2}(M_i/M_k)^{1/4}]^2/[8(1 + M_i/M_k]^{1/2} \quad (12)$$

The above two equations are equivalent to the approximation of Mason and Saxena.[16]

Proposed Approximation

Equations (3), (8), and (9) are the governing relations for the proposed approximation. The translational component of the thermal conductivities for the pure components are assumed to be known and the magnitude of A_{ij}^*, F_{ij}, and B_{ij} are needed in order to utilize these relations. Detailed discussion of these parameters and a procedure for approximating their magnitudes was described by the authors.[21] The final conclusions are repeated here:

1) The off-diagonal member in Eq. (2), L_{ij}, was assumed to be negligible without enforcing the magnitude of A_{ij}^* to be equal to 5/2. Exact results[6,22] indicate that indeed L_{ij} is significantly smaller than L_{ii} and that the magnitude of A_{ij}^* could be significantly different than 5/2, especially for some ion-atom interactions.

2) $A_{ij}^* = A_{ji}^* = 1.25$ for all interactions except atom interactions with its own ion. For that case, the following values are recommended:

$$A_{H-H^+}^* = 0.18,\ A_{He-He^+}^* = 0.025,\ A_{C-C^+}^* = 1.10$$

For other atom interactions with their own ions, the value can be assumed to be equal to that of C-C^+ interactions.

3) $F_{ij} = F_{ji} = 1.0$ for all interactions.

4) $B_{ij} = B_{ji}$ = as outlined below:

= 0.2 for atom and molecule interactions with electrons

= 0.15 for atom and molecule interactions with ions

= 0.78 for atom and molecule interactions with atoms and molecules

= 1.0 for all other interactions

The above were deduced from a detailed study of the available exact interaction potentials and the exact viscosity data of Jupiter atmosphere[6] and were used by the authors[21] in the approximation of mixture viscosity. The same values are used in this study to approximate the mixture thermal conductivity.

Results and Discussion

The proposed approximate method consists of utilizing the stated values for the three constants, A^*_{ij}, F_{ij}, and B_{ij}, with Eqs. (3), (8), and (9) to evaluate the conductivity of gas mixtures. The values of these constants are identical to the ones used by the authors[21] for calculating viscosity of gas mixtures. The translational component of the thermal conductivity for the pure-gas species λ_i, which is needed for calculating the mixture's conductivity, was calculated from available viscosity data[21] and the following relation

$$\lambda_i - C_1 \eta_i / M_i \qquad (13)$$

where C_1 is a dimensional constant and η_i the viscosity. Calculations were made for various mixtures and the results are compared with exact values, when possible, and with results obtained by using the approximation of Mason and Saxena.[16]

The conductivity of dissociated hydrogen and of ionized atomic hydrogen are presented in Figs. 1 and 2, respectively, as a function of temperature. The results are for an equilibrium composition at a pressure of 1 atm. It is clear from these figures that the proposed and existing[16] approximations predict with good accuracy the

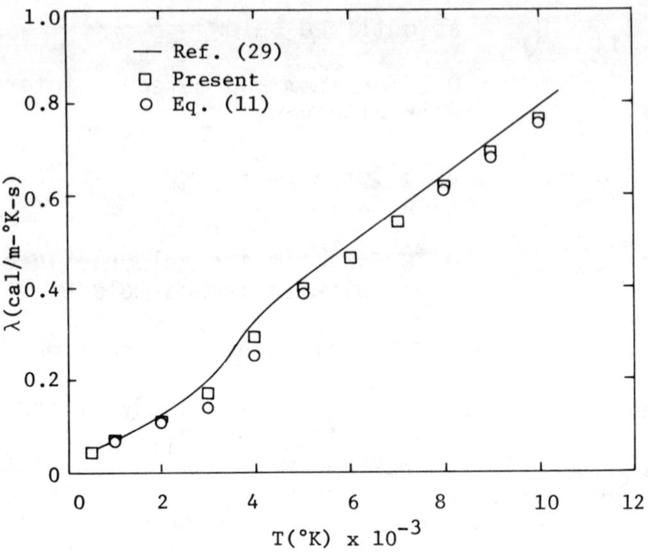

Fig. 1 Thermal conductivity of dissociated hydrogen.

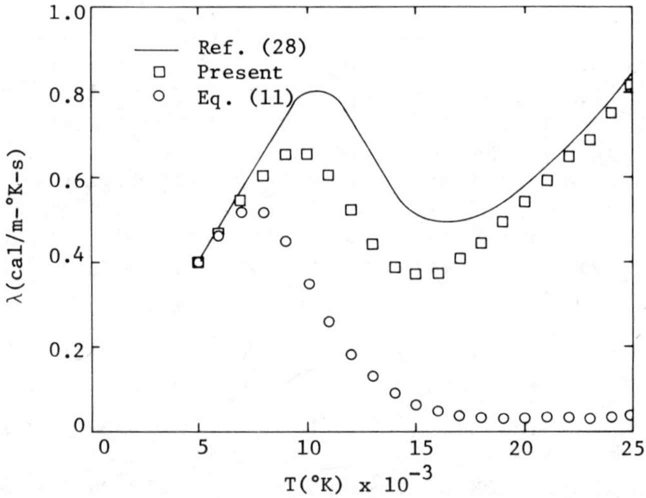

Fig. 2 Thermal conductivity of ionized atomic hydrogen.

thermal conductivity of dissociated hydrogen, but the latter starts to deviate significantly from exact values for ionized hydrogen. Similar to the Wilke[11] approximation for viscosity, this approximation fails when ionization starts to appear in the mixture. The proposed approximation offers a significant improvement in that region.

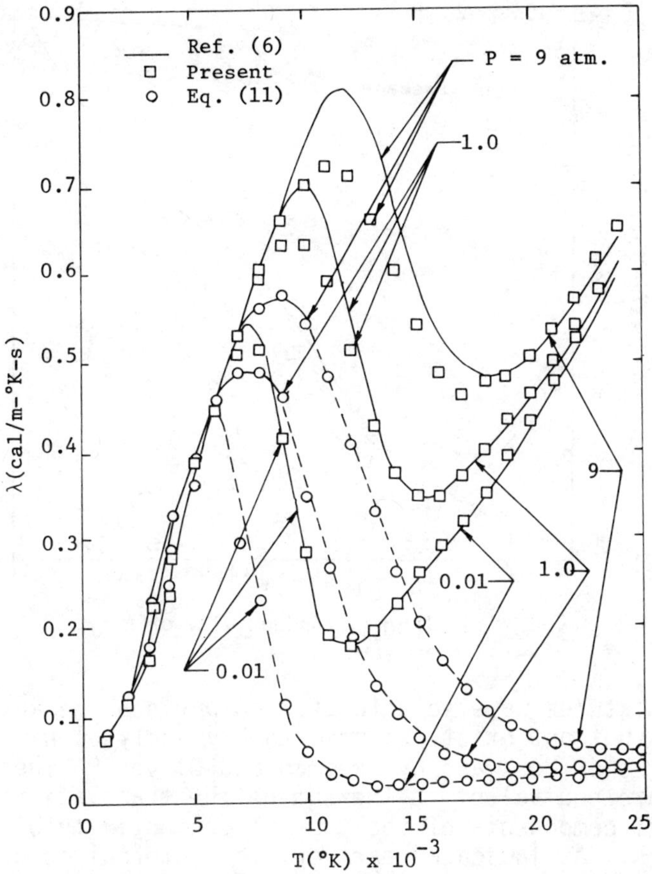

Fig. 3 Thermal conductivity of 0.89 H_2 - 0.11 He mixture by volume.

The conductivity of a hydrogen-helium (0.89 H_2-0.11 He, by volume) gas mixture is presented as a function of temperature in Fig. 3. The results are for equilibrium compositions at the indicated pressures. The Mason and Saxena[16] approximations deviates significantly from exact values[6] when ionization begins to occur in the mixture. The proposed approximation provides good agreement with exact values. This mixture received significant attention recently in the open literature because it represents the atmospheric composition of Jupiter.

The proposed approximation was applied to gas mixtures representative of the Earth atmosphere (air) and a Venus atmosphere model (0.50 CO_2-0.50 N_2 by volume).

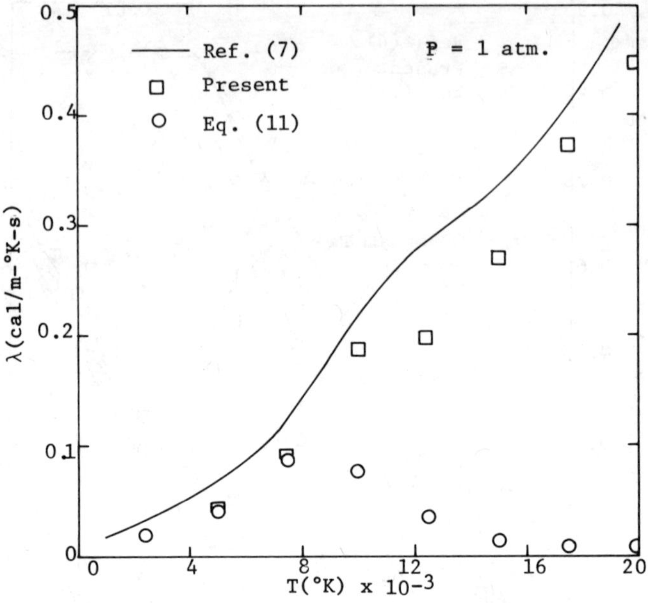

Fig. 4 Thermal conductivity of air.

These mixtures were selected for comparisons because detailed (but not exact) kinetic theory analyses have been performed by Yos[7] and by Freeman and Oliver.[8] Their results are equivalent to the sum of the translational and internal components of the thermal conductivity of these mixtures. As indicated earlier, the internal component is negligible at high temperatures. The comparisons are presented as a function of temperature for an equilibrium composition at 1 atm in Figs. 4 and 5. They demonstrate again the weakness of the Mason and Saxena[16] approximation and the improvements which are obtained by using the proposed approximation.

The present method was also applied to predict the thermal conductivity of a mixture consisting of 92% carbon, 6% oxygen, and 2% hydrogen by elemental mass fraction (the composition of a carbon-phenolic heat shield) and of a mixture consisting of 90% N_2 and 10% CH_4 by volume (approximate composition of Titan atmosphere). The results are presented as a function of temperature at an equilibrium pressure of 1 atm in Figs. 6 and 7. Exact results for these mixtures are not available for comparison and the results demonstrate the difference between the proposed and the existing[16] approximation. The conductivity of the

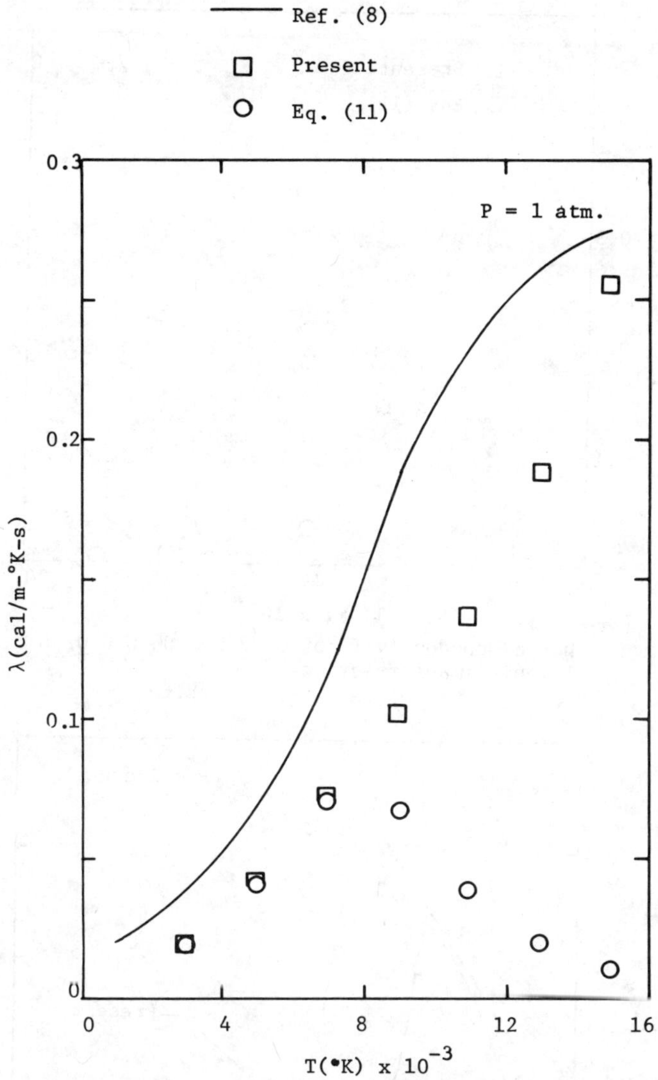

Fig. 5 Thermal conductivity of 0.5 CO_2-0.5 N_2 mixture by volume.

Titan atmosphere is one order of magnitude smaller than that of the Jupiter, Venus, and Earth atmospheres.

Conclusion

A new approximate method is proposed for evaluating the translational component of the thermal conductivity of

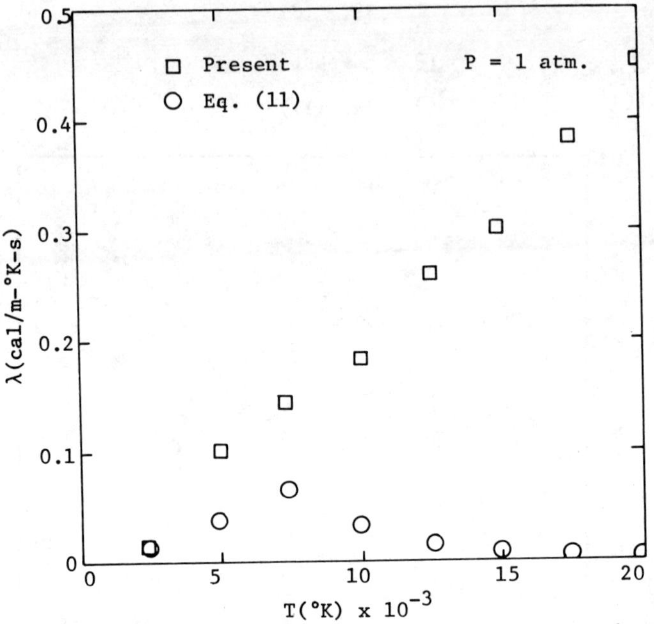

Fig. 6 Thermal conductivity of 0.92 C-0.06 O-0.02 H mixture by elemental mass fractions.

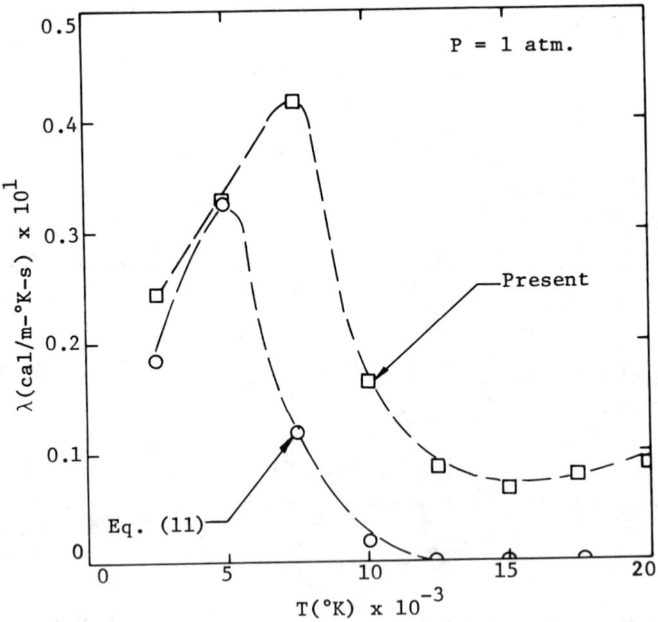

Fig. 7 Thermal conductivity of 90% N_2-10% CH_4 mixture by volume.

partially ionized gas mixtures. Results calculated with the method compare reasonably well with results calculated by exact and detailed kinetic theory analyses. For ionized gas mixtures, the proposed approximate method provides a significant improvement over the widely used Mason and Saxena approximation without significantly increasing the complexity of the calculations. It is recommended that the present method be considered for use in flowfield calculations where ionization occurs in the gas mixture.

References

[1] Moss, J. N., "Stagnation and Downstream Viscous Shock Layer Solution with Radiation and Coupled Ablation Injection," AIAA Paper 74-73, Jan. 1974.

[2] Kumar, Ajay, Tiwari, S. N., and Graves, R. A., Jr., "Effect of Small Angle of Attack on the Radiating Viscous Shock Layer Solutions for Jovian Entry," Outer Planet Entry Heating and Thermal Protection, Progress in Astronautics and Aeronautics, Vol. 64, edited by R. Viskanta, AIAA, New York, 1969, pp. 147-164.

[3] Sutton, Kenneth, "Characteristics of Coupled Nongray Radiating Gas Flow with Ablation Product Effects about Blunt Bodies During Planetary Entries," Ph.D. Dissertation, North Carolina State University, Raleigh, N.C., 1973.

[4] Moss, James N., Anderson, E. C., and Bolz, C. W., Jr., "Aerothermal Environment for Jovian Probes," AIAA Paper 76-469, July 1976.

[5] Anderson, E. C., Moss, James N., and Sutton, Kenneth, "Turbulent Viscous Shock Layer Solution with Strong Vorticity Interaction," AIAA Paper 76-120, Jan. 1976.

[6] Biolsi, L., "Transport Properties in the Jovian Atmosphere," Journal of Geophysical Research, Vol. 83, No. A3, 1978, p. 1125.

[7] Yos, J. M., "Transport Properties of Nitrogen, Hydrogen, Oxygen, and Air to 30,000°K," AVCO Corp., Wilmington, Mass., Tech. Memo. RAD-TM-63-7, 1963.

[8] Freeman, G. N. and Oliver, C. C., "High Temperature Thermodynamic and Transport Properties of Planetary CO_2 - N_2 Atmosphere," AIAA Journal, Vol. 8, Sept. 1970, p. 1687.

[9] Hirschfelder, J. O., Curtis, C. F., and Bird, R. B., Molecular Theory of Gases and Liquids, John Wiley & Sons, New York, 1954.

[10] Reichenberg, D., "New Simplified Methods for the Estimation of the Viscosities of Gas Mixtures at Moderate Pressures," NPL Report Chem. 53, May 1977.

[11]Wilke, C. R., "A Viscosity Equation for Gas Mixtures," Journal of Chemical Physics, Vol. 18, April 1950, p. 517.

[12]Saxena, S. C. and Tanzman, A., "A Note on the Calculation of Viscosities for Multicomponent Gas Mixtures," High Temperature Science, Vol. 6, No. 3, 1974, p. 215.

[13]Svehla, R. A., "Estimated Viscosities and Thermal Conductivities of Gases at High Temperatures," NASA TR R-132, 1962.

[14]Brokaw, R. A., "Approximate Formulas for the Viscosity and Thermal Conductivity of Gas Mixtures," Journal of Chemical Physics, Vol. 29, No. 2, 1958, p. 391.

[15]Ulybin, S. A., "A Simple Formula for the Thermal Conductivity of Dilute Gas Mixtures," Proceedings of XIII International Conference on Thermal Conductivity, University of Missouri-Rolla, Rolla, Mo., Nov. 1973, p. 124.

[16]Mason, E. A. and Saxena, S. C., "Approximate Formula for the Thermal Conductivity of Gas Mixtures," Physics of Fluids, Vol. 1, No. 5, 1958, p. 361.

[17]Brokaw, R. S., "Thermal Conductivity of Gas Mixtures in Chemical Equilibrium II," Journal of Chemical Physics, Vol. 32, No. 4, 1960, p. 1005.

[18]Muckenfuss, C. and Curtiss, C. F., "Thermal Conductivity of Multicomponent Gas Mixtures," Journal of Chemical Physics, Vol. 29, No. 6, 1958, p. 1273.

[19]Butler, J. H. and Brokaw, R. S., "Thermal Conductivity of Gas Mixtures in Chemical Equilibrium," Journal of Chemical Physics, Vol. 26, No. 6, 1957, p. 1636.

[20]Blake, L. H., "Approximate Transport Calculations for High-Temperature Air," AIAA Journal, Vol. 8, Sept. 1970, p. 1698.

[21]Armaly, B. F. and Sutton, K., "Viscosity of Multicomponent Partially Ionized Gas Mixtures," AIAA Paper 80-1495, July 1980.

[22]Biolsi, L., "Transport Properties in the Atmosphere of Jupiter," NASA Progress Report, Jan. 1977-Dec. 1978, CR-158094, 1978.

[23]Yun, K. S. and Mason, E. A., "Collision Integral for the Transport Properties of Dissociating Air at High Temperatures," Physics of Fluids, Vol. 5, No. 4, 1962, p. 380.

[24]Cubley, S. J. and Mason, E. A., "Atom-Molecule and Molecule-Molecule Potentials and Transport Collision Integrals for High Temperature Air Species," Physics of Fluids, Vol. 18, No. 9, 1975, p. 1109.

[25]Amdur, I. and Mason, E. A., "Properties of Gases at Very High Temperatures," Physics of Fluids, Vol. 1, No. 5, 1958, p. 370.

[26]Brezing, D., "Transport Properties of Hydrogen," AIAA Journal, Vol. 3, Aug. 1965, p. 1422.

[27]Devoto, R. S., "Transport Coefficients of Partially Ionized Argon," Physics of Fluids, Vol. 10, No. 2, 1967, p. 354.

[28]Grier, N. T., "Calculation of Transport Properties of Ionized Atomic Hydrogen," NASA TN D-3186, 1966.

[29]Grier, N. T., "Calculation of Transport Properties and Heat Transfer of Dissociating Hydrogen," NASA TN D-1406, 1962.

Optical Absorption of Carbon and Hydrocarbon Species from Shock-Heated Acetylene and Methane in the 135-220 nm Wavelength Range

Judy L. Shinn*
NASA Langley Research Center, Hampton, Va.

Abstract

Shock-tube absorption spectroscopy of carbon and hydrocarbon species behind an incident shock wave has been performed for the spectral range 135-220 nm to obtain information needed for calculating radiation blockage ahead of a planetary entry probe. Two test gas mixtures, one with acetylene and the other with methane, were used to provide reliable and distinct variation of the C_3/C_2H concentration ratio. Comparison of test results using these two mixtures in the temperature range 3750 ± 100 K determined the main absorbing species to be C_3. The possibility of significant absorption contribution from C_2 for the spectral range was eliminated by comparing test results at 4050 and 3650 K. The measured absorption band was identified to be due to the transition $X^1\Sigma_g^+ \rightarrow {}^1\Sigma_u^+$ of C_3 by comparison with the recent theoretical predictions. The location of the maximum near 158 nm agrees well with the theoretical values of 7.68 and 8.03 eV for the vertical excitation energy. A value of 0.90 for the electronic oscillator strength obtained from the measurement is also in agreement with the predicted value of 0.92.

Introduction

Entry into the atmospheres of the outer planets is accompanied by large radiative heating of the entry probe forebody. Use of a carbon-phenolic heat shield for the probe produces an ablation layer of carbon and hydrocarbon

---

Presented as Paper 81-1189 at the AIAA 16th Thermophysics Conference, Palo Alto, Calif., June 23-25, 1981.

*Aero-Space Technologist, Aerothermodynamics Branch, Space Systems Division.

species which absorbs a substantial amount of radiation directed toward the probe body from the shock layer.[1] To determine the net radiative flux reaching the probe wall, the knowledge of the radiative properties of these species is needed for calculating radiation blockage.

Several recent investigations have been conducted to improve the knowledge of radiative properties of carbon and hydrocarbon species. Jones and Cooper[2,3] reported new absorption data for the C_3 Swings band. Prakish and Park[4] measured absorption for shock-heated acetylene in the ultraviolet (uv) and vacuum uv range, which they attributed to C_2H. However, the nature of the absorption curve represented in Ref. 4, which showed continuously increasing absorption as the wavelength decreases from 300 toward 160 nm, left questions as to whether the absorption would have decreased if the data had been obtainable below 160 nm and whether this continuously increasing absorption might have been due to Mie scattering of soot particulates. The latter proposition seems plausible because if one were to calculate Mie scattering cross section, one would essentially obtain a monotonically increasing value toward lower wavelength.[5] For the experiment performed in Ref. 4, soot particulates are predicted to form behind the rather weak incident shock and to vaporize as they are processed by the reflected shock. The rate of vaporization is, however, difficult to predict and, since data were not taken in the form of a time history, it is not certain that the soot particulates had vaporized when the absorption measurements were made. Thus, the result reported in Ref. 4 for C_2H absorption warranted further investigation.

In the present study, absorption measurements were made behind an incident shock wave which raised the test gas to a temperature several hundred degrees above the limit for soot formation. Two different test gas mixtures, one with acetylene plus argon and the other with methane plus argon, were used. Absorption time histories were obtained for the spectral range 135-220 nm by using a hydrogen discharge lamp and photomultiplier detectors. The temperature and the number density of carbon and hydrocarbon species behind the incident shock wave were calculated by using the measured shock velocity and assuming thermochemical equilibrium. A comparison of absorption cross sections from these two test gas mixtures was made to identify absorbing species.

Experimental Apparatus

Shock Tube

The shock tube used in the present study is the same facility as in Ref. 2. It is an all-stainless-steel tube 15.2 cm in internal diameter with a 15.4 m long driven tube connected to a dump tank on one end and a driver section on the other. The length from the stainless-steel diaphragm to the test section centerline was 14.9 m. The driven tube was evacuated to a pressure of 0.03 Pa prior to the injection of test gas. The test gas used was either a mixture of 1% C_2H_2 and 99% argon or a mixture of 2% CH_4 and 98% argon. Unheated helium was used as the driver gas for the present study. A complete description of this facility and its operation can be found in Ref. 6.

Optical System

A schematic drawing of the optical arrangement and shock tube is shown in Fig. 1. The light source was a hydrogen discharge lamp operated at a pulsed, high-energy mode to produce a nearly constant light intensity for about 1 ms. This pulse was initiated about 200 μs prior to the arrival of the incident shock wave at the window location. The windows mounted on the shock tube were of M_gF_2 and all of the mirrors were also coated with M_gF_2 for the vacuum uv region. A concave mirror was used to focus the beam at the entrance slit of a 0.5 m, f/8.7 spectrograph which used a 2400 line/mm grating with a dispersion of 0.83 nm/mm. Five photomultiplier tubes coated with sodium salicylate for detecting light in the vacuum uv region were placed behind the exit slits of the spectrograph. The range of photomultiplier linearity was established prior to the experiment for selecting the slit width. Two diffusion pumps were used to evacuate the total enclosure.

Shock-Tube Wall Instrumentation

Instrumentation placed at 12 different locations along the shock-tube wall consisted of high-frequency response pressure transducers, thin-film heat-transfer gages, or photomultipliers coupled by light pipes. The incident shock velocity was determined using the time-of-arrival method between successive instrumented stations. Approximate test time was indicated by photomultipliers, while the steadiness of the test gas slug was inferred from pressure transducers. The thin-film heat-transfer gages

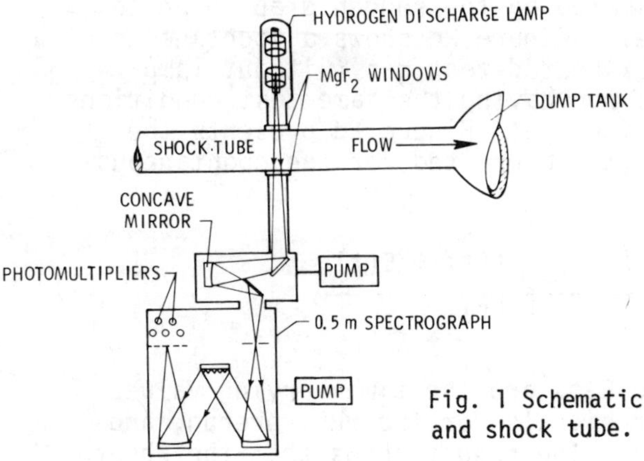

Fig. 1 Schematic of optical arrangement and shock tube.

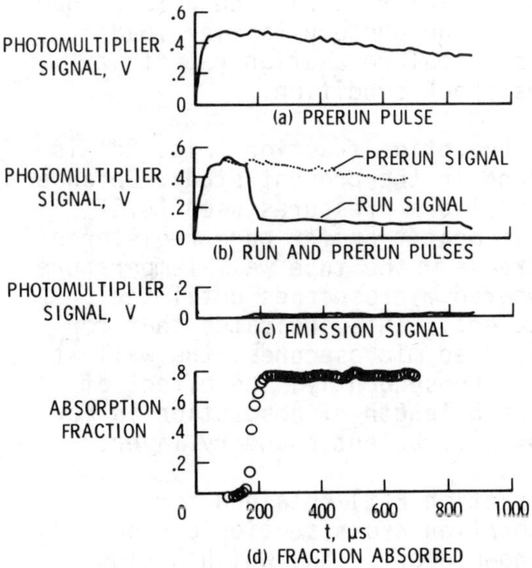

Fig. 2 Sample spectrograph channel record and reduced absorption fraction.

provided a clue as to whether the flow boundary layer was laminar.

Measurement and Analysis

A sample record from a spectrograph channel and the reduced absorption fraction are shown in Fig. 2. Figure 2a presents a prerun record of a nearly constant light pulse. Figure 2b shows both absorption run and prerun records where the prerun pulse has been normalized to match the absorption run record for the duration prior to shock arrival, which is indicated by the sudden drop in photomultiplier signal level. Figure 2c shows a spontaneous emission level of shock-heated test gas (without lamp pulse) from a separate run having the same test conditions as that of the absorption run. Figure 2d presents the absorption fraction, F(t), corrected for the spontaneous emission.

$$F(t) = \frac{I_o(t) - I(t) + E(t)}{I_o(t)} \tag{1}$$

where I(t), $I_o(t)$, and E(t) are the time-varying values of the photomultiplier signals for the run, prerun, and emission, respectively. The result shows that the absorption fraction remains almost constant after an initial rise of 60 μs or so. This long initial rise was found to be due to the large time constant of the photomultiplier load impedance rather than the chemical relaxation rate that the flow adjusts to the postshock condition.

The constant level of absorption fraction (Fig. 2d) is typical for all the runs made in the present study. Because of the fairly low initial test gas pressures used (0.62 and 0.76 kPa), the thin-flim heat-transfer gage registered a nearly constant step increase in the tube wall temperature which persisted several hundred microseconds until the boundary layer became turbulent. This indicates that for the test time of several hundred microseconds, the wall boundary layer was laminar. Consequently, the effect of the boundary layer on the path length of absorption is far less than that expected for a turbulent boundary layer.

Using the absorption fraction F(t) obtained for each channel, spectral absorption cross section can be determined from the well-known Beer's law, which having been corrected for spontaneous emission of heated gas, is written as

$$\sigma_\lambda = -\frac{1}{N\ell} \ln \frac{I(t) - E(t)}{I_o(t)} = \frac{1}{N\ell} \ln \frac{1}{1 - F(t)} \tag{2}$$

where

$\sigma\lambda$ = absorption cross section at wavelength λ, cm^2
N = number density of absorbing species, cm^{-3}
ℓ = path length of absorption, cm

The number density of absorbing species N was determined by a free-energy-minimization calculation assuming thermochemical equilibrium in the test region behind the incident shock. The chemical equilibrium computer program of Ref. 7 was used for this calculation. The thermodynamic state properties ahead of the shock wave and the shock velocity are required inputs to the program. It also requires a value for heat of formation and a definition of the temperature-dependent thermodynamic variables for each species which is to be formed behind the incident shock wave.

Because heat of formation and thermodynamic variables of some dominant species such as C_3 were not accurately known, extensive treatment and comparisons were made[2] among the values that existed in the literature for C_3. The resultant values recommended[2] were used in the present study. A literature search was also made for C_2H for the present study, showing a wide spread in the values for heat of formation. The JANAF tables[8] adopted a value of 114 kcal/mole for heat of formation of C_2H at room temperature from a range of 105-123 kcal/mole that existed in the literature prior to 1967. A more recent study of photon impact measurement made by Okabe and Dibeler[9] resulted in a value of 127 kcal/mole (at zero temperature) which was in good agreement with a value of 129 ± 3 kcal/mole (at zero temperature) obtained by Wyatt and Stafford[10] from the high-

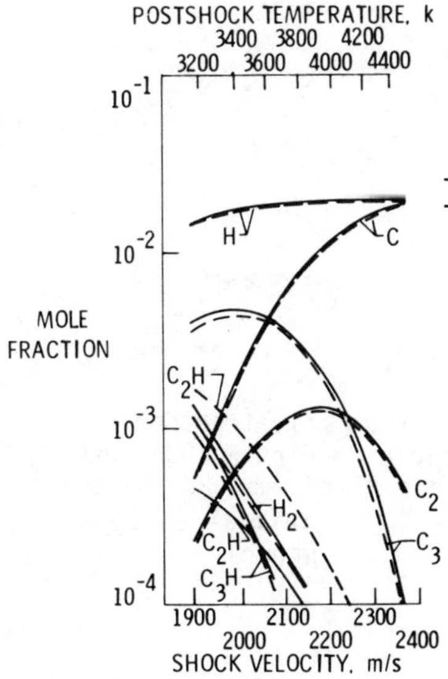

Fig. 3 Equilibrium mole fraction of principal species behind a shock wave into test gas mixture of 1% C_2H_2 and 99% argon; 0.76 kPa initial test gas pressure (argon not shown).

temperature reaction of graphite with hydrocarbons. For the present study, to show effects of varying the heat of formation, selected values of 114 and 123 kcal/mole at room temperature were used.

Figures 3 and 4 present the comparison for the mole fraction of the principal species (excluding argon) which were predicted to be present behind the incident shock wave as a function of shock velocity using different values for heat of formation of C_2H. The solid lines were generated by using a value of 123 kcal/mole and the dashed lines by using a value of 114 kcal/mole. The test gas used was a mixture of 1% C_2H_2 plus 99% argon for Fig. 3 and a mixture of 2% CH_4 plus 98% argon for Fig. 4. The initial test gas pressures were 0.76 and 0.62 kPa for Figs. 3 and 4, respectively. The result shows that there is a drastic decrease in predicted mole fraction of C_2H with an increase of merely 9 kcal/mole in the heat of formation for both test gas mixtures. The general trend of the velocity-dependent curve remains almost unchanged.

Results and Discussion

In Ref. 4, identification of absorbing species was achieved by comparing test results of various temperature conditions for one test gas. By changing the postshock temperature (or shock speed), species proportion behind

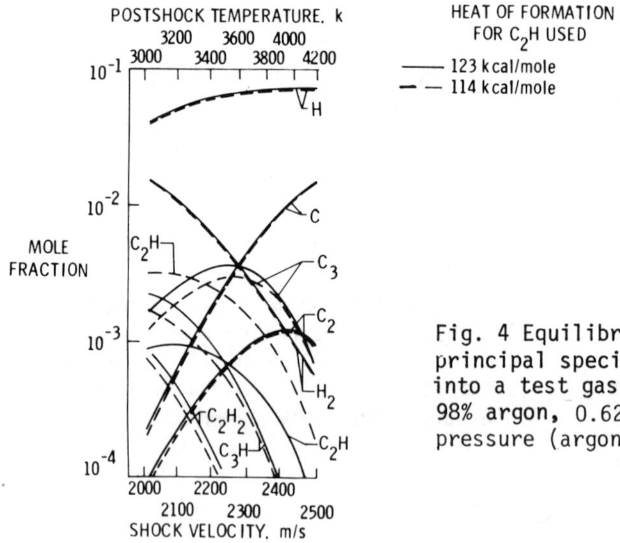

Fig. 4 Equilibrium mole fraction of principal species behind a shock wave into a test gas mixture of 2% CH_4 and 98% argon, 0.62 kPa initial test gas pressure (argon not shown).

the incident shock wave varied according to the thermodynamic properties of these species. For the present study, tests were conducted with two different test gas mixtures possessing different proportions of hydrogen and carbon to facilitate different ratios of C_3 and C_2H (or dominant carbon and hydrocarbon species) concentrations behind the incident shock at a given temperature. This can be easily seen by comparing Figs. 3 and 4. Thus, the identification of contributing absorbers between C_3 and C_2H does not rely heavily, as it did in Ref. 4, on the temperature-dependent thermodynamic properties of these species which may not be accurately known.

In Figs. 5 and 6, comparisons of absorption cross section are made between tests performed with acetylene and methane test gas mixtures for the spectral range 135-220 nm. The test conditions for all runs were at a temperature of approximately 3750 ± 100 K, while soot particulates were not predicted to form at temperatures above 3000 K. Figure 5 shows strong disagreement between the results of these two test gas mixtures when C_2H is assumed to be the absorber. However, when C_3 is the assumed absorber, good agreement is reached except in the region below 140 nm as revealed in Fig. 6. The discrepancy below 140 nm is probably due to the contribution from hydrogen or hydrocarbon species that are more abundant for the methane test gas mixture. The proposition that soot particulates were the main contributor for absorption[5] is

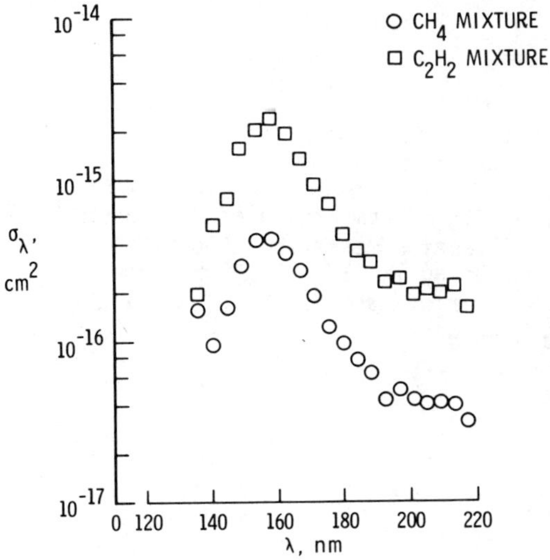

Fig. 5 Comparison of C_2H cross section measured at 3650-3850 K using 1% C_2H_2 plus 99% argon mixture (heat of formation for C_2H is assumed to be 123 kcal/mole).

invalidated since the present results show an increase of absorption from longer wavelength to a maximum around 158 nm and a decrease toward lower wavelength.

The ratios of C_3 and C_2 concentration behind the incident shock wave are almost identical for these two different test gas mixtures at a given temperature, as revealed in Figs. 3 and 4. Thus, the results obtained in Fig. 6 cannot conclusively exclude the possible significant contribution of absorption from C_2. For example, Freymark, Fox-Herzberg, and Mulliken were among the absorption band systems calculated[11] that overlap with the spectral range considered in this study. To determine if the contribution from C_2 is significant or not, additional tests were made for the methane test gas mixture at a slightly higher temperature of 4050 K, where the ratio of C_3 and C_2 concentration is substantially different from that at a temperature of 3650 K. Figures 7 and 8 present comparisons of absorption cross section between the results obtained at these two different temperatures, assuming the absorbing species to be C_2 and C_3, respectively. The results indicate that much better agreement is reached when C_3 is taken to be the absorber. It should be pointed out that although the elimination of C_2 is made based on the thermodynamic properties of C_2 and C_3, the difference in the result indicated in Fig. 7 is probably far beyond the uncertainties in the thermodynamic properties of these species.

Possible absorption contributions from CO(4+) was also examined because of some unavoidable contamination of

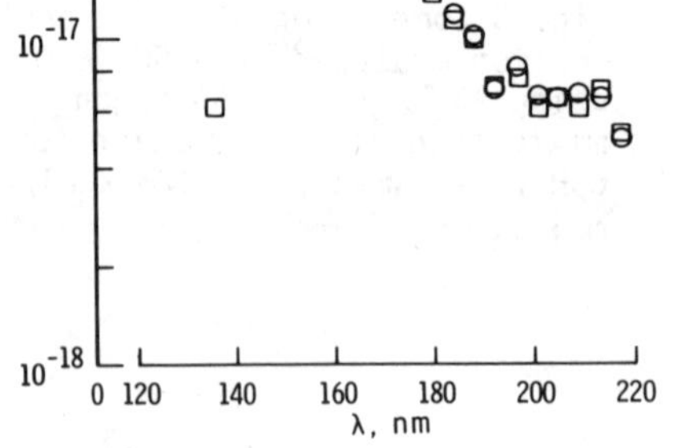

Fig. 6 Comparison of C_3 cross section measured at 3650-3850 K using 1% C_2H_2 plus 99% argon mixture and 2% CH_4 plus 98% argon mixture.

air or water vapor inside the driven section. If an initial pressure reading of 0.03 Pa in the driven section prior to the injection of test gas and a total assumed air leak of 0.06 Pa prior to the diaphragm rupture are considered, the number density of CO molecules is estimated to be, at most, 1% of that of C_3. Therefore, any contribution from CO has been neglected.

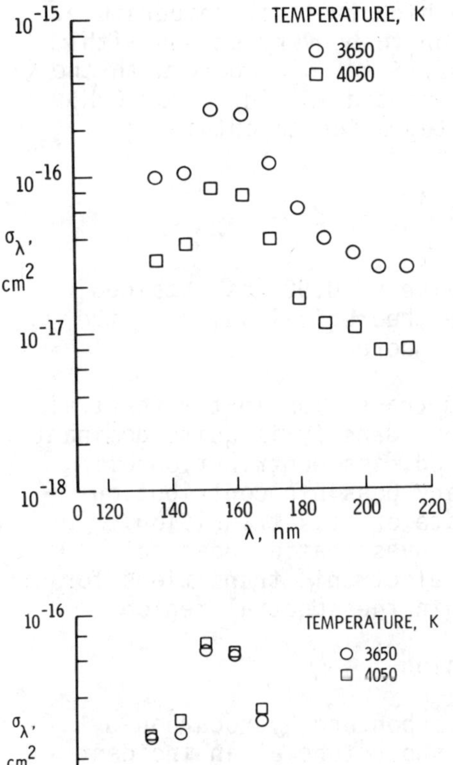

Fig. 7 Comparison of C_2 absorption cross section between tests performed at two different temperatures with a mixture of 2% CH_4 plus 98% argon.

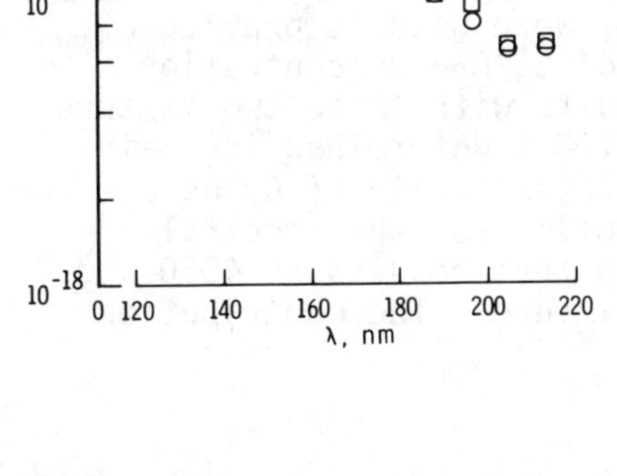

Fig. 8 Comparison of C_3 absorption cross section between tests performed at two different temperatures with a mixture of 2% CH_4 plus 98% argon.

Although there exist no other experimental data for the electronic transition of C_3 in the vacuum uv range for comparison, the result obtained in this study is found associated with the transition $X\,^1\Sigma_g^+ \rightarrow {}^1\Sigma_u^+$ predicted by Arnold et al.,[5] Williams,[12] and Römelt et al.[13] Williams predicted the vertical excitation energy for this transition to be 7.68 eV (equivalent to 162 nm) while Römelt predicted it to be near 8.03 eV (equivalent to 155 nm). These values agree well with the maximum absorption observed around 158 nm in this study, despite the fact that the maximum might be shifted somewhat due to high test gas temperature. They also predicted the transition to be very strong with the oscillator strength $f = 0.92$.[12,13] To compare with the predicted f value, the result presented in Fig. 6 for C_2H_2 test gas mixture was used for integration to obtain an f value according to the relation

$$f = 1.13 \times 10^{12} \; \sigma_{\bar{\nu}} \, d\bar{\nu}$$

where $\bar{\nu}$ is the wavenumber. A value of 0.90 was obtained that compares very well with the theoretical value if the difference in C_3 temperature is ignored.

Because C_3 has such a strong transition in the spectral region investigated and its number density is quite dominant for the test gas conditions tested, its contribution of absorption tends to overshadow any possible contribution from other species with a moderate or weak transition. Thus, the result of the present investigation does not preclude the existence of other electronic transitions for species such as C_2, C_2H, or C_3H in the spectral region.

Conclusion

Absorption spectroscopy of carbon and hydrocarbon species has been performed in a shock tube at an incident shock condition for a wavelength range of 135-220 nm. Two test gas mixtures, one with acetylene and the other with methane, both diluted with argon, were used to provide a reliable and distinct variation of C_3/C_2H concentration ratio. A comparison of test results with these two mixtures in the temperature range 3750 ± 100 K determined the main absorbing species to be C_3. The possibility of C_2 as a significant contributor of absorption for the spectral range was eliminated by comparing test results at 4050 and 3650 K with a methane test gas mixture. The contribution for CO(4+) due to air contamination of test gas was also estimated to be negligible. The result showing an absorption decrease from a maximum near 158 nm toward lower wavelength invalidates the speculation that an increasing trend toward lower wavelength observed by Prakash and Park[4] is due to soot particulates rather than C_2H as concluded by them.

The measured absorption band is identified to be due to the transition $X^1\Sigma_g^+ \rightarrow {}^1\Sigma_u^+$ of C_3 when comparison is made with the theoretical prediction made by Williams[12] and Römelt et al.[13] The wavelength for maximum absorption agrees well with the theoretical values of 7.68 and 8.03 eV for the vertical excitation energy, and a value of 0.90 for the electronic oscillator strength obtained from the measured absorption band is also in good agreement with the predicted value of 0.92.

References

[1]Moss, J. N., Anderson, E. C., and Bolz, C. W., Jr., "Viscous-Shock-Layer Solution with Radiation and Ablation Injection for Jovian Entry," AIAA Paper 75-671, May 1975.

[2]Jones, J. J., "The Optical Absorption of Triatomic Carbon C_3 for the Wavelength Range 260 to 560 nm," NASA TP-1141, March 1978.

[3]Cooper, D. M. and Jones, J. J., "An Experimental Determination of the Cross Section of the Swings Band System of C_3," Journal of Quantitative Spectroscopy and Radiative Transfer, Vol. 22, No. 2, Aug. 1979, pp. 201-208.

[4]Prakish, S. G. and Park, C., "Shock Tube Spectroscopy of C_3 + C_2H Mixture in the 140-700 nm Range," AIAA Paper 79-0094, Jan. 1979.

[5]Arnold, J. O., Cooper, D. M., and Park, C., "Line-By-Line Transport Calculations for Jupiter Entry Probes," AIAA Paper 79-1082, June 1979.

[6]Nealy, J. E., "Performance and Operating Characteristics of the Arc-Driven Langley 6-Inch Shock Tube," NASA TN D-6922, 1972.

[7]Powars, C. A. and Kendall, R. M., "User's Manual - Aerotherm Chemical Equilibrium (ACE) Computer Program," Aerotherm Corp., May 1969.

[8]JANAF Thermochemical Tables, NSRDS-NBS 37, June 1971.

[9]Okabe, H. and Dibeler, V. H., "Photon Impact Studies of C_2HCN and CH_3CN in the Vacuum Ultraviolet; Heat of Formation of C_2H and CH_3CN," Journal of Chemical Physics, Vol. 59, No. 5, Sept. 1973.

[10]Wyatt, J. R. and Stafford, F. E., "Mass Spectrometric Determination of the Heat of Formation of Ethynyl Radical, C_2H, and of Some Related Species," Journal of Physical Chemistry, Vol. 76, No. 13, 1972.

[11]Sutton, K. and Moss, J. N., "Radiative Absorption by C_2 Band Systems for Jupiter Entry Conditions," AIAA Paper 79-0033, Jan. 1979.

[12]Williams, G. R. J., "The C_3 Radical: A Theoretical Study of the Electronic Spectrum of C_3 Via the Equations of Motion Method," Chemical Physical Letters, Vol. 33, No. 3, June 1975.

[13]Römelt, J., Peyerimhoff, S. D., and Buenker, R. J., "Ab Initio MRD CI Calculations for the Electron Spectrum of the C_3 Radical," Chemical Physics Letters, Vol. 58, No. 1, Sept. 1978.

Chapter II Aerothermodynamics

Nondimensional Parameters in Radiation Gasdynamics

R. Goulard*
George Washington University, Washington, D.C.

Abstract

Nondimensional numbers (e.g., Re and Pe) allow for concise flux expressions that are valid over a wide range of properties. However, temperature dependence is a frequent restriction to the validity of these closed-form expressions. In 1974, a new approach was proposed by a Soviet group (headed by Biberman) for the case of radiation from re-entry shock layers. It consists of taking the rate of change of traditional nondimensional numbers near their design (reference) point. In this paper, a fluid dynamic interpretation of this approach is shown to give a good correlation of numerical results.

Introduction

In recent years Biberman and his co-workers have introduced a new dimensionless radiation-convection parameter which seems to be particularly appropriate.[1,2] Instead of using the ratio of the "adiabatic" shock-layer radiation flux over the incoming energy flux — as has been done traditionally[3] — Biberman introduced the rate at which the radiative flux changes for a given small change in the incoming energy flux. Because this new parameter is based on the thermodynamic properties in the range of temperatures relevant to the problem, a much better fit has been obtained by Biberman than has been possible with the traditional ratio.[3]

It would be interesting to test this method against the many numerical and experimental results that have been accu-

---

Presented as Paper 81-1191 at the AIAA 16th Thermophysics Conference, Palo Alto, Calif., June 23-25, 1981.

*Professor, School of Engineering and Applied Science.

mulated in a wide range of atmospheres and re-entry conditions.[4] Also, a similar treatment of the Peclet number might reduce the proliferation of ad hoc formulas in convection heat transfer. Since such work has not yet been done, this is a short report on what appears to be a clever physical insight into a complicated problem.

Re-entry Radiative Heat Transfer

In a recent paper,[2] Biberman et al. considered the numerical calculations of radiative heat transfer corresponding to high-speed re-entry under a wide range of density and velocity conditions. They showed that all of the data could be fitted into a single curve in terms of a dimensionless parameter κ. This parameter κ (see below) is characteristic of the change of thermodynamic properties during cooling from stagnation conditions. Figure 1 shows that it is better suited to the analysis of radiation losses than the parameters used earlier[5] which were expressed in terms of the properties themselves.

Figure 1 represents the ratio of the radiative flux q_c from a hypersonic shock layer over the radiative flux q_0 which would have occurred if such a flux had not cooled the layer (the adiabatic case). The difference is due to the progressive loss of enthalpy that is incurred by the elementary radiating layers as they travel from the shock toward the surface of the body (Fig. 2). This effect is important at high temperatures where a larger fraction of the energy

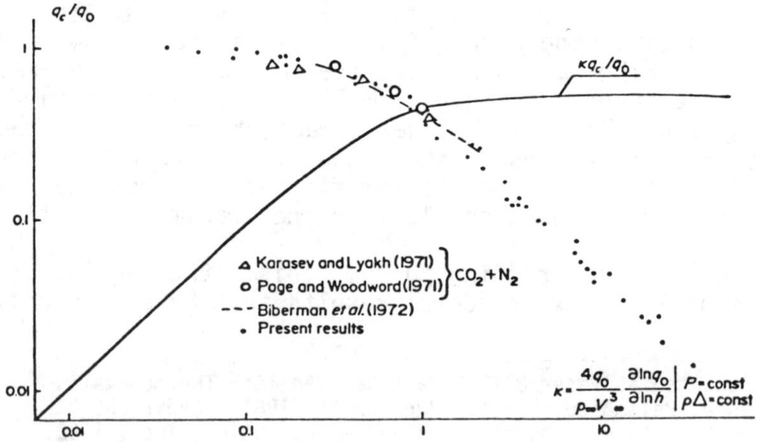

Fig. 1 Approximate similarity relation for radiative fluxes incident on nonablating body surface.

is stored in excited stages which are likely to de-excite radiatively. Hence, the ratio q_c/q_0 decreases as the entry velocities increase (Fig. 1).

Biberman's numerical results show that a single curve can account for all cases, (i.e., different trajectories, atmospheres, etc.) if a new variable κ is introduced in abcissa

$$\kappa \equiv \frac{4q_0}{\rho_\infty V^3} \frac{\delta(\ln q_0)}{\delta(\ln h_0)}\Bigg|_{\substack{p_0=\text{const}\\ \rho_s\Delta=\text{const}}} \tag{1}$$

where the conditions of constant stagnation pressure p_0 and of constant shock-layer mass per unit area $\rho_s\Delta$ correspond to the radiation cooling of a thin layer as it moves from the shock toward the wall (Fig. 2). Also, the product log $\kappa q_c/q_0$ tends to the limit log κ at low velocities and to the limit 0.5 at high velocities (Fig. 1).

The Biberman Number

It is interesting to rewrite κ by using the relation $x d(\ln x) = dx$. Then Eq. (1) becomes

$$\kappa \equiv \frac{2}{\rho_\infty V_\infty} \frac{\delta q_0}{\delta h_0}\Bigg|_{\substack{p_0=\text{const}\\ \rho_s\Delta=\text{const}}} \tag{2}$$

If one considers the motion of a thin layer as it moves from the shock to the stagnation area, it is clear that the characteristic time t available for radiation losses is

$$t = \Delta/\bar{v} \equiv \Delta/\frac{1}{a} V_s \tag{3}$$

where v is the time-averaged velocity of an elementary layer as it moves from the shock ($v=V_s$) to the boundary-layer edge (see Fig. 2). In the lower velocities case, radiation will be lost entirely near the shock, $\bar{v} \simeq V_s$ and $a \simeq 1$. At high velocities, radiation losses will persist throughout the shock layer, $\bar{v} < V_s$ and $a > 1$. Since at the shock[5] Eq. (3) can be written as

$$t = \Delta a \rho_s / \rho_\infty V_\infty \tag{4}$$

thus, Eq. (2) can be rewritten as

$$\kappa = \frac{1}{a}\, t\, \frac{\delta\left(\frac{2q_0}{\rho_s\Delta}\right)}{\delta h_0}\Bigg|_{\substack{p_0=\text{const}\\ \rho_s\Delta=\text{const}}} \tag{5}$$

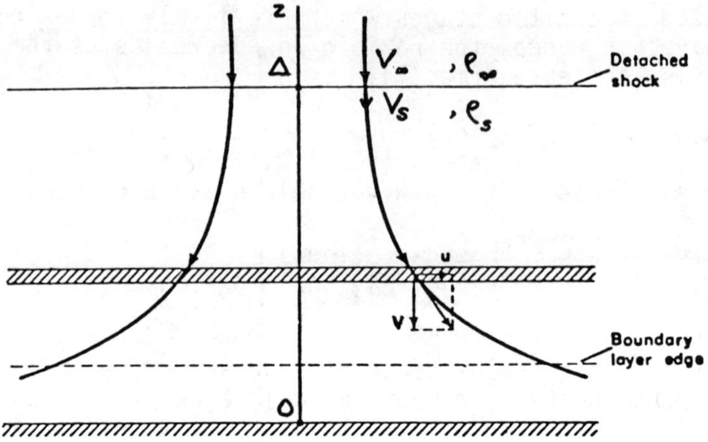

Fig. 2 Detached shock-layer flow at stagnation.

Since Biberman assumed optically thin shocks, the quantity $2q_0/\rho_S\Delta$ corresponds to the flux emitted by a thin elementary layer per unit mass, both toward the wall and away from it (hence, the factor 2).

Clearly,

$$r \equiv \delta(\frac{2q_0}{\rho_s\Delta})/\delta h_0 \tag{6}$$

is the time rate per unit mass at which a thin elementary layer of enthalpy h_0 loses its enthalpy by radiation as it moves past the stagnation area, if the adiabatic conditions (q_0, h_0), which are exact at the shock, are considered characteristic of the rest of the shock layer. Therefore, the product tr is characteristic of the fraction of the shock enthalpy h_0 radiated during the exposure time t. This is the physical meaning of the coefficient $\kappa \equiv (1/a)tr$.

With this in mind, it is possible to relate κ to the difference δq_C between the adiabatic flux q_0 (no loss) and the actual reduced flux q_C. This difference corresponds to half of the energy lost by the streamlines in the actual case $(1/2)\delta(\rho_S\Delta h_0)$. According to Eq. (6)

$$\delta q_c = q_0 - q_c = (1/2)r \cdot \delta(\rho_s\Delta h_0) \tag{7}$$

But this loss of energy $\delta(\rho_S\Delta h_0)$ is precisely equal to the total flux ($2q_C$) radiated during the time interval t

$$\delta(\rho_s\Delta h_0) = 2q_c \ . \ t \tag{8}$$

Therefore, Eq. (7) becomes

$$\delta q_c = r q_c t \tag{9}$$

and

$$\delta q_c / q_c = a\kappa \tag{10}$$

If one wishes to revert to the ratio q_c/q_0

$$\frac{q_c}{q_0} = \frac{q_c}{q_c + \delta q_c} = \frac{1}{1 + \dfrac{\delta q_c}{q_c}} = \frac{1}{1 + a\kappa} \tag{11}$$

This simple model appears to fit well the two asymptotic cases shown by Biberman, within the accuracy limits of Fig. 1. As discussed earlier, the constant a = 1 fits the low velocity cases

For $\kappa < 0.5$: $\quad q_c/q_0 = 1 - \kappa$ (12)

At very high speed, Fig. 1 shows an asymptotic value $a \simeq 2$

For $\kappa > 2$: $\quad q_c/q_0 = 1/2\kappa = 0.5/\kappa$ (13)

This model could be somewhat refined by following the elementary shock layer as it cools at the rate $2q_c/\rho_s\Delta$. In this case, Eq. (9) takes the differential form, $dq_c = rq_c dt$. It can be integrated for the time of migration t of an elementary shock layer from the shock to the boundary-layer edge. As the layer cools, the characteristic flux q_c diminishes from a maximum q_0. Also, the characteristic time rate r [Eq. (6)] corresponds to lower temperatures than T_0 and is also reduced. If one assumes that it changes in proportion to q_c, the solution is identical to Eq. (11).

Thus, a simplified model appears to be able to accommodate many numerical correlations into a simple function B of a physically meaningful thermodynamic coefficient κ

$$B = q_c/q_0 = 1/(1 + a\kappa) \tag{14}$$

Because of the important insights provided by Ref. 2 and earlier publications, it is proposed to name this dimensionless ratio B the Biberman number. The nondimensional radiative coefficient C_{R-C} can also be written

$$C_{R-C} \equiv \frac{q_c}{\frac{1}{2}\rho_\infty V_\infty^3} = \frac{q_0}{\frac{1}{2}\rho_\infty V_\infty^3} \cdot \frac{q_c}{q_0} = \frac{q_0}{\frac{1}{2}\rho_\infty V_\infty^3} \cdot B \qquad (15)$$

The Ablative Surface Case

The flux q_c discussed so far is the flux which reaches the outside of the ablation layer, if there is one. In Ref. 2, Biberman also discusses the effect of an ablation layer on the wall flux q_w. He shows that the radiation-layer flux q_c is attenuated by a transmittance factor K

$$K \equiv \frac{q_w}{q_c} \qquad (16)$$

which is shown to be a function of the properties of the gaseous ablation products only.

Thus, it is possible to obtain the overall radiation flux to the wall in the nondimensional form

$$C_{Rw-C} \equiv \frac{q_w}{\frac{1}{2}\rho_\infty V_\infty^3} = \frac{q_w}{q_c} \cdot \frac{q_c}{q_0} \cdot \frac{q_0}{\frac{1}{2}\rho_\infty V_\infty^3} \qquad (17)$$

If one keeps in mind that Biberman calls the ratio $q_0/(\frac{1}{2})\rho_\infty V_\infty^3$ the Goulard number G, it follows from Eqs. (15-17) that the nondimensional radiation number C_{Rw-C} is actually a KGB product. This should no doubt facilitate its acceptance in those aerospace circles that still have a viable space program.

References

[1] Biberman, L. M., "Radiation Heat Transfer at High Temperatures," Proceedings of 5th International Heat Transfer Conference, Sec. IL8, 1974, pp. 105-122.

[2] Biberman, L. M., Bronin, S. Ya., and Brykin, M. V., "Moving of a Blunt Body Through the Dense Atmosphere Under Conditions of Severe Aerodynamic Heating and Ablation," Acta Astronautica, Vol. 7, 1980, pp. 53-65.

[3] Olstad, W. B., "Correlations for Stagnation Point Radiative Heat Transfer," AIAA Journal, Vol. 7, Jan. 1969, pp. 170-172.

[4]Moss, J. N., "Flow Field Analyses for Future Planetary Entry," AIAA Paper 79-0918, 1979.

[5]Goulard, R., "The Coupling of Radiation and Convection in Detached Shock Layers," Journal of Quantitative Spectroscopy and Radiation Transfer, Vol. 1, 1961, pp. 249-257.

Blunt-Body Turbulent Boundary-Layer Parameters Including Shock Swallowing Effects

B.J. Griffith* and B.M. Majors†
Arvin/Calspan, Arnold Air Force Station, Tenn.
and
J.C. Adams Jr.‡
Sverdrup Technology, Inc., Tullahoma, Tenn.

Abstract

The primary objective of the present study is the development of correlation parameters which express turbulent flow quantities under perfect- and real-gas conditions, including effects of entropy-layer swallowing. Simple, yet accurate correlation charts of the freestream Stanton number ST_∞, the skin friction coefficient C_{F_∞}, and the boundary-layer thickness δ for a blunted cone at zero angle of attack are developed. The charts express ST_∞, C_{F_∞}, and δ for turbulent flow under perfect- and real-gas conditions, including the effects of entropy-layer swallowing. The charts also permit perfect-gas wind-tunnel data to be properly scaled to equilibrium real-gas flight conditions. The Stanton number ST_∞ and local skin friction coefficient CF_∞ from the methodology charts are accurate within 10-15%, depending on level of wall temperature. The boundary-layer thickness is accurate to within 15-20%. The charts are easy to use.

Nomenclature

C_{F_∞} = freestream skin friction coefficient, $2\,\tau_W/\rho_\infty U_\infty^2$
C_p = surface pressure coefficient, $[(p_W/p_\infty) - 1]/(\gamma/2)M_\infty^2$
F_{CF} = methodology parameter for C_{F_∞} correcting for wall temperature and entropy swallowing (Fig. 5b), turbulent flow

---

Presented as Paper 81-1089 at the AIAA 16th Thermophysics Conference, Palo Alto, Calif., June 23-25, 1981.

*Senior Scientist, Calspan Field Services, Inc./AEDC Division.
†Scientific Programmer, Calspan Field Services, Inc./AEDC Division.
‡Director, Sverdrup Computer Service Center.

F_{ST} = methodology parameter for ST_∞ correcting for wall temperature and entropy swallowing (Fig. 5a), turbulent flow
F_δ = methodology parameter for δ correcting for wall temperature and entropy swallowing (Fig. 5c), turbulent flow
G_{CF} = methodology parameter for C_{F_∞} correcting for Mach number and pressure coefficient (Fig. 5b), turbulent flow
G_{ST} = methodology parameter for ST_∞ correcting for Mach number and pressure coefficient (Fig. 5a), turbulent flow
G_δ = methodology parameter for δ correcting for Mach number and pressure coefficient (Fig. 5c), turbulent flow
H_{CF} = methodology parameter for C_{F_∞} correcting for flight conditions (Fig. 5b), turbulent flow
H_O = total enthalpy
H_{ST} = methodology parameter for ST_∞ correcting for tunnel conditions (Fig. 5a), turbulent flow
H_W = wall static enthalpy
H_δ = methodology parameter for δ correcting for tunnel conditions (Fig. 5c), turbulent flow
M_∞ = freestream Mach number
p_W/p_∞ = wall to freestream pressure ratio
$\dot{q}_W$ = wall heat-transfer rate
r = radius as defined in Fig. 2
R_{CF} = methodology parameter correcting C_{F_∞} for real-gas effects (Fig. 6), turbulent flow
Re_∞/FT = freestream unit Reynolds number, $\rho_\infty U_\infty/\mu_\infty$
$Re_{\infty R_N}$ = freestream Reynolds number based on nose radius
R_N = nose radius, in. or ft as noted
R_{ST} = methodology parameter correcting ST_∞ for real-gas effects (Fig. 6), turbulent flow
S = surface distance from the stagnation point
$\bar{S}$ = entropy-layer swallowing similarity parameter, Eq. (2)
ST_∞ = freestream Stanton number, $\dot{q}_W/[\rho_\infty U_\infty(H_O - H_W)]$
T_O = stagnation or total temperature
$T_{O,\infty}$ = freestream stagnation or total temperature
T_W = wall static temperature
T_∞ = freestream temperature, °R
U_∞ = freestream velocity
u = tangential velocity in boundary layer
y = distance normal to the body surface
y_e = normal distance from body surface to edge of boundary layer

y_s = radius of shock for mass balance
α = angle of attack
γ = ratio of specific heats, $\gamma = 1.4$
δ = boundary-layer thickness (at 0.995 $T_{0,\infty}$)
θ_c = cone half-angle
μ_∞ = freestream viscosity
ρ = mass density
ρ_∞ = freestream mass density
τ_w = total shear stress

Introduction

Four of the most important aerodynamic parameters needed in the design phase of a blunt-nosed re-entry vehicle are the inviscid surface pressure, the surface heat transfer, the surface shear stress, and the thickness of the boundary layer. The inviscid surface pressure is readily available from a number of sources, e.g., Refs. 1 and 2. Calculating the other parameters is time consuming and often difficult even for the engineer who is proficient in the use of digital computers. A series of design graphs would be most helpful in this regard. The present work extends the shock-swallowing concepts developed in Ref. 3 to the blunt-nosed re-entry vehicle body problem in order to develop a simple methodology that could be easily used in preliminary design work.

The present work includes the development of simple yet accurate correlations of the freestream Stanton number ST_∞, the skin friction coefficient C_{F_∞}, and the boundary-layer thickness δ under perfect- and real-gas conditions for a blunted cone at zero angle of attack. The primary objective of the present study is the development of correlation parameters which express turbulent flow quantities under perfect- and real-gas conditions, including effects of entropy-layer swallowing. A secondary objective is to demonstrate the effects of Mach number, cone angle, and other parameters on Stanton number, skin friction coefficient, and boundary-layer thickness under turbulent flow conditions. The resulting work will also enable perfect-gas wind-tunnel data to be properly scaled to equilibrium real-gas flight conditions, including effects of entropy-layer swallowing.

Flowfield Model

The model used in the present work consisted of a spherically blunted cone at zero angle of attack (Fig. 1).

The geometric variables and aerodynamic parameters are also presented and defined in Fig. 1.

Figure 2 shows a sketch of the entropy-layer swallowing process, and the following section presents a discussion of the analytical treatment of this process under turbulent flow conditions. A complete discussion of entropy-layer swallowing is given by Mayne and Adams.[4]

Flowfield Calculation Methods

Approach

The approach consisted of the following steps:

1) The selection of a realistic flowfield model amenable to existing analysis techniques. The computer code selected was the Spalding-Patankar code.[5] This code (see following subsection) has been validated and found to be one of the better tools for calculating the surface parameters of spherically blunted cones at zero angle of attack.

2) The running of a large number (120) of Spalding-Patankar solutions under turbulent flow conditions for a range of blunt cone configurations. These flow solutions (including entropy-layer swallowing) were ob-

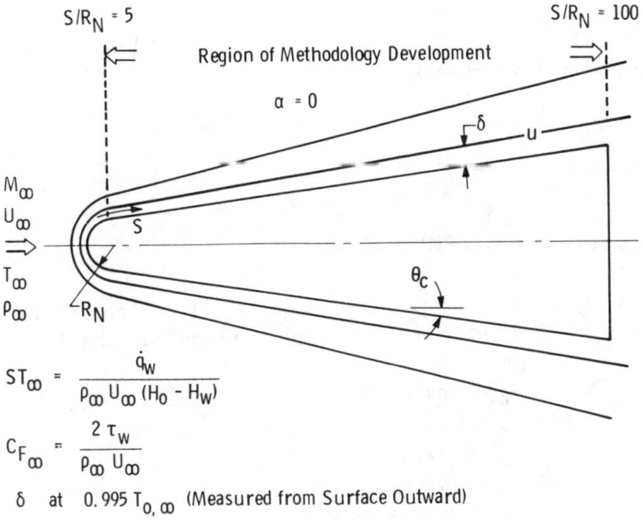

Fig. 1 Flowfield model.

tained for a range of perfect-gas wind-tunnel and flight conditions and equilibrium real-gas flight conditions.

3) A methodology was developed that expresses the results of the Spalding-Patankar solutions in terms of free-stream flow conditions and known geometry parameters (see Fig. 3). A regression analysis[6] was used in order to define the salient parameters (see section on methodology below).

A perfect-gas inviscid pressure coefficient is needed as an additional input to the methodology charts. These

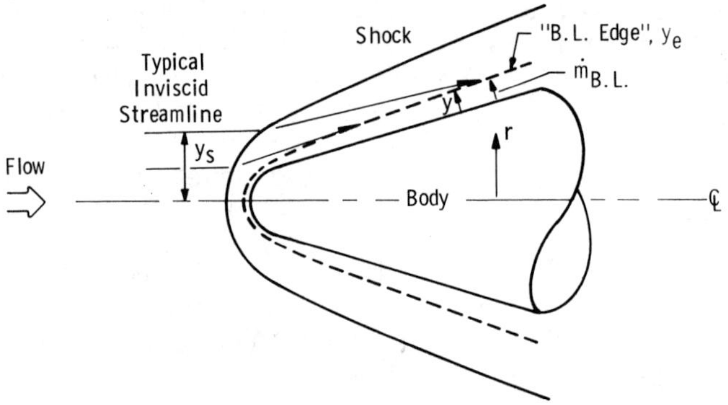

Fig. 2 Entropy-layer swallowing.

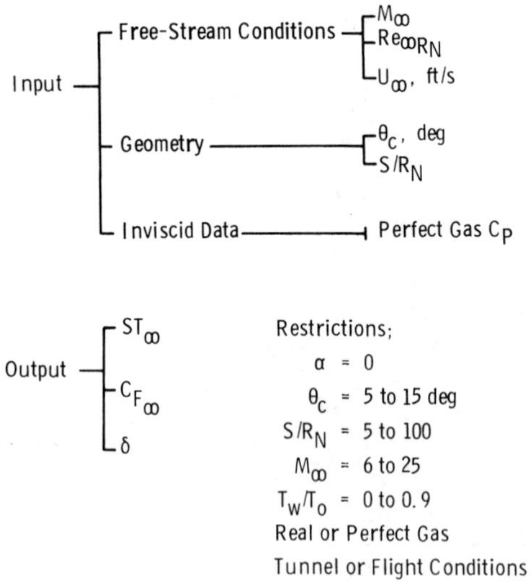

Fig. 3 Input and output parameters for methodology charts.

are available from a number of sources, for example, Refs. 1 and 2. The inviscid pressure coefficient for an 8 deg blunted cone is presented in Fig. 4.

Description of Numerical Scheme

The theory and numerical scheme used in obtaining the present boundary-layer results are based on the work of Patankar and Spalding.[5] In this approach the classical boundary-layer equations, with the addition of the transverse curvature terms, are expressed in a normalized von Mises coordinate system and solved by a marching, implicit finite-difference procedure. The computer program used in the present work is written in FORTRAN IV for use on an IBM 370/165 digital computer and is essentially a highly modified version of the code originally devised by Patankar and Spalding.[5] This program, formulated by Mayne and Dyer,[7] embodies a number of significant modifications and extensions to the basic technique of Patankar and Spalding. These include elimination of the Couette flow analysis at the body surface and the so-called slip value scheme in favor of applying the basic finite-difference scheme across the entire boundary layer.

Calculations for the real-gas flight conditions in the present study treat the flow as dissociating air in thermodynamic equilibrium using the thermodynamic and transport property correlation formulas of Cohen.[8] Under hypersonic

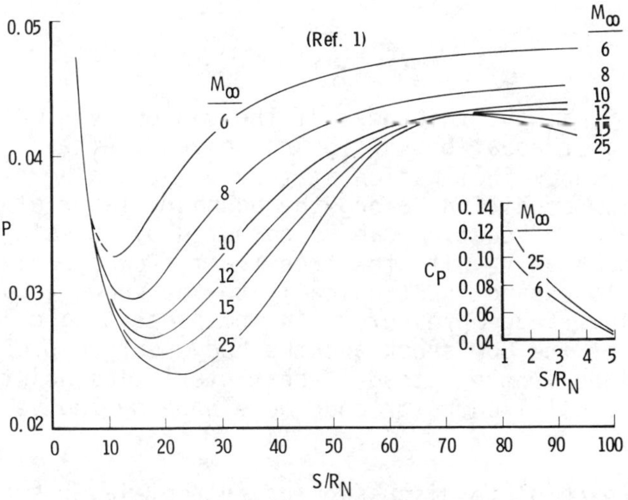

Fig. 4 Perfect-gas pressure distribution for 8 deg blunt cone.

wind-tunnel conditions, the flow is taken to be thermally and calorically perfect air with a constant Prandtl number of 0.7 in conjunction with Sutherland's viscosity law.

Since an item of primary interest in the present work is the effect of entropy-layer swallowing on turbulent boundary-layer parameters, a discussion of the swallowing process and its analytical treatment is in order. The specification of the conditions along the outer edge of the boundary layer is reasonably straightforward for bodies such as flat plates and sharp cones; however, for the case of a blunt body the problem is complex. As the flow proceeds along the body and the boundary layer grows because of entrainment of mass, the high-entropy portion of the flow which crossed the essentially normal portion of the bow shock is swallowed by the boundary layer. The flow along the edge of the boundary layer on the aft portions of the body will then have passed through an oblique part of the bow shock and will be in a different state than had it passed through a normal shock (see Fig. 2).

The determination of the local edge flow conditions may be improved by taking into consideration the inclination of the bow shock where the local flow streamline crossed the shock. The point at which the flow along the edge of the boundary layer crossed the shock can be determined by matching the mass flow in the boundary layer at a given location to the freestream mass flow in a cylinder with radius extending out to the location to be determined. Referring again to Fig. 2, this may be expressed as

$$\rho_\infty U_\infty \pi y_s^2 = \int_0^{y_e} 2\pi r \rho u \, dy \qquad (1)$$

Note that y_e is the thickness of the region over which the boundary-layer equations are being solved. After y_s is found, the shock inclination at that point can be determined and the flow conditions along the boundary layer at the corresponding body location can be computed by crossing the oblique shock at y_s with the freestream flow and allowing that flow to expand isentropically to the known local boundary-layer edge pressure. In the present calculations, the shape of the bow shock and the body surface pressures were obtained from a method-of-characteristics solution using the digital computer code developed by Inouye, Rakich, and Lomax.[9]

A more complete discussion of entropy-layer swallowing by laminar boundary layers, including comparisons of results from boundary-layer analyses which include and exclude entropy-swallowing effects relative to a fully viscous shock-layer method, is given by Mayne and Adams[4] and Adams, Martindale, Mayne, and Marchand.[3]

Methodology

The definitions of Stanton number, skin friction coefficient, and boundary-layer thickness are given in Fig. 1. Figure 2 shows a sketch depicting inviscid streamline swallowing. Streamline swallowing makes the definition of both surface and outer edge boundary-layer parameters very difficult (see Refs. 3 and 4). Therefore the present work circumvents this problem by expressing all variables in terms of freestream parameters and geometrical terms. This results in some loss of accuracy and leads to replacing some fundamental concepts with empirical considerations. However, by relying on a large theoretical data base, careful modeling considerations, and a thorough evaluation using a regression program,[6] simple charts that are quite accurate over the flight and tunnel range of interest can be generated. The 120 Spalding-Patankar solutions provided over 1800 data points (various S/R_N) that were used in the regression analysis.

The similarity parameter (denoted hereafter as $\overline{S}$) for the entropy-layer-swallowing process as developed by Adams et al.,[3] based upon the work presented in Refs. 10-13, can be expressed as

$$\overline{S} = \frac{S/R_N}{\left(Re_{\infty_{R_N}}\right)^{1/9}} \tag{2}$$

Expressing ST_∞, C_{F_∞}, and δ as functions of all known freestream and geometry parameters (see Refs. 14-16) that are likely to influence them, we have

$$ST_\infty, C_{F_\infty}, \delta = f\left(\overline{S}, C_p, M_\infty, T_w/T_o, T_\infty, U_\infty, S/R_N, R_N, Re_{\infty_{R_N}}, \theta_c\right) \tag{3}$$

Methodology Development. Treating Eq. (3) using the theoretical data base generated from Refs. 5 and 7 for turbulent flow and the regression program of Ref. 6, the variables fall into groups as follows:

$$ST_\infty, C_{F_\infty}, \delta = f\left(G, F, H, R, Re_{\infty_{R_N}}\right) \tag{4}$$

where

$$G = f\left(M_\infty, C_p\right)$$

$$F = f\left(\overline{S}, T_W/T_O\right)$$

$$H = f\left(\overline{S}, T_\infty, T_W/T_O\right)$$

$$R = f\left(\overline{S}, U_\infty, \theta_c\right)$$

The above functions as developed for the Stanton number, skin friction coefficient, and boundary-layer thickness are presented in Figs. 5 and 6. The Stanton number functions are restricted to $T_W/T_O \leq 0.70$ because of the recovery temperature problem. As the wall enthalpy approaches the recovery enthalpy, the heat-transfer rate ($\dot{q}_W$) approaches zero; therefore, the Stanton number becomes meaningless. The following subsection gives the estimated accuracy of the methodology plots. The resulting equations are as follows:

$$ST_\infty = \frac{G_{ST} * F_{ST} * H_{ST} * R_{ST}}{\left(Re_{\infty_{RN}}\right)^{0.1805}} \tag{5}$$

$$C_{F_\infty} = \frac{G_{CF} * F_{CF} * R_{CF}}{H_{CF} * \left(Re_{\infty_{RN}}\right)^{0.1805}} \tag{6}$$

$$\delta = \frac{F_\delta * R_N}{G_\delta * H_\delta * \left(Re_{\infty_{RN}}\right)^{0.0917}} \tag{7}$$

where

G_{ST}, F_{ST}, and H_{ST} are given in Fig. 5a
R_{ST} = 1.0 for perfect gas (Fig. 6 for real gas)
H_{ST} = 1.0 for flight conditions
G_{CF}, F_{CF}, H_{CF}, and R_{CF} are given in Fig. 5b.
R_{CF} = 1.0 for perfect gas (Fig. 6 for real gas)
H_{CF} = 1.0 for tunnel conditions
G_δ, F_δ, and H_δ are given in Fig. 5c
R_δ = 1.0 for real and perfect gas
H_δ = 1.0 for flight conditions

Estimated Accuracy and Sample Calculations. Considering the computer code data base and the regression analysis,

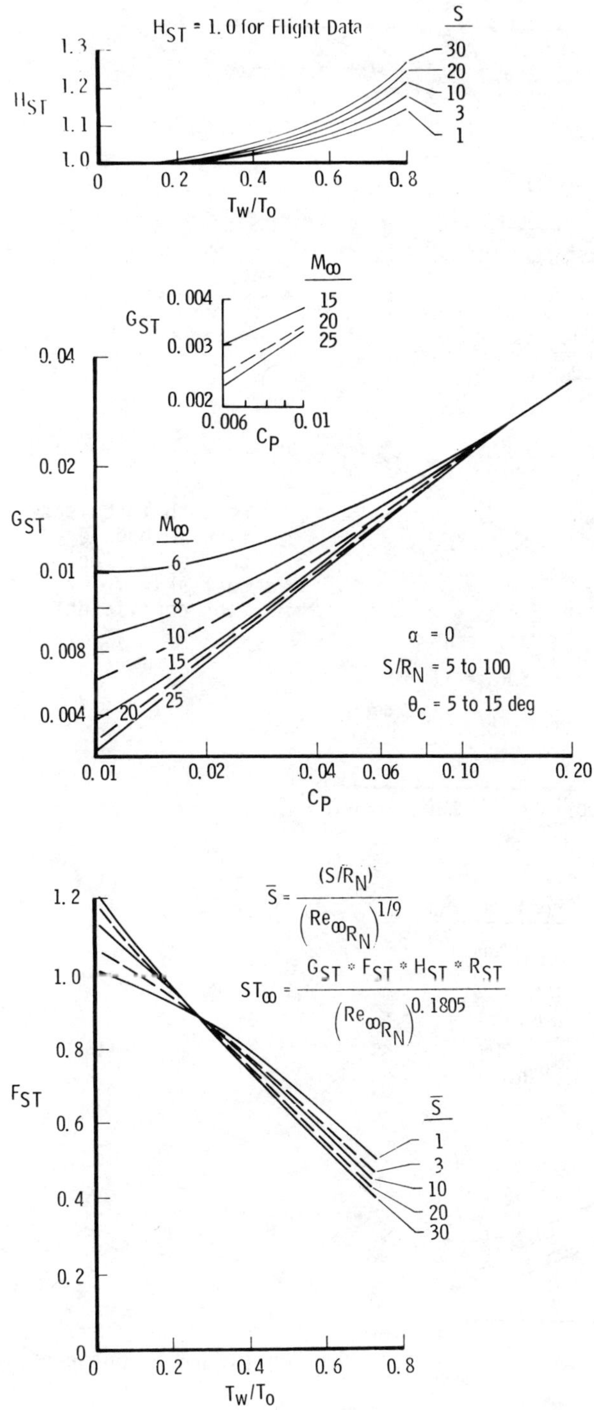

Fig. 5a Turbulent flow methodology charts, perfect gas: Stanton number, ST_∞.

(Figure 5 continued)

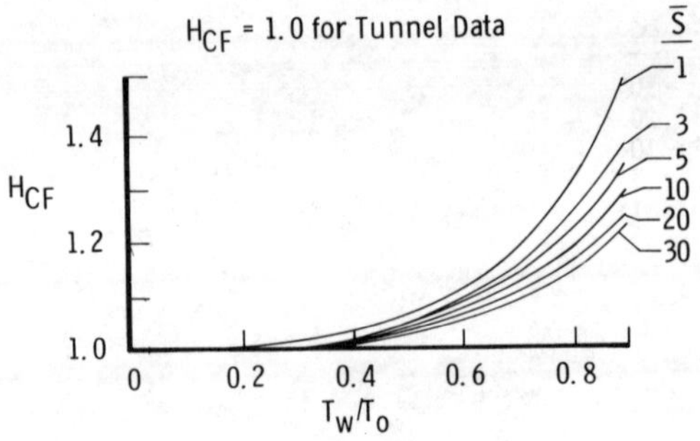

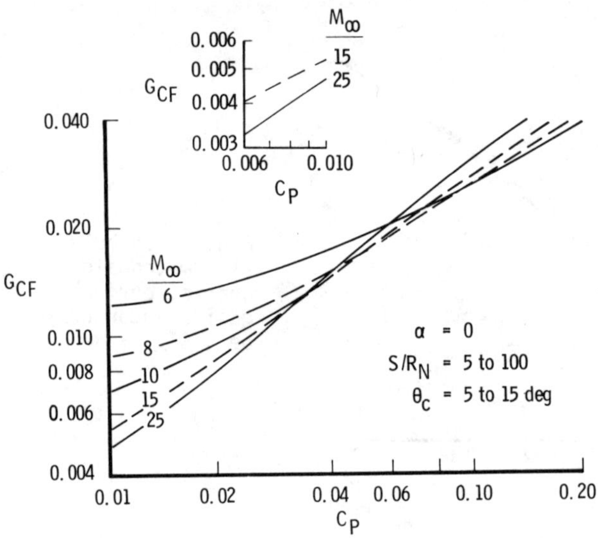

Fig. 5b Turbulent flow methodology charts, perfect gas: skin friction coefficient, C_{F_∞}.

$$C_{F_\infty} = \frac{G_{CF} * F_{CF} * R_{CF}}{H_{CF} * \left(Re_{\infty R_N}\right)^{0.1805}}$$

$$\bar{S} = \frac{\left(S/R_N\right)}{\left(Re_{\infty R_N}\right)^{1/9}}$$

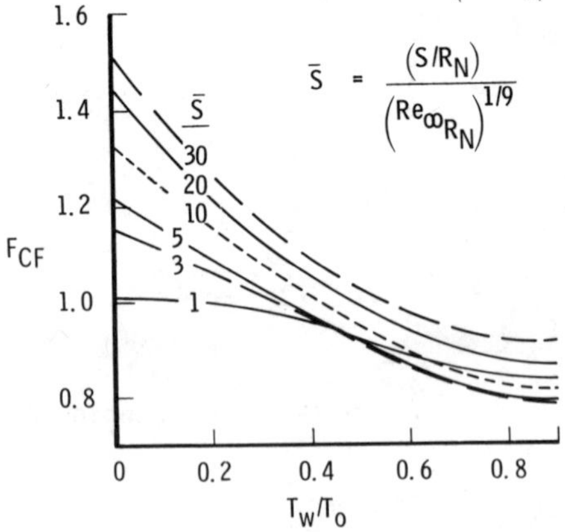

(Figure 5 continued)

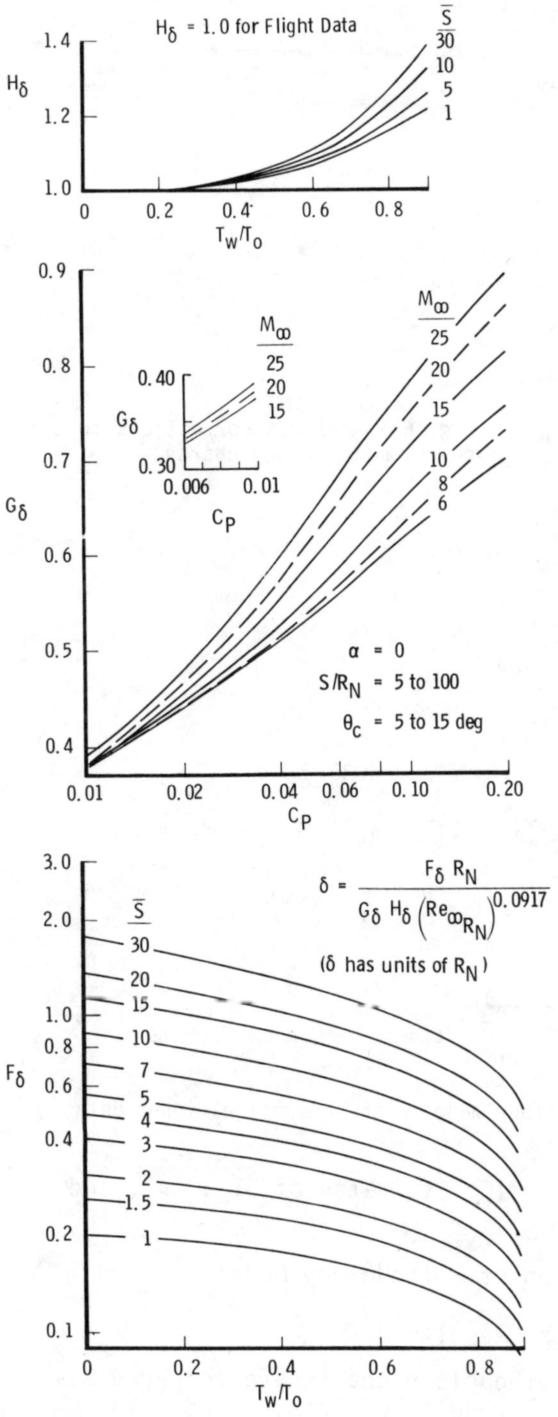

Fig. 5c Turbulent flow methodology charts, perfect gas: boundary-layer thickness, δ.

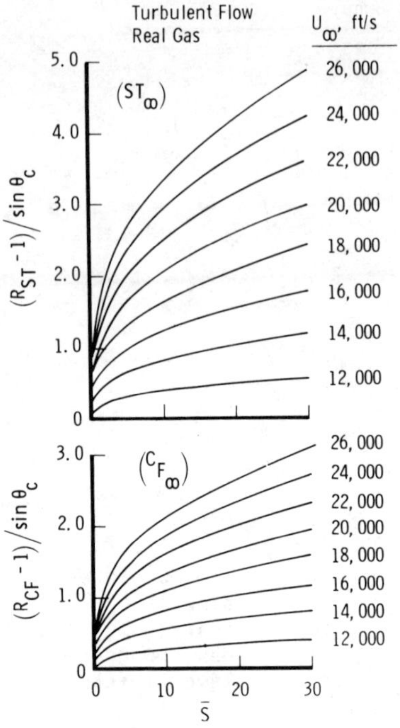

Fig. 6 Real-gas corrections to turbulent methodology charts.

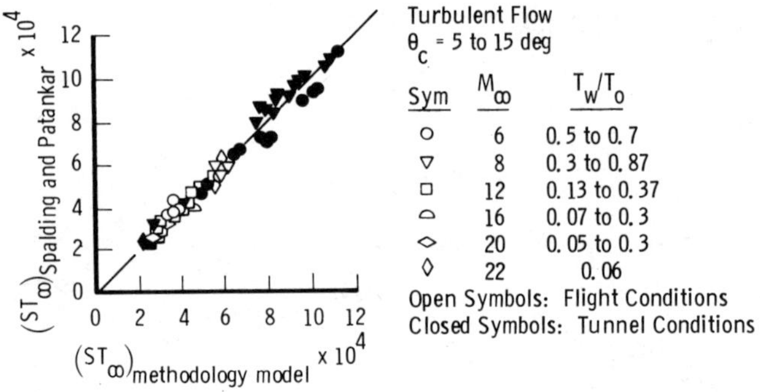

Fig. 7 Comparison of methodology model with theoretical input data.

the accuracy of the engineering estimates of ST_∞, C_{F_∞}, and δ are shown in Table 1.

Verification of Methodology Model

Comparison with Theoretical Results

Solutions from the methodology charts are compared directly with the theoretical results of Refs. 5 and 7 in

Table 1 Estimated accuracy and sample calculations

	Engineering Estimates				
	Perfect gas (PG)				Real gas (RG)
T_w/T_o:	0-0.3	0.4-0.5	0.6-0.7	0.8-0.9	
ST_∞, %	±10	±11	±12	NA	±12
C_{F_∞}, %	±11	±12	±13	±14	±14
δ, %	±16	±17	±18	±20	±20

Sample Calculation

Input

$M_\infty = 22$ $\qquad T_w/T_o = 0.065$

$\theta_c = 5$ deg $\qquad Re_{\infty R_N} = 4.55 \times 10^6$

$S/R_N = 36$ $\qquad U_\infty = 21{,}430$ ft/s

$C_p = 0.0097$ (PG) $\qquad R_N = 0.1166$ ft

Output

$$\bar{S} = 36/\left(4.55 \times 10^6\right)^{1/9} = 6.56$$

$$ST_\infty(PG) = \frac{G_{ST} * F_{ST} * H_{ST}}{\left(Re_{\infty R_N}\right)^{0.1805}}$$

From Fig. 5a:

$G_{ST} = 0.0033$

$F_{ST} = 1.07$

$H_{ST} = 1.0$

$$= \frac{(0.0033)(1.07)(1.0)}{\left(4.55 \times 10^6\right)^{0.1805}}$$

$$= \frac{0.00022}{\text{(Charts)}} \text{ vs } \frac{0.00022}{\text{(Spalding-Patankar)}}$$

$$ST_\infty(RG) = ST_\infty(PG) * R_{ST}$$

$$= (0.000218)(1.180)$$

From Fig. 6:

$R_{ST} = 1.18$

$$= \frac{0.00026}{\text{(Charts)}} = \frac{0.00027}{\text{(Spalding-Patankar)}}$$

Following the same procedure we have

	Methodology	Spalding-Patankar
$C_{F_\infty}(PG)$	0.000356	0.000367
$C_{F_\infty}(RG)$	0.000400	0.000392
$\delta(PG)$, ft	0.04864	0.05210
$\delta(RG)$, ft	0.04864	0.0550

Fig. 7. A wide range of Mach numbers, cone angles, and wall temperatures are shown for both tunnel and flight conditions. Figure 7 presents the Stanton number (ST_∞) comparison. Comparisons showing the local skin friction coefficient and boundary-layer thickness would be very similar in nature. Only a few comparisons fall outside of ±10%.

Figure 8 gives a comparison of both perfect- and real-gas ST_∞ and δ using the methodology charts and the computer program of Refs. 5 and 7. The real-gas parameters from the methodology charts (ST_∞ and δ) are calculated assuming a perfect-gas pressure distribution. A more accurate answer could be obtained by using a real-gas pressure distribution which would be consistent with the computations of Ref. 5. However, the increased difficulty of obtaining a real-gas pressure distribution is not worth the small (2-3%) increase in accuracy.

Comparison with Ground Test Data

The methodology charts were developed from the theoretical results of Ref. 5 with no experimental data input. Figures 9 and 10 show comparisons of the methodology model with experimental data from AEDC Hypersonic Wind Tunnel B. The Stanton number (ST_∞) at Mach 5.95 (Fig. 9) is overpredicted using the methodology charts by 10-15%, whereas the local skin friction coefficient (C_{F_∞}) calculations show ex-

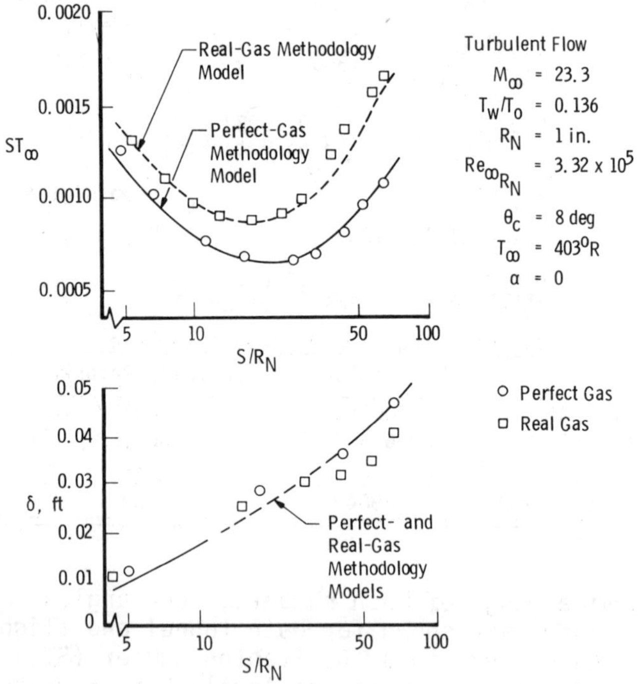

Fig. 8 Comparison of methodology model with real- and perfect-gas Spalding-Patankar solutions.

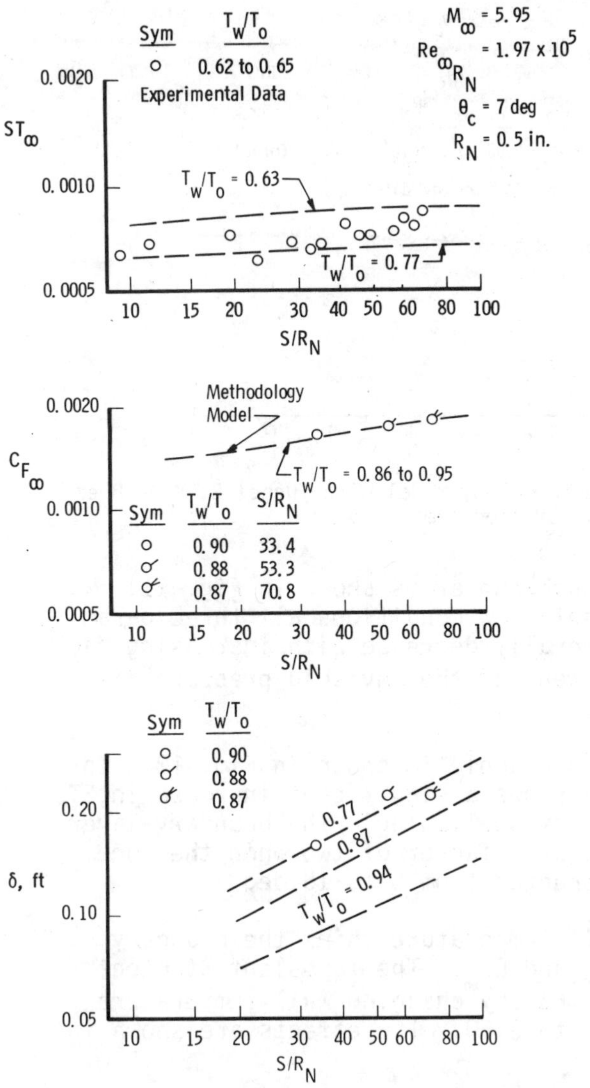

Fig. 9 Comparison of methodology model with Tunnel B, M_∞ = 5.95 experimental data, turbulent flow.

cellent agreement with the experimental data. In Fig. 10 excellent agreement is shown between the methodology model and Mach $\sim$ 8.0 experimental data.

Results

The methodology charts have been used to illustrate the influence of Mach number, cone angle, wall temperature, and real gas on ST_∞, C_{F_∞}, and δ. These results are presented in Figs. 11-14.

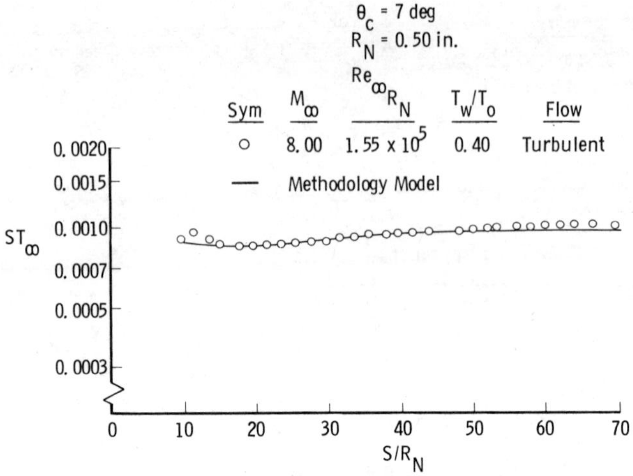

Fig. 10 Comparison of methodology model with Tunnel B, M_∞ = 8 experimental data, Stanton number, ST_∞.

The influence of Mach number is shown in Fig. 11. Note that under turbulent flow conditions all three parameters (ST_∞, CF_∞, δ) generally decrease with increasing Mach number and follow the trend of the inviscid pressure distribution.

The influence of cone angle is shown in Fig. 12. Increasing the cone angle shows a significant increase in ST_∞ and CF_∞ with early entropy swallowing. The boundary-layer thickness (δ) decreases by a factor of two when the cone angle (half-angle) is changed from 5 to 15 deg.

Increasing the wall temperature thins the boundary layer and decreases ST_∞ and CF_∞. The turbulent Stanton number is decreased by ~65% by changing wall temperature ratio (T_W/T_O) from 0.05 to 0.50. The effects are shown in Fig. 13.

Figure 14 shows the influence of real gas on ST_∞, CF_∞, and δ for an 8 deg cone. Under turbulent flow conditions ST_∞ is increased by 43% and CF_∞ is increased by 28% at an S/R_N of 50 at Mach 25 (U_∞ = 24,500 ft/s). The boundary-layer thickness did not change enough to obtain a correlation.

Concluding Summary

A methodology has been developed that has provided simple design charts for calculating the surface heat trans-

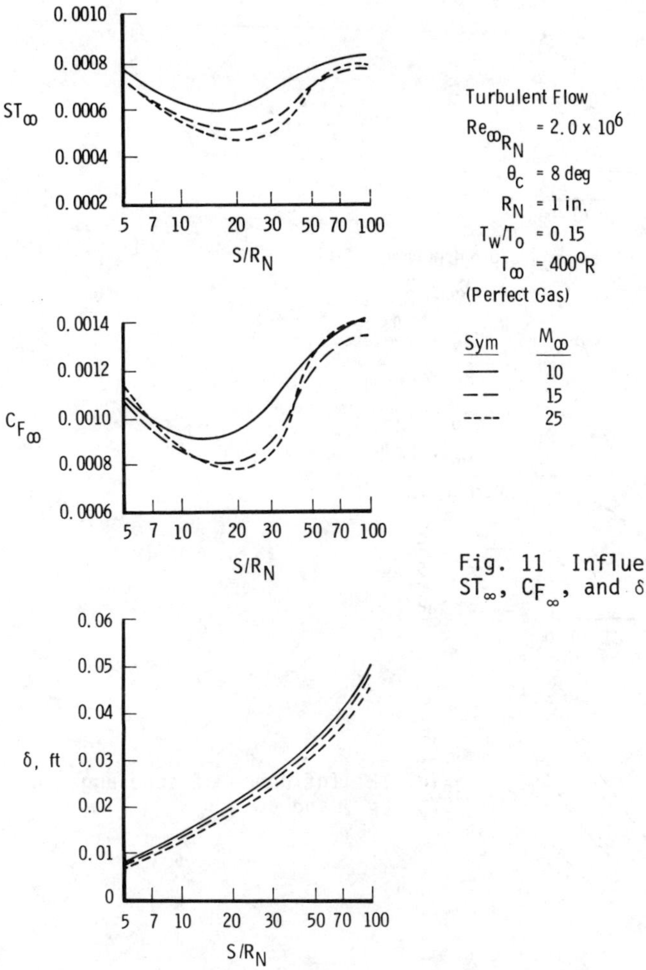

Fig. 11 Influence of Mach number on ST_∞, C_{F_∞}, and δ.

fer ST_∞, the skin friction coefficient C_{F_∞}, and the thickness of the boundary layer δ. The scope of the methodology includes necessary design charts for blunt nose re-entry vehicles at zero angle of attack under turbulent flow conditions, perfect- and real-gas conditions, and flight or wind-tunnel situations. The scope of the methodology is

1) Mach numbers of 6-25.
2) Angle of attack of zero, θ_C = 5-15 deg.
3) From 5 nose radii downstream to 100 nose radii (S/R_N = 5-100).

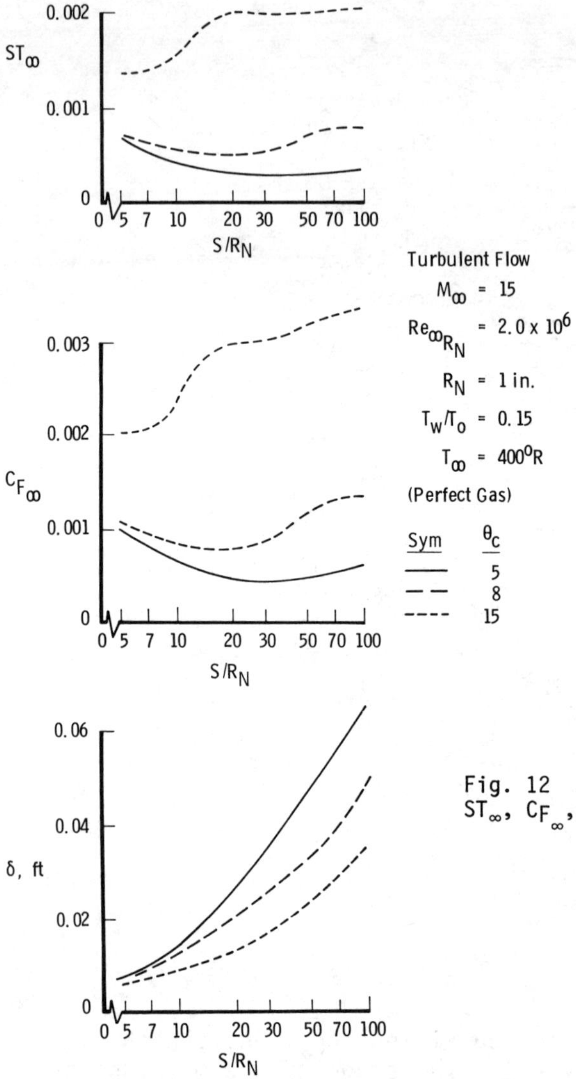

Fig. 12 Influence of cone angle on ST_∞, C_{F_∞}, and δ.

4) Flight and wind-tunnel conditions.

5) Wall temperature ratios of 0-0.9, except in the case of Stanton number where T_W/T_O = 0.7 is upper limit.

6) The Stanton number ST_∞ and local skin friction coefficient C_{F_∞} from the methodology charts are ac-

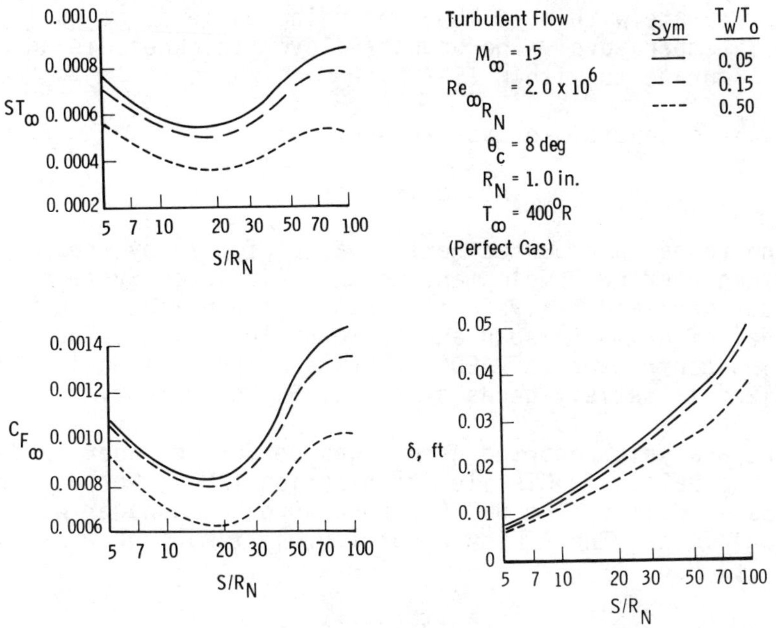

Fig. 13 Influence of wall temperature ratio on ST_∞, C_{F_∞}, and δ.

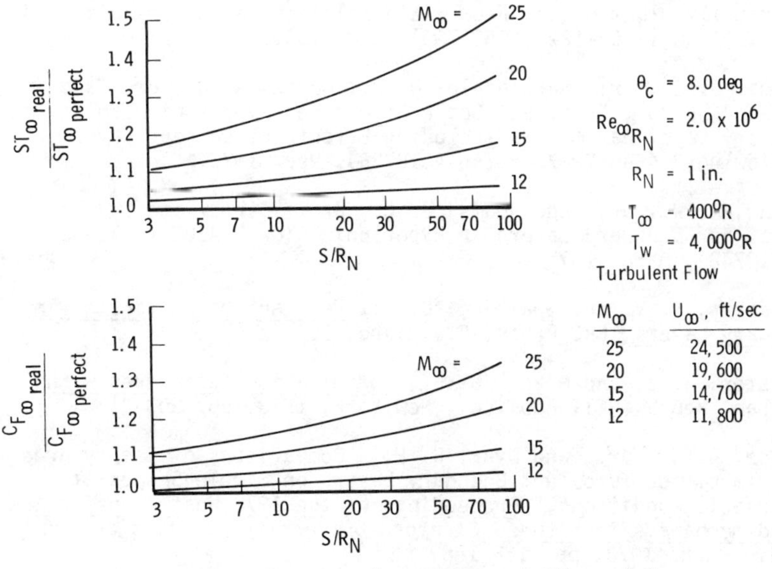

Fig. 14 Influence of real gas on ST_∞ and C_{F_∞}.

curate within 10-15%, depending on level of wall temperature. The boundary-layer thickness is accurate to within 15-20%.

7) The charts are easy to use.

Acknowledgments

The research reported herein was performed by the Arnold Engineering Development Center, Air Force Systems Command. Work and analysis for this research were done by personnel of Arvin/Calspan and Sverdrup Technology, Inc., operating contractors of AEDC. Further reproduction is authorized to satisfy needs of the U. S. Government.

The analysis reported herein was performed under funding by the Ballistic Missile Organization (BMO), Norton Air Force Base, Calif. The BMO Air Force Project Manager was Maj. Paul Toth. The AEDC Air Force Project Manager was Capt. Carl Bang.

References

[1]Morrison, A. M., Solomon, J. M., Ciment, M., and Ferguson, R. E., "Handbook of Inviscid Sphere-Cone Flow Fields and Pressure Distribution: Vol. I," NSWC/WOL/TR 75-45, Dec. 1975.

[2]Roberts, J. F., Lewis, Clark H., and Reed, Marvin, "Ideal Gas Spherically Blunted Cone Flow Field Solutions at Hypersonic Conditions," AEDC-TR-66-121 (AD637703), Aug. 1966.

[3]Adams, J. C., Jr., Martindale, W. R., Mayne, A. W., Jr., and Marchand, E. O., "Real Gas Scale Effects on Hypersonic Laminar Boundary-Layer Parameters Including Effects of Entropy-Layer Swallowing," AEDC-TR-75-2 (AD-A018755), Dec. 1975.

[4]Mayne, A. W., Jr., and Adams, J. C., Jr., "Streamline Swallowing by Laminar Boundary Layers in Hypersonic Flow," AEDC-TR-71-32 (AD719748), March 1971.

[5]Patankar, S. V. and Spalding, D. B., Heat and Mass Transfer in Boundary Layers, CRC Press, Cleveland, 1968.

[6]Kuester, J. L. and Mize, Joe H., Optimization Techniques with Fortran, McGraw-Hill Book Co., New York, 1973, pp. 203-217.

[7]Mayne, A. W., Jr., and Dyer, D. F., "Comparisons of Theory and Experiment for Turbulent Boundary Layers on Simple Shapes at Hypersonic Conditions," Proceedings of the 1970 Heat Transfer and Fluid Mechanics Institute, Stanford University Press, Stanford, Calif., June 1970, pp. 168-188.

[8]Cohen, N. B., "Correlation Formulas and Tables of Density and Some Transport Properties of Equilibrium Dissociating Air for Use in Solutions of the Boundary-Layer Equations," NASA TN D-194, Feb. 1960.

[9]Inouye, M., Rakich, J. V., and Lomax, H., "A Description of Numerical Methods and Computer Programs for Two-Dimensional and Axisymmetric Supersonic Flow over Blunt-Nosed and Flared Bodies," NASA TN D-2970, Aug. 1965.

[10]Rotta, N. R. and Zakkay, V., "Effects of Nose Bluntness on the Boundary-Layer Characteristics of Conical Bodies at Hypersonic Speeds," Astronautica Acta, Vol. 13, Nos. 5 and 6, 1968, pp. 507-516.

[11]Ferri, A., "Some Heat Transfer Problems in Hypersonic Flow," Aeronautics and Astronautics, Pergamon Press, New York, 1960, pp. 344-377.

[12]Zakkay, V. and Krause, E., "Boundary Conditions at the Outer Edge of the Boundary Layer on Blunted Conical Bodies," AIAA Journal, Vol. 1, July 1963, pp. 1671-1672.

[13]Rubin, I., "Shock Curvature Effect on the Outer Edge Conditions of a Laminar Boundary Layer," AIAA Journal, Vol. 1, Dec. 1963, pp. 2850-2852.

[14]Cheng, H. K., "Hypersonic Flow with Combined Leading-Edge Bluntness and Boundary-Layer Displacement Effect," Cornell Aeronautical Laboratory, Rept. AF-1285-A-4, 1960.

[15]Griffith, B. J. and Lewis, C. H., "Laminar Heat Transfer to Spherically Blunted Cones at Hypersonic Conditions," AIAA Journal, Vol. 2, March 1964, pp. 438-444.

[16]Hayes, W. D. and Probstein, R. F., Hypersonic Flow Theory, Academic Press, New York, 1959.

A Study of a Boundary-Layer Trip Concept at Hypersonic Speeds

D.E. Nestler*
General Electric Co., Philadelphia, Pa.
and
W.D. McCauley†
TRW Defense and Space Systems Group, Redondo Beach, Calif.

Abstract

Wind-tunnel tests of a spherically blunted 9 deg half-angle cone were conducted at AEDC Tunnel B at Mach 8 over a Reynolds number range of 1.0-3.5 x 10^6/ft to determine the tripping effectiveness, downstream augmented heating, and body forces for several three-dimensional trip configurations. Variations were made in trip height, circumferential spacing, number of rows, trip element shape, model nose radius, and angle of attack. A correlation for effective tripping was developed in terms of local flow properties that predicts the present as well as previous results for three-dimensional trips.

Nomenclature

C	= circumferential spacing of trip elements
C_A	= axial force coefficient
C_N	= normal force coefficient
C_M	= pitching moment coefficient
C_p	= pressure coefficient
$H_{0.9T}$	= heat-transfer coefficient based on recovery temperature of 0.9 T_O
h	= static enthalpy

Presented as Paper 81-1086 at the AIAA 16th Thermophysics Conference, Palo Alto, Calif., June 23-25, 1981.

*Staff Consultant, Re-entry Systems Division.

†Staff Engineer.

H_{FR} = Fay-Riddell stagnation point heat-transfer coefficient based on recovery temperature of T_0
H_{max} = maximum heat-transfer coefficient in interaction region
H_u = heat-transfer coefficient for undisturbed flow
K = trip height
K_A = downstream heating augmentation factor, $K_A = H_{0.9T}/(H_{0.9T})_{\text{predicted turbulent smooth body}}$
L = model length
M = Mach number
$\dot{q}$ = heat flux
R_N = nose radius
Re = Reynolds number
T = temperature
W = circumferential width of trip element
X_M = axial location measured from sharp apex
X = axial location measured from nose stagnation point
X_{CP} = axial distance to center of pressure
y = distance normal to cone surface
α = angle of attack
δ = boundary-layer velocity thickness
δ^* = boundary-layer displacement thickness
θ = boundary-layer momentum thickness
ϕ = circumferential angle
$\Delta\phi$ = circumferential angle between trip elements
λ = compressibility parameter; $= 1 + 0.90\,[(h_w/h_e) - 1] + 0.28\,[(h_R/h_e) - 1]$

Subscripts

B = base value
e = boundary-layer edge value
max = maximum value
R = recovery value
S = based on wetted length
T = turbulent value
TR = value at transition
W = wall value
0 = total value
∞ = freestream value

Introduction

Analyses of ballistic re-entry vehicle flight test data have shown that appreciable trajectory deflections develop as boundary-layer transition progresses forward over the vehicle. The cause of these deflections is probably the asymmetric transition process, possibly interacting with the antenna windows.[1] One method to obviate such effects may be to artificially induce a rapid, mostly symmetric occurrence of the transition process at a high altitude.

In a previous study[2] tripping effectiveness of active and passive trips was determined in wind-tunnel tests of a blunt 7.5 deg half-angle cone. The primary purpose of the present study was to obtain tripping effectiveness data for passive trips on a blunt 9.0 deg half-angle cone and to develop a generalized effective tripping correlation in terms of local boundary-layer properties.

It is well-known that protuberances such as boundary-layer trips cause increased heating in their immediate vicinity, as well as downstream. Sterrett et al.[3] showed that if the roughness is of the proper size, spanwise distortion is very slight and the measured turbulent heat-transfer rates agree with smooth body predictions. However, too large a roughness causes extensive spanwise variations in heat flux and deviation from theory. The same conclusion was reached by Stainback,[4] who measured 50% variations in heat-transfer circumferentially in tripped turbulent flow on a sharp 5 deg cone at $M_\infty = 8$. More recently, Thyson et al.[2] reported lateral variations of over 50% downstream of trips. Hence, a secondary objective of the present program was to measure circumferential distributions of heat-transfer downstream of the trip array.

Finally, the effect of trips on axial force and static stability must be considered. Previous tests[2] had addressed this question for a 7.5 deg half-angle cone. Therefore, a final objective was to perform force and moment tests on the 9 deg half-angle cone of the present study.

Experimental Program

An existing conical model was modified to incorporate nine different trip configurations. The test model consisted of a 9 deg half-angle cone of 10 in. base diameter, with sharp and blunt spherical nosetips. A schematic of the model is shown in Fig. 1, while a photograph of the model with a trip insert installed is shown in Fig. 2.

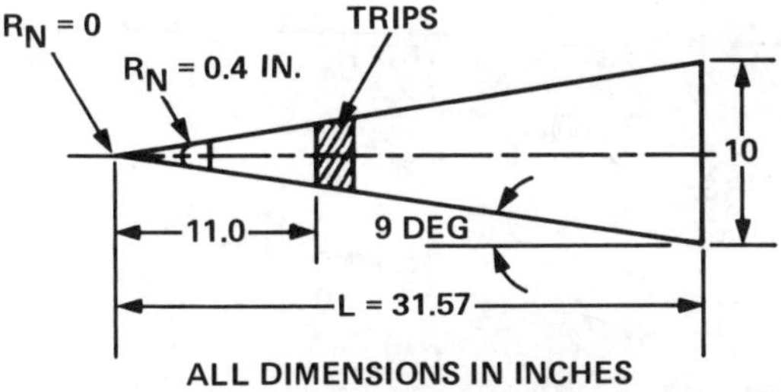

Fig. 1 Schematic of wind-tunnel test model.

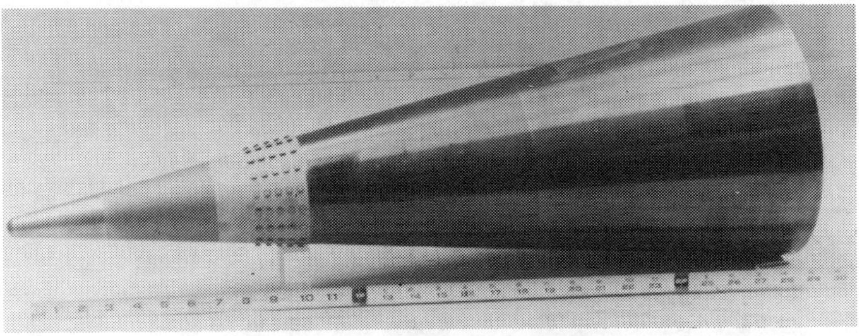

Fig. 2 Wind-tunnel test model.

Table 1 Trip geometries

Trip no.	Shape	No. of rows	$\Delta\phi$, deg	Height, in.
1	Rectangular	5	15	0.125
2	↓	↓	↓	0.100
3				0.075
4			22.5	0.125
5			↓	0.100
6				0.075
7			30	0.125
8		3	22.5	0.125
9	Ablated	5	22.5	0.065 - 0.125

The model body and its support accessories were fabricated from stainless steel, with the exception of one trip configuration which was made from lava to obtain phase-change paint data. Lava was selected based on its past performance in obtaining paint data and for its machinability.

The nine different trip configurations which were tested are summarized in Table 1 and Fig. 3. Trips 1-8 were of rectangular shape with variations in trip height, circumferential spacing, and number of rows. Trip 9

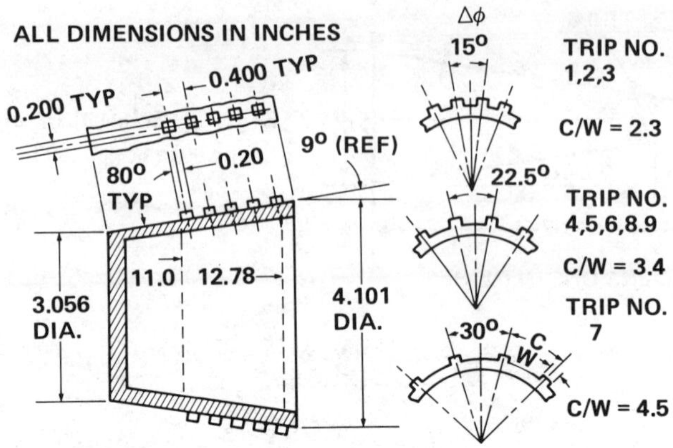

Fig. 3a Nonablated trip section details.

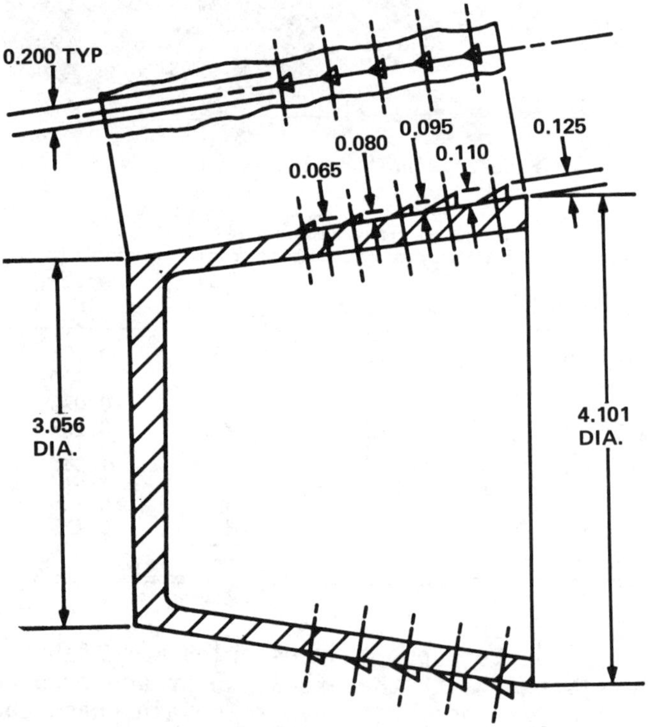

Fig. 3b Ablated trip section details.

represented an ablated shape as idealized from previous trip ablation tests. Photographs of trips 5 and 9 are shown in Fig. 4.

Model instrumentation consisted of coaxial heat gages installed at 97 locations. The gages were located along

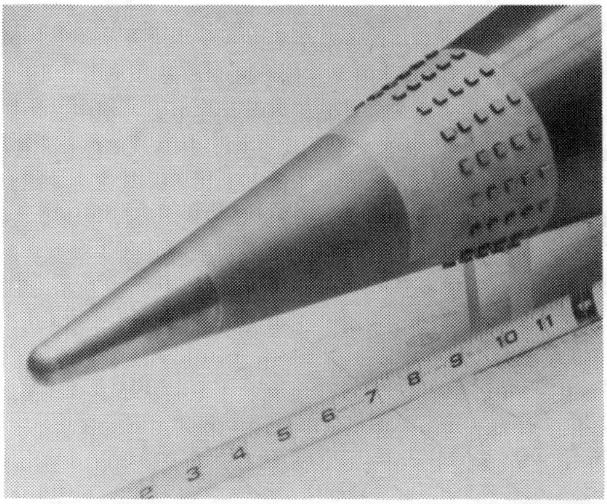

Fig. 4a Trip insert (trip 5, 0.100 in. high, $\Delta\phi = 22.5$ deg).

Fig. 4b Trip insert in ablated configuration (trip 9).

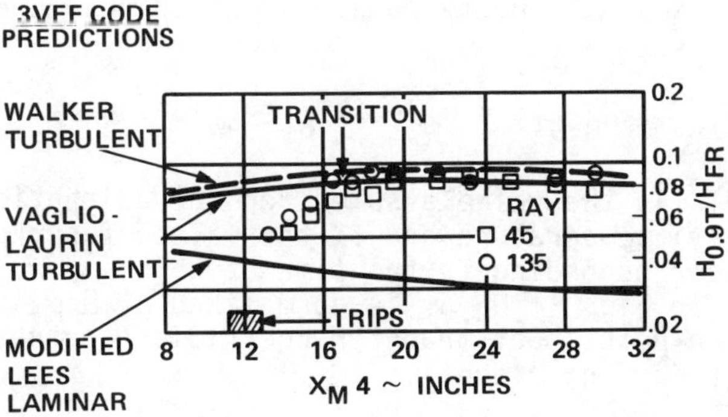

Fig. 5 Typical heat-transfer distributions (trip 5, $R_N = 0.4$ in., $Re_\infty/ft = 2.0 \times 10^6$, $\alpha = 0$).

several meridians to determine axial heat-transfer distributions, and were also installed in local arrays to obtain circumferential heat-transfer distributions between trip element meridians. Following the heat-transfer tests, the heat gages were removed, their holes were plugged and contoured, and force tests were conducted to determine the effect of the trips on axial and normal forces. Oil flow tests, boundary-layer surveys, shadowgraphs, and phase-change paint data were also obtained in support of the trip study.

Tests were conducted in AEDC Tunnel B at Mach 8 and $T_0 = 1350^oR$ over a freestream Reynolds number (Re_∞) range of 0.5-3.5 X 10^6/ft. A total of 208 heat-transfer runs were made, in which the freestream Reynolds number was varied over a sufficient range in an attempt to determine the effective trip condition for each trip configuration, including selected variations in nose radius (R_N = 0, 0.25, 0.40, and 0.75 in.) and angle of attack (α = 0, 1, and 2 deg). In the present paper, representative results will be presented to illustrate the effect of each of the major variables. More complete results are available in Ref. 5.

Determination of Effective Tripping

Definition of Transition

Typical results of the transition tests are shown in Fig. 5, in which the longitudinal heat transfer distribution is shown for two of the primary rays. The coaxial heat gage surface temperature response was reduced to wall heat flux $\dot{q}_W$ by standard relations for a one-dimensional semi-infinite solid.[6] Heat-transfer coefficient $H_{0.9T}$ is defined as

$$H_{0.9T} = \dot{q}_W/(0.9T_o - T_W) \quad (1)$$

in which $0.9T_0$ is the value assumed for the adiabatic wall (recovery) temperature and T_W the value of surface temperature corresponding to the time at which $\dot{q}_W$ was calculated. In Fig. 5, $H_{0.9T}$ is normalized by H_{FR}, the stagnation point heat-transfer coefficient computed by the method of Fay and Riddell.[7]

Figure 5 also includes plots of the normalized heat-transfer distribution predicted by the GE-RSD 3D viscous code (3VFF)[8] for laminar and turbulent boundary-layer

Table 2 Transition locations for α = 0
(inches from virtual apex)

R_N, in.	Trip No.	$Re_\infty/ft \times 10^{-6}$ 1.5	1.75	2.0	2.5	3.0	3.5
0	None	18.5	17/30	15/28	14/22	10/19	9/17
	4	16.5	16	15.5	15	-	-
	5	18.5	17	16.5	15.5	14.5	-
	6	21	19	18.5	16.5	15.5	14.5
0.25	5	24	20	18.5	16.5	16	-
	6	-	31	29	-	17.5	16.5
0.40	1	-	-	15	14.5	-	-
	2	-	-	17.5	15	-	-
	3	-	-	28	22	-	-
	4	19	16.5	15	14.5	-	-
	5	>31	23	17.5	15	14.5	14.5
	6	-	-	29	22	19	16.5
	7	-	-	19	19	-	-
	8	20	17	14	13	-	-
	9	>31	-	26	22	20	19
0.75	4	31	17	15	14	-	-
	5	-	31	17.5	15.5	14.5	-
	6	-	-	-	29	19.5	16.5

Note: For no trip, beginning and end of transition are shown. For all cases with trips, only end of transition is shown. indicates not tested.

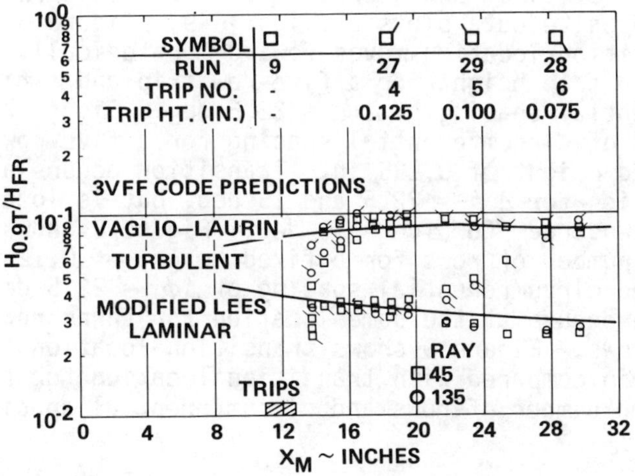

Fig. 6 Effect of trip height on transition location (R_N = 0.4 in., $Re_\infty/ft = 2.0 \times 10^6$, $\alpha = 0$, $\Delta\phi$ = 22.5 deg).

conditions. Two turbulent predictions are shown: Vaglio-Laurin[9] and Walker,[10] while the laminar prediction is the Lees method[11] modified by using properties based on the Eckert reference enthalpy.[12] The turbulent prediction assumed the virtual origin to be at the stagnation point.

The location of the completion of transition (defined herein as "transition") is identified by an arrow in Fig. 5, based on a judgment assessment of where the heat-transfer distribution merges into the predicted turbulent variation.‡ (Note that because of bluntness-induced entropy gradient effects, this does not correspond to a maximum in heat transfer, as in the classical method of identifying flat-plate or sharp-cone transition completion.) Some unavoidable uncertainty exists in identifying transition by this definition, due to the subjectivity involved, as well as data scatter of gage readings. However, this method of determining transition is consistent with previous evaluations.[2]

For completeness, transition locations for all tests are given in Table 2 for $\alpha = 0$ and Table 3 for $\alpha \neq 0$, including the sharp-cone results.

Effect of Trip Geometry (at Fixed Reynolds Number)

The effect of trip geometry on transition location for $\alpha = 0$, $R_N = 0.4$ in., and $Re_\infty/ft = 2.0 \times 10^6$ is shown in the composite data plots of Figs. 6-9. Figure 6 shows that transition location moves foward monatonically with increase in trip height for a five-row trip and a fixed circumferential spacing of $\Delta\phi = 22.5$ deg. Figure 7 shows the effect of circumferential spacing for a five-row trip and a fixed height of 0.125 in. Transition occurs at the same location for $\Delta\phi = 22.5$ and 15 deg, but is located further downstream for $\Delta\phi = 30$ deg. Figure 8 shows the effect of number of rows for a fixed height of 0.125 in. and a fixed circumferential spacing of $\Delta\phi = 22.5$ deg. Transition occurs at the same location for three rows as for five rows. Figure 9 shows transition location for the ablated trip compared with transition location for trips of the same number of rows and circumferential spacing,

‡In making this assessment, matching the slope of the predicted heat-transfer distribution was used as the criterion, rather than the absolute level.

Table 3 Transition locations for α = 0
(inches from virtual apex)

R_N, in.	Trip No.	α, deg	Re_∞/ft x10^6	ϕ, deg 0	45	90	135	180
0	None	1	1.5	20/-	22/-	22/-	18/-	18/-
			1.75	17/-	18/-	17/-	17/-	16/-
			2.0	14/24	15/26	17/26	16/30	17/-
			2.5	13/22	14/23	14/24	14/21	12/20
			3.0	10/18	*/19	*/19	*/17	*/18
			3.5	9/17	*/18	*/17	*/14	*/15
		2	2.0	9/16	16/27	20/-	19/-	-/-
			2.5	14/24	*/23	16/24	*/21	*/18
	5	1	1.5	18	17	18	20	20
			1.75	17	17	17	18	18
			2.0	16	15	16	17	20
			2.5	16	14	15	16	17
			3.0	14	14	15	15	16
		2	2.0	15	15	15	17	19
			2.5	14	14	15	16	14
	6	1	3.0	16	15	15	16	17
			3.5	14	14	14	14	14
0.25	5	1	2.0	16	17	19	20	24
		2	2.0	16	16	15	19	-
0.4	4	1	1.5	16	16	16	17	29
			1.75	16	15	15	16	24
		2	1.5	15	14	14	18	28
			1.75	15	14	14	17	22
	5	1	1.75	18	17	18	20	28
			2.0	16	16	17	20	24
0.4	5		2.5	15	15	15	17	21
			3.0	15	14	15	16	18
			3.5	15	15	15	15	15
		2	1.75	16	15	14	20	-
			2.0	15	14	14	16	27
			2.5	15	15	14	14	22
			3.0	15	15	14	14	19
			3.5	15	15	14	14	18
	6	1	3.0	16	15	17	17	22
			3.5	16	15	16	17	19
		2	3.0	16	15	14	16	22
			3.5	15	15	14	14	18
	8	1	1.5	16	16	15	18	-
			1.75	16	14	14	15	24
		2	1.5	15	14	14	15	26
			1.75	14	14	14	15	21
	9	1	1.5	26	28	25	-	-
		2	1.5	25	24	18	22	-
0.75	5	1	1.75	22	22	24	28	-
			2.0	16	16	16	17	26
			2.5	14	14	14	15	18
			3.0	14	14	14	14	15
		2	2.0	14	14	14	14	28
			2.5	14	14	14	14	18
	6	1	3.0	17	16	17	17	22
			3.5	15	14	14	16	18

* Transition occurred forward of first heat gage (X_M = 13.8 in.).
- Transition completion did not occur on model.

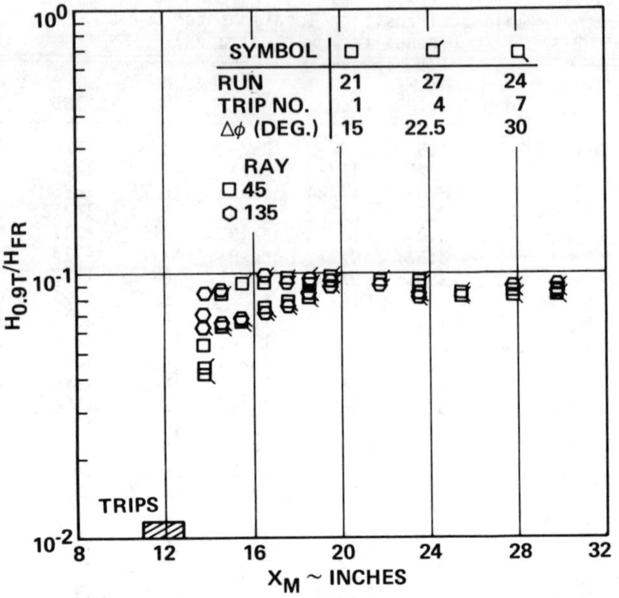

Fig. 7 Effect of circumferential spacing on transition location (R_N = 0.4 in., Re_∞/ft = 2.0 x 10^6, $\Delta\phi$= 0, 0.125 in. trip height).

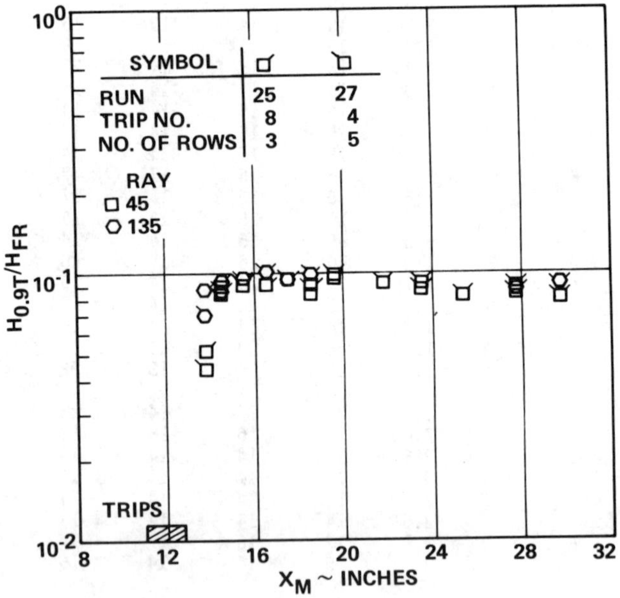

Fig. 8 Effect of number of rows on transition location (R_N = 0.4 in., Re_∞/ft = 2.0 x 10^6, α = 0, 0.125 in. trip height).

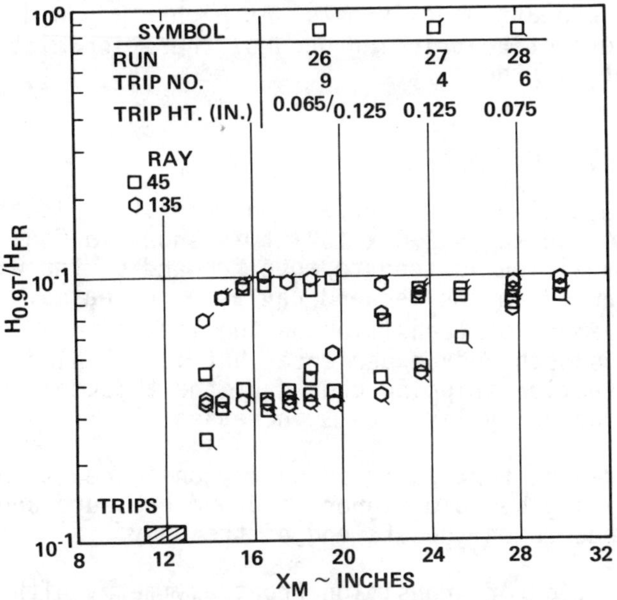

Fig. 9 Effect of ablated trip configuration on transition location (R_N = 0.4 in., Re_∞/ft = 2.0 x 10^6, α = 0).

but with heights of 0.075 and 0.125 in. The ablated trip array, which has heights increasing from 0.065 in. for the first row to 0.125 in. for the last (fifth) row, is seen to cause transition at a location intermediate to the other two trips.

Similar qualitative effects of trip geometry on transition location were observed at other Reynolds numbers.

Effect of Reynolds Number and Nose Bluntness

For each nose bluntness, the flow without trips was completely laminar for all Reynolds numbers. For the smallest nose bluntness (0.25 in.), significant roughness tripping occurs at Re_∞/ft = 1.5 x 10^6 while transition onset is just beginning for the larger bluntness. This is to be expected because as one approaches a sharp nose, smooth body transition occurs on the frustum for these freestream Reynolds numbers. Thus transition onset is most easily caused for small bluntness, with increasing roughness being required to overcome the normal transition delay (for no trips) being caused by increasing bluntness.

As Re_∞/ft increases from 1.5 x 10^6 to only 2.0 x 10^6, the effect of nose bluntness on the tripping to turbulent flow appears nonexistent. At higher Reynolds numbers where effective tripping occurs, there is relatively no effect of bluntness.

Effect of Angle of Attack

The effect of angle of attack on transition location for $R_N = 0.4$ and $Re_\infty = 2.0 \times 10^6$/ft is shown in Fig. 10 for trip 5. Transition is seen to move forward on the windward ray and rearward on the leeward ray as α increases. The rearward movement of transition on the leeward ray is opposite to smooth body experience, but is explained by the less effective tripping caused by the thickening of the leeward boundary layer as α increases.

Windward-side forward tripping is considered to be more stabilizing because higher windward pressure and shear are induced on the aft end of the body.

The variation of transition front asymmetry with Reynolds number is shown pictorially in Fig. 11a for trip 5, for $R_N = 0.4$ in. The degree of asymmetry is seen to decrease as Reynolds number increases and transition moves closer to the trip. Similar plots are given in Figs. 11b and 11c for trips 4 and 9, respectively. The character of the asymmetry is seen to vary, depending on the trip geometry: the ablated trip (No. 9) has transition most forward on the 90 deg meridian, while the highest trip (No. 4) has a uniform transition front for 0-90 deg, but delayed on the leeward side.

Crossplots of Transition Locations vs Freestream Reynolds Number

Crossplots of transition location vs freestream Reynolds number for trips 4-6 are given in Fig. 12 for α = 0 and for all nose radii tested (R_N = 0.75, 0.4, 0.25, and 0 in.). All of the configurations exhibit the same qualitative behavior of monatonic forward movement of transition location as Reynolds number increased for a fixed nose radius. For the smallest trip (No. 6), transition is still moving forward rather rapidly at the highest test Reynolds number available. The location of transition appears to be approaching a final forward position 2-4 in. downstream of the last row of trips, with R_N = 0.75 in. producing the most forward movement and R_N = 0.25 in. the least forward movement. Changing the

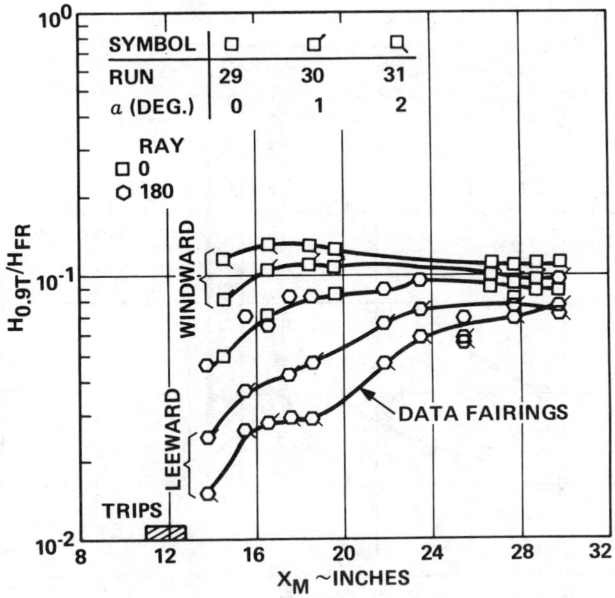

Fig. 10 Effect of angle of attack on transition location (trip 5, R_N = 0.4 in., Re_∞/ft = 2.0 x 10^6).

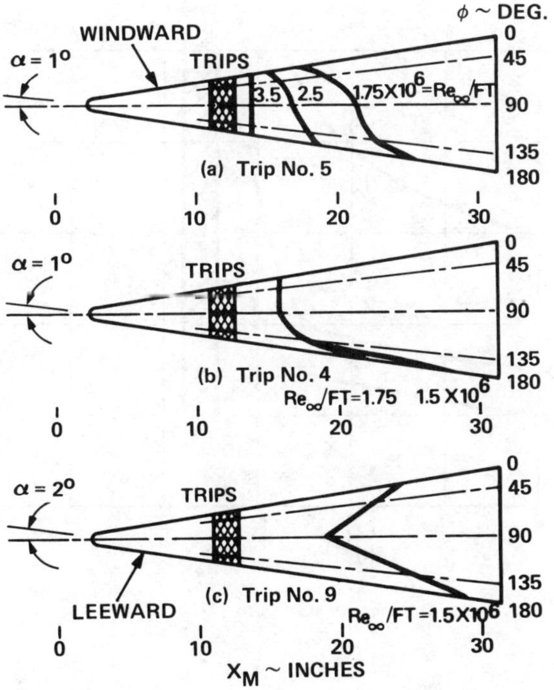

Fig. 11 Effect of angle of attack on transition front asymmetry (R_N = 0.4 in.).

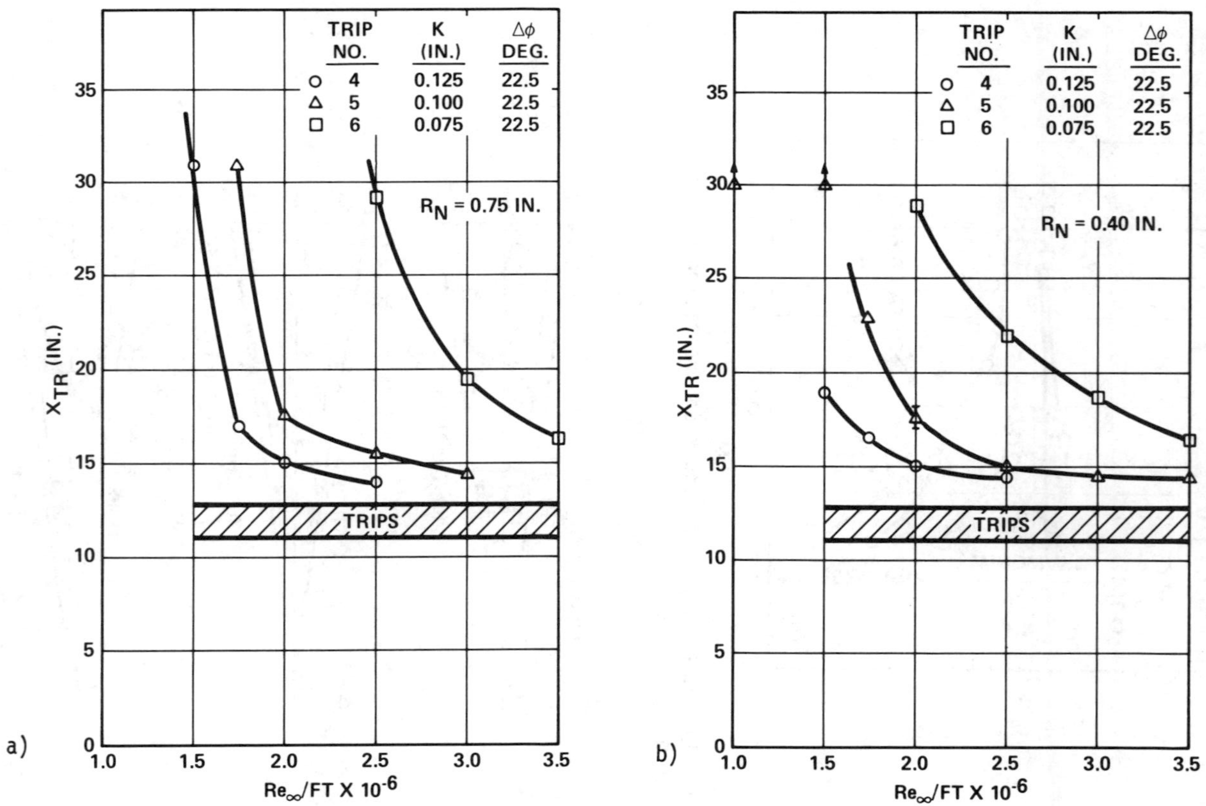

Fig. 12 Variation of transition with Reynolds number (trips 4-6, $\alpha = 0$).

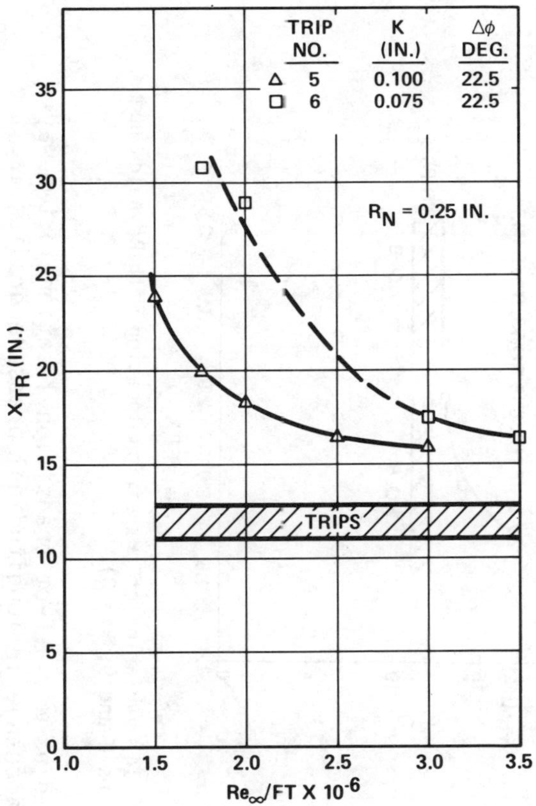

c)

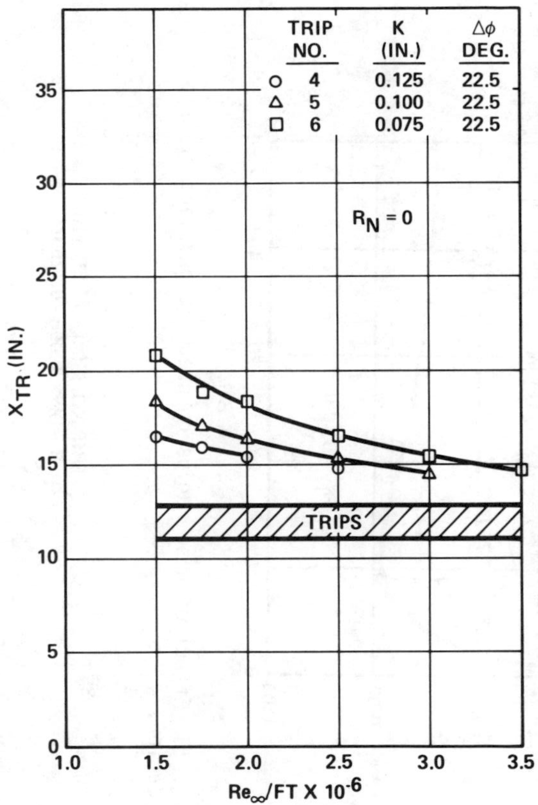

d)

Fig. 12 (cont.) Variation of transition with Reynolds number (trips 4-6, $\alpha = 0$).

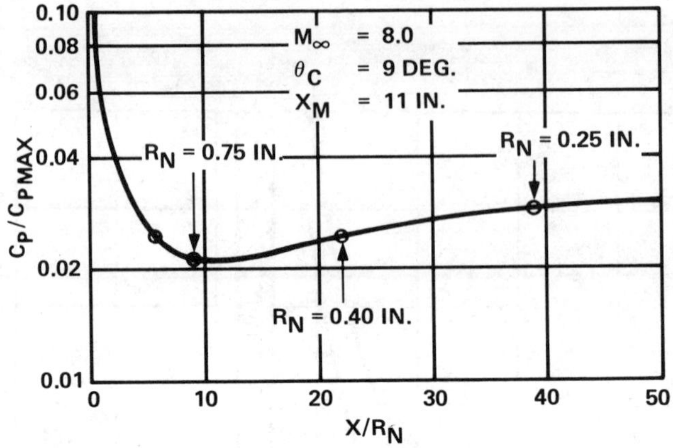

Fig. 13 Effect of nose radius on pressure gradient at trip location.

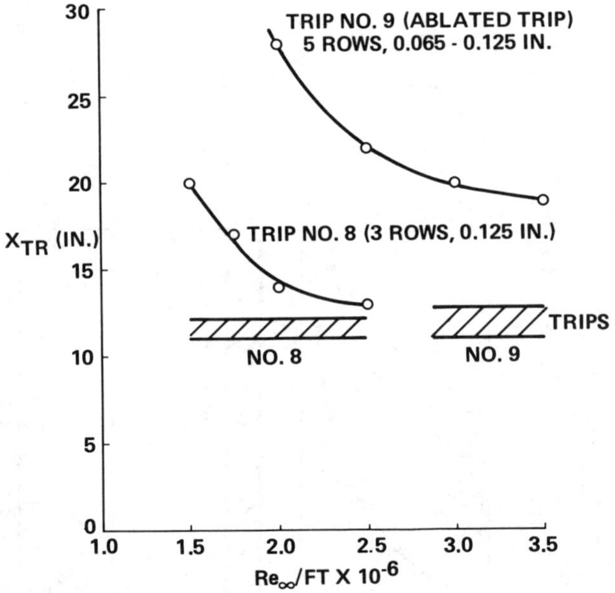

Fig. 14 Variation of transition location with Reynolds number (trips 8 and 9, $\alpha = 0$).

nose radius may affect transition location in two ways: 1) by a change in longitudinal pressure gradient at the trip location and 2) by a change in local Mach number, Reynolds number, and boundary-layer thickness at the trip location. The effect of nose radius on pressure gradient is shown in Fig. 13. An R_N of 0.75 in. is seen to produce a slightly negative pressure gradient at the trip

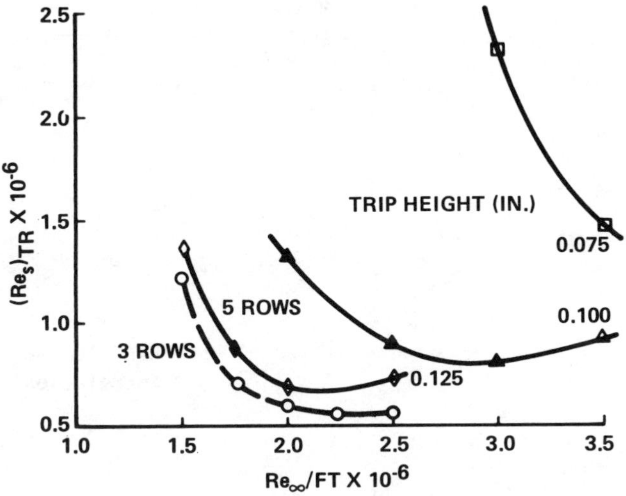

Fig. 15 Determination of effective tripping (R_N = 0.4 in., α = 0).

location, whereas R_N = 0.40 and 0.75 in. produce positive pressure gradients.

Crossplots of transition location vs Re_∞ for trips 8 and 9 are given in Fig. 14 for α = 0 and R_N = 0.4. The three-row trip (No. 8) is seen to perform similarly to the five-row trip of the same height and spacing (No. 4, Fig. 12b) but with a more forward final transition position. The ablated trip (No. 9) is seen to be much less effective than the unablated trip of the same final row height and spacing (No. 4, Fig. 12b), with a final transition position about 4 in. further downstream.

Determination of Effective Tripping Conditions

The definition of effective tripping which has been used frequently in supersonic studies is the Reynolds number at which transition ceases to move forward rapidly, as determined by a plot of transition location vs free-stream Reynolds number (e.g., Fig. 12). In the present case, this point is not always well defined, and the selection of the effective tripping condition becomes somewhat subjective. In order to provide a less subjective, more systematic procedure, the definition of effective tripping originally proposed by Van Driest and Blumer[13] was adopted: effective tripping corresponds to the minimum local value of $(Re_S)_{TR}$ (the wetted length

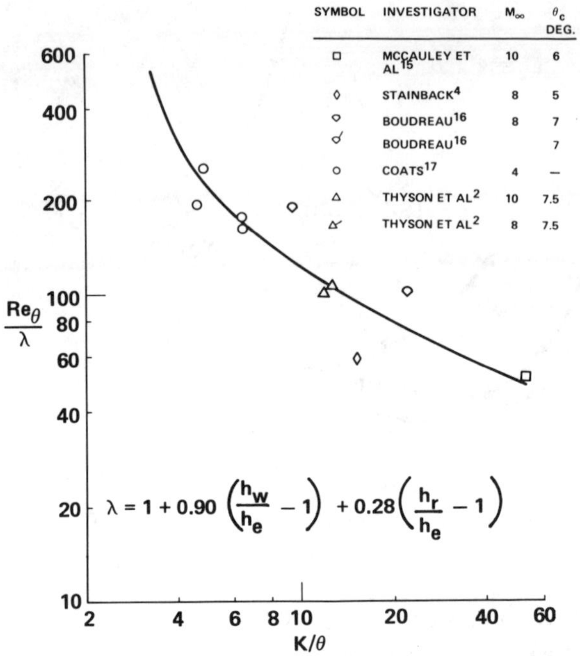

Fig. 16 Re_θ/λ vs K/θ correlation.

Table 4 Summary of effective tripping parameters for α = 0

R_N, in.	Trip No.	Trip height, in.	(Re_∞)eff x 10^{-6}/ft	θ, in.	Re_θ	λ	Re_θ/λ	k/θ
0	4	0.125	<1.5	>0.0027	<490	5.44	<90	<46
	5	0.100						<37
	6	0.075						<28
0.25	4	0.125	Not tested	-	-	-	-	-
	5	0.100	2.25	0.0054	187	1.95	96	18.5
	6	0.075	3.25	0.0047	222	1.86	119.5	16
0.40	4	0.125	2.16	0.0067	163	1.76	92.5	18.7
	5	0.100	2.85	0.0059	185	1.71	108.5	17
	6	0.075	>3.5	<0.0055	>200	1.63	>122.5	>13.6
	8	0.125	≈2.3	0.0065	168	1.75	96	19.1
	9	0.065/ 0.125	>3.5 3.7 (est.)	<0.0055	>200	1.63	>122.5	>22.7
0.75	4	0.125	2.05	0.0082	142	1.56	91	15.3
	5	0.100	2.35	0.0076	153	1.56	98	13.2
	6	0.075	>3.5	<0.0063	>185	1.53	>121	>11.9

Table 5 Summary of effective tripping parameters for α = 1 deg

R_N, in.	Trip No.	Trip height, in.	ϕ, deg	$(Re_\infty)_{eff}$ x 10^{-6}/ft	θ, in.	Re_θ	λ	Re_θ/λ	k/θ
0.4	5	0.100	0	2.05	0.00658	163	1.70	96	15.2
			180	3.0	0.00629	190	1.74	109	15.9

Reynolds number at the transition location), which occurs as Re_∞ is increased and transition moves forward. An example of the determination of effective tripping by this method for the present data is given in Fig. 15 for trips 4-6, R_N = 0.4 in., and α = 0. The values of Re_S plotted in this figure were obtained from the turbulent boundary-layer output of the GE-RSD 3D viscous code[8] (including entropy gradient effects) at the location of transition as plotted in Fig. 12. Values of Re_∞ corresponding to the minimum of the curves for K = 0.125 and 0.100 in. were defined as effective tripping conditions, $(Re_\infty)_{eff}$. The absence of a minimum for K = 0.075 in. implies that $(Re_\infty)_{eff} > 3.5 \times 10^6$/ft.

A summary of the value of $(Re_\infty)_{eff}$ determined by this method is given to Table 4 for α = 0, for all combinations of trip geometry and nose radius which were tested over a range of Reynolds numbers. Corresponding values of Re_θ, θ, λ, Re_θ/λ, and K/θ are also given, as obtained from the laminar boundary layer output of the 3VFF code at the location of the leading edge of the trip array. The computed values were based on a wall temperature of 540R. A similar summary for α = 1 deg is given in Table 5 for trip 5 for α = 0, 90, and 180 deg.

Correlation of Effective Tripping

An effective tripping correlation developed prior to the present tests[14] is shown in Fig. 16, with test data defined in Table 6, from Refs. 2, 4, and 15-17. The parameters used in this correlation are based on laminar boundary-layer properties at the location of the leading edge of the first row of trips, computed by the viscous code of Ref. 8.[§] The correlation allows for compressi-

§Since viscous codes for blunt bodies differ somewhat in computed local properties, application of the correlation of Fig. 16 to predict effective tripping should be made with the viscous code of Ref. 8 which was used to develop the correlation.

Table 6 Test data summary for Fig. 16

One or more rows of trip elements on axisymmetric bodies

Symbol	Investigator	Facility	M_∞	θ_c deg	R_N in.	H_W/H_R	$\left(\frac{X}{R_N}\right)_K$	K, in.	C/W	No. of rows
□	McCauley et al.[15][a]	AEDC-C	10	6	~0	0.29	>1000	0.094	2	1
◇	Stainback[4]	NASA-LRC	8	5	0.1	0.45	20	0.047	2.7	1
◇	Boudreau[16]	AEDC-B	8	7	0.589	0.47	6	0.109	4	1,2
◇	Boudreau[16]	AEDC-F	12	7	0.589	0.49	32	0.109	4	1,2
○	Coats[17]	AEDC-A	4	-	1.466 2.932	~1.0 ~1.0	7.5 1.6	0.025, 0.046 0.015, 0.025	?	1 1
△	Thyson et al.[2]	AEDC-C	10	7.5	0.833	0.31	3.9	0.100	2.5	1,3,10
△	Thyson et al.[2]	AEDC-F	8	7.5	0.833	0.27	3.9	0.100	2.5	5

Note: All bodies were sphere cones except Coats, who tested a sphere-cone-cylinder and a hemisphere-cylinder.

[a] These data for a sharp cone appear to correlate with the data from blunt cones; however, this conclusion must be viewed with caution, since sharp cone trip data may be more susceptible to the influence of tunnel disturbances.

bility effects by the same parameter λ used by Van Driest and Blumer[13] in their correlation for tripping due to a single row of spheres. The results of the present tests for α = 0 deg are compared with this correlation in Fig. 17 for R_N = 0.4 in and in Fig. 18 for R_N = 0.25 and 0.75 in., using values of Re_θ, λ, and θ from Table 4. It is seen that the results for R_N = 0.25 and 0.40 in. fall somewhat above the correlation curve, while the results for R_N = 0.75 in. fall quite close to the correlation. Referring to Fig. 13, it is seen that R_N = 0.75 in. provided a negative pressure gradient at the trip location, whereas R_N = 0.25 and 0.40 in. produced a positive pressure gradient at the trip location. Since a negative pressure gradient is usually regarded as stabilizing, the lower trip Reynolds number of R_N = 0.75 in. is contradictory to smooth body experience. Typical error bars are indicated in Fig. 18 for one point, based on an estimated ± 10% uncertainty in determining $(Re_\infty)_{eff}$ from plots such as Fig. 15. The three-row trip appears to be equally effective as the five-row trip. The ablated trip point falls well above the correlation, as expected from the plot of Fig. 14.

Comparison of Fig. 17 with Fig. 16 shows that the present data for R_N = 0.4 in. fall within the scatter of the data used in drawing the correlation curve (except for the ablated trip). Further, the values of effective tripping parameters derived from the present results are

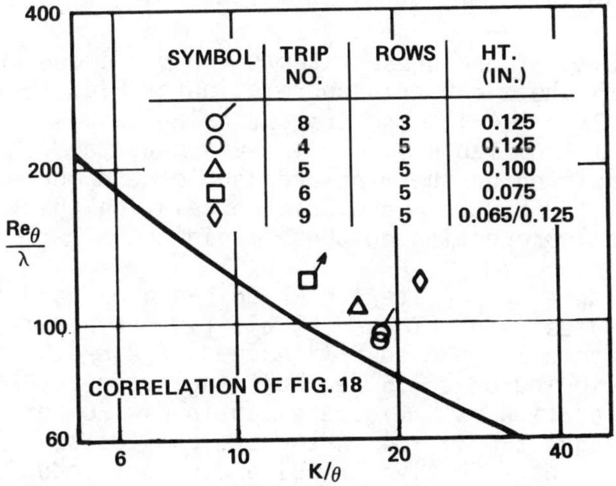

Fig. 17 Comparison of 9 deg cone data with trip correlation for R_N = 0.4 in. and α = 0.

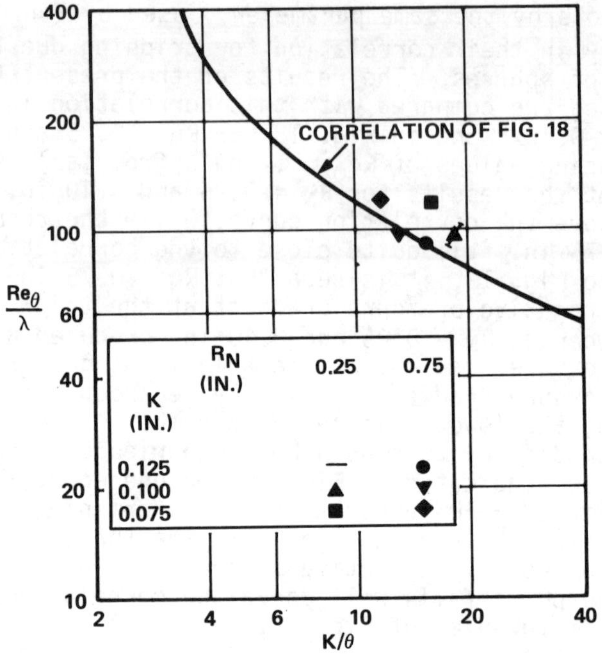

Fig. 18 Comparison of 9 deg cone data with trip correlation for R_N = 0.25 and 0.75 in. and α = 0.

regarded as more accurate than the previous data for two reasons: 1) closer increments of Re_∞ were employed in the present tests and 2) the minimum $(Re_S)_{TR}$ method used herein should give more consistent results.

The results of the present tests for α = 1 deg are compared with the α = 0 trip correlation of Fig. 16 in Fig. 19 for R_N = 0.4 in. and trip 5, using values of Re_θ, λ, and θ from Table 5. It is seen that the effective tripping condition on the windward and leeward meridians can be predicted by the α = 0 trip correlation using computed local properties on these meridians.

An alternate trip criterion which has been used in previous studies (e.g., Thyson et al.[2]) is the parameter K/δ^*. Reference 2 reported K/δ^* values of 2.5-4.0 for effective tripping on a 7.5 deg half-angle cone. Plots of transition location vs K/δ^* are given in Fig. 20 for trips 4-6, which suggest $K/\delta^* \approx 4$ in the present tests. Figure 20 indicates that effective tripping does not occur at a single value of K/δ^*, but is also a weak function of trip height.

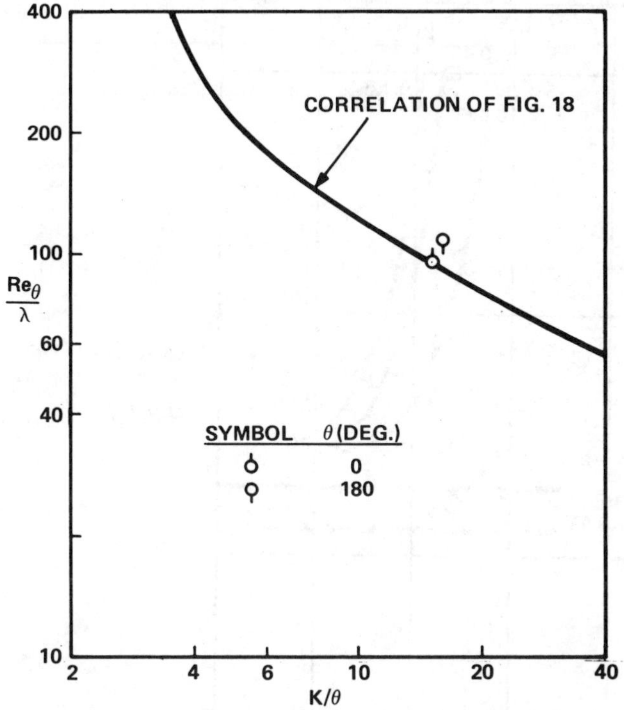

Fig. 19 Comparison of 9 deg cone data with trip correlation for trip 5 (α = 1 deg).

Finally, some words are in order concerning the applicability of wind-tunnel transition data to flight. It is well known that smooth body transition in wind tunnels is dominated by radiated aerodynamic noise from the tunnel side wall boundary layer. However, Pate[18] showed experimentally that transition due to roughness elements is trip disturbance dominated and is essentially independent of tunnel disturbance level. Hence the correlation of Fig. 16 should be applicable to flight.

Heat-Transfer Augmentation Due to Trips

Downstream Augmented Heating

Figure 21 is a typical map of heat-transfer coefficients downstream of a trip array, as constructed from the coaxial gage data. Significant circumferential nonuniformity is evident, with values exceeding the theroretical turbulent level by as much as 20%. A plot of the

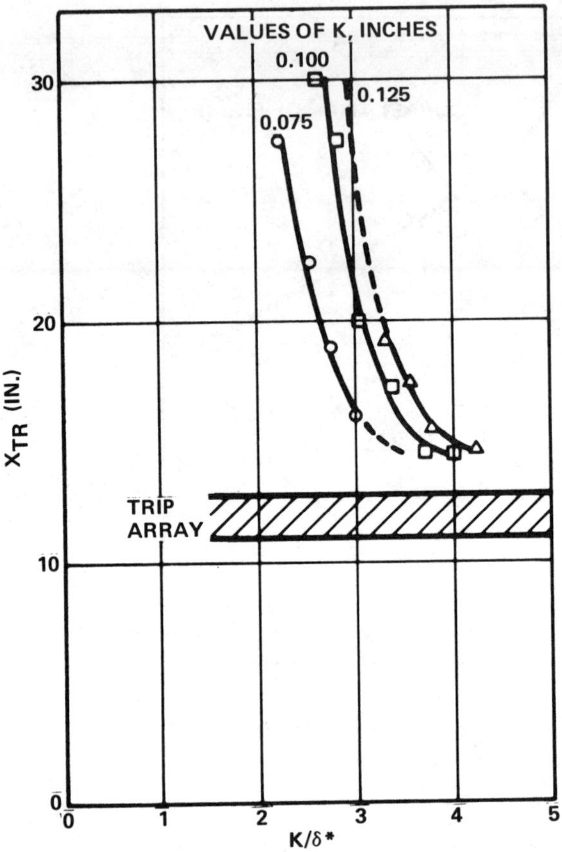

Fig. 20 Transition location vs K/δ^* for $R_N = 0.4$ in. and $\alpha = 0$.

variation in augmented heating factor computed from Fig. 21 is given in Fig. 22. Similar results obtained for other trip geometries and nose bluntness showed that the maximum augmented heating factor was independent of trip circumferential spacing, number of rows, and nose radius, but decreased from 1.2 to 1.1 as trip height decreased from 0.125 to 0.075 in.

Augmented Heating within Trip Array

Augmented frustum heating within the trip array itself was determined for trip 4, using the phase-change paint technique. Figure 23 shows the paint melt patterns at different times, for the maximum Reynolds number tested. The heat-transfer coefficient ratio $H_{0.9To}/H_{FR}$ corresponding to the melt line location is shown for each time, as computed by ARO-AEDC using one-dimensional semi-infinite

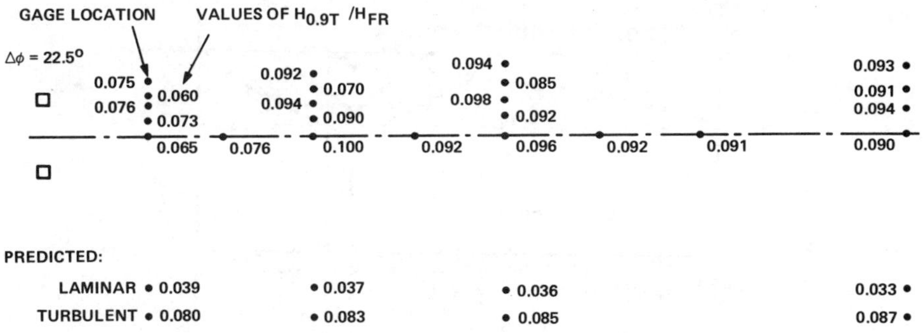

Fig. 21 Effect of trip height on downstream heating of trip 4 ($K = 0.125$ in., $R_N = 0.4$ in., $Re_\infty/ft = 2.0 \times 10^6$, $\alpha = 0$).

slab solutions.[6] The maximum frustum heating occurs immediately upstream of the first row of trip elements, suggestive of the hot spots reported by many investigators in studies of protuberance interaction heating. Since melt occurred in the first frame after model insertion, the maximum value of $H_{0.9T}/H_{FR}$ is greater than 0.49, which corresponds to an interaction heating factor of 5.5 times greater than the predicted turbulent value of 0.0887. This is consistent with the correlation of Nestler[19] for unswept cylindrical protuberances

$$H_{max}/H_u = 22\ M_e/(Re_s)^{0.15} \tag{2}$$

which predicts an increase factor H_{max}/H_u of 10 for the local flow properties of the test ($M_e = 3.15$, $Re_s = 3.43 \times 10^5$).

Another notable feature of the paint photographs is the extensive melt region at early times on the top face of the first row trip elements. The top faces of the second through fifth row trip elements appear to have about the same heating, on the order of half the first row. The top face heating values should be considered as qualitative, since the one-dimensional semi-infinite slab assumption is known to be in error for the trip elements.

Oil Flow Patterns

Oil flow tests were conducted for a few selected runs to obtain visual data on surface flow patterns to supplement the interpretation of transition and heat-transfer data. Due to the secondary nature of these tests, they

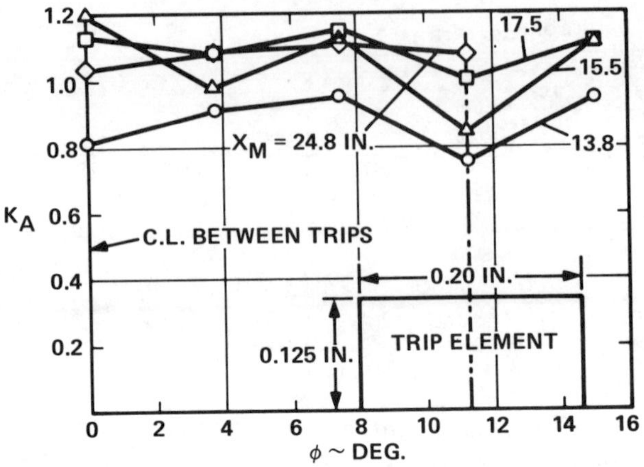

Fig. 22 Augmented heating factors downstream of trip 4 (Re_{∞} = 2.0 x 10^6/ft, α = 0; data from Fig. 21).

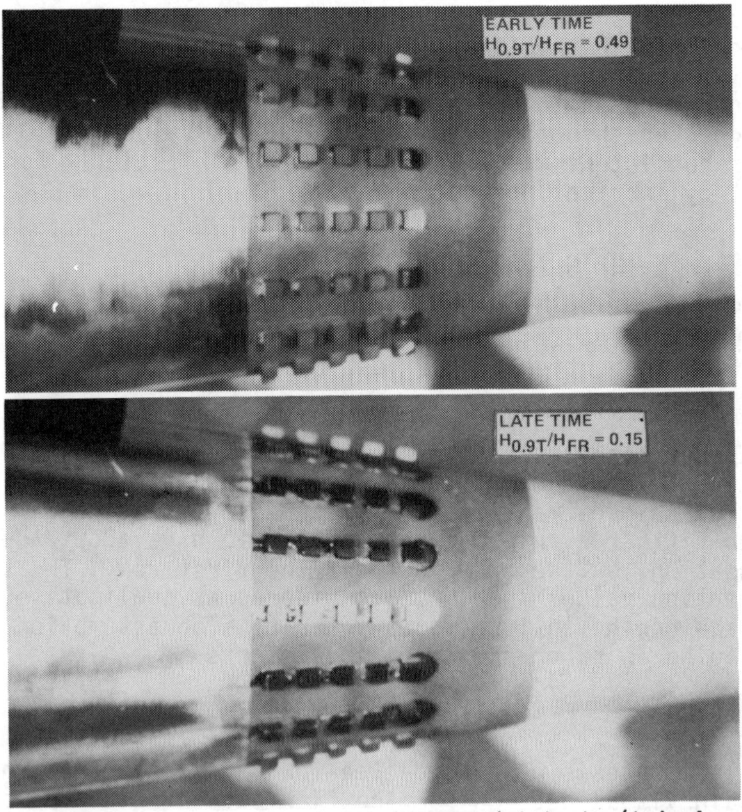

Fig. 23 Melt contours for phase-change paint tests (trip 4, R_N = 0.4 in., α = 0, Re_{∞}/ft = 3.5 x 10^6).

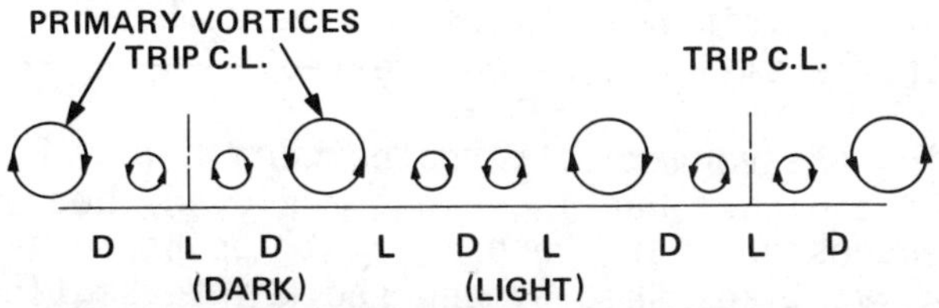

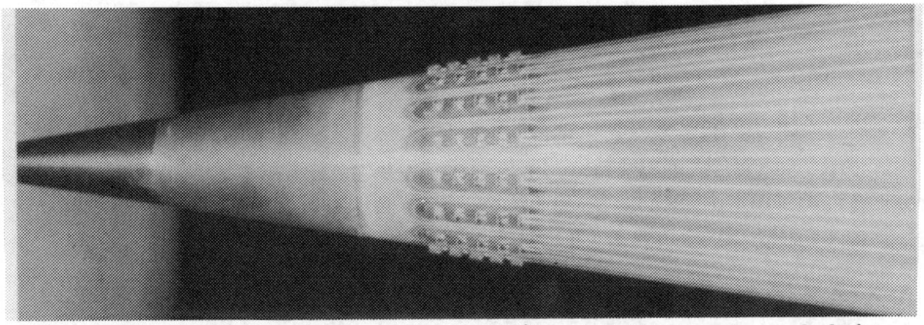

Fig. 24 Oil flow patterns of trip 5 (K = 0.100 in., R_N = 0.4 in., Re_∞/ft = 2 x 10, α = 0).

were confined to trip arrays of one circumferential spacing ($\Delta\phi$ = 22.5 deg). Variations were made in trip height, ablated shape, and Reynolds number.

Figure 24 is a photograph of oil flow patterns at $Re_\infty/ft = 2 \times 10^6$ for a 0.100 in. trip height. In this photograph, dark regions are regions in which the oil has been removed. Two dark trails, indicative of vortices, exist behind each column of trip elements. An additional dark trail developed midway between the elements. The number of vortex pairs for each case can be determined by counting either the light or dark regions, since, of course, they are equal in number. The hypothesized vortex pattern for Fig. 24 is shown in the accompanying sketch in which three pairs of vortices were inferred per trip meridian. The dark trails correspond to local heat-transfer maximum in the circumferential heating distributions of Fig. 22.

Similar oil flow trails behind spherical and cylindrical trip elements have been observed by many investigators, as summarized by Sedney[20] and more recently behind square cross-section trips by Thyson et al.[2] The striking feature of the present data is the additional dark region midway between trip meridians. Similar evidence of additional vortices has been reported previously (e.g., Van Driest and McCauley[21]). The intensity of the additional trail decreased as the trip height was reduced, almost disappearing at the smallest height tested.

Regarding the vortex patterns between trips, Ref. 2 also noted the higher heating along meridians between the trips and suggested that the primary tripping mechanism is longitudinal vortices shed by bow shock intersections between the trip elements. With this hypothesis, one can see that tripping becomes a function not only of trip height and Reynolds number, but that the $\Delta\phi$ and number of rows also play an important role.

At $Re_{\infty}/ft = 2.0 \times 10^6$ for a 0.075 in. trip height (trip 6), four dark trails between trip meridians were seen, instead of three as in Fig. 24. The number of trails downstream of trips was found to be a function of Reynolds number and trip height. In all cases, the trails persisted the full length of the model, even though the flow had become fully turbulent at about half the model length.

As Reynolds number was decreased, the trails become fainter, taking longer to develop. However, they still persisted at the lowest Reynolds number tested, indicating the presence of vorticity prior to the onset of transition.

Forces and Moments

The objective of the force tests was to measure the normal and axial forces and pitching moment of the 9 deg cone model with various trip geometries.

The aerodynamic force and moment coefficients are presented in the body axes. Pitching and yawing moment coefficients are referenced to a point on the model centerline which was 20.484 in. from the model nose. Model body length (31.514 in.) and base area (78.54 in.^{2}) were used as the reference length and area for the aerodynamic coefficients.

The variation of C_A with Reynolds number for trip 5 is plotted in Fig. 25. Also shown are the theoretical values of C_A for laminar and turbulent flow, as predicted by the GE 3D viscous code.[8] The tripped data are seen to agree reasonably well with the turbulent prediction, indicating that the principal effect of the trip on drag is the increased viscous contribution due to turbulent flow. The measured laminar values of C_A for the body with no trip are also shown in Fig. 25, in which excellent agreement with prediction is evident.

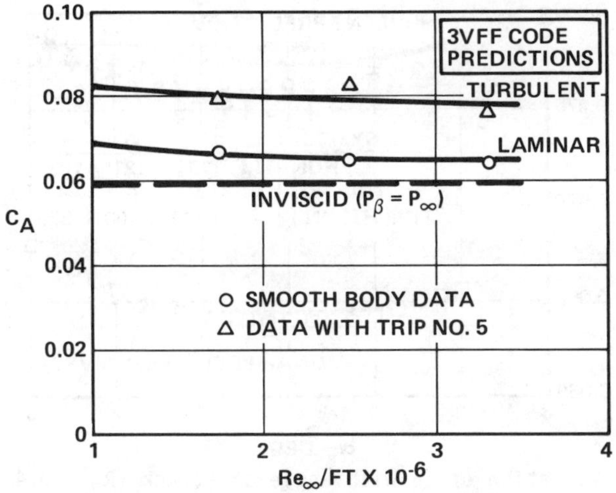

Fig. 25 Variation of C_A with Reynolds number (R_N = 0.4 in.).

Plots of C_A, C_N, and C_M vs α for trips 4-6, as well as for the body with no trip, are shown in Figs. 26-28. The predicted variations of C_A, C_N, and C_M with α are also shown in these figures. The agreement of prediction and experiment is again good. Figures 27 and 28 show that the trips have a small effect on static stability. For the largest trip tested, the decrease in $dC_M/d\alpha$ corresponds to a forward movement of X_{CP}/L of 0.5% (less stable), which may be within the accuracy of the data considering sting base pressure effects.

Discussion of Circumferential Spacing Effects

For isolated single element roughness, it can be shown[21] that the two primary vortices that wrap around the roughness proceed downstream to some point at which a transverse contamination starts to occur. (In Ref. 21, the process appeared to be caused by additional induced vortices.) The transverse contamination angle has been correlated by Fisher.[22] The angle describes turbulent regions (as from Emmons[23] turbulent-type spots) which are regions having a constant angle to each meridian around the body such that

$$X_{vo} = X_{so}\, e^{\Delta\theta(\sin\theta_c/\tan\gamma)} \tag{3}$$

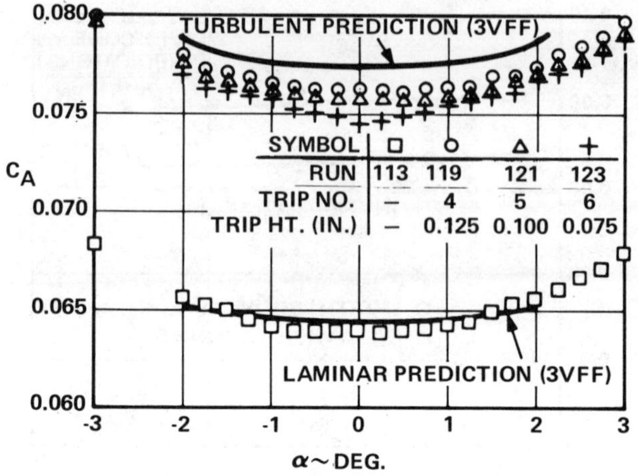

Fig. 26 Variation of C_A with angle of attack (R_N = 0.4 in., Re_∞/ft = 3.3 x 10^6).

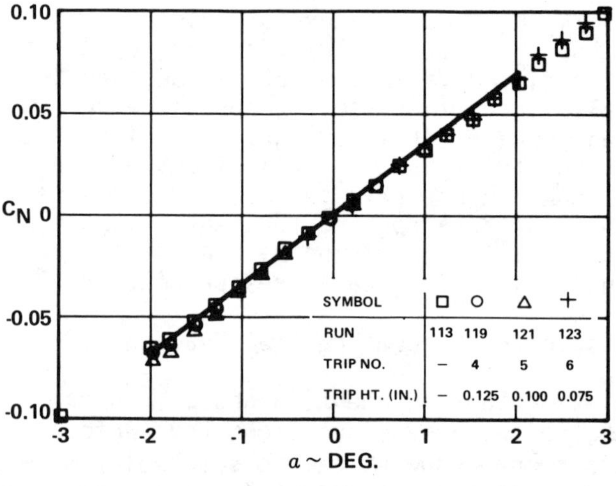

Fig. 27 Variation of C_N with angle of attack (R_N = 0.4 in., Re_∞/ft = 3.3 x 10^6).

where

X_{so} = turbulent spot origin from the sharp tip (virtual origin)

$\Delta\phi$ = change in meridian angle from the spot origin to meridian angle desired

θ_c = cone half-angle

γ = Fischer lateral spread angle

SYMBOL	□	○	△	×
RUN	113	119	121	123
TRIP NO.	–	4	5	6
TRIP HT. (IN.)	–	0.125	0.100	0.075

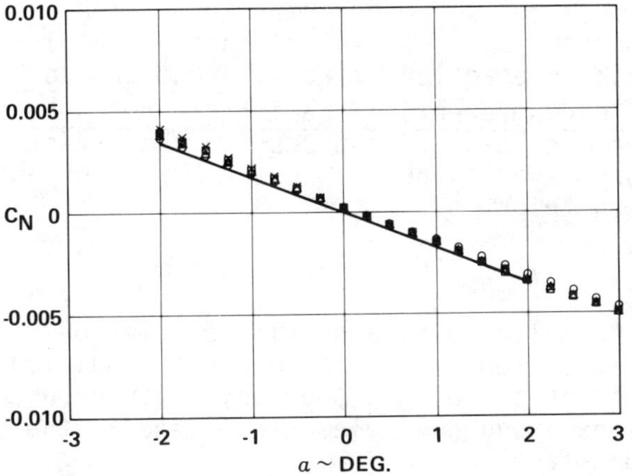

Fig. 28 Variation of C_M with angle of attack (R_N = 0.4 in., $Re_\infty/ft = 3.3 \times 10^6$).

X_{vo} = distance from the virtual origin to the spot on the $\Delta\phi$ meridian desired (i.e., assuming a sharp tip)

The basic difference between the statistical occurrence of Emmons spots and roughness-induced transition is that the primary vortices which start the transverse contamination always occur at the same spot. Thus, the element roughness-induced transverse contamination region can be considered completely turbulent. For effective tripping when similar adjacent roughnesses are placed (in a single row) circumferentially around the body, a pattern of transverse contamination regions emerges[21] which adequately describes the downstream turbulent regions. Moving roughnesses closer together moves the average downstream transition forward, as demonstrated by Morrisette et al.[24] However, placing trips too close to one another can cause a two-dimensional tripping effect to occur. For relatively wide trips, one can also show that since transverse contamination must occur behind the trip, spacing should never be less than two times the trip width.

The preceding discussion does not even consider the complex bow shock interactions that were noted for the large trips of this study and those of Ref. 2 where downstream heating and oil flow pattern intensity between the trips is most intense. As also indicated in Ref. 2, the heating and vortex patterns between the trips are most likely caused by the roughness-induced bow shock interactions. For each trip type, height, number of rows, and a given Mach number and Reynolds number condition, there is probably an optimum circumferential spacing to maximize these bow shock interaction effects. From ground tests, the effect of circumferential spacing can not be easily scaled and one must make allowances for the effect of flight boundary-layer thickness on the bow shock interactions between trips.

Conclusions

The wind-tunnel results and trip correlation described herein are felt to represent a significant addition to the important topic of tripping a boundary-layer by use of three-dimensional roughness elements. Some of the principal test results are

1) Tripping effectiveness was about equal for the two smaller circumferential spacings (15 and 22.5 deg) but was less for the larger spacing (30 deg).

2) The three-row trip was equally effective as the five-row trip.

3) The ablated trip configuration was less effective than an unablated trip of the same maximum height.

4) Significant transition front asymmetry existed at α = 1 deg at lower Reynolds numbers, with windward side forward. The asymmetry disappeared as the Reynolds number increased and transition approached the trip.

5) Circumferential heating variations of up to 20% in turbulent flow occurred downstream of the trips. Trip height and spacing had little effect on these variations. The oil flow tests disclosed twin streaks downstream of each trip column, indicative of vortices, which are related to the heating variations.

6) Force data showed that the axial force coefficient increased with trip height, spacing, and number of rows. The measured values with trips agreed with viscous code predictions based on a smooth wall turbulent boundary-

layer. Moment data showed a small destabilizing effect of trips on static stability.

7) For large frontal area trips, bow shock interactions play an important role in transition tripping such that a large lateral or too small spacing can cause a significant reduction in trip effectiveness.

8) Freestream Reynolds number for effective tripping was the same for R_N of 0.25-0.75 in.

9) The effective tripping results of these and previous tests using three-dimensional roughness elements were correlated well by a momentum thickness Reynolds number parameter vs a relative roughness parameter.

Acknowledgments

This work was performed as part of the Performance Technology Program-G II, under Air Force Contract F04701-78-C-0104.

The author acknowledges the cooperation of the staff of the Von Karman Gas Dynamics Facility of the Arnold Engineering Development Center in conducting the test program described herein, particularly A. Boudreau, D. Fikes, J. Hahn, J. Noble, and L. Jahnke. The support and suggestions of Maj. K. Yelmgren, Maj. M. Sabin, and Capt. R. Chambers (BMO), as well as H. Dyner (Aerospace Corporation), are also acknowledged. The efforts of F. George (GE-RSD) in model design and C. Harris (GE-RSD) in monitoring the tests are also appreciated.

References

[1]Nestler, D. E., "An Experimental and Analytical Study of Antenna Window Effects on Aerothermodynamics," AIAA Paper 81-0335, Jan. 1981.

[2]Thyson, N., Todisco, A., Reeves, B., and McCauley, W., "Active and Passive Tripping of Frustum Transition at Mach Numbers of 8 and 10," AIAA Paper 78-1128, July 1978.

[3]Sterrett, J. R., Morrisette, E. L., Whitehead, A. H., Jr., and Hicks, R. M., "Transition Fixing for Hypersonic Flow," NASA TN D-4129, Oct. 1967.

[4]Stainback, P. C., "Effect of Unit Reynolds Number, Nose Bluntness, Angle of Attack and Roughness on Transition on a 5° Half-Angle Cone at Mach 8," NASA TN D-4961, Jan. 1969.

[5]Nestler, D. E., "An Experimental and Analytical Study of the Feasibility of Boundary Layer Trips ...," Vol. 1, Wind Tunnel Tests, BMO-TR-80-46 (GE RSD 80SDR2077) Jan. 1980.

[6]Noble, J. A., "Wind Tunnel Tests of the PTP-G II Vehicle at Mach 8 in AEDC-VKF Tunnel B (Phase 3: Heat Transfer Test)," AEDC-TSR-79-V57, Oct. 1979.

[7]Fay, J. A. and Riddell, F. R., "Theory of Stagnation Point Heat Transfer in Dissociated Air," Journal of the Aero/Space Sciences, Vol. 25, No. 2, Feb. 1958, pp. 73-85, 121.

[8] Hecht, A. M. and Nestler, D. E., "A Three-Dimensional Boundary-Layer Computer Program for Sphere-Cone Type Re-entry Vehicles," Vol. I, Engineering Analysis and Code Description; Richbourg, D. H., Vol., II, User's Manual, AFFDL-TR-78-67, Vol. I and II, June 1978.

[9]Vaglio-Laurin, R., "Turbulent Heat Transfer on Blunt Nose Bodies in Two-Dimensional and General Three-Dimensional Hypersonic Flow," Journal of the Aero/Space Sciences, Vol. 27, No. 1, Jan., 1960, pp. 27-36.

[10]Walker, G. K., "A Particular Solution to the Turbulent Boundary-Layer Equations," Journal of the Aero/Space Sciences, Vol. 27, No. 9, Sept. 1960, pp. 715-716.

[11]Lees, L., "Laminar Heat Transfer Over Blunt-Nosed Bodies at Hypersonic Flight Speeds," ARS Journal, Vol. 26, No. 4, April 1956, pp. 259-269.

[12]Eckert, E. R. G., "Engineering Relations for Heat-Transfer and Friction in High-Velocity Laminar and Turbulent Boundary-Layer Flows over Surfaces with Constant Pressure and Temperature," Transactions of ASME, Vol. 78, 1956, pp. 1273-1283.

[13]Van Driest, E. R. and Blumer, C. B., "Boundary-Layer Transition on Cones and Spheres at Supersonic Speeds-Effects of Roughness and Cooling," AFOSR Scientific Rept. 67-2048, July 1967 (also Boundary-Layer Transition Study Group Meeting, Vol. II, Aug. 1967).

[14]Nestler, D. E., "A Parametric Study of Boundary-Layer Trip Heights to Induce Transition of MK 12A RV," SAMSO TR-78-98, Vol. VI.

[15]McCauley, W. D., Saydah, A., and Bueche, J., "The Effect of Controlled Three Dimensional Roughness on Hypersonic Laminar Boundary-Layer Transition," AIAA Paper 66-26, Jan. 1966.

[16]Boudreau, A. H., "Artifically Induced Boundary-Layer Transition on Blunt-Slender Cones Using Distributed Roughness and Spherical-Type Tripping Devices at Hypersonic Speeds," AEDC-TR-77-120, Feb. 1978.

[17]Coats, Jack D., "Investigation of the Effects of Nose Bluntness on Natural and Induced Boundary-Layer Transition on Axisymmetric Bodies in Supersonic Flow," AEDC-TR-73-36, Feb. 1973.

[18]Pate, S. R., "Dominance of Radiated Aerodynamic Noise on Boundary-Layer Transition in Supersonic-Hypersonic Wind Tunnels: Theory and Application," Ph.D. Dissertation, University of Tennessee, Knoxville, March 1977 (also AEDC-TR-77-107, March 1978).

[19]Nestler, D. E., "Results of Heat Transfer Tests of a Blunt Cone with Protuberances at Mach 10," Proceedings of 3rd International Heat Transfer Conference, Vol. 2, The Science Press, Ephrata, Pa., 1966, pp. 241-250.

[20]Sedney, R., "The Effects of Steady, Three-Dimensional Perturbations in Boundary Layers," AIAA Paper 72-713, June, 1972.

[21]Van Driest, E. R. and McCauley, W. D., "The Effect of Controlled Three-Dimensional Roughness on Boundary-Layer Transition at Supersonic Speeds," Journal of Aero/Space Sciences, Vol. 27, No. 4, April 1960, pp. 261-271, 303.

[22] Fischer, M. C., "Spreading of Turbulent Disturbances," AIAA Journal, Vol. 10, July 1972, pp. 957-959.

[23]Emmons, H. W., "The Laminar-Turbulent Transition in a Boundary Layer-Part I," Journal of the Aero/Space Sciences, Vol. 18, July 1951, pp. 490-498.

[24]Morrisette E., Stone, D., and Whitehead, A., "Boundary Layer Tripping with Emphasis on Hypersonic Flows," Proceedings of Symposium on Viscous Drag Reduction, Plenum Press, New York, 1969.

Low-Temperature Ablator Tests for Shape-Stable Nosetip Applications on Maneuvering Re-entry Vehicles

W.S. Kobayashi* and J.L. Saperstein†
Acurex Corporation, Mountain View, Calif.

Abstract

A series of low-temperature ablator (LTA) tests, designed to assess the applicability of the shape-stable nosetip (SSN) core concept to maneuvering re-entry vehicles (MaRV) were performed at Mach 5 in Tunnel 8 at the Naval Surface Weapons Center, White Oak Laboratory. The objectives of this test series were to acquire fundamental shape-change phenomenology data on cored nosetips in variable angle of attack and variable freestream environments, and to compare five candidate SSN core designs for MaRV applications. Force and moment data as well as photographic (shape) data were obtained. The LTA test results, data analysis, and conclusions are presented. Overall, for the conditions tested the circular core concept provides a more symmetric shape history and lower lateral loads than either the uncored nosetip or the noncircular cored nosetips that were tested.

Nomenclature

C_{M_α} = slope of pitch moment coefficient
C_{N_α} = slope of normal force coefficient

Presented as Paper 81-1061 at the AIAA 16th Thermophysics Conference, Palo Alto, Calif., June 23-25, 1981.

*Staff Engineer, Aerotherm Division.
†Project Engineer, Aerotherm Division.

F_A = total axial force
F_N = total normal force
$F_{N,body}$ = normal force due to angle of attack
$F_{N,asym}$ = normal force due to nosetip asymmetry
F_T = resultant force due to F_N and F_Y
F_Y = total yaw force
k = characteristic wall roughness height
L_{ref} = reference length (model base diameter)
M_p = total pitch moment
$M_{p,asym}$ = pitch moment due to nosetip asymmetry
$M_{p,body}$ = pitch moment due to angle of attack
NFASYM = normal force asymmetry coefficient
NFC = normal force coefficient = $F_N/q_\infty S_{ref}$
PMASYM = pitch moment asymmetry coefficient
PMC = pitch moment coefficient = $M_p/q_\infty S_{ref} L_{ref}$
q_∞ = freestream dynamic pressure
Re_θ = local Reynolds number based on θ
R_N = initial nose radius
S_{ref} = reference area = $\pi(R_N)^2$
t = time
T_e = local boundary-layer edge temperature
T_w = local wall temperature
TP = transition parameter defined in Eq. (1)
$\bar{z}$ = sonic point recession normalized by R_N
α = angle of attack
$\Delta\alpha$ = $\alpha_{max} - \alpha_o$
θ = boundary-layer momentum thickness

Superscript

* = local values at the sonic point of the initial nosetip contour

Introduction

During ballistic re-entry, relatively small nosetip shape asymmetries, resulting from asymmetric boundary-layer transition and material ablation, can cause significant impact dispersions, primarily through the interaction between vehicle lift and vehicle roll.[1-3] Consistent and symmetric nosetip shaping is necessary for improved vehicle accuracy and reduced structural loads. The shape-stable nosetip (SSN) concept was initially advanced as a means to minimize the effects of asymmetric boundary-layer transition on the nosetip and to promote symmetric shaping during ballistic re-entry by using a modified material region in the center (or core) of the nosetip. The SSN core concept has been successfully demonstrated in ground tests at zero angle of attack and in severe ballistic re-entry tests.[4]

The asymmetric nosetip shape effects on maneuvering re-entry vehicles (MaRV) are more complicated. Asymmetric nosetip shape change on a MaRV will result from angle of attack as well as asymmetric nosetip boundary-layer transition. A maneuvering re-entry vehicle can use its control system to compensate for the lift and torque resulting from nosetip asymmetry. Unfortunately, as the magnitude of the asymmetry increases, weight and performance penalties accumulate. In addition, the orientation of the nosetip asymmetry is significant in the MaRV design -- asymmetries in the vehicle pitch plane (the maneuver plane) usually require greater control system size and capability while asymmetries in the yaw plane may require design modifications to increase the vehicle stability in that plane.

Interest in the severity of nosetip asymmetry during maneuvering angle-of-attack trajectories has prompted the investigation of different SSN core concepts at angle of attack. In assessing nosetip shape development, it is necessary to understand the interaction of the re-entry flowfield and the material response of the nosetip. Unfortunately, it is impossible to duplicate both the flowfield and the material in ground test facilities. Ablation tests of real nosetip materials in arc plasma generators cannot simulate all of the pertinent flowfield parameters (i.e., Mach number, Reynolds number, enthalpy). While there are several materials response properties (transition resistance, steady-state recession) that can be successfully investigated in the low Mach number flows provided by high-pressure, high-enthalpy arc heater facilities (AFWAL 50 MW RENT, AEDC HEAT, etc.), these facilities are less suitable to investigate those characteristics of nosetip behavior, particularly shaping, that depend on the correct simulation of hypersonic pressure and heat flux distributions. On the other hand, hypersonic pressure and heat flux distributions can be simulated in wind-tunnel facilities; but the enthalpies are much too low to investigate the ablation characteristics of real nosetip materials. Therefore, test simulation procedures have been developed using low-temperature ablator (LTA) materials in high Mach number wind tunnels in order to provide information on Mach number dependent ablation phenomena. In particular, camphor has been used successfully in the PANT programs[5,6] as an analog for carbon-based nosetip materials, since it is a sublimer which recedes at appropriate rates under the normal operating ranges of pressures and enthalpies. Though surface morphology differences exist between camphor and carbonaceous nosetip materials, particularly in turbulent flow regimes, LTA test shaping is quite similar to ground and recovered flight test shapes of real materials (Fig. 1).

LTA simulations of the SSN were performed at Mach 5 in Tunnel 8 at the Naval Surface Weapons Center (NSWC), White Oak Laboratory, Silver Spring, Md. The test series consisted of 21 runs, performed during the period of April 18-30, 1980. The objectives of the test program were twofold: acquire fundamental shape-change phenomenology data on cored nosetips in variable angle-of-attack and flow environments, and compare five candidate SSN core designs for MaRV applications. Photographic and force moment data were obtained to characterize boundary-layer transition, nosetip ablation and recession, and nosetip loads during the test program. Photographic coverage provided data on both the shaping and the

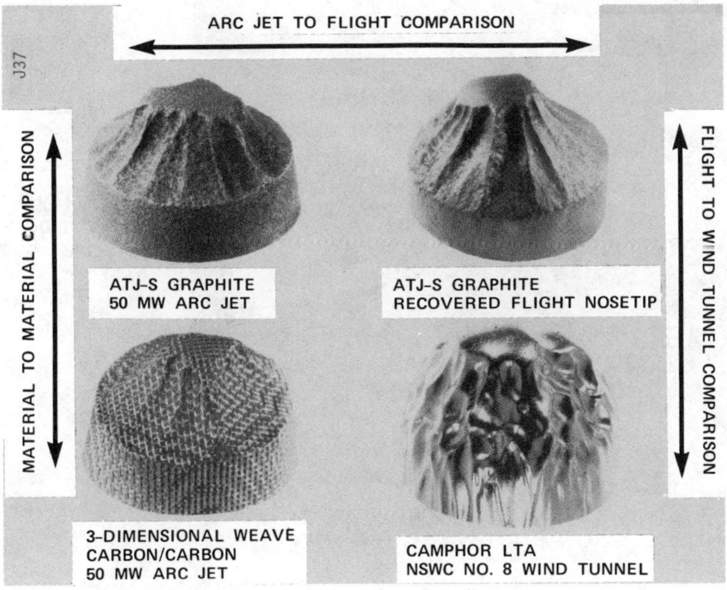

Fig. 1 Comparison of nosetip shapes for several materials and test environments.

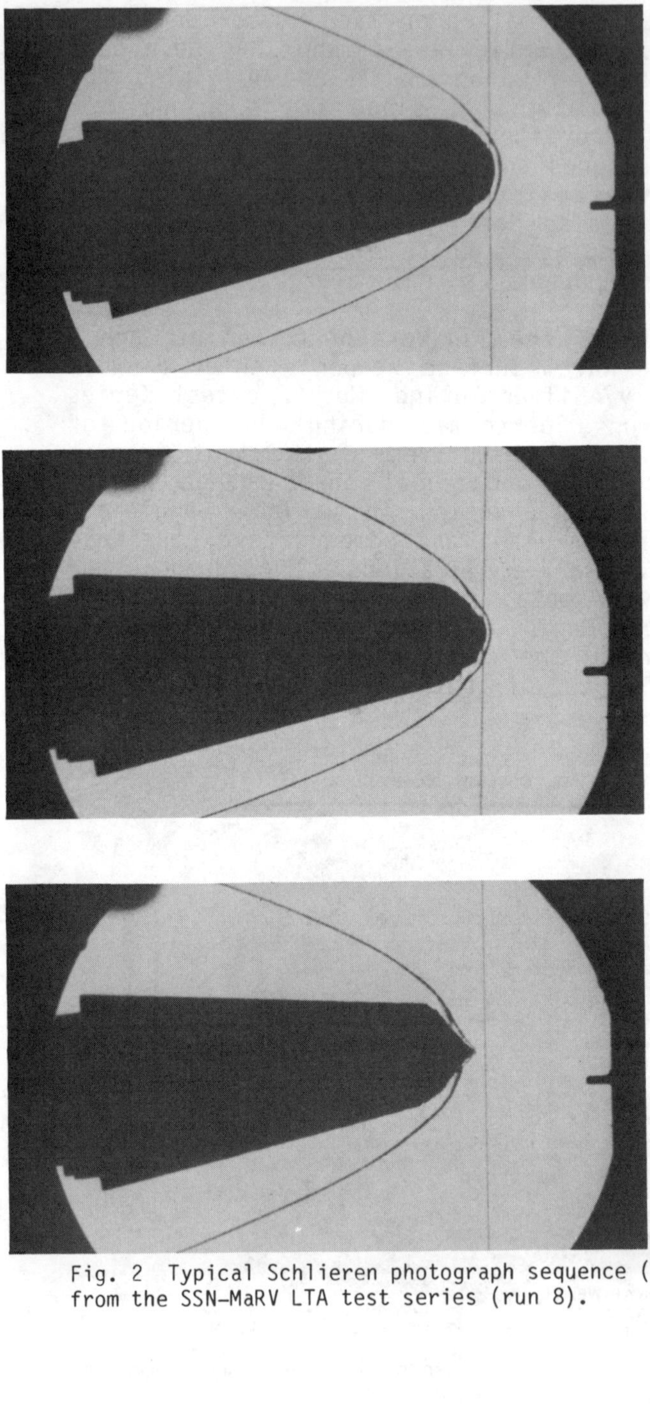

a) Early transitional shape (t = 63.6 s).

b) More developed transitional shape (laminar cap with turbulent sidewalls, t = 66.4 s).

c) Fully turbulent shape with embedded shock pattern (t = 74.7 s).

Fig. 2 Typical Schlieren photograph sequence (uncored model) from the SSN-MaRV LTA test series (run 8).

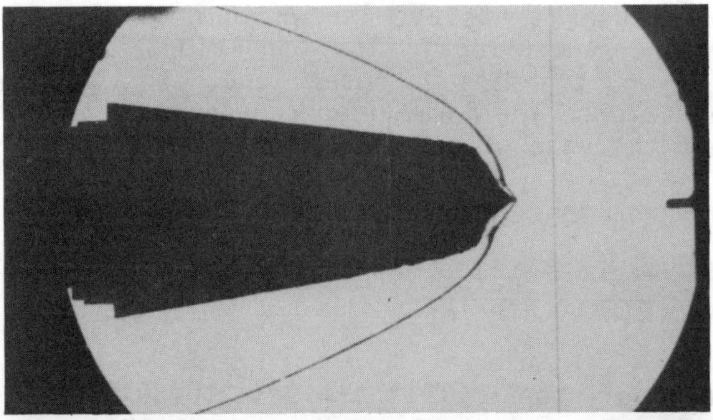

d) Indented turbulent shape with separation region (t = 85.4 s).

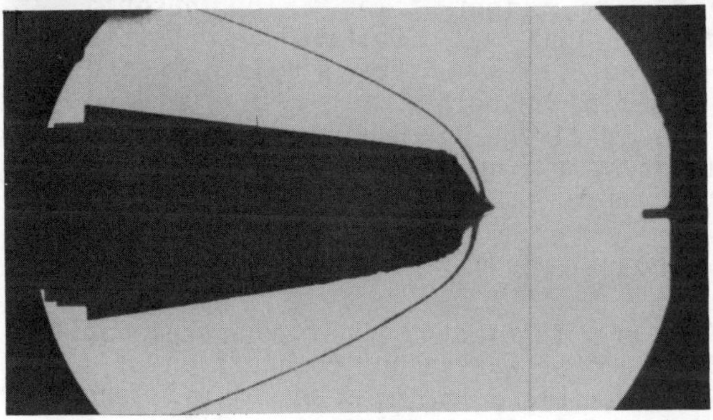

e) Indented turbulent shape with a return to an embedded shock pattern (t = 85.9 s).

Fig. 2 (cont.) Typical Schlieren photograph sequence (uncored model) from the SSN-MaRV LTA test series (run 8).

flowfield (Fig. 2). Four re-entry trajectories for a typical carbon-carbon nosetip were simulated.

The LTA simulation criteria used in this test are identical to those used in earlier PANT tests[5]; that is, the environment should be changing at the same rate, relative to the rate of nosetip shape change, in the wind tunnel as in flight. It is impossible to simulate the variation of all environmental and shape parameters in the wind tunnel; but based on experience in previous LTA testing, the environment has been characterized with a nondimensional sonic-point transition parameter (TP^*) and the nosetip shape change has been characterized by $\overline{z}$, the sonic recession normalized by R_N.

The transition parameter is defined by

$$TP^* = Re^*_\theta \left(\frac{k}{\theta^*} \frac{T^*_e}{T^*_w} \right)^{0.7} \tag{1}$$

TP* has been chosen because it has been shown to combine material surface roughness and local flow conditions to correlate boundary-layer transition onset and, to a certain degree, progression. The normalized recession $\bar{z}$ is a good measure of the rate of nosetip sharpening.

The trajectory then is simulated by matching the ratio

$$\frac{d\bar{z}}{dTP^*} = \frac{d\bar{z}/dt}{dTP^*/dt} \tag{2}$$

in flight and in the wind tunnel. Test times in the wind tunnel are different from flight times. The nosetip recovery vehicle (NRV) flight was simulated using this approach[5] and the LTA nosetip was very similar to the recovered graphite nosetip in both total recession and shape (Fig. 1). The desired trajectory simulations (pressure and angle-of-attack histories) are shown in Figs. 3–6.

Nosetip Models

All test models were fabricated from pure camphor powder ($C_{10}H_{16}O$). The camphor was vacuum compression molded at 20,000 psi into 10.75 in. o.d. by 13 in. long billets and then machined into the model configuration. A centered aluminum insert molded into the base of the model was attached to a steel baseplate that was mated to either an externally mounted five-component balance or a fixed-sting adapter.

All the models were 8 deg half-angle cones with spherically blunted noses having a radius of 1.50 in. After initial machining into the model configuration, the candidate SSN cores were drilled out and repacked with low-density camphor. The five candidate SSN design concepts (Fig. 7) tested were: circular core (the baseline SSN design), large circular core, elliptic core, large elliptic core, and teardrop core.

Test Matrix and Run Schedule

The test matrix (Table 1) subjected the various candidate SSN designs to the four trajectories. Trajectory A simulated a zero angle-of-attack ballistic re-entry. Trajectory B simulated the two-reverse dive

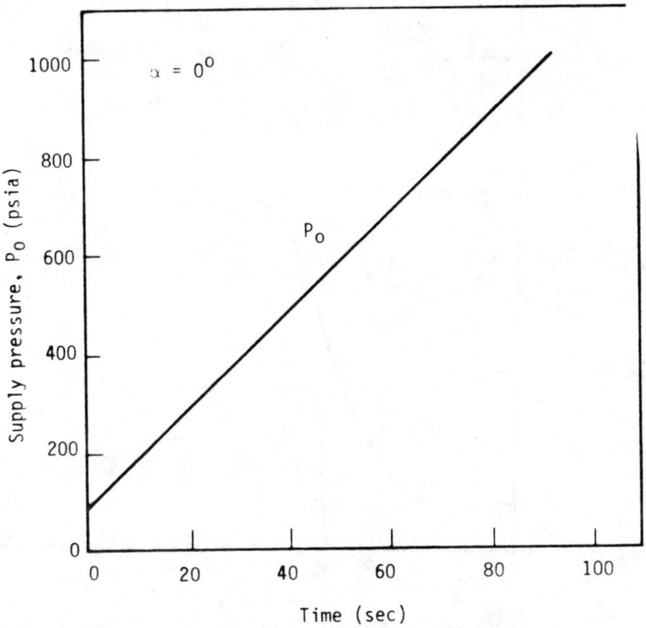

Fig. 3 Trajectory A: ramped trajectory (10 psi/s).

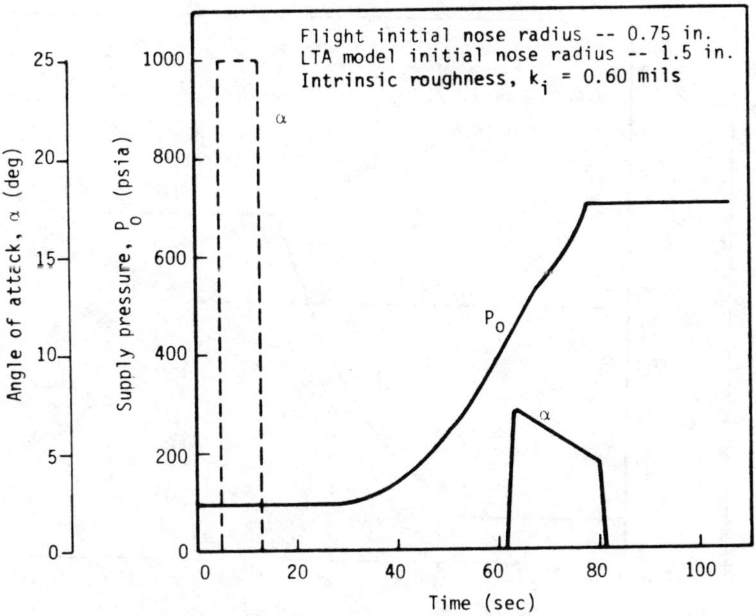

Fig. 4 Trajectory B: two-reverse dive simulation flight (R_N = 0.75 in.).

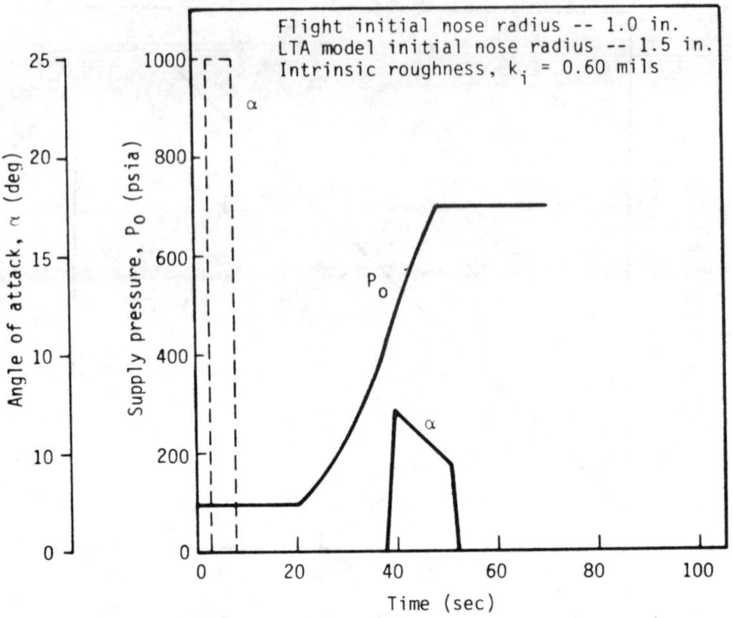

Fig. 5 Trajectory C: two-reverse dive simulation (flight R_N = 1.00 in.).

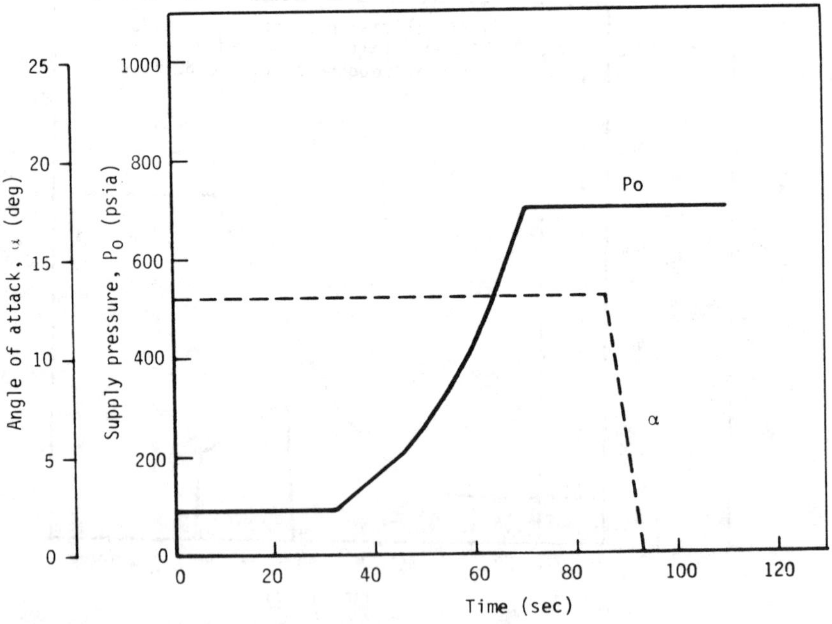

Fig. 6 Trajectory D: MPP simulation.

Table 1 SSN/LTA test matrix

	Trajectory			
	A	B	C	D
	α = 0 deg	R_N = 0.75 in.	R_N = 1.00 in.	
Homegeneous, no core	1	8	13	17
SSN circular core, BRV baseline	2	7, 16	14	21
SSN large circular core	--	11	--	20
SSN elliptic core	3	9	15	18
SSN large elliptic core	4	10	--	--
SSN teardrop core	5	12	--	19

Note: No data on run 6.

maneuvering angle-of-attack trajectory with an initial flight nose radius of 0.75 in. Trajectory C also simulated the two-reverse dive trajectory, but with an initial flight nose radius of 1.0 in. Trajectory D simulated the bent nose (constant angle of attack) trajectory of a Multiple Payload Program (MPP) vehicle.

Five runs were made at zero angle of attack, and the remaining 16 runs were made with varying angle of attack. The maximum angle of attack reached during these runs was 13 deg. Due to structural problems with the camphor models, some of the runs were not completed. During the testing, it was decided to omit the early α maneuver in the two-reverse dive trajectory simulations (B and C) because of the structural problems that were being experienced with the camphor models.

Previous LTA test series[5,6] have investigated the repeatability of camphor LTA simulations. In general, the shaping is repeatable at high Reynolds numbers for turbulent shaping and less repeatable for low Reynolds numbers. This reflects the statistical nature of transition onset and progression on changing nosetip shapes, one of the aspects of real re-entry shaping that can be simulated with LTA tests. Therefore, repeat runs are desirable; unfortunately, the scope of this test did not provide for repeat runs. Therefore, the test matrix was designed to identify comparative trends and demonstrate the feasibility of LTA testing with force moment instrumentation.

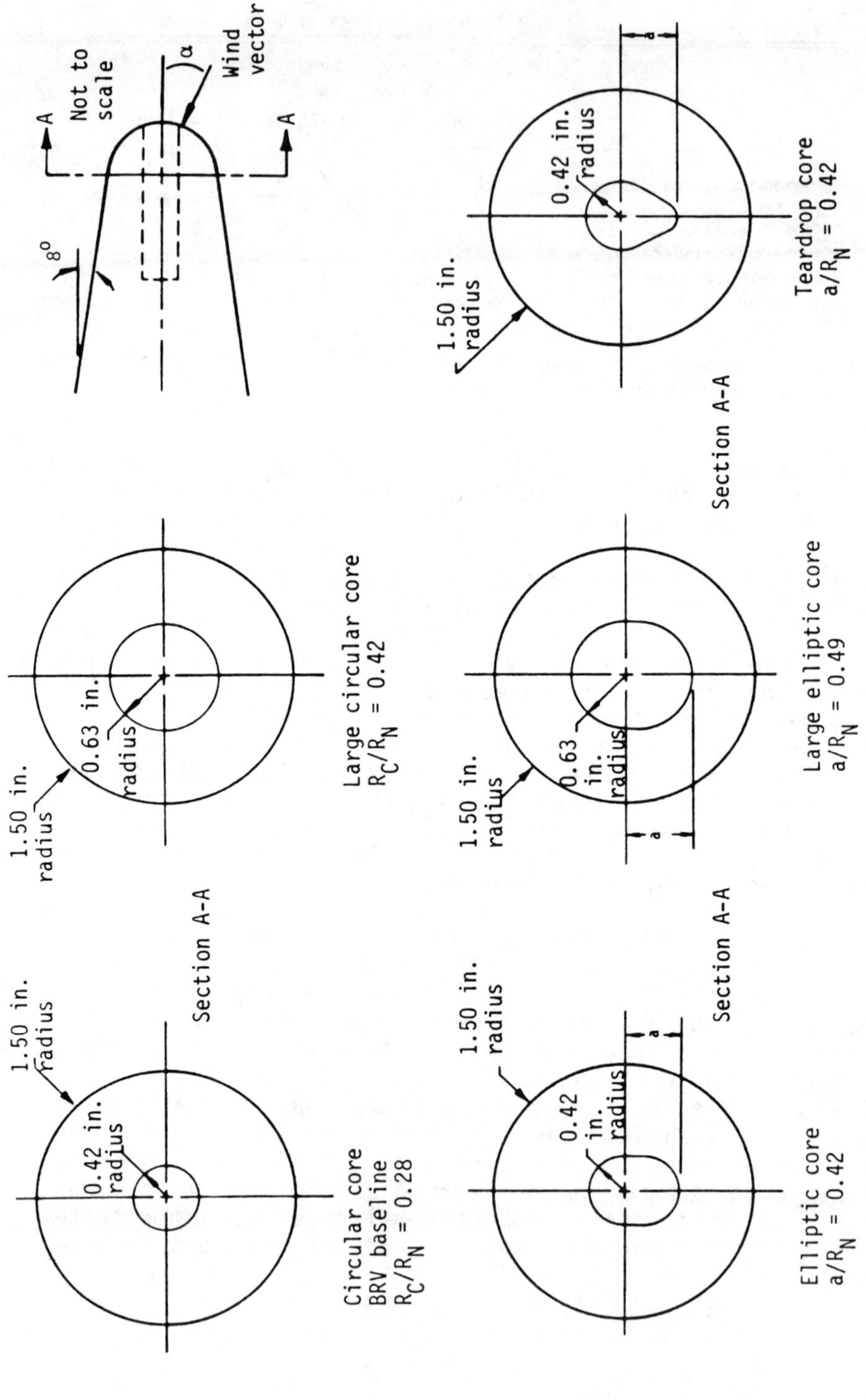

Fig. 7 SSN–MaRV core designs: LTA tests.

LTA Test Data Analysis

Centerline Recession Data

Centerline recession data have been analyzed to improve the understanding of the effect of an SSN core on nosetip shaping. Recession histories from the photographic data have been used to derive the recession rate as a function of total pressure for uncored models (Fig. 8a) and cored models (Fig. 8b). Several differences in recession peformance can be seen.

In the fully laminar flow regime, the recession rates are small, but the cored recession rates are about 50% greater than the uncored recession rates (0.09 vs 0.06 in./s at 200 psia). This result is consistent with previous LTA camphor tests and is explained by the relative densities at the model centerlines. The uncored model densities are 50% greater than the density of the camphor cores (64.2 vs 40.8 lb/ft^3).

In the fully turbulent regime, the cored nosetips maintain blunt shapes and the uncored nosetips develop sharpened, indented shapes. This difference in turbulent shaping has a significant effect on the recession rate. The shock angle is steeper on a blunt, cored shape than on a sharpened shape, and there is less entropy layer swallowing, lower heating, and lower recession rates.

Force and Moment Data Reduction

A five-component balance used during runs 1-16 (trajectory simulations A, B, and C) provided data on the total model normal force, side force, axial force, pitching moment, and yawing moment (Fig. 9). All force and moment data have been filtered using a fast Fourier transform. The 2 Hz, low-pass filtering procedure successfully eliminated high-frequency noise in the data and retained the character and detail necessary for the data analysis.

Because the interest is in analyzing small asymmetries in the presence of relatively large loads, it was necessary to correct for the small data biases due to the balance, model misalignment, flow angularity, camphor imperfections, etc. Therefore, these biases were corrected by averaging each balance component over an initial 4-6 s period (after the initial period of insertion tran-

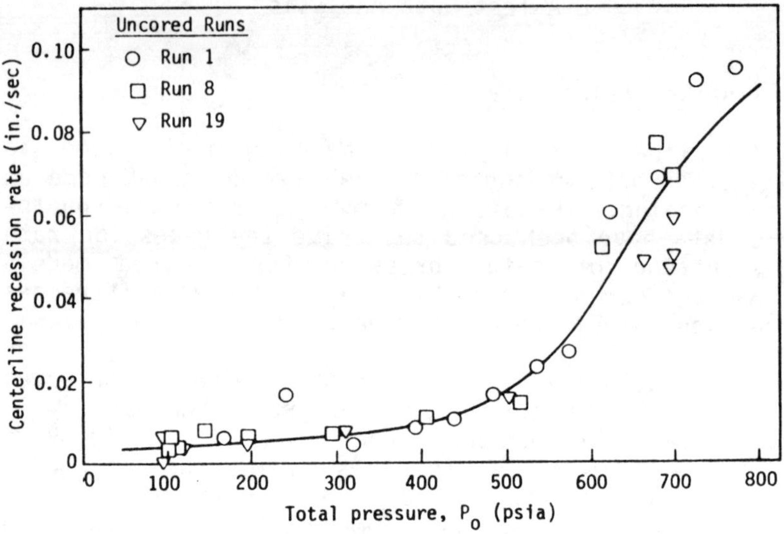

a) Uncored models.

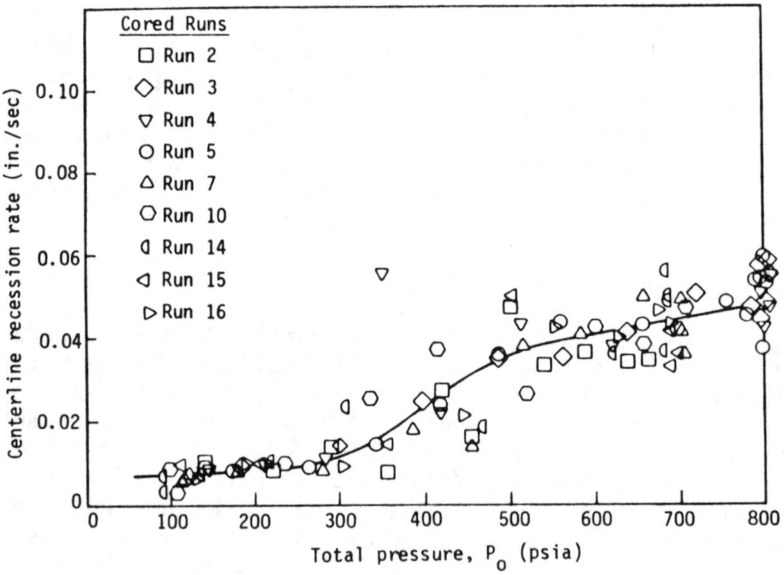

b) Cored models.

Fig. 8 Centerline recession rates.

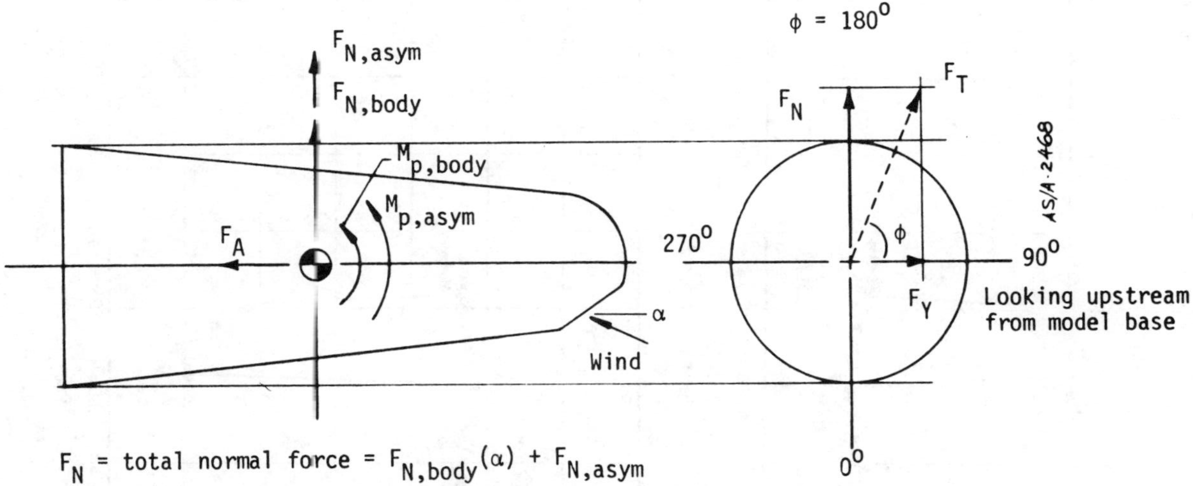

F_N = total normal force = $F_{N,body}(\alpha) + F_{N,asym}$

M_p = total pitch moment = $M_{p,body}(\alpha) + M_{p,asym}$

F_T = total resultant force = $\sqrt{F_N^2 + F_Y^2}$

ϕ = total force orientation

Fig. 9 Pitch-plane force and moment orientation.

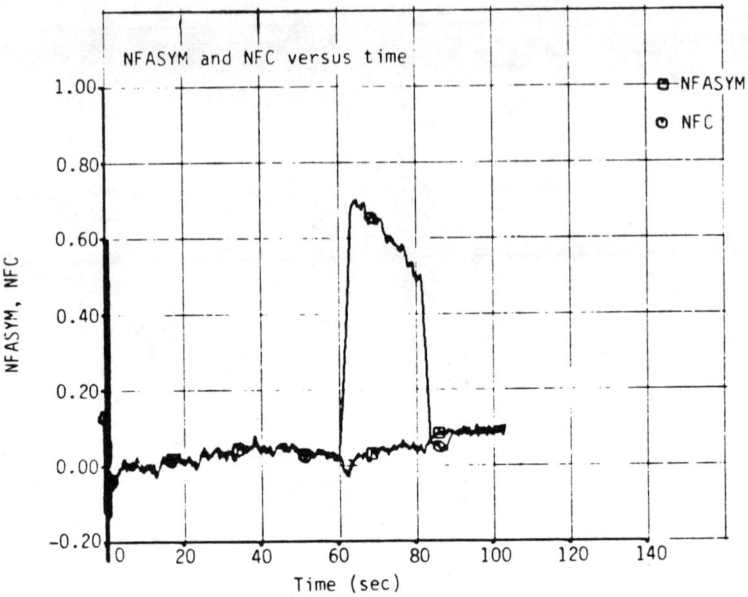

a) Normal force.

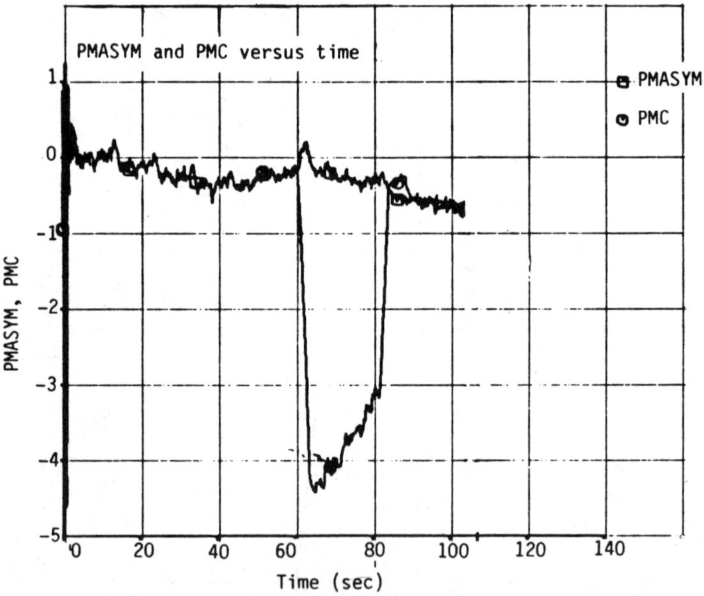

b) Pitching moment.

Fig. 10 Sample normal force correction for body aerodynamics angle of attack:

sients) and subtracting these averages from the balance component data.

The data as received from NSWC were corrected for the model tare taken before each run. Because these tares are taken without model ablation, they do not properly correct for the varying model weight during the actual run. An additional correction for changing model weight has been made using a correlation of the model recession rate as a function of the total pressure. This correlation, discussed in the previous section, was used to compute the recession and resulting approximate volume change of the model for each run. Using a mean camphor density of 62.4 lb/ft^3, the model mass loss was determined and subtracted from the pitch plane balance data. This correction was as large as 4 lb for some of the longer runs. This is not large compared to the maximum balance pitch load (100 lb) but it is quite significant compared to the small lateral loads experienced on the runs with relatively symmetric shaping, e.g., less than 20 lb.

At angle of attack, the balance measures the total of the body aerodynamics and the nosetip asymmetry component in the pitch plane. The body normal force and pitching moment must be subtracted to determine the nosetip asymmetry components. The normal force and pitching moment coefficient slopes (C_{N_α} and C_{M_α}) have been derived using the measured pitch plane loads before and after the first angle-of-attack pitch-up, that is

$$C_{N_\alpha} = (F_N \text{ at } \alpha_{max} - F_N \text{ at } \alpha = 0)/\Delta\alpha q_\infty S_{ref} \tag{3}$$

$$C_{M_\alpha} = (M_p \text{ at } \alpha_{max} - M_p \text{ at } \alpha = 0)/\Delta\alpha q_\infty S_{ref} L_{ref} \tag{4}$$

These empirically derived slopes have been used to remove the body aerodynamics at angle of attack from the total measured pitch plane loads. This approach is more appropriate for the ablated camphor shapes than using theoretical sphere-cone slopes. The effect of this correction is shown in Fig. 10, which compares the total normal force coefficient (NFC) and total pitch moment coefficient (PMC) to the derived normal force and pitching moment asymmetry coefficients (NFASYM and PMASYM) for a typical angle-of-attack run. The asymmetry of interest is a small part of the total pitch plane load at angle of attack. Figure 10 also shows the large transient loads at model insertion into the flow.

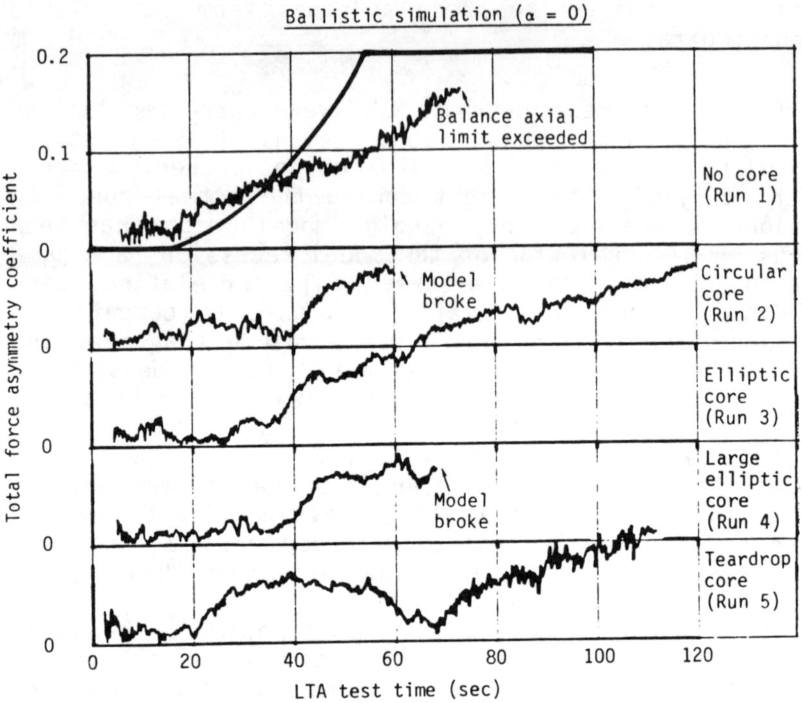

Fig. 11 Total force asymmetry comparison for trajectory A (ballistic simulation α = 0).

All of the force and moment data presented are in coefficient form referenced to the freestream dynamic pressure, the original nosetip area (πR_N^2), and the original nosetip radius (R_N).

Force and Moment Data Analysis

Total nosetip force and moment asymmetry coefficients and total axial force coefficient for trajectories A, B, and C were analyzed and compared. The different core concepts were compared to each other as well as to the uncored homogeneous case. Trajectories B and C were very similar, hence, for brevity, only the analyses for the ballistic re-entry simulation (trajectory A) and the 0.75 in nose radius maneuvering re-entry simulation (trajectory B) are presented here.

Trajectory A: Zero Angle of Attack, Ramped Simulation. The force asymmetry, moment asymmetry, and axial force coefficient histories for the zero angle of

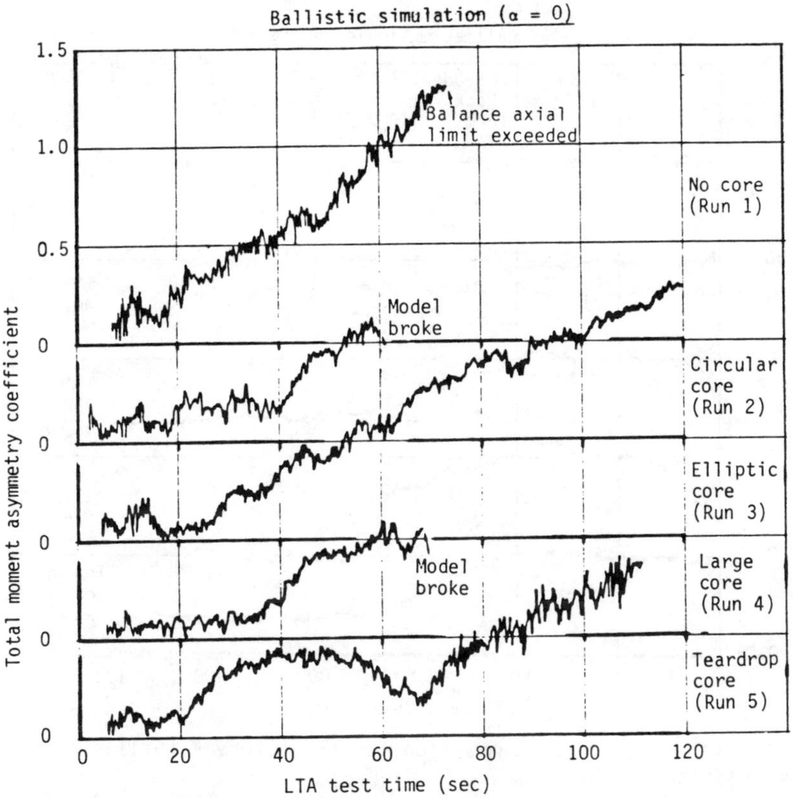

Fig. 12 Total moment asymmetry comparison for trajectory A (ballistic simulation α = 0).

attack, ramped simulations (i.e., the ballistic simulation) are presented in Figs. 11–13, respectively. Five different nosetip configurations are compared: no core (run 1), circular core (run 2), elliptic core (run 3), large elliptic core (run 4), and teardrop core (run 5). These data are from the initial test runs, before the models were reinforced, and several runs were not complete. Sufficient data are available to make useful comparisons with all of the runs.

The force asymmetry for the homogeneous model is compared to the ballistic flight correlation of Crenshaw[7] for nosetip normal force in Fig. 11. The general trends of the data and the correlations agree; that is, the force begins a steady increase at transition onset and reaches approximately the same level. Unfortunately, the uncored test run had to be terminated before any leveling was measured. There is a notable difference between the data

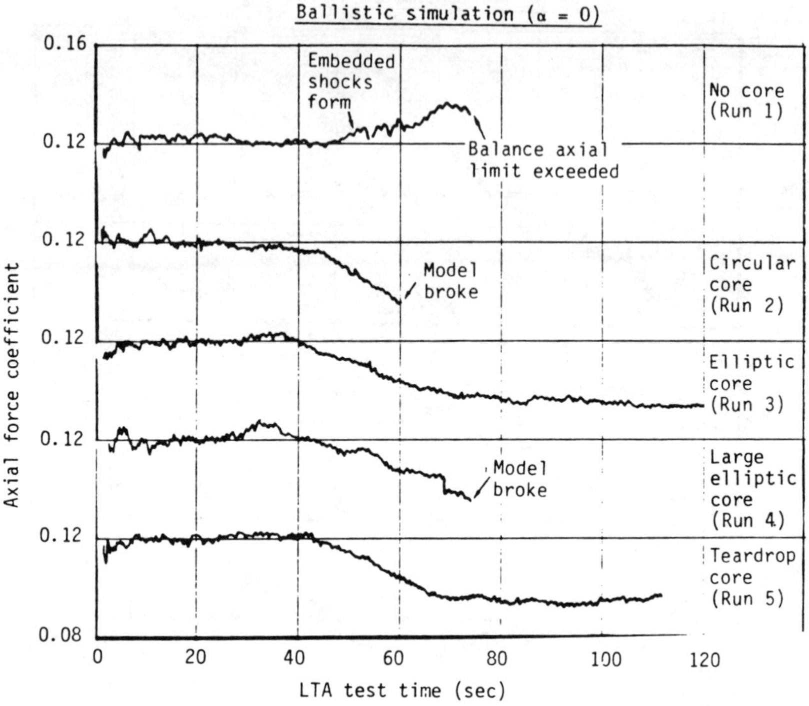

Fig. 13 Axial force comparison for trajectory A (ballistic simulation α = 0).

and the flight correlation; the data are increasing at a much slower rate than the correlation predicts.

The force asymmetries for the cored models are all somewhat lower than the force asymmetry of the uncored models. More significantly, all of the different cored models behave similarly. For this simulation, this observation is only tentative because of the incomplete runs (1, 2, and 4). As expected, the noncircular cores caused some shape and force asymmetry at zero angle of attack, but less than the uncored model. Based on a large data base from earlier LTA, arc, and flight tests, the circular core was expected to perform more symmetrically than the others at zero angle of attack. This was not the case and emphasizes the need for repeat testing with LTA simulations. The moment asymmetry coefficients are presented in Fig. 12, and they are consistent with the conclusions drawn from the force asymmetry comparison.

The axial force coefficients for the trajectory A simulation are compared in Fig. 13. The cored models show a consistent trend -- a drag decrease after transition. This is consistent with the observed shaping: the core region remains flat and the nosetip forecone angle decreases as the outerbody sharpens. The uncored drag behavior is quite different. The drag begins to increase slightly at approximately 50 s. The shape is observed to sharpen between 30 and 50 s, although there is no significant decrease in the measured drag. At 52 s, embedded shocks, oscillating with separated flow, are observed on the sharp, indented shape. This explains the variations in the uncored drag and its increasing trend.

Trajectory B: Two-Reverse Dive Simulation (Flight R_N = 0.75 in.). The force asymmetry history for the uncored configuration (Fig. 14) agrees well with the flight correlation of Crenshaw. This is the only uncored run that lasted long enough to observe the asymmetry growth level off; the asymmetry reaches a level close to that of the correlation. The effect of angle of attack on the uncored asymmetry growth is ambiguous; there appears to be some leveling off during the middle of the α maneuver, but this may be a result of the approximations in removing the body aerodynamics to derive the asymmetry component. In any case, the asymmetry resumes its growth before the end of the maneuver. After the uncored model has returned to $\alpha = 0$ deg, the asymmetry remains constant, possibly because it has already reached the maximum as predicted by the ballistic flight correlation, and does not decrease.

Generally, all of the cored models performed similarly and better (with respect to maximum asymmetry) than the uncored model. The maximum force asymmetries for cored models ranged from approximately 0.075 (the large elliptic core) to 0.125 (the elliptic core). These values are 30-60% lower than the uncored maximum force asymmetry coefficient (0.18). Repeat simulations are needed to establish the statistical significance of this result. The differences between cored models are small and the same magnitude as the early-time (laminar) variations in measured force. The moment asymmetry data (Fig. 15) are consistent with the conclusions drawn from the force asymmetry data.

The trends of the axial force data (Fig. 16) for this two-reverse dive simulation are similar to the trends of the $\alpha = 0$ deg simulation: the drag decreases slightly after transition on the cored models (due to forecone

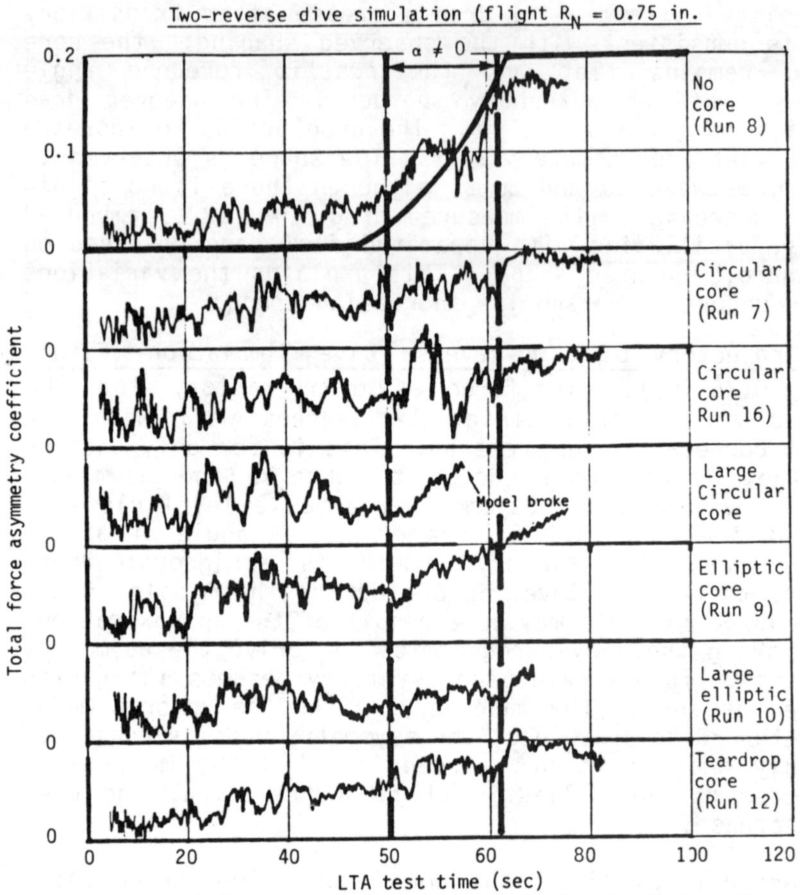

Fig. 14 Total force asymmetry comparison for trajectory B (two-reverse dive simulation, flight R_N = 0.75 in.).

sharpening outside the core) and the drag increases for the uncored model (presumably due to shape indenting and embedded shocks). The measured rise in uncored axial force history correlates well with the observed times of embedded shock formation and oscillating shock and separated flow structure (Fig. 2). The drag increase measured between 45 and 70 s for the model with the large elliptic core is a result of a very blunt shape that developed due to an irregularity in the fabrication of the large core.

Shape Profile Data

The shape profile data for the different nosetip configurations in trajectories A, B, and D (Figs. 17–19)

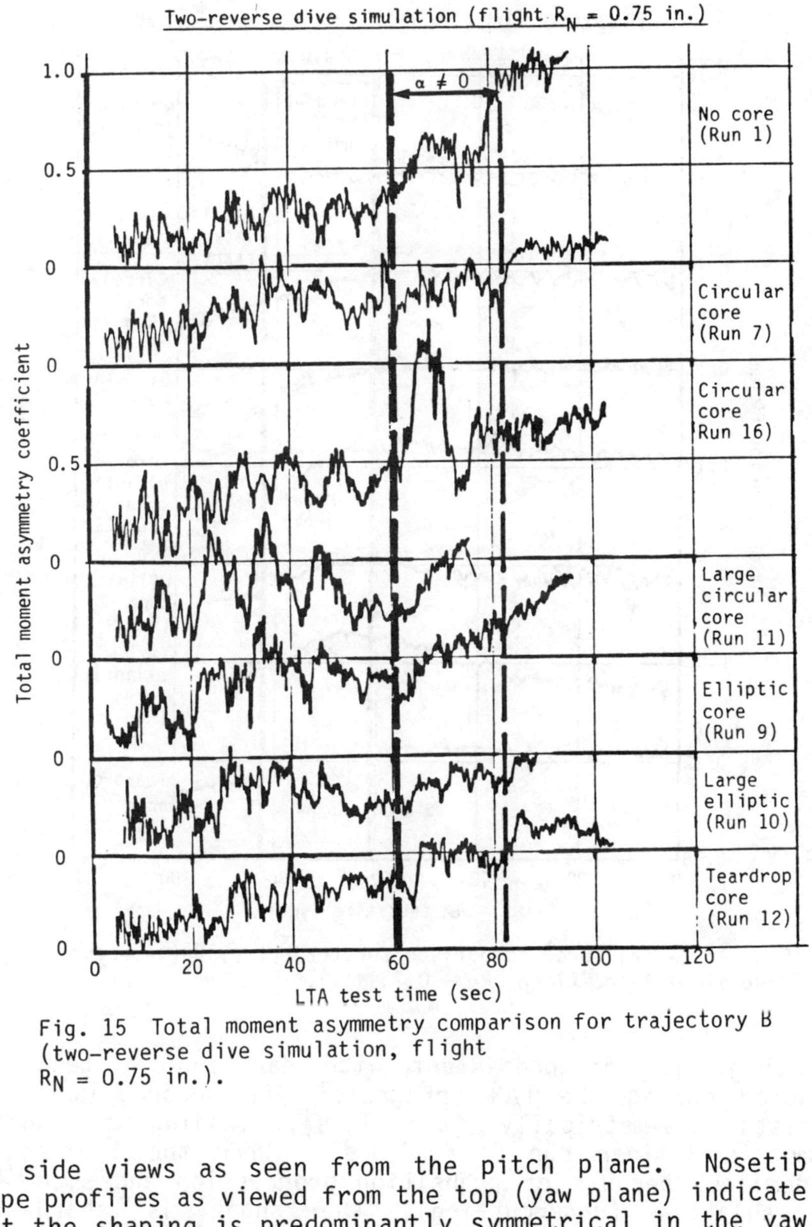

Fig. 15 Total moment asymmetry comparison for trajectory B (two-reverse dive simulation, flight R_N = 0.75 in.).

are side views as seen from the pitch plane. Nosetip shape profiles as viewed from the top (yaw plane) indicate that the shaping is predominantly symmetrical in the yaw plane.

Trajectory A: Zero Angle of Attack, Ramped Simulation

The observed shape performance of the uncored and circular cored models during the zero angle-of-attack

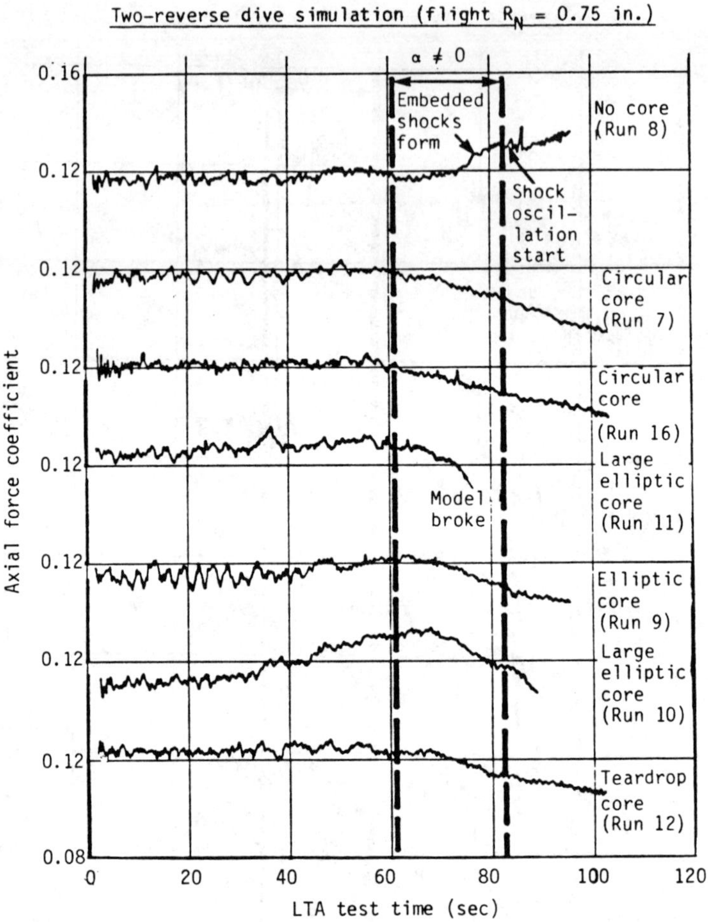

Fig. 16 Axial force comparison for trajectory B (two-reverse dive simulation, flight R_N = 0.75 in.).

ramp (Fig. 17) is consistent with earlier LTA tests conducted during the PANT program.[5] The uncored model transists asymmetrically (t ~ 29 s), leading to an asymmetric laminar cap (t ~ 38.6 s). When the laminar cap demises, because of transition progression and side-wall ablation (undercutting), the shape is sharp, asymmetric (although the stagnation point may still be on the centerline), and slightly indented (t ~ 53 s). Embedded shocks form in the indented regions, and low-frequency flowfield oscillations (embedded shock to separated flow and back again) are observed.

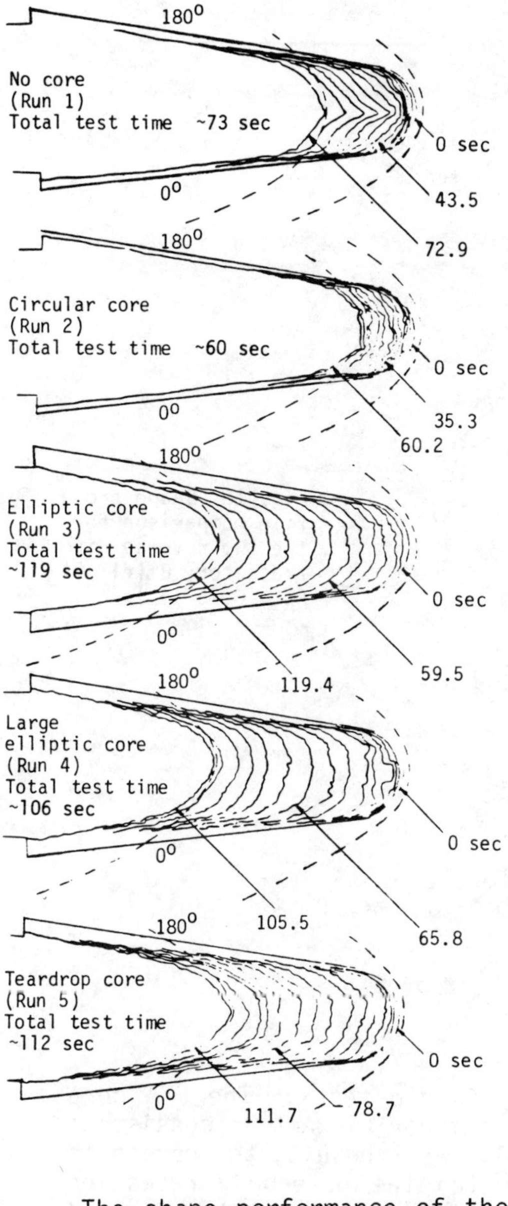

Fig. 17 Shape profile comparison for the zero angle-of-attack simulation trajectory A.

The shape performance of the circular core (run 2), similar to the baseline SSN design, is blunt and symmetric, but not as good as would be expected in this zero angle-of-attack environment. The ideal SSN core is designed to recede at the same rate as the outerbody at the core edge. This has been demonstrated with real nosetip materials (three-dimensional fine-weave carbon-

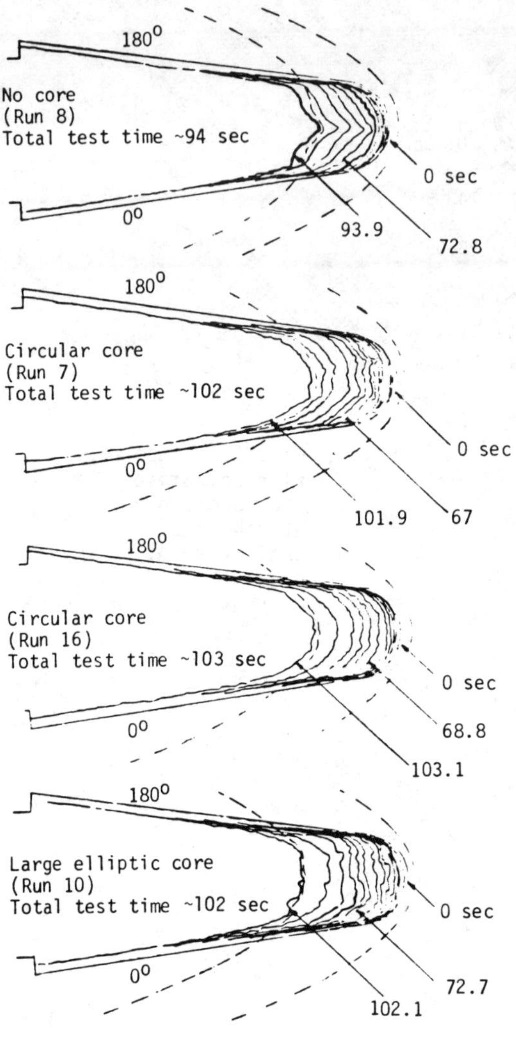

Fig. 18 Shape profile comparison for the two-reverse dive, trajectory B (flight R_N = 0.75 in.).

carbons with modified cores) in arc tests. The LTA core analog (lower density packed camphor) is not as consistent as the real SSN core material. As a result, the recession rate in the core did not match the outerbody recession rate in run 2; the core receded faster than outerbody and formed a lip. While the shape remained blunt and symmetric, the lip receded in a slightly irregular manner (the wiggles at $t \approx 35.3$ s). The circular cored model did not perform as well as it has in earlier LTA test, but it performed as well as the other cored models and somewhat better than the uncored model.

The noncircular cored models were tested at $\alpha = 0$ deg, a case of interest for maneuvering vehicles. The shape data for the elliptic, large elliptic, and teardrop cores all show small asymmetries in the expected direction at $\alpha = 0$ deg (the noncircular cores were aligned with the core asymmetry at $\phi = 0$ deg). However, the cores performed very well; the nosetip shape performance in all cases, as indicated by the profile data and as measured by the force asymmetry data, is better than the cored configuration and as good as the circular core.

Trajectory B: Two-Reverse Dive (Flight $R_N = 0.75$ in.)

The shaping performance for this simulation, which experiences only mild angle of attack (Fig. 18), is remarkably similar to the shaping performance at zero angle of attack (trajectory A). The uncored model goes through the same shape regimes identified at $\alpha = 0$ deg and does not exhibit any effects of the angle-of-attack maneuver. The circular and large elliptic cores remained blunt and relatively symmetric. The two models with circular cores show good repeatability. The observed shaping for these runs are consistent with the force and moment data, which provide a more quantitative measure of shape asymmetry. Note that there is an observable difference in the windward and leeward frustum recession.

Trajectory D: Bent-Nose ($\alpha = 13$ deg) MPP Simulation

These simulations are at the largest angle of attack (13 deg±) of this test series, and, therefore, the most severe nosetip shape asymmetries are observed on these runs (Fig. 19).

There are differences between the uncored shape change at $\alpha = 13$ deg and that at $\alpha = 0$ deg. The shape, notably the laminar cap at 70.3 s, tends to align with the wind vector. The sharpened shape of run 17 does not indent, and neither strong embedded shocks nor flow oscillations are observed. There is a very large difference in the windard and leeward frustum recession; this is true of all models tested with the MPP trajectory simulation.

± The MPP nosetip is tentatively planned to have a 10 deg bent nose, which will cause approximately 2.5 deg of angle of attack and 0.5 deg of nose deflection. Therefore, the total nosetip incidence to the flow will be approximately 13 deg.

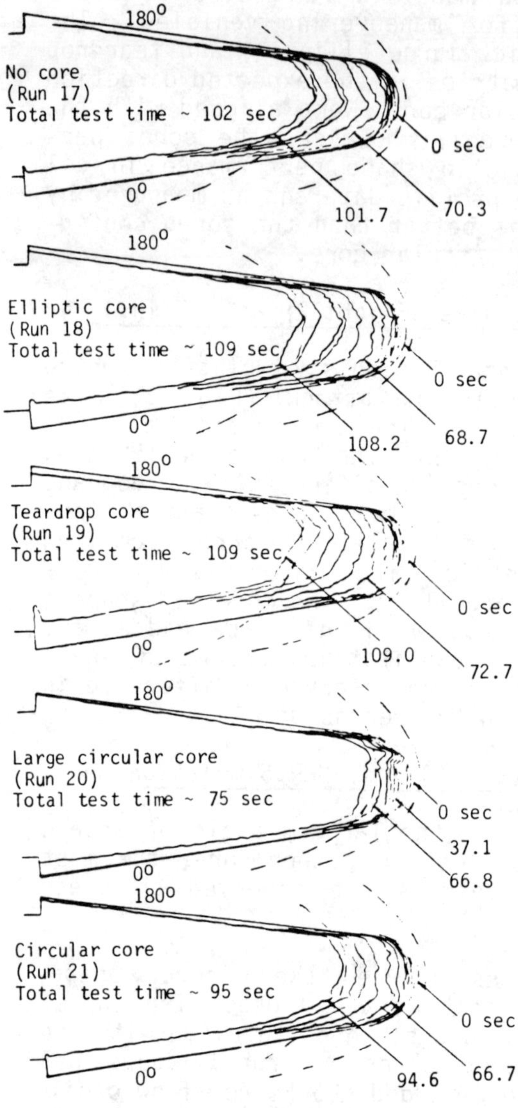

Fig. 19 Shape profile for the MPP simulation (trajectory D).

The circular core (run 21) appears to provide the best shaping performance in this group. The nosetip shape is blunt and very symmetric. It appears that the only component of asymmetry is the differential frustum recession. The large circular core (run 20) also provides blunt, symmetric shaping, but the large core region appears to be aligning normal to the flow after 66.8 s. The outer edge of the large circular core is somewhat irregular, particularly on the lee side. Without repeat runs, it is uncertain whether this is a characteristic of large cores at large angles of attack or if this is a random core irregularity.

The elliptic and teardrop cores provided similar shaping: blunt, symmetric shapes at early times ($t \lesssim 70$ s) and asymmetric, sharp shapes at late times. There are not enough runs, particularly repeats, to conclusively characterize these cores at large angles of attack, but these data suggest that sustained higher heating on the windward side of the outerbody can asymmetrically expose the core and result in the leeside outerbody protruding.

LTA Test Conclusions

This LTA test series has demonstrated the feasibility and value of testing camphor analogs in a varying freestream environment at angle of attack. This is particularly useful when both photographic (shape) and balance (force and moment) data are obtained.

The conclusions about the relative performance of the five core concepts depend on the angle-of-attack regime of interest:

1) At zero angle of attack, all of the cored concepts tested provide better symmetry, lower nosetip lateral loads, and a more stable nosetip flowfield than the uncored nosetip. This is consistent with data from previous $\alpha = 0$ deg LTA tests on cored models.[5]

2) At moderate angles of attack ($\alpha < 7$, trajectory simulations B and C), the uncored nosetip performance is very similar to its performance at $\alpha = 0$ deg. All of the cored concepts tested provide better shape symmetry, lower lateral loads, and a more stable nosetip flowfield, particularly the circular core.

3) At large sustained angles of attack ($\alpha \approx 13$ deg, the MPP simulation), shape profile data indicate that asymmetric cores and large cores may not offer any improvement over conventional homogeneous nosetips (because of large windward sidewall recession), but the circular core provides very symmetric nosecap shapes. (Unfortunately, no balance data were taken for these cases.) Asymmetric frustum recession existed for all

models, with or without cores, tested in this angle of attack regime.

Overall, the circular core concept, similar to the SSN baseline for ballistic re-entry, provides the best performance in all angle-of-attack regimes tested.

References

[1]Crenshaw, J. P., "Some Effects of Trim Angle of Attack on Reentry Vehicle Dispersion (U)," The Aerospace Corp., Rept. TOR-0066 (S5307)-1, 1973 (secret).

[2]Hall, D. W. and Nowlan, D. T., "Aerodynamics of Ballistic Re-entry Vehicles with Asymmetric Nosetips," Journal of Spacecraft and Rockets, Vol. 15, Jan.-Feb. 1978, pp. 55-61

[3]Dirling, R. B., Jr., "Asymmetry Nosetip Shape Change During Atmospheric Reentry," AIAA Paper 77-779, June 1977.

[4]Saperstein, J. L., et al., "The ICBM Flight Test Demonstration of the Shape Stable Nosetip (U)," presented at AIAA Strategic/Tactical Missile Systems and Space Sciences Meeting, San Diego, Sept. 1981. Sept. 1980 (Secret)

[5]Holtz, T., Derbidge, T. C., and Sandhu, S. S., "Low-Temperature Ablator Experiments on Shape Change Technology (Series K) and Shape Stable Nosetip Design (Series L)," Acurex Corp./Aerotherm Div., Aerotherm Rept. TM-76-128, Aug. 1976.

[6]Baker, D. L., Wool, M. R., Powars, C. A., and Derbidge, T. C., "Definition of Shape Change Phenomenology from Low Temperature Ablation Experiments -- Experimental Data, Series C (Preliminary Test Series)," Passive Nosetip Technology (PANT) Interim Final Report, Acurex Corp./Aerotherm Div., Mountain View, Calif., Vol. V, Pt. 1, 1974.

[7]Crenshaw, J. P., "Materials Requirements Definition (MRD), Handbook of Methods and Material Properties, Appendix IV," PDA Rept. TR-1046-01-15, Sept. 30, 1976.

The Hypersonic Flowfield over a Re-entry Vehicle Indented-Nose Configuration

A.M. Morrison,* W.J. Yanta,† and R.L.P. Voisinet‡
Naval Surface Weapons Center, White Oak, Silver Spring, Md.

Abstract

Data from an experimental program are provided that will allow an understanding of turbulent, indented, re-entry body nosetip flowfield physics and validation of numerical arbitrary nose shape flowfield modeling techniques. Results from the experiment include the first noninterference probe mapping of the region between the model surface and the bow shock wave using a two-dimensional laser Doppler velocimeter. Flowfield density profiles and surface pressure measurements are also obtained. The test matrix includes six angle-of-attack positions and seven roll orientations. These Mach 5 results provide the most extensive indented-nose flowfield definitions available to date.

Nomenclature

C_ρ	= pressure coefficient, $P-P_\infty/\frac{1}{2}\rho_\infty Q^2$
L	= reference length
M_∞	= freestream Mach number
P	= pressure, psi
P_∞, P_{inf}	= freestream pressure
Q	= freestream velocity, ft/s
R	= model radius, in.
$R_{e/ft}$	= Reynolds number per foot

Presented as Paper 81-1060 at AIAA 16th Thermophysics Conference, Palo Alto, Calif., June 23-25, 1981.

*Head, Re-entry Systems Branch.
†Aerospace Engineer.
‡Head, Aerodynamic Instrumentation Group.

S = arc length
U = velocity component along model longitudinal axis, ft/s
V = velocity component perpendicular to model longitudinal axis, ft·s
X = axial coordinate, in.
Y = radial coordinate, in.
α = angle of attack, deg
∞ = angular pressure tap position, deg
δ = density
δ_{Θ} = freestream density
ϕ = roll angle, deg

Introduction

Understanding, modeling, and predicting the accuracy of high-performance re-entry systems have been the goals of a number of recent programs throughout the strategic community. Results of such studies indicate that nosetip shape change is a significant re-entry contributor to system inaccuracy.[1-3] Asymmetric nosetip shapes generate aerodynamic forces and moments that can act in the nosetip region or can induce significant forces and moments by distorting the afterbody flowfield. Such resultant forces and moments can produce steady-state trim growth, transient trims, roll torques and subsequent roll excursions, and roll/trim imbalance[4,5] resulting in a dispersion of the re-entry vehicle from its predicted ballistic impact point.[6] Attempts to model these phenomena as subelements of weapon system models have included the use of similarity expressions[1] as well as "full up" solutions of the inviscid flow equations.[7] Such techniques have been limited to mild degrees of nosetip asymmetry and indentation.

Successful recoveries of high-β re-entry vehicles have yielded nosetips which have asymmetric, highly indented shapes.[8] Such shapes have also been noted in low-temperature ablator wind-tunnel tests and predicted using shape change codes. Attempts to model the flowfields over classes of arbitrary nosetip shapes, including the severely indented shapes, have recently been undertaken[9] in response to the flight test re-entry body recoveries, the wind-tunnel tests, and the predictions.

The flow over indented noses can be very complicated. Separation, reattachment, and embedded shock waves have been observed in ground tests[10] of indented-nose models.

The prime output of these ground tests have been shock shape, surface pressure and heat transfer, and aerodynamic forces and moments. Attempts to measure the flowfield between the nose surface and the shock have been largely unsuccessful as, to date, the use of temperature or pressure probes has been required. Inserting probes into the flowfield distorts the flow, inducing uncertainty as to whether the probe is measuring the actual flow or the flowfield the probe itself is generating. Thus, the existing indented-nose experimental data base does not provide sufficient information to allow an understanding of the basic flowfield physics. Information from this data base which can be used for flowfield prediction validation is, for the most part, limited to surface pressure and shock shape, two of the least sensitive of the governing flowfield parameters.

The objective of the current effort is to provide a data base for use in conjunction with arbitrary nose shape flowfield modeling programs. Data are provided which will allow an understanding of the indented nosetip flowfield physics and validation of numerical arbitrary nose shape flowfield modeling techniques. Prediction of nosetip aerodynamics with confidence will permit inclusion of the nosetip effects in total strategic system models and an assessment of the system impacts of this important phenomenon.

The approach taken involves conducting a ground experiment in which an indented-nose flowfield is mapped using noninterference probes to provide the most extensive indented-nose flowfield definition which is available to date.

Description of the Experiments

The experiments were conducted in the Naval Surface Weapons Center Supersonic Tunnel No. 2. Tunnel 2 can be operated in either a continuous flow mode or a blowdown mode and has a multiple Mach/Reynolds number capability. For the current study, a continuous mode of operation at a Mach number of 5 and a Reynolds number of 8×10^6/ft was selected. A 6 in. base diameter nosetip model was used (Fig. 1). The model contour was representative of a class of indented shapes which had been determined from nosetip shape change predictors for realistic re-entry configurations and environments. The model was mounted in the wind tunnel to allow angle-of-attack excursions of 0-4 deg and roll angle excursions of 0-180 deg. The model was instrumented with 38 surface pressure transducers along a meridian extending from the centerline to the base of the model. Great care was taken in the selection and calibration of these transducers to insure high-quality results. Surface pressure data were gathered at six angle-of-attack positions and at seven roll orientations.

Flowfield velocity measurements were made at 100 grid locations between the model surface and the bow shock wave using a two-dimensional laser Doppler velocimeter (LDV) system as described in Ref. 11. Polystyrene latex particles, 0.312 μm in diameter were injected into the supply plenum upstream of the model. An argon-ion laser was used to track the particles, thereby determining two components of velocity in a forward scatter mode. LDV measurements were made at angles of attack of 0 and 4 deg and at six roll orientations.

Flowfield density measurements were made using a pulsed ruby laser, dual-plate, optical holography system as described in Ref. 12. Holography data were gathered at angles of attack of 0 and 4 deg.

LDV data have also been gathered along a plane perpendicular to the model, near the model's base. This information can be used to start a supersonic afterbody

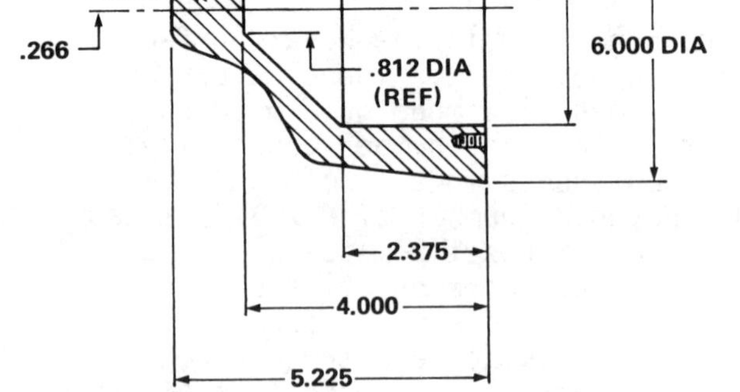

Fig. 1 Model configuration.

code to assess the effects of the indented nose flowfield on a re-entry vehicle afterbody.

Experimental Results

The experimental results derived from the Mach 5 wind-tunnel program are presented in this section. Normalized values of surface pressure, flowfield density, bow shock shape, and nose flowfield velocity are presented in tabular form in order to facilitate comparison with theoretical predictions and to allow detailed study of the experimental data to yield a basic understanding of the flowfield physics. Such a basic understanding of the physics is essential if adequate arbitrary body flowfield solvers are to be developed.

Surface Pressure Measurements

Great care was taken in the selection and calibration of the transducers for the pressure measurements and in acquiring and processing the pressure data. As a result, final pressure values have been analyzed to have an experimental uncertainty of $\pm$ 1%. The surface pressure values were obtained at 38 locations along a meridian extending from the stagnation point to the base of the model. Surface pressure values are also presented for six stations, located about the model's circumference in the stagnation region.

Values of pressure coefficient angle of attack and roll position are presented in Figs. 2 and 3. The stagnation region (forward flat face), the expansion corner (first model corner) region, the compression turn (indentation) region, and the conical afterbody region are evident from the behavior of the pressure coefficient variation from the model's most forward point to its base. Pressure coefficients at angle of attack suggest the presence of an embedded shock located on the windward side just below the expansion shoulder.

A comparison of surface pressure results at zero angle of attack with analytical predictions from Ref. 13, using techniques developed in Refs. 14 and 15 is given in Fig. 4. In general, these two-dimensional inviscid predictions were found to provide satisfactory agreement with the experimental pressure data in the stagnation region only.

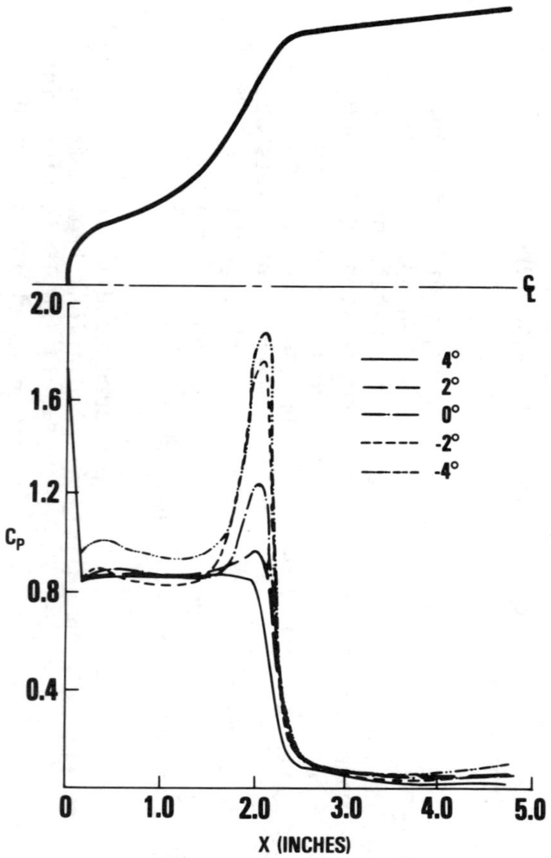

Fig. 2 Pressure coefficient for various angles of attack.

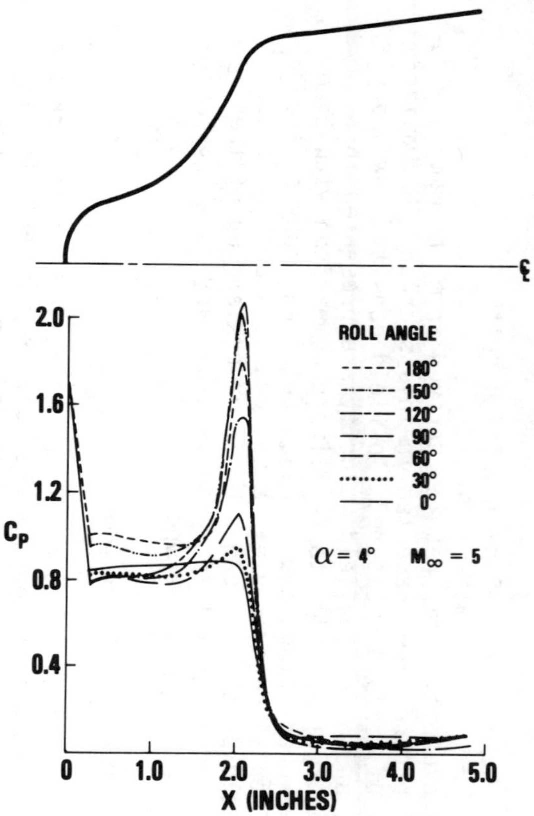

Fig. 3 Pressure coefficient for various roll angles.

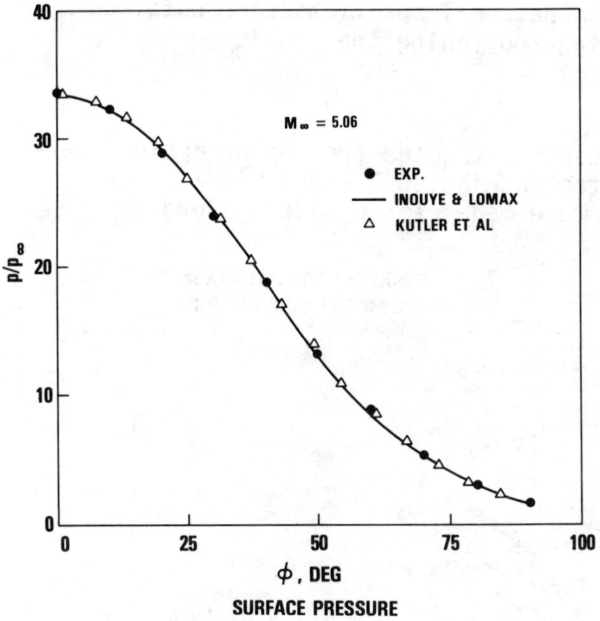

Fig. 4 Comparison of pressure results with predictions.

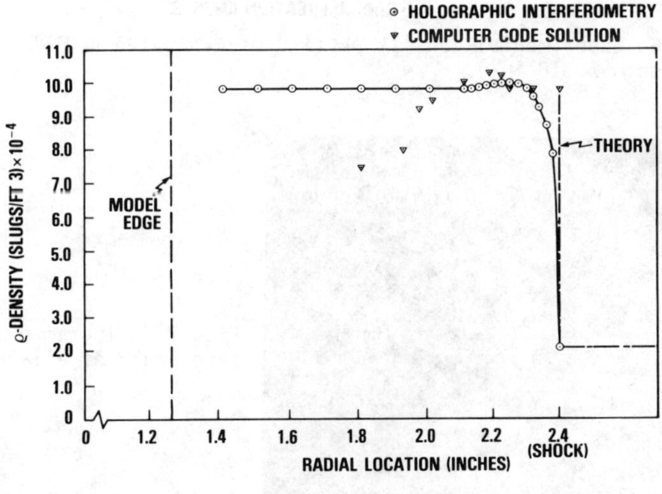

Fig. 5 Experimental density profile of separated region (axial station = 1.39 in.).

A summary of the surface pressure results is given in Table 1. Equivalent model roll positions were obtained by simultaneously pitching and yawing the model.

Density Measurements

Density measurements extracted from holographic interferograms are presented in Figs. 5-7 and Table 2. The density measurements were made at two sites, along rays

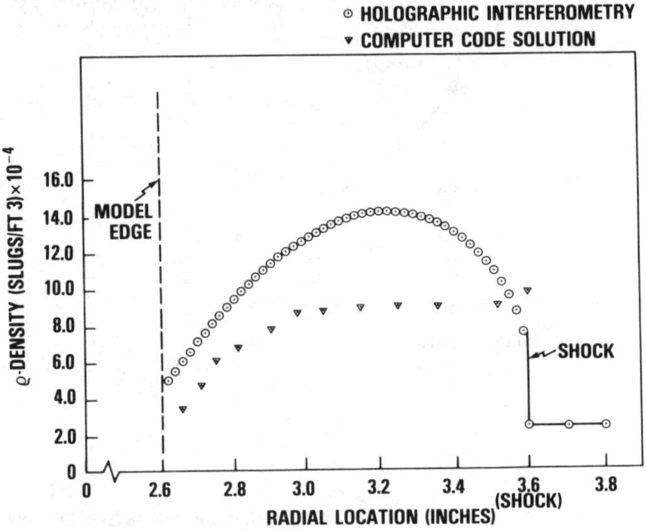

Fig. 6 Experimental density profile of expansion region (axial station = 2.318 in.).

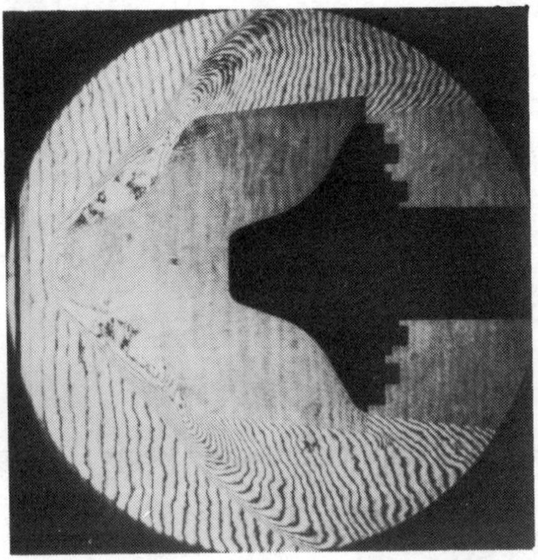

Fig. 7 Hologram of zero angle of attack.

Table 1 Experimental surface pressure summary, α/ϕ

X	Y	θ	0°/0°	4°/0°	4°/30°	4°/60°	4°/90°	4°/120°	4°/150°	4°/180°
In.	In.	Deg.	P/P_∞	P/P_∞	P/P_∞	P/P_∞	P/P_∞	P/P_∞	P/P_∞	P/P_∞
0.000	0.000	0.	33.3252	32.5447	32.3304	32.5634	32.6615	32.5155	33.7467	32.3342
0.000	.133	0.	33.1976	32.2140	32.0083	32.3527	32.5697	32.4893	33.7954	31.9385
0.000	.266	0.	32.6695	31.4595	31.2474	31.6952	32.0377	32.0736	33.5176	31.2256
.073	.495	0.	26.3256	24.4547	24.4124	25.0960	25.8993	26.4291	28.1798	24.2221
.263	.643	3.	16.5387	15.9409	15.4519	15.3397	15.0405	14.1597	18.6555	15.9087
.388	.670	0.	17.0927	16.2715	15.5820	15.5135	15.3153	14.4949	18.8181	16.0959
.513	.734	3.	17.2788	16.3034	15.6153	15.5555	15.5074	14.7996	18.7139	16.1097
.638	.779	0.	17.2798	16.3136	15.5844	15.5259	15.5140	15.0626	18.4075	16.1331
.763	.824	0.	17.2627	16.3636	15.5645	15.4589	15.5140	15.3683	18.2011	16.1880
.888	.871	0.	17.2500	16.4120	15.5978	15.3440	15.4911	15.5589	18.1714	16.2459
1.008	.926	0.	17.0812	16.4741	15.5374	15.1348	15.3558	15.7519	18.0476	16.2987
1.123	.991	0.	17.0014	16.5052	15.4876	15.0349	15.3462	16.1809	18.0264	16.3286
1.232	1.067	0.	16.9376	16.5506	15.5790	15.0648	15.5054	16.6929	18.1308	16.4187
1.335	1.166	0.	16.8646	16.6062	15.5478	15.2010	15.7875	17.3473	18.3001	16.4456
1.427	1.242	0.	16.9438	16.6599	15.6626	15.4623	16.2520	18.0464	18.6767	16.5574
1.516	1.342	0.	16.9459	16.7018	15.7353	15.7945	16.8843	18.8862	19.0348	16.6031
1.595	1.448	0.	17.1637	16.8536	15.9478	16.3157	17.7459	19.9317	19.6869	16.7420
1.665	1.562	0.	17.4592	16.8661	16.1049	16.8177	18.5951	21.0037	20.3844	16.7701
1.732	1.697	0.	17.9582	16.8267	16.3715	17.4736	19.5591	22.1966	21.3995	16.7500
1.793	1.797	0.	18.7749	16.8298	16.6533	18.1225	20.5529	23.6283	22.8836	16.7613
1.844	1.915	0.	19.9511	16.8459	17.0892	19.0181	22.0163	25.9694	25.6918	16.7818
1.915	2.033	0.	21.2014	16.7540	17.4541	19.8351	23.4576	26.2579	27.3630	16.6989
1.977	2.151	0.	22.4027	16.0392	17.4539	20.3777	25.3017	33.5011	35.6657	16.1908
2.038	2.269	0.	23.7130	16.1740	17.8504	21.1688	27.0448	37.9268	39.3617	16.0963
2.101	2.387	0.	23.5821	14.7732	16.8134	20.2329	27.0163	39.0561	38.4901	14.6493
2.177	2.496	0.	18.6631	11.4999	13.2252	15.9641	22.0971	26.4167	26.6078	11.3098
2.279	2.582	0.	11.2410	7.7848	8.5967	10.0321	14.1194	12.7873	14.2106	7.6130
2.398	2.641	0.	5.0921	4.2291	4.3455	4.8173	6.3215	4.5138	5.7703	4.1027
2.529	2.669	0.	2.9078	2.7183	2.6243	2.8126	3.3521	2.1348	3.0914	2.6268
2.661	2.686	0.	2.7287	2.4841	2.4359	2.6008	2.8009	1.7046	2.6807	2.5404
2.925	2.718	0.	2.5052	2.2662	2.1982	2.4459	2.1816	1.3633	2.2830	2.3489
3.189	2.751	0.	2.2924	1.9898	1.9644	2.3078	1.7849	1.1745	2.0048	2.1020
3.453	2.783	0.	2.1968	1.8280	1.8462	2.1957	1.6325	1.1623	1.9334	1.9550
3.717	2.816	0.	2.0075	1.7407	1.6510	1.9328	1.4458	1.2079	1.9322	1.8823
3.981	2.848	0.	1.9426	1.6623	1.6463	1.8032	1.4419	1.4576	2.1866	1.6699
4.245	2.880	0.	1.8747	1.5921	1.6015	1.6512	1.4209	1.7168	2.4056	1.6454
4.509	2.913	0.	1.8466	1.5532	1.6217	1.5845	1.4561	2.0060	2.6506	1.6574
4.773	2.945	0.	1.8444	1.5212	1.6128	1.5287	1.5304	2.2350	2.8182	1.5637
3.189	2.751	90.	2.2014	1.3968	.6783	1.9169	2.1932	2.5721	1.7274	1.5115
3.189	2.751	180.	2.1450	1.6336	2.9925	1.3617	1.5942	2.4115	2.1563	1.6902
3.189	2.751	270.	2.3184	1.0973	2.2612	1.9022	2.0550	2.0699	2.1489	1.1394
4.773	2.945	90.	1.7293	1.3967	2.2563	2.5441	2.7141	2.6794	2.2441	1.4568
4.773	2.945	180.	1.8611	2.8674	2.6942	2.2791	1.5594	1.5328	1.6899	2.0253
4.773	2.945	270.	1.8788	1.5664	1.5115	1.5833	1.5188	1.6224	1.8028	1.5757

extending from the model's surface through the model's bow shock. One set of measurements is from the compression turn region where a separated flow was found to exist, while a second set is presented from the expansion shoulder region. Holographic interferograms are presented for the 0 and 4 deg angle-of-attack cases in Figs. 7 and 8. Bow shock-wave position data determined from the interferograms are presented in Table 3. Comparisons of the

Table 2 Density profiles at zero angle of attack

X=2.318 INCHES		X=2.318 INCHES		X=2.318 INCHES	
Y (INCHES)	ρ/ρ_∞	Y (INCHES)	ρ/ρ_∞	Y (INCHES)	ρ/ρ_∞
2.62	2.2776	3.07	6.3100	3.52	4.7790
2.63	2.4108	3.08	6.3493	3.53	4.6126
2.64	2.5421	3.09	6.3862	3.54	4.4317
2.65	2.6716	3.10	6.4205	3.55	4.2339
2.66	2.7992	3.11	6.4523	3.56	4.0160
2.67	2.9250	3.12	6.4815	3.57	3.7731
2.68	3.0489	3.13	6.5081	3.58	3.4980
2.69	3.1710	3.14	6.5321	3.59	1.0000
2.70	3.2912	3.15	6.5534		
2.71	3.4095	3.16	6.5720		
2.72	3.5259	3.17	6.5879	X=1.390 INCHES	
2.73	3.6404	3.18	6.6010		
2.74	3.7530	3.19	6.6113	Y (INCHES)	ρ/ρ_∞
2.75	3.8637	3.20	6.6187		
2.76	3.9724	3.21	6.6232		
2.77	4.0791	3.22	6.6248		
2.78	4.1839	3.23	6.6233	2.10	4.6206
2.79	4.2867	3.24	6.6188	2.11	4.6143
2.80	4.3876	3.25	6.6112	2.12	4.6117
2.81	4.4864	3.26	6.6003	2.13	4.6124
2.82	4.5833	3.27	6.5862	2.14	4.6159
2.83	4.6781	3.28	6.5688	2.15	4.6217
2.84	4.7708	3.29	6.5479	2.16	4.6293
2.85	4.8616	3.30	6.5236	2.17	4.6382
2.86	4.9503	3.31	6.4957	2.18	4.6478
2.87	5.0369	3.32	6.4641	2.19	4.6573
2.88	5.1214	3.33	6.4287	2.20	4.6663
2.89	5.2038	3.34	6.3894	2.21	4.6738
2.90	5.2841	3.35	6.3462	2.22	4.6791
2.91	5.3623	3.36	6.2987	2.23	4.6813
2.92	5.4383	3.37	6.2470	2.24	4.6795
2.93	5.5122	3.38	6.1908	2.25	4.6725
2.94	5.5839	3.39	6.1300	2.26	4.6592
2.95	5.6534	3.40	6.0643	2.27	4.6381
2.96	5.7207	3.41	5.9936	2.28	4.6077
2.97	5.7857	3.42	5.9175	2.29	4.5663
2.98	5.8486	3.43	5.8359	2.30	4.5115
2.99	5.9092	3.44	5.7484	2.31	4.4410
3.00	5.9674	3.45	5.6547	2.32	4.3516
3.01	6.0234	3.46	5.5543	2.33	4.2393
3.02	6.0771	3.47	5.4468	2.34	4.0988
3.03	6.1285	3.48	5.3316	2.35	3.9231
3.04	6.1775	3.49	5.2080	2.36	3.7016
3.05	6.2241	3.50	5.0754	2.37	3.4176
3.06	6.2683	3.51	4.9328	2.38	1.0000

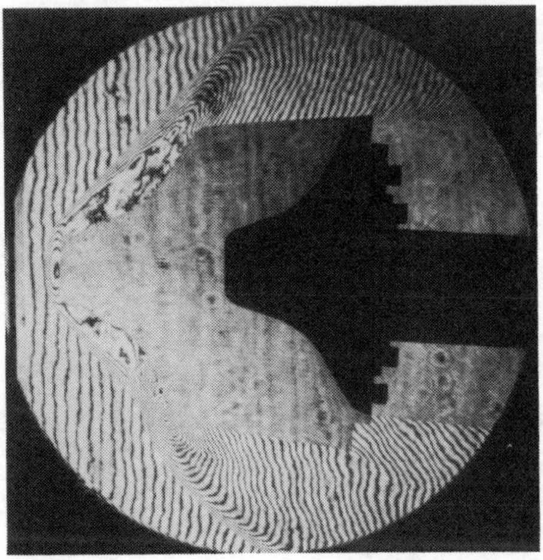

Fig. 8 Hologram of 4 deg angle of attack.

Table 3 Experimental shock shape summary

α=0°				α=4°			
X/L	Y/L	X/L	Y/L	X/L	Y/L	X/L	Y/L
0.42715	-0.66677	-0.04832	-0.00975	0.38058	-0.63860	-0.05166	-0.02336
0.40737	-0.63846	-0.04877	0.01333	0.36409	-0.60906	-0.05138	0.00069
0.39294	-0.61373	-0.04781	0.03971	0.34747	-0.58502	-0.05084	0.02391
0.37879	-0.59230	-0.04493	0.06224	0.33276	-0.55877	-0.04644	0.05468
0.36670	-0.57238	-0.04245	0.08230	0.32081	-0.53212	-0.03971	0.09109
0.36011	-0.56056	-0.03462	0.10634	0.31367	-0.50794	-0.02981	0.12228
0.34568	-0.54201	-0.02899	0.12310	0.30721	-0.48884	-0.01772	0.14193
0.33180	-0.52539	-0.01717	0.14330	0.30144	-0.46480	-0.00343	0.16432
0.31655	-0.50808	-0.00522	0.16006	0.29745	-0.43636	-0.01511	0.18603
0.29704	-0.48582	0.00866	0.17820	0.29636	-0.40874	-0.03091	0.20705
0.27616	-0.46549	0.02748	0.19702	0.29800	-0.38813	0.05152	0.23013
0.25541	-0.44529	0.04548	0.21708	0.29773	-0.36381	0.06471	0.24868
0.23604	-0.32399	0.06032	0.23302	0.29855	-0.34417	0.08175	0.27080
0.21433	-0.40077	0.07529	0.24964	0.30268	-0.32012	0.09284	0.29141
0.19908	-0.38305	0.08834	0.26448	0.31051	-0.29471	0.11816	0.31573
0.17957	-0.36052	0.10565	0.28454	0.30584	-0.47304	0.13863	0.33812
0.16391	-0.34362	0.12613	0.30776	0.29732	-0.46301	0.15663	0.35969
0.14756	-0.32452	0.14866	0.33221	0.26448	-0.43237	0.16473	0.36821
0.13011	-0.30666	0.16391	0.34733	0.24332	-0.41080	0.18246	0.39198
0.10936	-0.28426	0.18136	0.36629	0.21983	-0.38662	0.20444	0.42097
0.08972	-0.26036	0.20004	0.38703	0.18740	-0.35571	0.22477	0.44804
0.07103	-0.24044	0.21955	0.40737	0.15580	-0.32273	0.25050	0.47895
0.05413	-0.21955	0.23467	0.42633	0.12777	-0.29553	0.27451	0.50670
0.03902	-0.20375	0.25170	0.44543	0.10483	-0.27272	0.29265	0.52814
0.02446	-0.18699	0.27245	0.46672	0.08092	-0.24731	0.31023	0.55108
0.00797	-0.16817	0.29327	0.48513	0.05070	-0.21351	0.32768	0.57320
-0.00453	-0.15347	0.31064	0.50478	0.02844	-0.18974	0.34870	0.60082
-0.01896	-0.13093	0.33180	0.53020	0.00921	-0.17037	0.36409	0.62033
-0.02899	-0.11541	0.35076	0.55314	-0.01346	-0.14399	0.37948	0.64135
-0.03462	-0.09645	0.36316	0.57952	-0.02872	-0.11637	0.39528	0.66058
-0.04081	-0.07982	0.38717	0.59532	-0.03847	-0.08463		
-0.04424	-0.05894	0.40490	0.61827	-0.04685	-0.05317		
-0.04822	-0.02624						

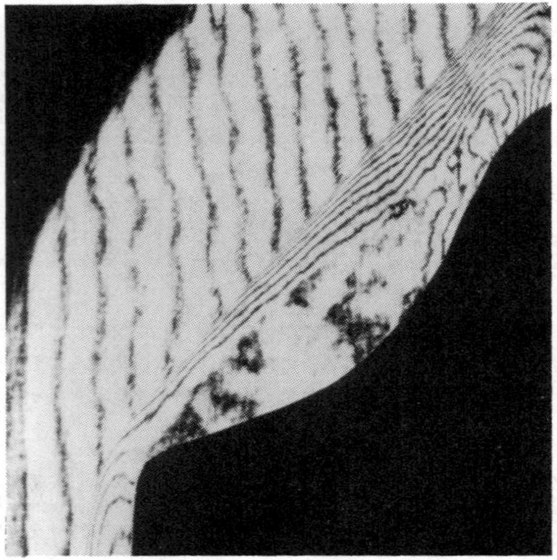

a)

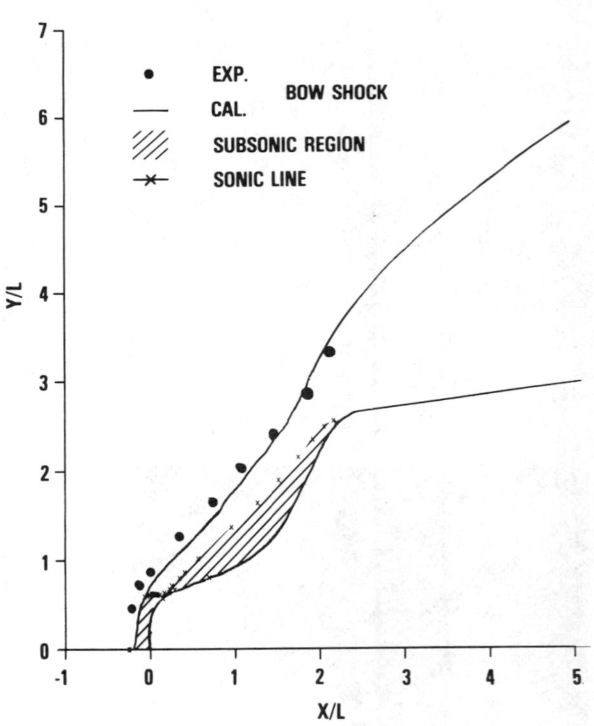

b)

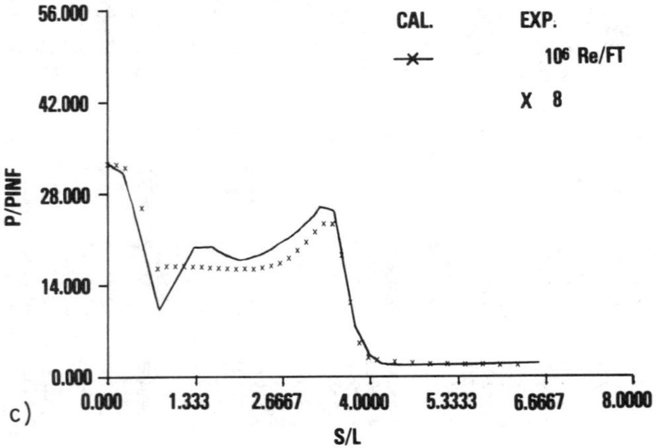

c)

Fig. 9 Comparison of experimental and calculated data at zero angle of attack: a) flow picture; b) shock shape; and c) surface pressure.

experimental data with results from Ref. 15 are given in Fig. 9. An enlargement of the zero angle-of-attack interferogram (Fig. 9a) shows that the flow is separated immediately at the beginning of the expansion corner and the separation region extends all the way to the expansion shoulder. Comparison of the inviscid solution with the experimental surface pressure data as shown in Fig. 9c indicates that good agreement can be found only in the stagnation region. The comparison with shock shape location is also poor (Fig. 9b). This poor agreement is due to the effects of the separation region on the flow.

Inviscid solutions do not provide accurate modeling of the viscous separation influences for this severely indented-nose configuration. Full viscous flowfield solvers or inviscid solvers modified to account for the separation effects (approximate body techniques) would probably be required to predict indented-nose flowfields.

The 4 deg angle-of-attack interferogram (Fig. 8) shows the windward meridian embedded shock suggested by

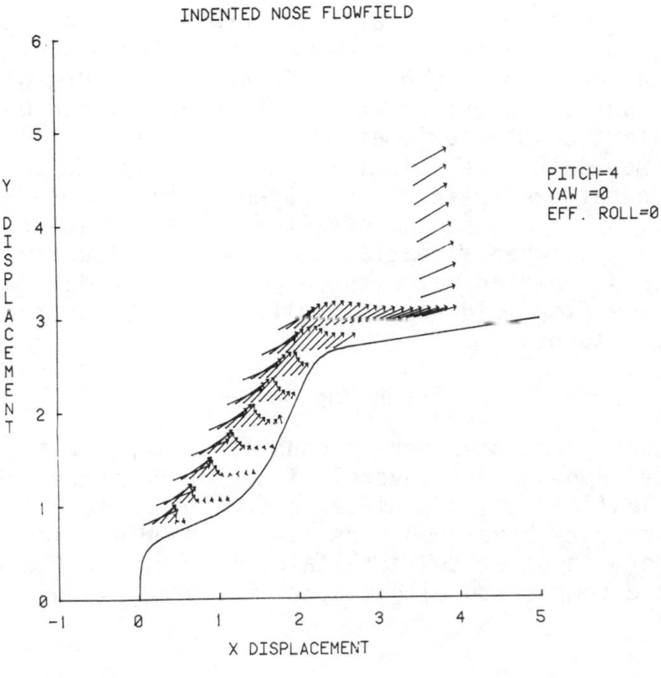

Fig. 10 Experimental velocity flowfield at 4 deg angle of attack.

the pressure data. The shock is generated by the reattachment of the separated flow region.

Velocity Measurements

Velocity profiles generated using a two-dimensional LDV system are presented in Tables 4-11.[§] A velocity vector plot of the flow, presented in Fig. 10, clearly shows the separation region which corresponds to the separation region shown in the interferogram (Fig. 8). The velocity profiles have been analyzed to have an experimental uncertainty band of $\pm$ 5%. Velocities measured in the region immediately adjacent to the bow shock may have a slightly higher degree of uncertainty. This is due to a lag effect experienced between the LDV tracer particles and the actual flowfield velocity as the tracer particles cross the bow shock. This effect quickly dies out away from the shock. Investigations are currently underway at Naval Surface Weapons Center to quantify the uncertainties associated with the lag effect.

Velocity measurements are also presented along a ray extending outward from the model's base. The associated vectors can be seen in Fig. 10. These data, along with the local surface pressure and shock location, can be used to start a supersonic afterbody calculation. The results from such a calculation would give an indication of the "downstream" effects generated by an indented-nose flowfield. This downstream effect would be greatest at angle of attack when a reattachment shock is located on the windward meridian below the expansion shoulder, while the lee-side flowfield remains fully separated over the compression turn.

Conclusion

The data presented herein constitute the first comprehensive mapping of a severely indented re-entry vehicle nosetip flowfield using noninterference probe techniques. Detailed results have been presented in nondimensional tabular form in order to facilitate study of the flowfield physics and comparison with theoretical predictions.

---

§Tables 4-11 appear on pages 193-200.

The severely indented nose investigated was found to have a separated flow region extending from the model's expansion corner to its expansion shoulder (zero angle of attack). An embedded reattachment shock occurred at angle of attack on the windward side, below the expansion shoulder.

It is hoped that the data base contained in this report will contribute to the advance of the understanding of and the ability to model the complex indented-nose flowfield.

Acknowledgments

This work was conducted under the Reentry Technology Program Office of the Strategic Systems Department of the Naval Surface Weapons Center, Dr. W. Carson Lyons, Manager. The authors wish to acknowledge the contributions of J. Marshall, Richard Westermeyer, Mark Roberts, Donald Ausherman, and Mary Ellen Falusi in the processing and reduction of the data contained in this report.

References

[1]Morrison, A. M., "Effects of Nosetip Shape Change on Reentry Vehicle Dispersion," Journal of Spacecraft and Rockets, Vol. 12, Oct. 1973, pp. 633-635.

[2]Hall, D. W. and Nowlan, D. T., "Aerodynamics of Ballistic Reentry Vehicles with Asymmetric Nosetips," AIAA Paper 77-701, June 1977.

[3]Swain, C. E., "Aerodynamics of Reentry Vehicles with Asymmetric Nosetip Shape Change," AIAA Paper 77-782, June 1977.

[4]Platus, D. M., "Dispersion of Spinning Missiles Due to Lift Nonaveraging," AIAA Journal, Vol. 15, July 1977, pp. 909-915.

[5]Nicolaides, J. D., "On the Free Flight Motion of Missiles Having Slight Configurational Asymmetries," Ballistic Research Laboratories, Aberdeen Proving Grounds, Rept. BRL-858, Oct. 1952.

[6]Krumins, M. V., Morrison, A. M., and Jobe, M. D., "Aerodynamics of Re-entry Bodies with Blunt, Indented Nosetips," AIAA Paper 81-0147, Jan. 1981.

[7]Solomon, J. M., Ciment, M., Ferguson, R. E., Bell, J. B., and Wardlaw, A. B., "A Program for Computing Steady Inviscid Three-Dimensional Supersonic Flow on Reentry Vehicles, Vol. 1: Analysis and Programming," Naval Surface Weapons Center Tech. Rept. NSWC.WOL, TR 77-28, Feb. 1977.

[8]Otey, G. R. and English, E. A., "High-B Reentry Vehicle Recovery," Journal of Spacecraft and Rockets, Vol. 14, May 1977, pp. 290-293.

[9]Kutler, P. and Shankar, V. S., "Diffraction of a Shock Wave by a Compression Corner, I: Regular Reflection," AIAA Journal, Vol. 15, Feb. 1977, pp. 197-203.

[10]Holden, M. S., "Studies of the Effects of Transitional and Turbulent Boundary Layers' on the Aerodynamic Performance of Hypersonic Re-entry Vehicles in High Reynolds Number Flows," Calspan Rept. AG-5834-A-2, Dec. 1978.

[11]Yanta, W. J., "A Three-Dimensional Laser Doppler Velocimeter (LDV) for Use in Wind Tunnels," Paper IEEE 79 CH 1500-8AES, presented at International Congress on Instrumentation in Aerospace Facilities, Monterey, Calif., Sept. 1979.

[12]Hannah, B. W. and King, W. L., "Extensions of Dual-Plate Holographic Interferometry," AIAA Journal, Vol. 15, May 1977, pp. 725-727.

[13]Hsieh, T., "Numerical Investigation of Flowfield about a Series of Indented Nosetips," AIAA Paper 81-0077, Jan. 1981.

[14]Kutler, P., Chakravartly, S. R., and Lombard, C. P., "Supersonic Flow over Ablated Nosetips Using an Unsteady Implicit Numerical Procedure," AIAA Paper 78-213, Jan. 1978.

[15]Inonye, M. and Lomax, H., "Comparison of Experimental and Numerical Results for the Flow of a Perfect Gas about Blunt-Nosed Bodies," NASA TN D-1426, Sept. 1962.

Table 4 Velocity profiles at zero pitch, yaw, and roll

X INCHES	Y INCHES	U/Q	V/Q	X INCHES	Y INCHES	U/Q	V/Q
0.4920	0.8000	-0.0601	0.0663	1.5490	2.2500	0.1848	0.0997
0.0940	0.8000	0.2097	0.0665	1.6490	2.2500	0.1636	0.1195
0.1940	0.8000	0.1385	0.0853	1.7490	2.2500	0.1481	0.1242
0.2940	0.8000	0.0683	0.0742	1.8490	2.2500	0.1083	0.0951
0.3940	0.8000	-0.0060	0.0098	1.9490	2.2500	0.0690	0.0634
0.2600	1.0000	0.2308	0.0622	1.6410	2.5000	0.2194	0.0651
0.3590	1.0000	0.1699	0.0858	1.7410	2.5000	0.1878	0.0945
0.4590	1.0000	0.1074	0.0951	1.8410	2.5000	0.1587	0.1201
0.5590	1.0000	0.0789	0.0649	1.9410	2.5000	0.1409	0.1285
0.6600	1.0000	0.0120	0.0156	2.0420	2.5000	0.1257	0.1194
0.7590	1.0000	-0.0173	0.0006	2.1410	2.5000	0.0969	0.0689
0.8600	1.0000	-0.0182	-0.0061	1.8400	2.7500	0.0872	0.1037
0.9590	1.0000	-0.0288	-0.0092	1.9390	2.7500	0.1842	0.0939
0.4790	1.2500	0.2440	0.0534	2.0400	2.7500	0.1539	0.1203
0.5780	1.2500	0.1955	0.0879	2.1390	2.7500	0.1444	0.1290
0.6780	1.2500	0.1378	0.1032	2.2400	2.7500	0.1413	0.1275
0.7780	1.2500	0.0971	0.0829	2.3400	2.7500	0.1484	0.1205
0.8790	1.2500	0.0659	0.0533	2.4400	2.7500	0.1511	0.1034
0.9780	1.2500	0.0218	0.0217	2.5400	2.7500	0.1642	0.0890
1.0780	1.2500	-0.0030	0.0058	2.6400	2.7500	0.1755	0.0716
1.1780	1.2500	-0.0160	-0.0048	2.7400	2.7500	0.1842	0.0542
1.2790	1.2500	-0.0233	-0.0104	2.0550	2.9990	0.2077	0.0684
0.7250	1.5000	0.2368	0.0581	2.1550	3.0000	0.1670	0.1029
0.8240	1.5000	0.1934	0.0920	2.2550	3.0000	0.1549	0.1214
0.9240	1.5000	0.1597	0.1125	2.3550	3.0000	0.1561	0.1267
1.0240	1.5000	0.1109	0.0960	2.4550	3.0000	0.1634	0.1273
1.1240	1.5000	0.0826	0.0662	2.5550	3.0000	0.1733	0.1224
1.2240	1.5000	0.0610	0.0415	2.6550	3.0000	0.1832	0.1151
1.3240	1.5000	0.0268	0.0200	2.7550	3.0000	0.1924	0.1045
1.4240	1.5000	0.0054	0.0045	2.8550	3.0000	0.1995	0.0943
1.5240	1.5000	-0.0098	-0.0061	2.9550	3.0000	0.2061	0.0858
0.9580	1.7500	0.2205	0.0692	3.0550	3.0000	0.2122	0.0759
1.0590	1.7500	0.1875	0.0986	3.1550	3.0000	0.2160	0.0678
1.1580	1.7500	0.1665	0.1177	3.2550	3.0000	0.2198	0.0611
1.2590	1.7500	0.1362	0.1124	3.3550	3.0000	0.2233	0.0566
1.3580	1.7500	0.1049	0.0831	3.4550	3.0000	0.2253	0.0530
1.4580	1.7500	0.0628	0.0509	3.5540	3.0000	0.2284	0.0406
1.5580	1.7500	0.0450	0.0321	3.6550	3.0000	0.2302	0.0453
1.6580	1.7500	0.0181	0.0096	3.6550	3.1990	0.2344	0.0629
1.1840	2.0000	0.2355	0.0603	3.6550	3.4000	0.2303	0.0749
1.2850	2.0000	0.1936	0.0945	3.6550	3.6000	0.2260	0.0836
1.3840	2.0000	0.1706	0.1145	3.6550	3.8000	0.2162	0.0897
1.4840	2.0000	0.1546	0.1258	3.6550	3.9990	0.2010	0.0881
1.5840	2.0000	0.1166	0.0968	3.6550	4.1990	0.1990	0.0914
1.6840	2.0000	0.0779	0.0636	3.6550	4.4000	0.2116	0.0965
1.7840	2.0000	0.0529	0.0395	3.6550	4.6000	0.2402	0.0764
1.4490	2.2500	0.2222	0.0719				

Table 5 Velocity profiles at pitch = 4, yaw = 0, and roll = 0

X INCHES	Y INCHES	U/Q	V/Q	X INCHES	Y INCHES	U/Q	V/Q
0.0718	0.8080	0.2142	0.0816	1.4197	2.1042	0.1477	0.1348
0.1716	0.8150	0.1471	0.0987	1.5194	2.1111	0.1142	0.1063
0.2713	0.8219	0.0992	0.1018	1.6192	2.1181	0.0809	0.0738
0.3711	0.8289	0.0497	0.0438	1.7190	2.1251	0.0587	0.0521
0.4708	0.8359	-0.0024	0.0073	1.8187	2.1321	0.0368	0.0396
0.5706	0.8429	-0.0151	-0.0044	1.9185	2.1390	0.0171	0.0340
0.2255	1.0192	0.2388	0.0785	1.3175	2.3476	0.2264	0.0804
0.3252	1.0262	0.1765	0.1045	1.4172	2.3546	0.1917	0.1056
0.4250	1.0332	0.1190	0.1074	1.5170	2.3616	0.1669	0.1256
0.5247	1.0401	0.0946	0.0874	1.6167	2.3686	0.1536	0.1366
0.6245	1.0471	0.0384	0.0371	1.7165	2.3755	0.1408	0.1321
0.7242	1.0541	0.0078	0.0114	1.8162	2.3825	0.1049	0.0984
0.8240	1.0611	-0.0096	-0.0009	1.9160	2.3895	0.0784	0.0714
0.9237	1.0680	-0.0145	-0.0037	2.0157	2.3975	0.0563	0.0612
1.0235	1.0750	-0.0125	-0.0075	1.5574	2.6150	0.2151	0.0866
1.1233	1.0820	-0.0150	-0.0089	1.6571	2.6220	0.1792	0.1146
0.4404	1.2849	0.2342	0.0804	1.7569	2.6290	0.1603	0.1288
0.5402	1.2918	0.1849	0.1116	1.8567	2.6359	0.1533	0.1364
0.6400	1.2988	0.1399	0.1146	1.9564	2.6429	0.1474	0.1377
0.7397	1.3058	0.0972	0.0943	2.0562	2.6499	0.1280	0.1193
0.8395	1.3128	0.0673	0.0598	2.1559	2.6569	0.1133	0.0979
0.9392	1.3197	0.0325	0.0295	2.2557	2.6638	0.1079	0.0873
1.0380	1.3266	0.0038	0.0108	2.3554	2.6708	0.1136	0.0835
1.1397	1.3338	-0.0015	0.0063	2.4552	2.6778	0.1358	0.0759
1.2375	1.3406	-0.0060	0.0001	1.7654	2.8802	0.2083	0.0926
1.3383	1.3476	-0.0062	-0.0018	1.8652	2.8871	0.1794	0.1139
1.4370	1.3545	-0.0075	-0.0050	1.9649	2.8941	0.1632	0.1289
0.6704	1.5516	0.2356	0.0768	2.0647	2.9011	0.1597	0.1345
0.7702	1.5585	0.1914	0.1091	2.1644	2.9081	0.1603	0.1379
0.8699	1.5655	0.1553	0.1254	2.2642	2.9150	0.1615	0.1348
0.9697	1.5725	0.1139	0.1069	2.3639	2.9220	0.1638	0.1236
1.0694	1.5795	0.0857	0.0751	2.4637	2.9290	0.1639	0.1137
1.1692	1.5864	0.0562	0.0486	2.5634	2.9360	0.1694	0.0999
1.2689	1.5934	0.0271	0.0259	2.6632	2.9429	0.1682	0.0884
1.3687	1.6004	0.0134	0.0190	2.7630	2.9499	0.1742	0.0803
1.4685	1.6074	0.0062	0.0122	2.8627	2.9569	0.1814	0.0737
1.5682	1.6143	0.0036	0.0091	2.9625	2.9639	0.1871	0.0664
1.6680	1.6213	0.0030	0.0120	3.0622	2.9709	0.1953	0.0606
0.8934	1.8178	0.2264	0.0837	3.1620	2.9778	0.1979	0.0574
0.9931	1.8247	0.1872	0.1073	3.2617	2.9848	0.2023	0.0549
1.0929	1.8317	0.1643	0.1263	3.3615	2.9918	0.2062	0.0520
1.1927	1.8387	0.1358	0.1243	3.4613	2.9988	0.2075	0.0494
1.2924	1.8457	0.0963	0.0886	3.5610	3.0057	0.2135	0.0475
1.3922	1.8526	0.0676	0.0625	3.5471	3.2052	0.2145	0.0626
1.4919	1.8596	0.0443	0.0401	3.5331	3.4048	0.2266	0.0816
1.5917	1.8666	0.0261	0.0243	3.5192	3.6043	0.2244	0.0968
1.6914	1.8736	0.0180	0.0207	3.5052	3.8038	0.2161	0.1033
1.7912	1.8805	-0.0153	0.0419	3.4913	4.0033	0.2105	0.1065
1.1214	2.0833	0.2189	0.0848	3.4773	4.2028	0.2061	0.1099
1.2202	2.0902	0.1855	0.1091	3.4634	4.4023	0.2102	0.1104
1.3199	2.0972	0.1642	0.1283	3.4494	4.6018	0.2185	0.1027

Table 6 Velocity profiles at pitch = 3.46, yaw = 2, and effective roll = 30

X INCHES	Y INCHES	U/Q	V/Q	X INCHES	Y INCHES	U/Q	V/Q
0.0186	0.8026	0.2452	0.0669	1.4905	2.0928	0.1205	0.1107
0.1183	0.8086	0.1816	0.0879	1.5903	2.0989	0.0843	0.0755
0.2181	0.8147	0.1181	0.0987	1.6900	2.1049	0.0617	0.0524
0.3179	0.8207	0.0900	0.0830	1.7898	2.1109	0.0417	0.0392
0.4176	0.8267	0.0118	0.0061	1.8895	2.1170	0.0293	0.0335
0.5174	0.8328	-0.0209	-0.0114	1.3696	2.3370	0.2086	0.0928
0.2360	1.0161	0.2303	0.0775	1.4694	2.3430	0.1794	0.1180
0.3357	1.0221	0.1711	0.0988	1.5691	2.3490	0.1614	0.1356
0.4355	1.0282	0.1112	0.1037	1.6689	2.3551	0.1441	0.1357
0.5352	1.0342	0.0812	0.0731	1.7687	2.3611	0.1173	0.1082
0.6350	1.0402	0.0208	0.0193	1.8684	2.3671	0.0884	0.0813
0.7347	1.0463	-0.0138	-0.0038	1.9682	2.3732	0.0655	0.0611
0.8345	1.0523	-0.0173	-0.0070	2.0679	2.3792	0.0499	0.0576
0.9343	1.0584	-0.0204	-0.0052	1.5282	2.5960	0.2250	0.0779
1.0340	1.0644	-0.0186	-0.0077	1.6279	2.6021	0.1924	0.1043
0.4483	1.2794	0.2396	0.0741	1.7277	2.6081	0.1700	0.1251
0.5481	1.2854	0.1907	0.1068	1.8275	2.6141	0.1567	0.1368
0.6478	1.2915	0.1354	0.1139	1.9272	2.6202	0.1477	0.1372
0.7476	1.2975	0.1023	0.0925	2.0270	2.6262	0.1373	0.1252
0.8474	1.3036	0.0666	0.0566	2.1267	2.6322	0.1163	0.1006
0.9471	1.3096	0.0176	0.0202	2.2265	2.6383	0.1090	0.0871
1.0469	1.3156	-0.0029	0.0062	2.3262	2.6443	0.1121	0.0824
1.1466	1.3217	-0.0092	0.0056	1.7864	2.8621	0.2093	0.0905
1.2464	1.3277	-0.0087	0.0054	1.8862	2.8681	0.1788	0.1140
1.3461	1.3337	-0.0050	0.0039	1.9860	2.8742	0.1639	0.1302
1.4459	1.3398	-0.0071	0.0337	2.0857	2.8802	0.1603	0.1370
0.6547	1.5424	0.2466	0.0680	2.1855	2.8862	0.1575	0.1362
0.7545	1.5484	0.2010	0.1021	2.2852	2.8923	0.1630	0.1324
0.8542	1.5544	0.1659	0.1259	2.3850	2.8983	0.1636	0.1192
0.9540	1.5605	0.1166	0.1087	2.4847	2.9043	0.1636	0.1078
1.0537	1.5665	0.0853	0.0772	2.5845	2.9104	0.1655	0.0950
1.1535	1.5725	0.0571	0.0443	2.6843	2.9164	0.1745	0.0861
1.2533	1.5786	0.0313	0.0236	2.7840	2.9225	0.1810	0.0773
1.3530	1.5846	0.0130	0.0203	2.8838	2.9285	0.1879	0.0693
1.4528	1.5906	0.0066	0.0171	2.9835	2.9345	0.1958	0.0628
1.5525	1.5967	0.0086	0.0124	3.0833	2.9406	0.2019	0.0578
1.6523	1.6027	0.0054	0.0070	3.1830	2.9466	0.2061	0.0534
0.8941	1.8063	0.2373	0.0761	3.2828	2.9526	0.2099	0.0502
0.9938	1.8123	0.1930	0.1083	3.3826	2.9587	0.2121	0.0480
1.0936	1.8184	0.1686	0.1278	3.4823	2.9647	0.2142	0.0448
1.1933	1.8244	0.1286	0.1175	3.5821	2.9707	0.2173	0.0434
1.2931	1.8304	0.0925	0.0810	3.5700	3.1704	0.2194	0.0600
1.3929	1.8365	0.0676	0.0569	3.5579	3.3700	0.2256	0.0791
1.4926	1.8425	0.0461	0.0375	3.5459	3.5696	0.2233	0.0963
1.5924	1.8485	0.0265	0.0288	3.5338	3.7693	0.2205	0.1014
1.6921	1.8546	0.0201	0.0222	3.5218	3.9689	0.2133	0.1049
1.7919	1.8606	0.0116	0.0227	3.5097	4.1675	0.2087	0.1083
1.0915	2.0687	0.2378	0.0739	3.4977	4.3672	0.2109	0.1091
1.1912	2.0747	0.2020	0.1002	3.4855	2.0000	0.0282	0.0008
1.2910	2.0808	0.1702	0.1270	71.0000	2.0000	0.0282	0.0008
1.3908	2.0868	0.1518	0.1337				

Table 7 Velocity profiles at pitch = 2, yaw ▪ 3.46, and effective roll = 60

X INCHES	Y INCHES	U/Q	V/Q	X INCHES	Y INCHES	U/Q	V/Q
-0.6683	0.7771	0.3100	0.2649	1.4230	2.3012	0.2046	0.0905
-0.6683	0.7771	0.2978	0.2116	1.5227	2.3046	0.1854	0.1096
0.1178	0.8046	0.1809	0.0816	1.6225	2.3081	0.1629	0.1270
0.2185	0.8081	0.1151	0.0950	1.7222	2.3116	0.1452	0.1280
0.3173	0.8116	0.0965	0.0873	1.8220	2.3151	0.1267	0.1124
0.4180	0.8151	0.0302	0.0481	1.9218	2.3186	0.0824	0.0790
0.2555	1.0096	0.2217	0.0794	2.0215	2.3221	0.0525	0.0589
0.3552	1.0130	0.1601	0.0957	1.6367	2.5588	0.2111	0.0817
0.4550	1.0165	0.1106	0.0990	1.7365	2.5623	0.1826	0.1064
0.5547	1.0200	0.0824	0.0755	1.8362	2.5658	0.1620	0.1249
0.6545	1.0235	0.0382	0.0462	1.9360	2.5693	0.1463	0.1329
0.7532	1.0270	0.0022	0.0222	2.0357	2.5727	0.1352	0.1273
0.8540	1.0305	-0.0170	0.0124	2.1355	2.5762	0.1144	0.1039
0.9538	1.0340	-0.0212	0.0068	2.2353	2.5797	0.0932	0.0957
1.0535	1.0375	-0.0268	0.0698	1.8674	2.8170	0.1850	0.1007
0.4961	1.2691	0.2198	0.0880	1.9672	2.8205	0.1712	0.1138
0.5959	1.2726	0.1812	0.1054	2.0669	2.8240	0.1578	0.1268
0.6956	1.2761	0.1233	0.1039	2.1667	2.8275	0.1519	0.1337
0.7954	1.2796	0.1039	0.0840	2.2665	2.8310	0.1532	0.1331
0.8941	1.2830	0.0639	0.0600	2.3662	2.8345	0.1552	0.1224
0.9949	1.2866	0.0351	0.0388	2.4660	2.8380	0.1651	0.1112
1.0937	1.2900	0.0048	0.0235	2.5657	2.8414	0.1717	0.0998
1.1944	1.2936	-0.0066	0.0162	2.6655	2.8449	0.1821	0.0862
1.2932	1.2970	-0.0090	0.0103	2.7652	2.8484	0.1889	0.0766
1.3939	1.3005	-0.0886	0.0806	2.8650	2.8519	0.1962	0.0680
0.7019	1.5265	0.2346	0.0788	2.9647	2.8554	0.2033	0.0596
0.8026	1.5300	0.1967	0.1015	2.1101	3.0747	0.1823	0.0993
0.9014	1.5335	0.1608	0.1208	2.2099	3.0781	0.1634	0.1176
1.0021	1.5370	0.1159	0.1010	2.3096	3.0816	0.1618	0.1262
1.1009	1.5404	0.0930	0.0737	2.4094	3.0851	0.1677	0.1306
1.2017	1.5440	0.0567	0.0527	2.5092	3.0886	0.1739	0.1272
1.3004	1.5474	0.0321	0.0398	2.6089	3.0921	0.1807	0.1224
1.4012	1.5509	0.0187	0.0252	2.7087	3.0956	0.1912	0.1167
1.5009	1.5544	0.0023	0.0151	2.8084	3.0991	0.1968	0.1089
1.6007	1.5579	-0.0063	0.1152	2.9082	3.1026	0.2042	0.0995
0.9496	1.7843	0.2303	0.0788	3.0079	3.1061	0.2090	0.0901
1.0493	1.7878	0.1918	0.1045	3.1077	3.1096	0.2141	0.0835
1.1491	1.7913	0.1725	0.1216	3.2075	3.1130	0.2189	0.0777
1.2489	1.7948	0.1360	0.1146	3.3072	3.1165	0.2182	0.0717
1.3486	1.7983	0.1022	0.0865	3.4070	3.1200	0.2233	0.0660
1.4484	1.8017	0.0715	0.0601	3.5067	3.1235	0.2263	0.0638
1.5481	1.8052	0.0506	0.0433	3.6065	3.1270	0.2300	0.0598
1.6479	1.8087	0.0118	0.0260	3.5995	3.3269	0.2327	0.0752
1.7476	1.8122	0.0132	0.0556	3.5926	3.5268	0.2256	0.0873
1.1633	2.0419	0.2360	0.0761	3.5856	3.7266	0.2190	0.0941
1.2631	2.0454	0.1959	0.0993	3.5786	3.9265	0.2109	0.0962
1.3628	2.0489	0.1730	0.1202	3.5717	4.1264	0.2014	0.0977
1.4626	2.0524	0.1586	0.1305	3.5647	4.3263	0.2024	0.1011
1.5624	2.0559	0.1263	0.1100	3.5577	4.5272	0.1934	0.1021
1.6621	2.0594	0.0870	0.0745	3.5507	4.7270	0.2139	0.0757
1.7619	2.0629	0.0581	0.0553	3.5437	4.9269	0.1988	0.1102
1.8616	2.0664	0.0291	0.0505				

Table 8 Velocity profiles at pitch = 0, yaw = 4, and effective roll = 90

X INCHES	Y INCHES	U/Q	V/Q	X INCHES	Y INCHES	U/Q	V/Q
-0.9078	0.8010	0.3151	0.0899	1.6410	2.2500	0.1619	0.1185
-0.9078	0.8010	0.3038	0.0098	1.7408	2.2500	0.1375	0.1272
0.0828	0.8010	0.1747	0.0691	1.8405	2.2500	0.0972	0.0931
0.1826	0.8010	0.1434	0.0859	1.9403	2.2500	-0.0040	0.0562
0.2823	0.8010	0.0990	0.0917	1.6759	2.5000	0.1439	0.0865
0.3821	0.8010	0.0258	0.0351	1.7757	2.5000	0.1367	0.1092
0.4818	0.8010	-0.0080	0.0312	1.8754	2.5000	0.1469	0.1258
0.2624	1.0000	0.2278	0.0630	1.9752	2.5000	0.1275	0.1266
0.3621	1.0000	0.1667	0.0875	2.0749	2.5000	0.1039	0.1054
0.4619	1.0000	0.1188	0.0934	2.1747	2.5000	0.0979	0.1000
0.5616	1.0000	0.0784	0.0664	1.7956	2.7500	0.1461	0.0547
0.6614	1.0000	0.0465	0.0389	1.8954	2.7500	0.1272	0.0933
0.7611	1.0000	0.0259	0.0219	1.9951	2.7500	0.1407	0.1154
0.8609	1.0000	0.0101	0.0083	2.0949	2.7500	0.1385	0.1225
0.9607	1.0000	-0.0049	-0.0022	2.1946	2.7500	0.1348	0.1268
0.4998	1.2500	0.2459	0.0578	2.2944	2.7500	0.1428	0.1214
0.5995	1.2500	0.1941	0.0907	2.3942	2.7500	0.1521	0.1089
0.6993	1.2500	0.1309	0.1036	2.4939	2.7500	0.1590	0.0921
0.7990	1.2500	0.1010	0.0807	2.5937	2.7500	0.1731	0.0784
0.8988	1.2500	0.0640	0.0537	2.6934	2.7500	0.1833	0.0621
0.9986	1.2500	0.0495	0.0371	2.7932	2.7500	0.1898	0.0479
1.0983	1.2500	0.0327	0.0209	1.9433	3.0000	0.1555	0.0537
1.1981	1.2500	0.0050	0.0042	2.0430	3.0000	0.1114	0.0807
1.2978	1.2500	-0.0102	-0.0091	2.1428	3.0000	0.1281	0.1001
0.7422	1.5010	0.2420	0.0569	2.2425	3.0000	0.1261	0.1145
0.8419	1.5010	0.2007	0.0939	2.3423	3.0000	0.1505	0.1227
0.9407	1.5010	0.1616	0.1120	2.4420	3.0000	0.1595	0.1221
1.0415	1.5010	0.1173	0.0964	2.5418	3.0000	0.1700	0.1172
1.1402	1.5010	0.0872	0.0696	2.6416	3.0000	0.1850	0.1117
1.2410	1.5010	0.0564	0.0440	2.7413	3.0000	0.1932	0.1006
1.3397	1.5010	0.0351	0.0227	2.8411	3.0000	0.2008	0.0914
1.4405	1.5010	0.0121	0.0048	2.9408	3.0000	0.2047	0.0832
1.5392	1.5010	-0.0779	0.0637	3.0406	3.0000	0.2113	0.0749
0.9746	1.7510	0.2358	0.0582	3.1403	3.0000	0.2140	0.0690
1.0744	1.7510	0.1974	0.0928	3.2401	3.0000	0.2162	0.0632
1.1741	1.7510	0.1743	0.1178	3.3398	3.0000	0.2214	0.0575
1.2739	1.7510	0.1344	0.1147	3.4396	3.0000	0.2237	0.0536
1.3736	1.7510	0.0968	0.0799	3.5394	3.0000	0.2279	0.0504
1.4734	1.7510	0.0630	0.0488	3.6391	3.0000	0.2284	0.0484
1.5732	1.7510	0.0377	0.0247	3.7389	3.0000	0.2300	0.0441
1.6729	1.7510	-0.0347	0.0539	3.7389	3.2000	0.2299	0.0613
1.1891	2.0000	0.2434	0.0550	3.7389	3.4000	0.2361	0.0729
1.2889	2.0000	0.2027	0.0863	3.7389	3.6000	0.2270	0.0803
1.3886	2.0000	0.1761	0.1120	3.7389	3.8000	0.2128	0.0824
1.4884	2.0000	0.1561	0.1265	3.7389	4.0000	0.1822	0.0787
1.5881	2.0000	0.1130	0.1035	3.7389	4.2000	0.1862	0.0852
1.6879	2.0000	0.0684	0.0603	3.7389	4.4000	0.1778	0.0931
1.7876	2.0000	0.0196	0.0350	3.7389	4.6000	0.1464	0.0968
1.4415	2.2500	0.2302	0.0595	3.7389	4.8000	0.2050	0.0791
1.5412	2.2500	0.1850	0.0949				

Table 9 Velocity profiles at pitch = 2, yaw = 3.46, and effective roll = 120

X INCHES	Y INCHES	U/Q	V/Q	X INCHES	Y INCHES	U/Q	V/Q
0.0927	0.7982	0.2348	0.0537	1.5777	2.1962	0.1820	0.0937
0.1915	0.7948	0.1727	0.0732	1.6775	2.1927	0.1534	0.1144
0.2923	0.7913	0.1150	0.0886	1.7772	2.1892	0.1261	0.1074
0.3910	0.7878	0.0567	0.0426	1.8770	2.1857	0.0838	0.0792
0.2843	0.9917	0.2439	0.0504	1.6713	2.4441	0.1868	0.0654
0.3840	0.9882	0.1806	0.0799	1.7710	2.4406	0.1689	0.0949
0.4838	0.9847	0.1210	0.0871	1.8708	2.4371	0.1389	0.1117
0.5835	0.9812	0.0885	0.0644	1.9705	2.4336	0.1202	0.1104
0.6833	0.9777	0.0355	0.0277	2.0703	2.4301	0.0807	0.0807
0.7830	0.9742	-0.0094	0.0023	1.8076	2.6884	0.1901	0.0417
0.8828	0.9707	-0.0289	-0.0068	1.9074	2.6850	0.1336	0.0707
0.5404	1.2329	0.2336	0.0577	2.0071	2.6815	0.1313	0.1051
0.6401	1.2294	0.1897	0.0892	2.1069	2.6780	0.1241	0.1148
0.7399	1.2259	0.1309	0.0917	2.2067	2.6745	0.1208	0.1097
0.8396	1.2224	0.0944	0.0701	2.3064	2.6710	0.1330	0.1069
0.9394	1.2189	0.0639	0.0424	1.9400	2.9340	0.1441	0.0595
1.0392	1.2154	0.0333	0.0210	2.0398	2.9305	0.1013	0.0725
1.1389	1.2119	-0.0030	0.0054	2.1396	2.9270	0.1128	0.0938
1.2387	1.2084	-0.0156	-0.0065	2.2393	2.9235	0.1320	0.1134
0.7456	1.4758	0.2465	0.0477	2.3391	2.9200	0.1461	0.1245
0.8454	1.4723	0.2043	0.0788	2.4388	2.9165	0.1562	0.1235
0.9451	1.4689	0.1676	0.1060	2.5386	2.9130	0.1693	0.1156
1.0449	1.4654	0.1238	0.0909	2.6383	2.9095	0.1810	0.1080
1.1446	1.4619	0.0938	0.0671	2.7381	2.9060	0.1917	0.0982
1.2444	1.4584	0.0662	0.0439	2.8379	2.9026	0.2026	0.0850
1.3441	1.4549	0.0359	0.0217	2.9376	2.8991	0.2088	0.0736
1.4439	1.4514	0.0096	0.0081	3.0374	2.8956	0.2124	0.0634
1.5437	1.4479	0.0001	0.0160	3.1371	2.8921	0.2132	0.0544
1.0017	1.7160	0.2319	0.0607	3.2369	2.8886	0.2146	0.0484
1.1014	1.7125	0.1970	0.0855	3.3366	2.8851	0.2179	0.0435
1.2012	1.7090	0.1745	0.1054	3.4364	2.8816	0.2194	0.0407
1.3009	1.7056	0.1415	0.1010	3.5362	2.8781	0.2215	0.0358
1.3997	1.7021	0.1001	0.0726	3.6359	2.8746	0.2233	0.0334
1.5005	1.6986	0.0628	0.0464	3.7357	2.8711	0.2219	0.0308
1.6002	1.6951	0.0280	0.0228	3.7426	3.0710	0.2338	0.0484
1.7000	1.6916	0.0148	0.0146	3.7496	3.2709	0.2349	0.0605
1.2488	1.9575	0.2270	0.0644	3.7566	3.4708	0.2325	0.0698
1.3486	1.9540	0.1929	0.0902	3.7635	3.6707	0.2240	0.0732
1.4483	1.9506	0.1674	0.1098	3.7705	3.8705	0.1966	0.0691
1.5481	1.9471	0.1410	0.1091	3.7775	4.0704	0.1967	0.0737
1.6478	1.9436	0.1006	0.0762	3.7844	4.2703	0.2058	0.0806
1.7476	1.9401	0.0440	0.0444	4.0000	71.0000	0.0016	0.0282
1.4780	2.1997	0.2235	0.0626	71.0000	4.0000	0.0282	0.0016

Table 10 Velocity profiles at pitch = 3.46, yaw = 2, and effective roll = 150

X INCHES	Y INCHES	U/Q	V/Q	X INCHES	Y INCHES	U/Q	V/Q
0.1011	0.7953	0.2339	0.0489	1.8086	2.1437	0.0697	0.0637
0.2009	0.7893	0.1744	0.0667	1.9083	2.1377	0.0485	0.0346
0.3006	0.7833	0.1259	0.0741	1.5543	2.4095	0.2429	0.0558
0.4004	0.7772	0.0458	0.0325	1.6541	2.4035	0.1767	0.0898
0.5002	0.7712	0.0619	0.0400	1.7538	2.3975	0.1486	0.1030
0.2389	0.9874	0.2486	0.0523	1.8536	2.3914	0.1231	0.1079
Ø.3885	0.9783	0.1583	0.0777	1.9533	2.3854	0.0891	0.0903
0.4893	0.9722	0.1140	0.0723	1.7619	2.6464	0.0010	0.2202
0.5880	0.9663	0.0807	0.0528	1.8616	2.6404	0.0613	0.0979
0.6888	0.9602	0.0366	0.0285	1.9614	2.6344	0.1312	0.1104
0.7875	0.9542	0.0021	0.0071	2.0611	2.6283	0.1257	0.1098
0.4604	1.2234	0.2685	0.0340	2.1609	2.6223	0.1207	0.1024
0.5602	1.2174	0.2076	0.0779	2.2606	2.6163	0.1211	0.0677
0.6599	1.2114	0.1443	0.0857	1.9096	2.8880	0.0689	0.1168
0.7597	1.2053	0.1014	0.0674	2.0094	2.8819	0.1328	0.0788
0.8594	1.1993	0.0799	0.0487	2.1091	2.8759	0.1358	0.1039
0.9592	1.1933	0.0259	0.0212	2.2089	2.8698	0.1419	0.1113
1.0589	1.1872	-0.0109	-0.0031	2.3086	2.8638	0.1505	0.1097
1.1587	1.1812	-0.0286	-0.0093	2.4084	2.8578	0.1579	0.1039
0.7069	1.4590	0.2362	0.0576	2.5082	2.8517	0.1667	0.0959
0.8057	1.4530	0.1894	0.0859	2.6079	2.8457	0.1805	0.0862
0.9064	1.4469	0.1421	0.0962	2.7077	2.8397	0.1934	0.0772
1.0052	1.4409	0.1079	0.0745	2.8074	2.8336	0.1974	0.0667
1.1059	1.4348	0.0672	0.0450	2.9072	2.8276	0.2071	0.0575
1.2047	1.4289	0.0184	0.0187	3.0069	2.8216	0.2049	0.0512
1.3055	1.4228	-0.0097	0.0014	3.1067	2.8155	0.2102	0.0445
1.4042	1.4168	-0.0293	-0.0058	3.2065	2.8095	0.2116	0.0391
0.9425	1.6952	0.2285	0.0608	3.3062	2.8035	0.2149	0.0359
1.0402	1.6893	0.1788	0.0890	3.4121	2.8982	0.2218	0.0450
1.1410	1.6832	0.1509	0.1032	3.4121	2.8982	0.2246	0.0441
1.2407	1.6771	0.1178	0.0838	3.5118	2.8922	0.2232	0.0407
1.3405	1.6711	0.0833	0.0552	3.6116	2.8862	0.2247	0.0369
1.4392	1.6651	0.0356	0.0264	3.7113	2.8801	0.2266	0.0349
1.5400	1.6590	-0.0522	0.0057	3.7234	3.0798	0.2326	0.0497
1.6398	1.6530	-0.0199	-0.0135	3.7354	3.2784	0.2300	0.0568
1.1341	1.9340	0.2300	0.0577	3.7474	3.4770	0.2289	0.0613
1.2329	1.9281	0.1862	0.0858	3.7595	3.6767	0.2202	0.0680
1.3336	1.9220	0.1629	0.1016	3.7715	3.8753	0.1999	0.0633
1.4324	1.9160	0.1413	0.1065	3.7835	4.0750	0.2025	0.0709
1.5331	1.9099	0.1064	0.0777	3.7956	4.2746	0.2115	0.0792
1.6329	1.9039	0.0588	0.0445	3.8076	4.4732	0.2208	0.0838
1.7326	1.8978	-0.0264	0.0594	3.8197	4.6729	0.2439	0.0658
1.4095	2.1678	0.2034	0.0751	71.0000	3.0000	0.0282	0.0012
1.5093	2.1618	0.1667	0.1003	71.0000	3.0000	0.0282	0.0012
1.6091	2.1558	0.1487	0.1088	71.0000	3.0000	0.0282	0.0012
1.7088	2.1497	0.1252	0.0999				

Table 11 Velocity profiles at pitch = -4, yaw = 0, and effective roll = 180

X INCHES	Y INCHES	U/Q	V/Q	X INCHES	Y INCHES	U/Q	V/Q
-0.0306	0.7530	0.2201	0.0487	1.3698	2.1567	0.1775	0.0923
0.0692	0.7460	0.1392	0.0741	1.4685	2.1498	0.1505	0.1099
0.1690	0.7390	0.1211	0.0787	1.5693	2.1428	0.1305	0.1063
0.1645	0.9899	0.2545	0.0467	1.6691	2.1358	0.1057	0.0794
0.2652	0.9829	0.1768	0.0761	1.7688	2.1288	0.0636	0.0486
0.3640	0.9760	0.0834	0.0718	1.5129	2.3973	0.1905	0.0759
0.4647	0.9689	0.0268	0.0220	1.6127	2.3903	0.1538	0.1017
0.6642	0.9550	-0.0349	-0.0114	1.7114	2.3834	0.1311	0.1111
0.4612	1.2198	0.1981	0.0831	1.8122	2.3764	0.1090	0.0974
0.5620	1.2128	0.1196	0.0914	1.9119	2.3694	0.0823	0.0717
0.6617	1.2058	0.0629	0.0484	1.6969	2.6351	0.0388	0.1218
0.7615	1.1988	0.0026	0.0095	1.7967	2.6281	0.1066	0.1001
0.8612	1.1918	-0.0278	-0.0080	1.8965	2.6211	0.1234	0.1070
0.9610	1.1849	-0.0365	-0.0118	1.9962	2.6141	0.1110	0.1029
1.0608	1.1779	-0.0383	-0.0123	2.0960	2.6071	0.1041	0.0905
1.1605	1.1709	-0.0407	0.0333	1.7962	2.8787	-0.0270	0.1940
0.6313	1.4585	0.2213	0.0622	1.8959	2.8717	0.1067	0.0713
0.7310	1.4515	0.1649	0.1011	1.9957	2.8648	0.1269	0.1009
0.8298	1.4446	0.1181	0.0923	2.0955	2.8578	0.1295	0.1119
0.9305	1.4376	0.0602	0.0477	2.1952	2.8508	0.1368	0.1104
1.0293	1.4307	0.0022	0.0136	2.2950	2.8438	0.1421	0.1041
1.1301	1.4236	-0.0292	-0.0027	2.3947	2.8369	0.1516	0.1028
1.2288	1.4167	-0.0349	-0.0087	2.4945	2.8299	0.1674	0.0888
1.3296	1.4097	-0.0284	0.0703	2.5942	2.8229	0.1795	0.0723
1.4283	1.4028	-0.0318	0.0059	2.6940	2.8159	0.1863	0.0545
0.8333	1.6950	0.1971	0.0623	2.7938	2.8090	0.1894	0.0414
0.9330	1.6880	0.1806	0.0918	2.8112	3.0584	0.1980	0.0857
1.0328	1.6811	0.1490	0.1095	2.9109	3.0514	0.2042	0.0781
1.1325	1.6741	0.0980	0.0753	3.0107	3.0444	0.2103	0.0676
1.2323	1.6671	0.0524	0.0389	3.1105	3.0374	0.2156	0.0608
1.3321	1.6601	0.0055	0.0106	3.2102	3.0305	0.2185	0.0547
1.4318	1.6532	-0.0212	0.0024	3.3100	3.0235	0.2203	0.0493
1.5316	1.6462	-0.0242	-0.0055	3.4097	3.0165	0.2207	0.0449
1.1040	1.9247	0.1877	0.0840	3.5095	3.0095	0.2227	0.0415
1.2037	1.9177	0.1623	0.1053	3.6092	3.0026	0.2206	0.0393
1.3035	1.9107	0.1381	0.1094	3.6232	3.2021	0.2256	0.0490
1.4032	1.9038	0.0968	0.0810	3.6371	3.4016	0.2267	0.0567
1.5030	1.8968	0.0599	0.0490	3.6511	3.6011	0.2226	0.0626
1.6027	1.8898	0.0114	0.0201	3.6650	3.7996	0.2010	0.0612
1.7025	1.8828	-0.0121	0.0133	3.6790	4.0001	0.1940	0.0661
1.8022	1.8759	-0.0080	0.0183	3.6930	4.1990	0.0024	0.0282
1.2700	2.1637	0.1561	0.0730	71.0000	6.0000	0.0282	0.0024

Ablation and Deceleration of Mass Driver-Launched Projectiles for Space Disposal of Nuclear Wastes

Chul Park*
NASA Ames Research Center, Moffett Field, Calif.
and
Stuart W. Bowen†
Beam Engineering, Sunnyvale, Calif.

Abstract

Ablation and deceleration characteristics are analyzed for a hemisphere-cylinder-shaped projectile protected by a graphite nose tip and launched vertically upward with a velocity in excess of 17 km/s. It is shown that ablation and deceleration of the projectile are in tolerable ranges for a scheme in which such projectiles are packed with nuclear wastes and launched with a mass driver to dispose of nuclear wastes outside the solar system.

Nomenclature

C_D = drag coefficient
$|fw|$ = blowing parameter, see Eq. (2)
H = enthalpy
ΔH = energy of vaporization of heat-shield material
h = altitude
M = mass of projectile
$\dot{m}$ = stagnation-point ablation rate
n = exponent for temperature in Eq. (6)
p = pressure
q = radiative heat flux
R = nose radius
Re = Reynolds number based on freestream conditions and nose radius
T = temperature
t = time
u,v = tangential and normal components of flow velocity

Presented as Paper 81-0355 at the AIAA 19th Aerospace Sciences Meeting, St. Louis, Mo., Jan. 12-15, 1981.

*Research Scientist, Entry Technology Branch.
†Research Scientist, Beam Engineering.

V_j = exhaust velocity
V_f = final velocity
V_∞ = projectile velocity
x,y = circumferential and normal distance
α = Rosseland mean opacity, see Eq. (7)
η = reduced normal coordinate
ρ = density

Subscripts

f = final value (40 km altitude)
i = interface
o = reference value
s = shock wave
w = wall
+ = air shock layer
- = ablation-product (blowing) layer
∞ = freestream

Introduction

A device that accelerates a projectile by the use of electromagnetic forces is referred to as a mass driver. Clarke[1] first suggested such a device for use as a means of space transportation. O'Neill[2] and Arnold et al.[3] have pointed out potential uses of a mass driver for constructing large-scale space structures. Kolm et al.[4] discussed the use of such a device as an Earth-based launcher of nuclear wastes. According to their proposal, a mass driver could be made to accelerate a projectile containing nuclear waste materials vertically upward with a muzzle velocity of about 20 km/s. The projectile would fly through the atmosphere in a ballistic mode. If the projectile survived the atmospheric flight and had a velocity when above the atmosphere greater than 16.6 km/s, it would escape the solar system, provided the launch was properly timed and aimed. Kolm et al. speculate that construction of such a device may not be entirely unfeasible.

The concept of disposing of radioactive nuclear wastes in space has been studied by various groups in recent years (see e.g., Refs. 5 to 7). Such studies assume the use of existing chemical rockets as the means of propulsion. This is a reasonable and prudent assumption, at least in the near term, because the technology of rocket propulsion is developed to a high degree of sophistication and reliability. For the long term, however, rocket propulsion may not necessarily be the most desirable method because of its relatively poor energy efficiency.

In rocket systems, fuel consumption increases exponentially with final velocity. The fuel mass consumed in a rocket system is related to the final velocity V_f by

$$\text{rocket fuel mass} = \text{final vehicle mass} \times [\exp(V_f/V_j) - 1]$$

where V_j is the rocket engine exhaust velocity. Since V_j is typically 4 km/s, the exponential factor is overwhelming when V_f approaches 20 km/s. In comparison, a mass driver is highly energy efficient because the efficiency of conversion of electrical energy into kinetic energy is usually nearly constant in most electrical devices. Assuming an efficiency of 75% in converting electrical energy to projectile kinetic energy, and assuming an electrical energy cost of $0.05/kWh, the energy cost for the launch would be $4.2 per kilogram of projectile mass. This is two orders of magnitude less than the fuel cost of typical rocket systems.

The present paper examines the problems of atmospheric flight of a mass driver accelerated projectile. For the scheme to be feasible, ablation and deceleration of the projectile must be within certain tolerable limits. Therefore, the paper concentrates on the problems of ablation and deceleration. The calculations show that it is possible for the projectiles to survive the flight through the atmosphere.

General Description of Mass Driver Scheme

There are several possible mass driver schemes.[3,4] Figure 1 is a schematic of a projectile suited for one such scheme; its cylindrical shape and length minimize drag. The nose of the projectile is assumed to be hemispherical, thus simplifying the analysis in the present work. Superconducting niobium alloy rings, which form belts around the projectile, interact with the pulsed drive coils to provide traction forces during launching; they are needed only for the mass driver scheme shown. To maintain attitudinal and directional stability, the projectile will have to be spun. As shown in Fig. 1, the nuclear waste materials are packed in the core of the projectile.

Since nuclear waste materials generate heat, they must be stored in a cooling environment separate from other heat-sensitive components. The projectile would be assembled shortly before launching and the niobium alloy bucket coils cooled cryogenically after the projectile was loaded into the launcher.[3] The insulating layers shown in Fig. 1 would shield the superconducting coils from the rest of the projectile, which would be heated by the nuclear waste material.

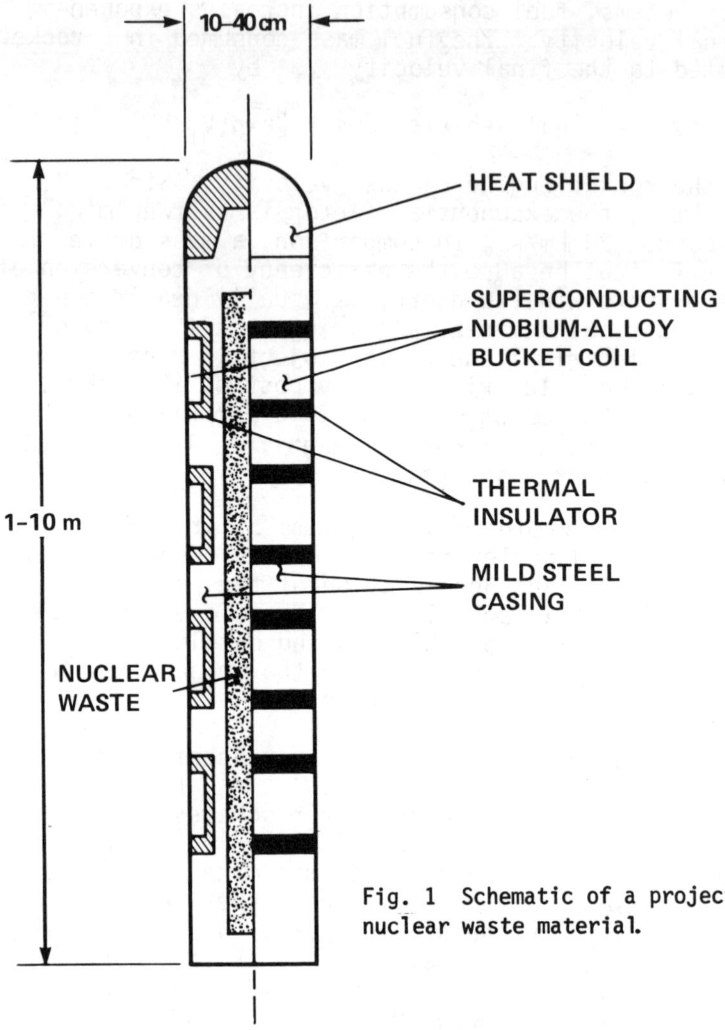

Fig. 1 Schematic of a projectile containing nuclear waste material.

Kolm et al.[4] estimate that the mass driver would be about 2 km long; it would be positioned vertically within a high mountain. The launch tube would be evacuated before launch to facilitate cryogenic cooling of the superconducting coils and to reduce the drag of the projectile during launch. A thin diaphragm, used to cover the muzzle of the launch tube, would be broken by impact of the projectile.

Electrical energy for the launch, generated by a nearby power plant, would be stored in a large energy storage system (e.g., a system of flywheels). Assuming the efficiency of conversion of electrical to kinetic energy of the projectile to be

75%, launching a 1 ton projectile would require 30,000 MJ. An energy storage system with this capacity is possible.[3,4] One small power plant operating continuously could supply enough energy for one launch per day.

In the axial mass driver configuration suitable for accelerating the projectile shown in Fig. 1, the driver coils are sequentially pulsed to produce an attractive force to incoming superconducting bucket coils as they approach the drive coil plane, and a repulsive force after the bucket coil passes through the drive coil plane.[3] Kolm et al.[4] speculate that an acceleration of about 10,000 g could be attained by this means. A constant acceleration of 10,000 g over a distance of 2 km would produce a muzzle velocity of 20 km/s.

Because each of the thousands of drive coil circuits operates almost independently of the others, the overall reliability is much greater than that of a rocket system. To ensure a successful launch even with some drive coil circuit failures, the mass driver will be made longer than normally needed to house the extra coils. The projectile's acceleration can be monitored and the extra coils can be used as needed if other coils fail. In case of an early abort, the phasing of the remaining coils can be reversed to stop the projectile within the tube.

The launch must take place at sunrise and the projectile must be aimed in a direction normal to the Sun to best utilize Earth's orbital velocity. If the direction of launch is in the ecliptic plane, a velocity of 16.6 km/s is needed for escape from the solar system. Otherwise, the required launch velocity becomes higher. For this reason, the launcher should be located at or near the equator.

The launch will produce a strong overpressure, that is, a sonic boom. From the calculated drag of the projectile, one can estimate that a 1 ton projectile will produce a boom roughly equivalent to that from five Concorde supersonic transports combined. One can surmise that the region within about 50 km of the launch site will not be suitable for human habitation. An uninhabited area at least 100 km in diameter must therefore be found for this purpose if the system is to be located on land.

Ideally, however, the mass driver could be located on an artificial island floating on an ocean. The tubing that houses the mass driver would be immersed vertically in the water and the entire facility related to the mass driver operation — the nuclear waste processing and fabricating plant, power plant, energy storage, etc. — could be located alongside the launcher on the floating platform. Safety and environmental constraints, as well as operational and maintenance problems, would be eased greatly by this means. In addition, equatorial positions requiring the minimum launch velocity become accessible with little restriction. The technology related to submerging a long tube into a deep ocean and maintaining its alignment, similar to that used currently in deep-sea oil drilling, is being improved further for deep-ocean mineral mining.

Analysis

Principles of Ablation

For the hemisphere-cylinder configuration under study, ablation is expected to occur almost entirely in the hemispherical nose region. Ablation characteristics of a hemisphere have been studied for the flow conditions of the entry flights into large planets; the flow conditions of the present case are similar.[8,9] Such studies indicate that when the flow is laminar, the ablation rates over the hemispherical surface are proportional to $\cos(x/R)$, where x is the circumferential distance measured from the stagnation point and R is the nose radius.[8] An approximate total ablation rate value can be obtained for such a case, therefore, by multiplying the stagnation-point value by the cross-sectional area of the hemisphere-cylinder. In reality, however, turbulence is likely to develop a short distance away from the stagnation point. The turbulence will tend to increase the ablation rates in the downstream region and hence the cosine distribution will be modified; however, studies indicate that the increase in the ablation rate due to turbulence is small for the severe entry environments[9] because energy transport occurs mostly in the radiative mode rather than the convective mode. Hence, ablation will be calculated here assuming the cosine distribution, under the implicit assumption that the flow is laminar everywhere. The question of turbulence is discussed later (see "Discussion").

The qualitative description of the ablation behavior is as follows: almost 80% of the ablation occurs below an altitude of 25 km. The temperature behind the normal shock wave formed in front of the nose varies from about 40,000 K at sea level to about 25,000 K at an altitude of 25 km. The stagnation-point pressure varies from about 5000 to 100 atm in the same altitude range. Air heated to such temperatures at such pressures is

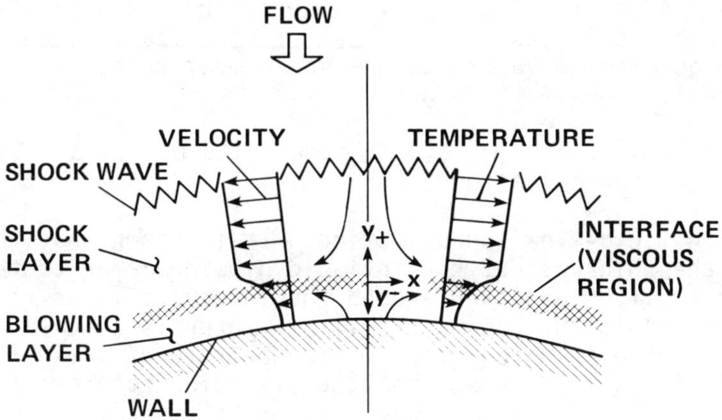

Fig. 2 Stagnation region flowfield.

expected to radiate immensely. Initially, when the nose is not ablating, its exterior surface will be heated by both radiation and convection. The tremendous heat-transfer rate from the air to the projectile's surface will induce ablation quickly, and a layer of ablation-product gases will be formed (see Fig. 2). Although the layer of gases will completely block the convective heat from reaching the wall, radiative heat will penetrate the gas layer; however, the radiative intensity will be diminished because of absorption by the layer. The higher the opacity of the layer, the lower the radiative heat flux reaching the wall.

The rate of ablation at the stagnation-point $\dot{m}$ is governed by the energy conservation relation at the wall

$$q_w = \Delta H \dot{m} \tag{1}$$

where q_w is the wall heat flux and ΔH is the heat of vaporization of the heat-shield material. Spallation is neglected for the reason to be described later (see "Discussion"). In the regime of interest, the ablation rate is very high, so the flow field of the ablation-product gases behaves as an inviscid gas. The viscous-flow phenomena are confined only to the narrow region where the two inviscid layers interface. The flowfield is referred to as a "massive blowing" flow.[10,11] The layer of the ablation-product gases is termed "blowing layer."[11]

To minimize the ablation rate $\dot{m}$ in Eq. (1), the heat-shield material must have a high vaporization energy and high opacity in its gaseous state. In addition, it must be sublimable so as not to flow away in liquid form. Graphite — which is sublimable, has a high heat of vaporization, and is highly opaque in its vapor state — would meet these requirements.

Analysis of the flowfield is simplified by use of the following two approximations:

1) Inviscid-flow approximation: It is known that the ratio of the thickness of the inviscid blowing layer to that of the viscous-diffusion layer around the interface is of the order of the "blowing parameter" $|f_w|$, which is a measure of the ratio of the ablation rate $\dot{m}$ to the rate of laminar diffusion.[10] It is known also that the diffusion rate is of the order of

$$\text{diffusion rate} \cong \rho_\infty V_\infty / \sqrt{Re}$$

where ρ_∞ is the freestream density, V_∞ is the freestream velocity, and Re is Reynolds number based on freestream properties and nose radius. Hence

$$\text{blowing layer thickness/viscous layer thickness} = |f_w| \cong (\dot{m}/\rho_\infty V_\infty)\sqrt{Re} \qquad (2)$$

As shown later, this ratio is of the order of 100 in the present problem. For this reason, the viscous-diffusion layer can be neglected; the entire flowfield is considered to be inviscid, and energy transport is assumed to occur in the radiative mode only.

2) Optically thick (Rosseland) approximation: Figure 3 shows the Rosseland mean opacity values α of air and carbon vapor as functions of temperature and pressure. In the figure, the reference density ρ_r and pressure p_r are 1.23×10^{-3} g/cm^3 and 1 atm, respectively. The opacity values of air are taken from the work of Aroeste and Magee;[12] the carbon values are calculated in the present work using the computer program described by Arnold et al.[13]

For carbon vapor, the Rosseland mean-free path $1/\alpha$ is of the order of 0.01 mm at 1000 atm. As will be shown later, the thickness of the blowing layer is of the order of 1 mm at this pressure, and hence the condition of an optically thick radiation field is satisfied at this pressure. At lower pressures, the Rosseland mean-free path becomes larger; but the thickness of the blowing layer also becomes larger, though less rapidly.

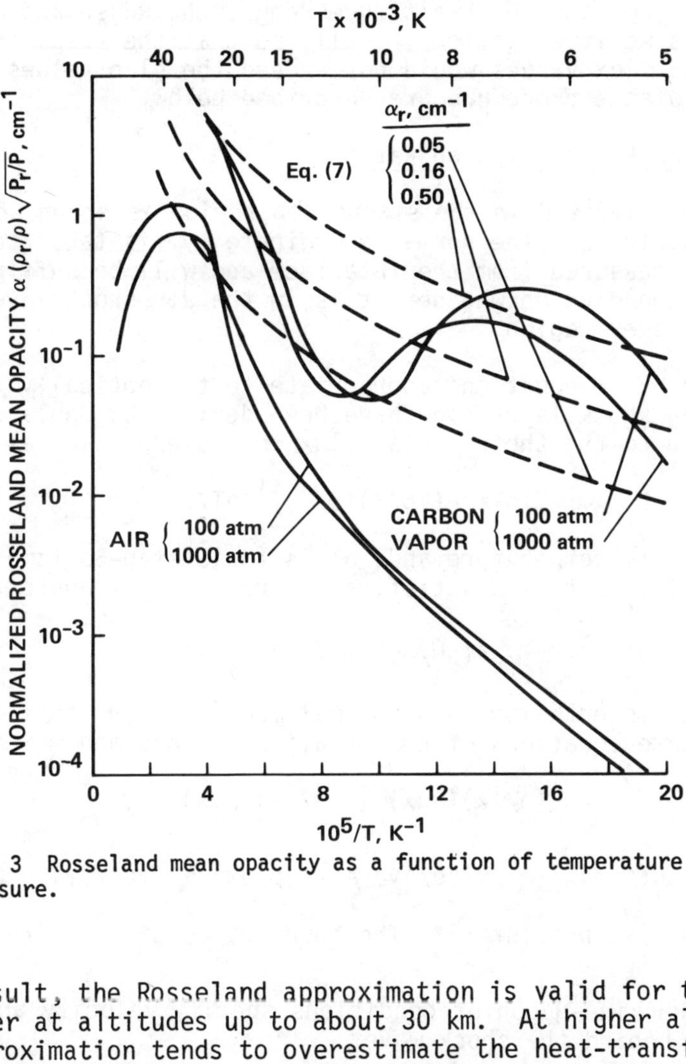

Fig. 3 Rosseland mean opacity as a function of temperature and pressure.

As a result, the Rosseland approximation is valid for the blowing layer at altitudes up to about 30 km. At higher altitudes, the approximation tends to overestimate the heat-transfer rates to the wall. However, since the absolute values of the heat-transfer rates are small in the high-altitude range, the approximation is used for altitudes above 30 km in the present work, that is, up to 40 km.

For the air layer, the Rosseland approximation is valid up to about 20 km. As will be shown later, the approximation substantially overestimates the heat fluxes at the interface at higher altitudes: the calculated interface heat fluxes become larger than the values for a uniform slab, the theoretical maximum. For this high-altitude range, the Rosseland approximation is modified semiempirically so that the resulting interface heat flux values would not exceed the slab values. The details of the procedure are described below.

Stagnation-Point Ablation Rates

The flowfield in the stagnation region is shown schematically in Fig. 2. The normal coordinate y is taken to be the distance measured from the interface and will be suffixed (+) or (-) depending on whether it is in the air shock layer or the blowing layer region.

The basic equations appropriate to the optically thick radiating shock layer flow have been derived by Goulard.[14] According to the theory, heat flux q is expressed as

$$q = -(16/3)(\sigma/\alpha)T^3(\partial T/\partial y)$$

where T is temperature and σ is the Stefan-Boltzmann constant. The heat flux satisfies the radiative transport equation

$$\rho v(\partial H/\partial y) + (\partial q/\partial y) = 0 \qquad (3)$$

where H is enthalpy. The normal momentum ρv satisfies the approximate equations of conservation of mass and momentum:

$$(\partial/\partial x)(\rho ux) + (\partial/\partial y)(\rho vx) = 0 \qquad (4)$$

$$\rho u(\partial u/\partial x) + \rho v(\partial u/\partial y) = -(\partial p/\partial x), \quad (\partial p/\partial y) = 0 \qquad (5)$$

where p is pressure.[14] The boundary conditions are as follows:

1) Rankine-Hugoniot conditions specify velocity and temperature behind the shock wave.

2) Temperature and heat flux are continuous across the interface.

3) At the interface, the blowing layer flow is accelerated in the tangential direction only by the tangential pressure gradient.

4) Tangential velocity is zero at the wall.

5) The energy balance condition, Eq. (1), is satisfied at the wall.

Solutions to the dynamic equations (4) and (5), subject to the boundary conditions (c) and (d), are obtained by setting the blowing parameter $|f_w|$ to infinity in the solutions given by Libby[10] and by Liu and Nachtsheim.[11] In terms of the reduced normal coordinates,

$$\eta_+ = \int_0^s \rho dy / \int_0^s \rho dy \ , \quad \eta_- = \int_0^w \rho dy / \int_0^w \rho dy$$

where s denotes the shock wave and w the wall; the normal momentum becomes approximately

$$\rho v_+ = \rho_s v_s \eta_+ \ , \quad \rho v_- = -2\eta_-[1 - (1/2)\eta_-]\dot{m}$$

With ρv thus expressed in terms of η, the energy transport Eq. (3) becomes a second-order nonlinear ordinary differential equation. It is made linear here by introducing the following approximations:

1) Enthalpy H is assumed to be related to T by

$$H = H_r(T/10{,}000)^n \tag{6}$$

The reference enthalpy H_r and the exponent n, which are functions of pressure only, are determined by first solving the equation of state at several thermodynamic conditions and then fitting a curve through the resulting data, for both air and carbon vapor. The thermodynamic conditions of air are taken from the work of Fenter,[15] and the carbon-vapor conditions are calculated in the present work using the thermodynamic functions given by the JANAF tables.[16] In the regime of interest, the exponent n is found to be about 1.5 for both air and carbon vapor. For air, H_r is found to be expressible as

$$H_r = 3.6 \times 10^{11}[1 - (1/6)\log(p/100)] \text{ erg/g}$$

where p is in atmospheres.

2) The Rosseland mean opacity α is expressed as

$$\alpha = \alpha_r(\rho/\rho_r)\sqrt{p/p_r}(T/10{,}000)^{4-n}, \text{ cm}^{-1} \tag{7}$$

In Fig. 3, Eq. (7) is shown for $n = 1.5$ and compared with the computed values of α. For air, the temperatures pertinent to the present work range between about 20,000 K and 40,000 K. For this range, the computed α values can be approximated fairly closely by Eq. (7) for $\alpha_r = 0.05$ cm^{-1}. The errors arising from this approximation for air are small, as will be shown later (see "Discussion"). For carbon vapor, the pertinent range of temperatures is between 5000 K and 25,000 K. The α values for carbon vapor lie approximately between those corresponding to $\alpha_r = 0.05$ and $0.5\ \mathrm{cm}^{-1}$. The present calculations are carried out using the α_r values varying between these two limits.

The linearization leads to the solution of Eq. (3) of the form:

$$H_+ = H_i + (H_s - H_i)\mathrm{erf}(\sqrt{A_1/2}\eta_+)/\mathrm{erf}(\sqrt{A_1/2})$$

$$H_- = H_i + (H_w - H_i)F_1(A_2,\eta_-)/F_1(A_2,1)$$

where

$$F_1 = \int_0 \exp\{-A_2[(1/2)\eta_-{}^2 - (1/6)\eta_-{}^3]\}d\eta_-$$

Subscript i here refers to the interface. The constant A_1 is a function only of the freestream properties, and A_2 is a function of the freestream properties, wall temperature T_w, interface temperature T_i, and ablation rate $\dot{m}$. After the linearization and after insertion of the known constants, the radiative heat fluxes at the interface are found to be expressible approximately, for the interested range, according to the Rosseland approximation as

$$q_{i_1} = 43.5(\rho_r/\rho_s)^{1/4}V_\infty{}^2[1 - (T_i/T_s)^{1.5}]/\{\sqrt{1 - \log(p/100)/6}\ \sqrt{R}\ \mathrm{erf}(\sqrt{A_1}/2)\}\mathrm{erg/cm^2\text{-}s} \quad (8)$$

where ρ_s and T_s are the after shock density and temperature given by the Rankine-Hugoniot relationships, and V_∞ is in cm/s. The interface temperature T_i in Eq. (8) is determined by imposing the boundary conditions at the interface and the wall, as explained below.

In Fig. 4, the heat fluxes given by Eq. (8), with T_i determined by the methods described below, are compared with the blackbody heat fluxes and with the values for a uniform slab thickness equal to that of the shock layer over a hemisphere of 10 cm radius. The values for the slab are obtained using the work of Aroeste and Magee.[12] All values are for the aftershock conditions corresponding to the flight velocity of 17 km/s. As seen in Fig. 4, Eq. (8) results in heat flux values lower than those of either the blackbody or the slab

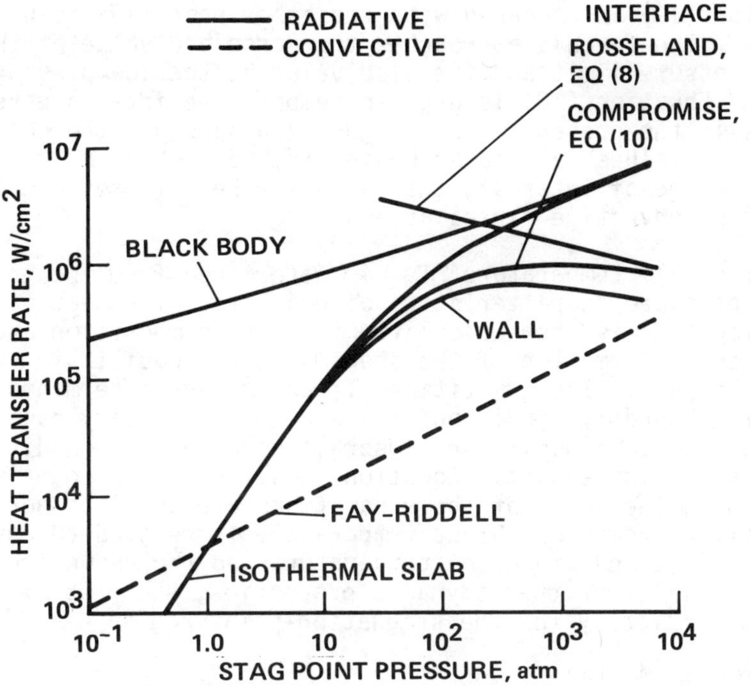

Fig. 4 Stagnation point heat-transfer rates: $R = 10$ cm, $V_\infty = 17$ km/s.

when the pressure is greater than about 300 atm. This is expected; according to the Rosseland approximation, a "thermal boundary layer" develops at the interface and the interface temperature T_i becomes substantially lower than the Rankine-Hugoniot temperature T_s, resulting in a low heat flux. At lower pressures, however, Eq. (8) exceeds the blackbody and the slab values, in obvious violation of thermodynamic laws. This anomaly is caused by the fact that the regime is no longer amenable to the Rosseland approximation. For this low-pressure region, therefore, a semiempirical modification is introduced as follows. The slab heat flux is expressed first in the form

$$q_{i_2} = 5.679 \times 10^{-5} T_s^4 \exp[-25.45(pR)^{-0.545}] \text{ erg/cm}^2\text{-s} \quad (9)$$

The expression agrees with the values of Ref. 12 to within about ±3% at pressures above 10 atm, which occurs at the altitude of 40 km. The interface heat flux q_i is taken to be the harmonic average of the Rosseland value, Eq. (8), and the slab value, Eq. (9):

$$q_i = q_{i_1} q_{i_2} / (q_{i_1} + q_{i_2}) \quad (10)$$

Equation (10) is compared with the other heat flux values in Fig. 4. The formula approaches the Rosseland value at the high-pressure limit and the slab value at the low-pressure limit. Equation (10) is used to compute the interface heat fluxes in the present work. Figure 4 also shows the convective heat flux values calculated by the method of Fay and Riddell.[17] In the range of interest, the radiative heat fluxes are larger than the convective fluxes as assumed.

The wall temperature T_w is determined from the known vapor pressure characteristics of graphite. The heat of sublimation ΔH is then calculated from the composition and the energies of formation of the species in the equilibrium vapor over the wall. The quantities T_i and $\dot{m}$ are determined by applying boundary conditions (2) and (5). Boundary condition (2) reduces to a nonlinear algebraic equation and condition (5) becomes a transcendental equation involving an exponential function. The roots of these equations are obtained here by an iterative procedure. Since temperature is related to enthalpy via Eq. (6), and since the temperature and the known pressure specify all other thermodynamic properties, the problem is solved completely for the stagnation-point region.

Equation of Motion

The motion of the projectile satisfies the equation

$$M(dV/dt) = -Mg - C_D(\rho_\infty/2)V_\infty{}^2\pi R^2 \tag{11}$$

where M is the mass of the projectile, q the gravitational constant, t is time, and C_D is the drag coefficient. For the hemisphere-cylinder under consideration, drag is caused almost entirely by the hemispherical nose, for which C_D is 1. Hence the present work adopts the C_D value

$$C_D = 1 \tag{12}$$

To determine the freestream density ρ_∞, one must know the altitude h; it is obtained by integrating the equation

$$(dh/dt) = V_\infty \tag{13}$$

As stated before, the present work assumes that the total ablation rate is a product of the stagnation-point ablation rate and the cross-sectional area of the projectile. Hence, the mass of the projectile satisfies the equation

$$(dM/dt) = -\pi R^2\dot{m} \tag{14}$$

Equations (1-3) are solved numerically on a CDC 7600 computer. Calculation is made backward in time, starting

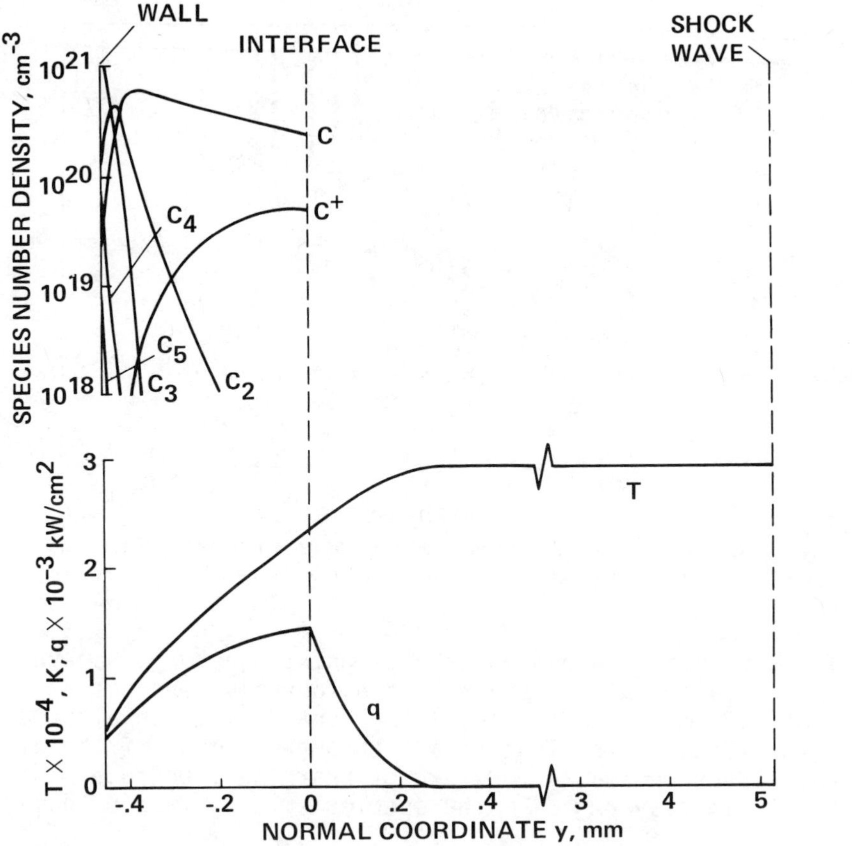

Fig. 5 Variation of preperties in stagnation region: $\alpha_r = 0.158$ cm^{-1}, R = 10 cm, M_f = 1 ton, h - 11.64 km, p_s = 1000 atm.

from the altitude of 40 km where the vertical velocity V_∞ is specified as 17 km/s, the velocity required for escape from the solar system. The integration is marched in the negative time direction from the 40 km altitude to sea level. The mass loss and velocity drop are then determined by subtracting the values at 40 km from the values at sea level.

Results

The results of the calculations are shown in Figs. 5 through 9. Figure 5 shows the variation of thermodynamic properties along the normal coordinate y in the stagnation region at one point on the flight trajectory. As shown, the thickness of the blowing layer and air shock layer are 0.45 and 5.2 mn, respectively, at the point. Examination of other calculated cases reveals that the air layer thickness remains

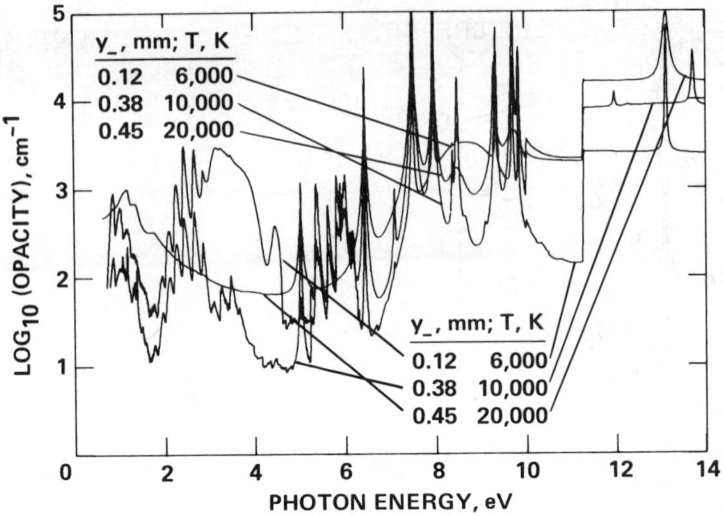

Fig. 6 Opacity of ablation product gases in blowing layer as a function of photon energy, at conditions of Fig. 5.

almost constant throughout the flight trajectory; however, the blowing layer thickness varies considerably, increasing from about 0.2 mm at the sea level to about 2 mm at an altitude of 30 km. Temperature is constant throughout most of the air shock layer region ($y > 0$) except near the interface. Within the blowing layer ($y < 0$), the species C_3, C_2, and C dominate.

Figure 6 shows the variation of opacity of the blowing layer with photon energy at three points within the layer shown in Fig. 5. At all three conditions, the photon energy range above 11.5 eV is highly opaque. At 6000 K, the continuous absorptions centered at 3.5 and 8.5 eV, which are due to C_3, provide high opacity. At 20,000 K, the nondistinct continuum resulting from atomic carbon provides high opacity. At 10,000 K, however, low opacity values occur at about 1.8, 4.5, and 6.5 eV because neither C_3 nor C is present in sufficient concentration. This is the cause of the dip in the Rosseland mean opacity value at about 10,000 K (Fig. 3).

Figures 7a through 7c show typical variations of properties along the flightpath. As shown in Fig. 7a, the projectile reaches the 40 km altitude in 2.3 s. This is a very short time compared with those in other entry flight problems. In the present problem, the short duration of the heating period is a contributing factor to the relatively small loss of mass due to ablation. Mass change occurs more or less linearly with time,

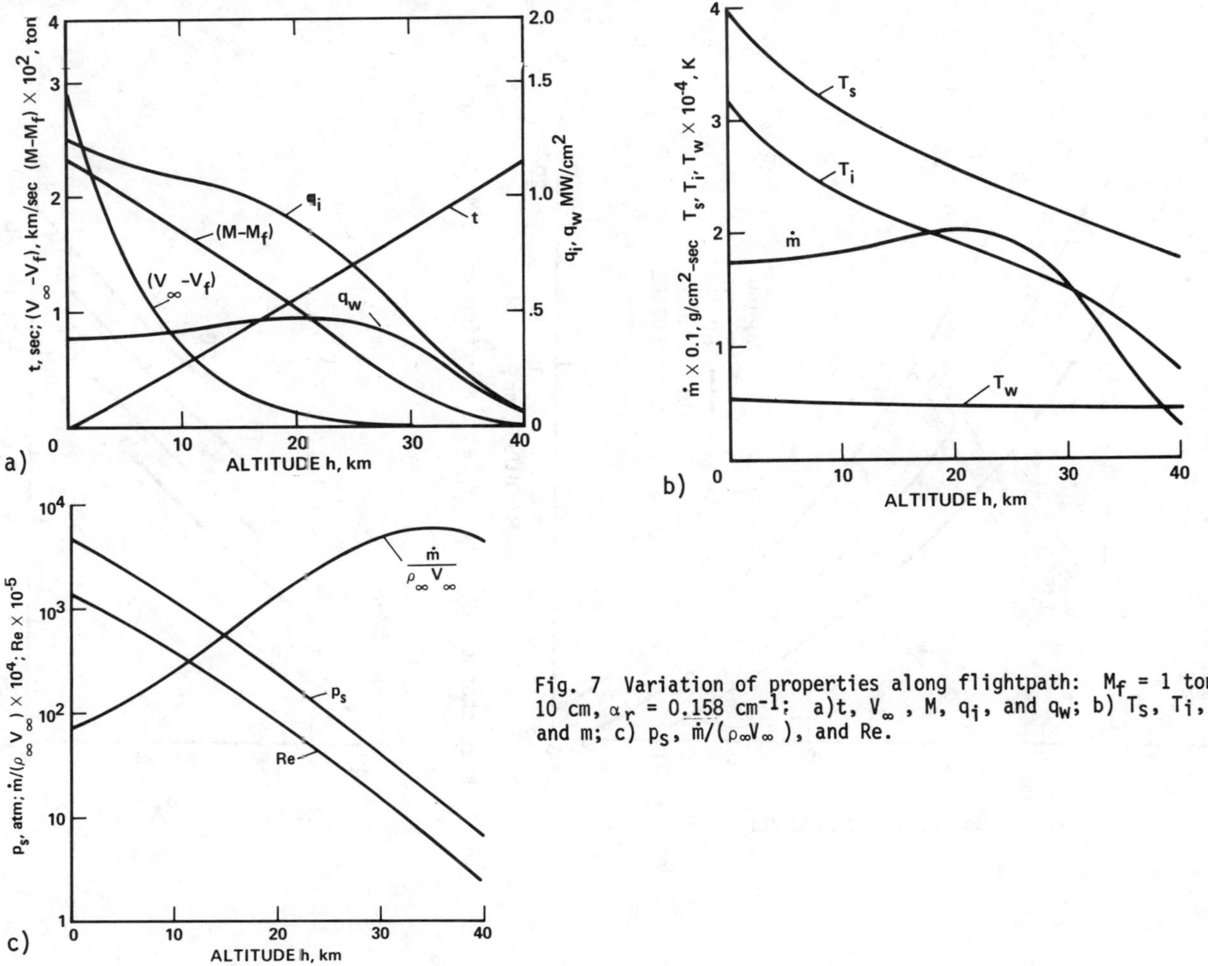

Fig. 7 Variation of properties along flightpath: M_f = 1 ton, R = 10 cm, α_r = 0.158 cm^{-1}; a) t, V_∞, M, q_i, and q_w; b) T_s, T_i, T_w, and m; c) p_s, $\dot{m}/(\rho_\infty V_\infty)$, and Re.

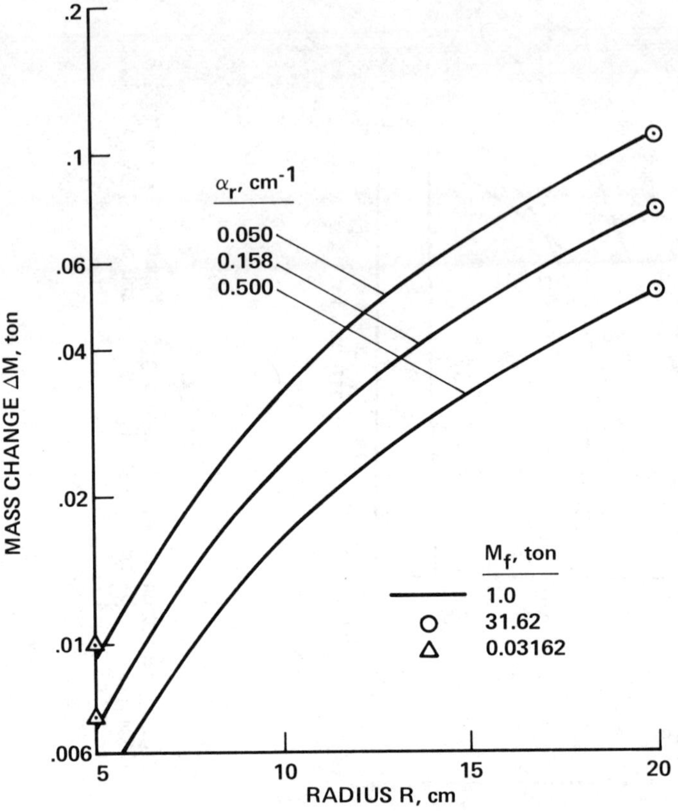

Fig. 8 Mass change $\Delta M - M_f - M_0$ as a function of R, M_f and α_r.

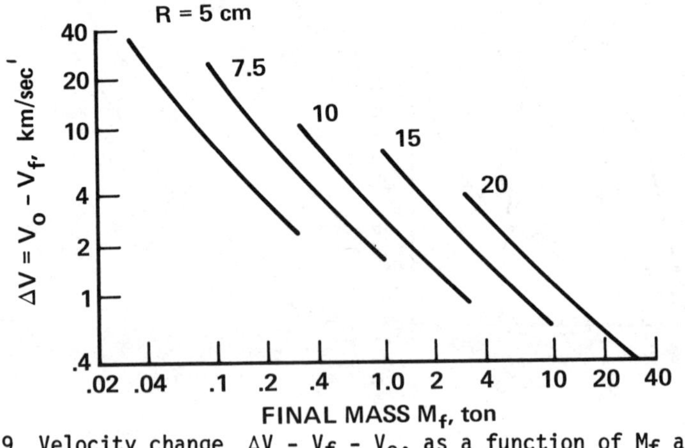

Fig. 9 Velocity change $\Delta V - V_f - V_0$, as a function of M_f and R; $\alpha_r = 0.158\ cm^{-1}$.

but velocity change is appreciably nonlinear; half the total velocity change occurs below an altitude of 6 km. One sees from this that there is a significant advantage in installing the launcher at the top of a high mountain. The heat fluxes q_i and q_w both approach zero at high altitudes as expected. It is interesting to note that the wall heat flux q_w increases with altitude below 20 km. This is because the radiative blockage by the blowing layer becomes less effective at low densities.

Figure 7b shows the temperatures T_s, T_i, T_w, and the stagnation point ablation rate $\dot{m}$. Temperatures T_s and T_i mostly range from about 20,000 to 40,000 K. Referring to Fig. 3, one sees that the air layer remains in the temperature range where Eq. (7) approximately reproduces the computed Rosseland mean opacity values. The ablation rate $\dot{m}$ increases with altitude below 20 km, corresponding to the increase in q_w.

Figure 7c shows the stagnation point pressure p_s, the ratio $\dot{m}/(\rho_\infty V_\infty)$, and Reynolds number Re based on freestream properties and nose radius. The stagnation-point pressure ranges from 10 to 5,000 atm and $\dot{m}/(\rho_\infty V_\infty)$ varies from about 0.01 to 1. Reynolds number is in the range of 3×10^5 to 10^8. The ratio $\sqrt{Re}\,\dot{m}/(\rho_\infty V_\infty)$ in Eq. (2) varies from about 30 to about 1,000; thus, neglecting the viscous-diffusion layer between the two inviscid layers is justified (see "Principle of Ablation").

In Fig. 8, the mass change ΔM is shown as a function of nose radius R, final mass M_f, and the reference opacity α_r. Large R and small α_r cause large mass change. This is understandable: a large radius provides a large frontal area subject to ablation, while a small α_r provides a small radiation blockage in the blowing layer. The mass change is virtually independent of the final mass M_f. In the range considered, the absolute mass change ΔM varies from about 0.005 to about 0.1 tons, or about 1% to 10% of the total mass for a projectile mass of 1 ton. The mass loss values are generally in the range tolerable for the purpose of nuclear waste disposal.

In Fig. 9, the velocity change ΔV is shown as a function of final mass M_f and radius R. Here, the effect of varying α_r is not shown because ΔV is found to be virtually independent of α_r. The velocity change is a fairly strong function of both M_f and R; a large M_f and a small R lead to a small velocity change. In the computed range, the ΔV value varies from about 0.4 to 30 km/s. Since the energy required for launch increases as the square of the launch velocity, the large ΔV values would penalize the operation quite heavily. The figure thus demonstrates the importance of properly selecting the two parameters in the design.

Discussion

As stated in the discussion of the principles of ablation, the present calculations implicitly assume laminar flow in the viscous-diffusion layer around the interface between the two inviscid layers. As shown in the preceding paragraphs, the thickness of the laminar viscous layer is less than 3% of the blowing layer thickness. Even if turbulence is present, the inviscid flow assumption should be valid because it is unlikely that the thickness of the layer could be increased by a factor greater than 30 by the turbulence. As long as the layer remains thinner than the blowing layer, the radiative transport phenomenon that dominates the energy transport process will not be affected.

The calculated mass loss due to ablation is small compared with losses expected in comparable problems of entry flights into the large planets. For instance, the entry flight into Jupiter is expected to cause a mass loss of about 25% due to ablation. The relatively small mass loss in the present problem can be attributed to the following:

1) The high pressures in the stagnation region provide a dense blowing layer of the ablation product gases which effectively absorbs the radiation from the hot air layer. The tendency is demonstrated in Fig. 7b, in which it is shown that the stagnation-point ablation rate tends to decrease with decreasing altitudes at altitudes below 20 km. The phenomenon is apparent also in Fig. 4, which shows q_w decreasing with pressure at pressures above 300 atm.

2) The duration of the heating period is quite short.

Figure 9 shows that velocity loss due to drag may be a serious problem in the design of the present system. One reason for such a large velocity loss is a result of the assumption of a spherical nose. Drag can be reduced by shaping the nose into a slender hemisphere-cone. Although the spherical nose tip of such a hemisphere-cone may ablate quickly and reduce the overall shape nearly to the hemisphere-cylinder considered in the present work, by then the projectile would have traveled through the lower atmosphere where the greatest velocity loss occurs.

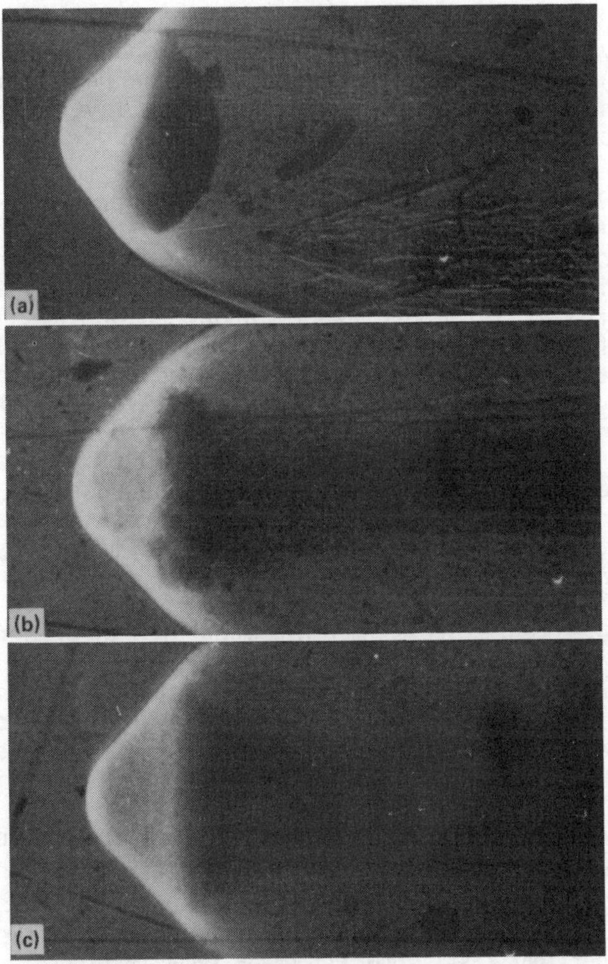

Fig. 10 Shadowgraphs of flow around a model flying through xenon where ambient pressure - 100 Torr; flight velocity ≅ 4.3 Km/s: a) graphite; b) reinforced carbon-carbon composite; c) carbon-phenolic.

Precursor heating, ignored in the present analysis, would also reduce drag. In the present flight regime, the free-stream air ahead of the bow shock wave will be irradiated strongly by the hot shock layer, causing a rise in temperature and pressure there. As a result, the air mass will move away laterally from the path of the projectile. The resulting flowfield will resemble that of a subsonic flow. The drag coefficient of the hemisphere will decrease as a result, perhaps to as low as 0.5, the subsonic limiting value. Velocity change will be lessened accordingly. One might note there that the precursor phenomenon will not affect ablation characteristics appreciably; the loss of energy from the shock layer through irradiation will tend to lower the ablation rate, but the accompanying drop in pressure in the shock layer may tend to increase it.

Of several approximations made in the present work, seemingly the least justifiable is the use of Eq. (7), with $\alpha_r = 0.05\ cm^{-1}$, for air. Figure 3 shows that the analytical approximation overestimates the computed values by as much as a factor of 3 at a temperature of about 20,000 K. To investigate the effect of this discrepancy, the present calculations were repeated with the α_r value changed by a factor of 10 to $\alpha_r = 0.005\ cm^{-1}$. The resulting mass-loss values were only 15% higher than the values shown in Figs. 5 through 9. Thus, the opacity of air has little effect on the ablation characteristics in the present flight environments because the air layer is close to being optically black. As a result, the radiation power emitted by the air layer is dictated mainly by the temperature and only weakly by its opacity.

Another important assumption made in the present analysis concerns spallation. Ablation rates are determined in the present work using Eq. (1), which neglects the possible contribution of spallation. Lundell and Dickey,[18] among others, studied the ablation characteristics of graphite, reinforced carbon-carbon composite, and carbon-phenolic under intense radiation, that is, of heat fluxes up to 1000 kW/cm^2. The study showed that at such high heat fluxes, graphite and reinforced carbon-carbon composite produce little spallation. The

Table 1 Calculated environmental parameters of the freeflight tests: ambient gas = xenon; ambient pressure = 100 Torr

Parameter	Run number		
	1543	1544	1545
Model material	Graphite	Carbon	Carbon-phenolic
Flight velocity, km/s	4.30	4.43	4.14
Stagnation pressure, atm	140	149	130
Interface radiative heat flux, kW/cm^2	388	429	331
Shadowgraph, figure	10a	10b	10c

contribution of spallation to the overall ablation rate for graphite was smaller than the uncertainties involved in the experiment.

Recently, Chul Park conducted an ablation experiment in the freeflight range. Hemisphere-cone shaped models, with nose radii of 0.476 cm, frustum diameters of 1.905 cm, and cones angles of 45 deg, made of graphite, reinforced carbon-carbon composite, and carbon-phenolic, were flown through xenon at an ambient pressure of 100 Torr. In Figs. 10a through 10c, the shadowgraph pictures of the flow around the models are shown for three typical cases. The calculated environmental parameters are listed in Table 1. As indicated in the table, the interface heat fluxes, which are calculated using the method described by Arnold et al.,[13] were about 400 kW/cm^2. Stagnation pressures were about 140 atm, which corresponds to an altitude of about 23 km in the flight regime of the mass driver launched projectiles. Figures 10b and 10c show a dark trail behind the model, which is attributed to the solid particulates produced by spallation. According to the figures, the carbon-carbon composite produces less spallation than carbon-phenolic, and graphite produces no spallation. The freeflight tests thus confirm, at least qualitatively, the findings of the laser experiments of Lundell and Dickey.[18]

One can conclude from this study that ablation is likely to be always in the range that can be tolerated for the proposed scheme for disposal of nuclear wastes outside the solar system. Deceleration can be in the tolerable range also, if the mass and diameter of the projectile are chosen properly. However, there remain many practical problems other than those of ablation and deceleration, even excluding problems associated with construction of the launcher itself. Included among those problems are the following: the enormous drag force applied to the nose region will tend to yaw the projectile; the sudden heating of the nose within the first milliseconds of flight will cause a severe thermal shock; and the cylindrical portion of the projectile, which will be heated by radiation (although not as intensely as the nose region), will require some thermal protection. Such problems await further study.

Conclusions

If a hemisphere-cylinder-shaped projectile protected thermally with a graphite nose is launched vertically upward to attain a velocity of 17 km/s at 40 km altitude, the mass loss due to ablation during atmospheric flight will be less than

0.1 ton, provided the radius of the projectile is under 20 cm and the mass of the projectile is of the order of 1 ton. The velocity loss due to drag will vary from about 0.4 to 30 km/s, depending on the mass and radius of the projectile, the smaller velocity loss corresponding to large mass and small radius. Ablation is always within a tolerable range for a scheme aimed at using a mass driver launcher to dispose of nuclear wastes outside the solar system. Deceleration can also be in the tolerable range if the mass and diameter of the projectile are properly chosen.

Acknowledgment

The authors thank Dr. Jin S. Chung of Lockheed Missile and Space Co., Ocean Minerals Program, Mountain View, California, for suggestions and discussions relating to positioning the mass driver system on a floating platform.

References

[1]Clarke, A. C., "Electromagnetic Launching as a Major Contribution to Space Flight," Journal of the British Interplanetary Society, Vol. 9, No. 6, Nov. 1960, pp. 261-267.

[2]O'Neill, G. K., "The Colonization of Space," Physics Today, Vol. 27, No. 9, Sept. 1974, pp. 32-40.

[3]Arnold, W. H., Bowen, S., Fine, K., Kaplan, D., Kolm, M., Kolm, H., Newman, J., O'Neill, G. K., and Snow, W. R., "Mass Drivers, I. Electrical Design," Space Resources and Space Settlements, edited by J. Billingham, W. Gilbreath, and B. O'Leary, NASA SP-428, 1979, pp. 87-100.

[4]Kolm, H., Fine, K., Mongeau, P., and Williams, F., "Electro Magnetic Propulsion Alternatives," paper presented at the 1979 Princeton Symposium on Space Manufacturing, Princeton, N.J., May 1979.

[5]Priest, C. C., Nixon, R. F., and Rice, E. E., "Space Disposal of Nuclear Wastes," Astronautics and Aeronautics, Vol. 18, No. 4, April 1980, pp. 26-35.

[6]Ruppe, H. O., Hayn, D., Braitinger, M., and Schmucker, R. H., "A Nuclear Waste Deposit in Space — The Ultimate Solution for Low-Cost and Safe Disposal," IAF Paper No. 79-A-15, 30th International Astroautical Congress, Sept. 17-22, 1979, Munich.

[7]Salkeld, R. and Beichel, R., "Nuclear Waste Disposal in Space — Implications of Advanced Space Transportation," IAF Paper No. 79-51, 30th International Astronautical Congress, Sept. 17-22, 1979, Munich.

[8]Sutton, K., Moss, J. M., Falanga, R. A., and Olstad, W. B., "Outer Planet Entry Probes Aerothermal Environment — Status of Prediction Methodology," AIAA Paper 75-1148, 1975, St. Louis.

[9]Moss, J. N., Anderson, E. C., and Simmonds, A. L., "The Impact of Turbulence on a Radiating Shock Layer," AIAA Paper 78-1186, 1978, Seattle.

[10]Libby, P. A., "The Homogeneous Boundary Layer at an Axisymmetric Stagnation Point with Large Rates of Injection," Journal of the Aerospace Sciences, Vol. 29, No. 1, Jan. 1962, pp. 48-60.

[11]Liu, T. M. and Nachtsheim, P. R., "Shooting Method for Solution of Boundary-Layer Flow with Massive Blowing," AIAA Journal, Vol. 11, Nov. 1973, pp. 1584-1586.

[12]Aroeste, H. and Magee, J. L., Thermal Radiation Phenomenon, Vol. 3, Tables of Radiative Properties of Air, AD 654-773, Washington, D.C., 1976.

[13]Arnold, J. O., Cooper, D. M., Park, C., and Prakash, S. G., "Line-by-Line Transport Calculations for Jupiter Entry Probes," AIAA Paper 79-1082, 1979, Orlando.

[14]Goulard, R., "Preliminary Estimates of Radiative Transfer Effects on Detached Shock Layers," AIAA Journal, Vol. 2, Mar. 1964, pp. 494-502.

[15]Fenter, F. W., "The Thermodynamic Properties of High Temperature Air," Chance Vought Research Center, Dallas, Texas, Rept. RE-IR-14, 1971.

[16]Stull, D. R. and Prophet, H., JANAF Thermochemical Tables, 2nd ed., NSRDS-NBS 37, National Bureau of Standards, Washington, D.C., 1971.

[17]Fay, J. A. and Riddell, F. R., "Theory of Stagnation Point Heat Transfer in Dissociated Air," Journal of the Aeronautical Sciences, Vol. 25, No. 2, Feb. 1958, pp.73-85, 121.

[18]Lundell, J. H. and Dickey, R. R., "Response of Heat-Shield Materials to Intense Laser Radiation," Outer Planet Entry Heating and Thermal Protection, Progress in Astronautics and Aeronautics, Vol. 64, edited by R. Viskanta, AIAA, New York, 1979, pp. 3-21.

Chapter III Space Shuttle Studies

Approximate Heating Analysis for the Windward Symmetry Plane of Shuttle-like Bodies at Large Angle of Attack

E.V. Zoby*
NASA Langley Research Center, Hampton, Va.

Abstract

An engineering method has been developed for computing the windward symmetry plane convective heat-transfer rates on Shuttle-like vehicles at large angles of attack. The engineering code includes an approximate inviscid flowfield technique, laminar and turbulent heating rate expressions, an approximation to account for the variable entropy effects on the surface heating, and the concept of an equivalent axisymmetric body to model the windward-ray flowfields of Shuttle-like vehicles at angles of attack of 25-45 deg. The engineering method is validated by comparing computed heating results with corresponding experimental data measured on the Shuttle and other advanced transportation models over a wide range of flow conditions and angles of attack of 25-40 deg and also with results of existing prediction techniques. The comparisons are in good agreement.

Nomenclature

c_1-c_5	=	defined by Eqs. (8-12)
C_f	=	skin friction coefficient
h	=	enthalpy
H	=	heat-transfer coefficient
H_{REF}	=	stagnation point heat-transfer coefficient to a 0.3048 m scaled nose radius

Presented as Paper 81-1042 at the AIAA 16th Thermophysics Conference, Palo Alto, Calif., June 23-25, 1981.

*Aero-Space Technologist, Aerothermodynamics Branch, Space Systems Division.

K = defined by Eq. (1)
L = model length
M = Mach number
m = exponent in friction law
N = reciprocal exponent in velocity profile power law
p = pressure
Pr = Prandtl number
$\dot{q}$ = heat-transfer rate
R = unit Reynolds number
R_θ = momentum thickness Reynolds number
R_N = nose radius
r = radius of body of revolution
S = body wetted length
T = temperature
U = velocity
X = axial measurement of model
Y = distance normal to wall
α = angle of attack
δ = boundary-layer thickness
δ^* = displacement thickness
μ = viscosity
ρ = density
θ = momentum thickness
θ_H = hyperboloid half angle

Subscripts and Superscripts

aw = adiabatic wall
c = convective heating rate
e = edge conditions
∞ = freestream conditions
L = laminar
* = conditions determined by reference enthalpy
s = stagnation point condition
t = turbulent
w = wall condition
REF = stagnation point value based on 0.3048 m scaled nose radius

Introduction

The development of the Space Shuttle as well as possible missions using advanced transportation systems have resulted in additional problems for convective heating analyses. Such problems as complex three-dimensional and/or variable entropy flows and large angle-of-attack (AOA) conditions are now encountered. Hence, the spacecraft designer needs some procedure to evaluate the impact of such problems on the vehicle design for parametric or actual design calculations.

Experimental wind-tunnel testing can be instrumental in defining the aerothermodynamic environment about Shuttle-like vehicles. For example, the Shuttle experimental heat-transfer data[1,2] have been used to verify the results of heating techniques[3-6] used in the Shuttle design studies and to complement the results of the heating methods in regions where the prediction techniques are not valid, for instance, the leeward surfaces.[7,8] However, experimental facilities cannot fully simulate the impact of certain entry flow conditions (real-gas effects and the influence of the inviscid entropy layer) on the test model. Thus computational analyses are needed.

Even with advances in high-speed computers and numerical techniques, recently developed viscous flowfield analyses[9-14] which have greatly enhanced the capabilities in computational fluid dynamics are still not capable of predicting the flowfields about Shuttle-like vehicles at rather large angles of attack. To date, the most detailed thermal analysis about such vehicles involves three-dimensional inviscid flowfield solutions coupled with laminar two-dimensional heating techniques.[15-20] The application of the two-dimensional heating methods[18-20] to three-dimensional flow conditions is based on the assumptions of the axisymmetric analog concept.[21] Several of these inviscid-viscous methods account for variable entropy edge conditions in surface heat-transfer calculations. The effect of variable entropy conditions on the heating rates is shown[19,20] to be more significant at flight conditions than at wind-tunnel conditions and to impact the predicted heating levels more than the assumption of a gas model. Unfortunately, these inviscid-viscous methods require a rather large amount of computational time and computer storage; consequently, these solution techniques are expensive to run and are not applicable generally to parametric or preliminary design studies. Engineering methods that have been substantiated by experimental data and/or detailed calculations are usually employed in the parametric-type studies.

The purpose of this paper is to present a rapid but reliable engineering method for predicting the windward-ray convective heat-transfer rates on vehicles similar to Space Shuttle at rather large angles of attack. Such a method should provide useful information for parametric or preliminary design studies. The engineering method incorporates a recently developed approximate equilibrium-air inviscid flowfield code[22] and a convective heating method[23] that includes an approximate technique to account for variable entropy effects on the surface heating rates. The windward ray of the desired body at a given angle of attack is modeled by an "equivalent axisymmetric body"[24] at zero angle of attack. A laminar heating analysis[24] which uses the equivalent body concept has been demonstrated to provide good comparison with experimental heat-transfer data obtained on a Shuttle model at a 30 deg AOA. The "equivalent axisymmetric body" concept is extended herein to other AOA conditions. The present engineering method is validated by comparing computed heating results with corresponding experimental data measured on Shuttle models[1,2] and advanced transportation models over a wide range of flow conditions and angles of attack of 25-40 deg and with results of existing prediction techniques.

Analysis

This section presents a discussion of the engineering heating method which has been developed to predict the convective heat transfer along the windward symmetry plane of Shuttle-like bodies at large angles of attack. Essential elements of the present engineering method, e.g., the inviscid flowfield calculation and the convective heating method, have been developed previously and were demonstrated to yield good comparisons with corresponding viscous shock layer (VSL)[25] calculations for Earth[22] and outer-planet conditions.[23] Those aspects of the present engineering method are outlined for completeness. The additional feature of the present method is the use of the "equivalent axisymmetric body" concept.[24] This concept assumes that the windward-ray flowfield of Shuttle-like bodies at large angles of attack can be modeled by an axisymmetric body at zero AOA. A discussion of the equivalent body concept is also presented.

Inviscid Flowfield Calculations

The inviscid flowfield technique is based on an inverse method of solution, i.e., analytic shock-shape equations are required such that the calculated body closely approximates the desired body configuration. Presently, two analytic expressions for bow shocks capable of generating body shapes closely approximating either spherically blunted cones or hyperboloids are used. The flowfield analysis is based on the Maslen[26] technique which permits an independent evaluation of the pressure through the shock layer. With the inclusion of radiative and convective heating procedures as well as correlations for the equilibrium thermodynamic and transport properties, the approximate flowfield technique has been demonstrated to provide reliable engineering calculations of the radiative and/or convective heat-transfer rates at Earth,[22] Venusian,[27] and outer-planet[23,28,29] entry conditions.

Approximate Heating Relations

Laminar and turbulent heating rate equations appropriate for engineering predictions of the convective heating rates about blunt re-entry spacecraft at hypersonic conditions have been reported.[23] The approximate methods were incorporated in the engineering flowfield code and were demonstrated to be applicable to both nonreacting and reacting gas mixtures for either constant or variable entropy edge conditions. In addition, the convective heating analysis has been extended to Earth entry conditions with recently developed equilibrium air thermodynamic correlations.[22]

For the stagnation point heat-transfer calculations, the equation of Sutton and Graves[30]

$$\dot{q}_{w,s} = K(p_s/R_N)^{\frac{1}{2}} (h_s - h_w) \tag{1}$$

is used. Stagnation point heat-transfer results computed with Eq. (1) are in good agreement with results computed by the Cohen method[31] at equilibrium-air conditions. Since the constant K in Eq. (1) can be determined by a simple but accurate method over a wide range of gas mixtures and is simple to program, the method of Sutton and Graves[30] is used in the approximate flowfield analysis rather than the more widely recognized Cohen[31] method.

The laminar heat-transfer distributions are computed by relating heat transfer to a skin friction relation based on R_θ through a modified form of Reynolds analogy. This approach was used because existing laminar heating methods are not applicable over wide ranges of gas mixtures. The approach was useful for approximating the variable entropy effects on the heat-transfer calculations and is used also in this investigation. The laminar heat transfer is computed using a standard incompressible skin friction relation with compressibility effects accounted for by

Eckert's reference enthalpy relation[32] and is given by

$$\dot{q}_{w,L} = 0.22(R_{\theta,e})^{-1}(\rho^*/\rho_e)(\mu^*/\mu_e)\rho_e U_e(h_{aw} - h_w)(Pr_w)^{-0.6} \quad (2)$$

where θ is computed by

$$\theta_L = 0.644\left(\int_0^S \rho^*\mu^* U_e r^2 \, dS\right)^{1/2} / (\rho_e U_e r) \quad (3)$$

The turbulent heat transfer is also computed by relating heat transfer to a skin friction relation based on R_θ through a modified form of Reynolds analogy. The skin friction relation is determined by assuming a velocity profile ($U/U_e = (Y/\delta)^{1/N}$) to compute the required information. Usually, the existing approximate turbulent methods, e.g., Refs. 33 and 34, assume a constant 1/7th power-law velocity profile. However, experimental results[35] show N to be a function of R_θ and the effect of a variable N on the skin friction has been demonstrated.[36] In order to incorporate this effect, the skin friction relation is written as

$$c_f/2 = c_1(R_\theta)^{-m} \quad (4)$$

With compressibility effects again accounted for by Eckert's reference enthalpy technique, the turbulent heating rate expression is

$$\dot{q}_{w,t} = c_1(R_{\theta,e})^{-m}(\rho^*/\rho_e)(\mu^*/\mu_e)^m \rho_e U_e(h_{aw} - h_w)(Pr_w)^{-0.4} \quad (5)$$

where θ can be written as

$$\theta_t = \left(c_2 \int_0^S U_e \rho^* \mu^{*m} r^{c_3} dS\right)^{c_4} / (\rho_e U_e r) \quad (6)$$

The coefficients and exponents (m, c_1, c_2, c_3, and c_4) in Eqs. (4-6) are functions of N and are given as

$$m = 2/(N + 1) \quad (7)$$

$$c_1 = (1/c_5)^{2N/(N + 1)}[N/(N + 1)(N + 2)]^m \quad (8)$$

$$c_2 = (1 + m)c_1 \quad (9)$$

$$c_3 = 1 + m \tag{10}$$

$$c_4 = 1/c_3 \tag{11}$$

$$c_5 = 2.24333 + 0.93N \tag{12}$$

A curve fit of the axisymmetric nozzle wall data[35] for N as a function of R_θ yielded

$$N = 12.67 - 6.5 \log (R_{\theta,e}) + 1.21[\log(R_{\theta,e})]^2 \tag{13}$$

and was used to calculate N in the present study.

Procedures for applying the heating methods to three-dimensional and variable entropy flow conditions were also discussed.[23] Based on the axisymmetric analog,[21] simplifying assumptions can be applied to the complex three-dimensional boundary-layer theory which allows equations such as those presented herein to be applied to three-dimensional heating rate calculations. As a result of these assumptions, the distance along a streamline is substituted for the distance S along the body surface and the metric coefficient for the surface coordinate normal to the streamline is used rather than the cross-sectional radius r. The calculated results of procedures using the axisymmetric analog have been shown to be in good agreement with measured three-dimensional heating data.[15,16,19,20,37] Second, as a result of inherent difficulties involved with applying mass balancing to three-dimensional flows and apparent discrepancies that result from applying approximate or classical boundary-layer methods to mass balancing in axisymmetric flows, the effect of variable entropy on the heat transfer was approximated by a somewhat different approach than mass balancing. An iterative process involving a known inviscid solution, the momentum equations [Eqs. (3) and (6)], and the ratios of boundary-layer thickness to momentum thickness is used to determine the local flow conditions. Thus, variable entropy effects are accounted for locally by moving out in the inviscid flowfield a distance equal to the boundary-layer thickness. It appears that similar procedures have been used in previous investigations.[18-20] The inviscid properties at this location are used as the boundary-layer edge properties. The boundary-layer-thickness to momentum-thickness ratios were given[23] as

$$(\delta/\theta)_L = 5.55 \tag{14}$$

and

$$(\delta/\theta)_t = N + 1 + \left\{[((N + 2)/N)(h_w/h_{aw}) + 1] \times [1 + 1.29(Pr_w)^{0.333} (U_e^2/2h_e)]\right\} \quad (15)$$

Equivalent Axisymmetric Body

The Space Shuttle and proposed advanced transportation systems (e.g., the control configured vehicle (CCV) described in Ref. 38) presented rather complex configurations for flowfield calculations. However, these vehicles have a rather flat bottom surface with a smoothly decreasing slope which essentially becomes constant over the aft portion of the vehicle. Adams et al.[24] proposed that an appropriate axisymmetric body at zero AOA which modeled the windward centerline flowfield over the Shuttle at a given angle of attack could be determined. The results of the investigation demonstrated good agreement between the experimental heat-transfer data on the windward ray of the Shuttle at 30 deg AOA and the resulting computed heating rates obtained from a flowfield calculation about a 31 deg half-angle hyperboloid. The "equivalent axisymmetric body" concept has been extended herein to Shuttle and Shuttle-like vehicles over a range of angles of attack of 25-45 deg. For a Shuttle-like vehicle with a known geometry (known longitudinal and cross-sectional measurements), nose radius, and angle of attack, the nose radius and asymptotic half

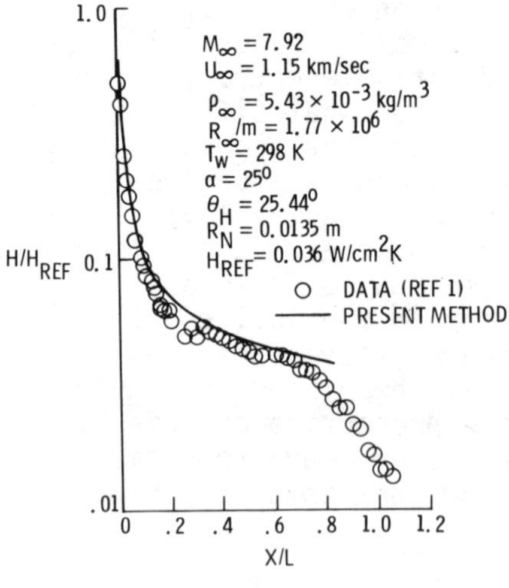

Fig. 1 Comparison of predicted and experimental windward-ray heating distributions on Shuttle model at α = 25 deg (OH4B, AEDC Tunnel B).

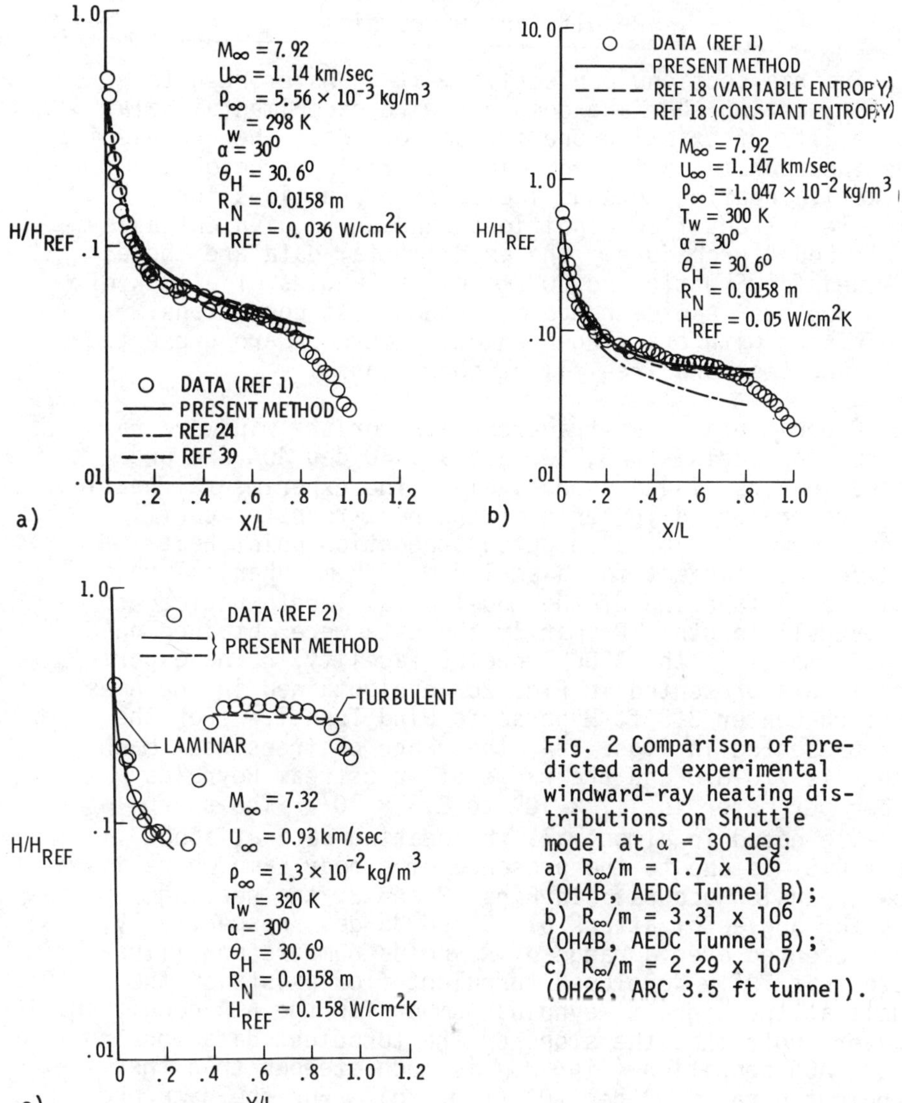

Fig. 2 Comparison of predicted and experimental windward-ray heating distributions on Shuttle model at α = 30 deg: a) $R_\infty/m = 1.7 \times 10^6$ (OH4B, AEDC Tunnel B); b) $R_\infty/m = 3.31 \times 10^6$ (OH4B, AEDC Tunnel B); c) $R_\infty/m = 2.29 \times 10^7$ (OH26, ARC 3.5 ft tunnel).

angle are determined herein for a hyperboloid which models the geometry of the windward ray of the desired vehicle at an angle of attack. Naturally, an analytic body will not provide a particularly good description of a body shape with discontinuous body slopes. Such a region is encountered on the Shuttle vehicle at the start of the body ramp located approximately at $X/L = 0.8$.

Results and Discussion

In this section, convective heat-transfer results of the engineering code are compared with experimental data and results of existing prediction methods. The engineering code incorporates an approximate flowfield technique, convective heating methods including corrections for variable entropy flow conditions, and an "equivalent axisymmetric body" procedure. The experimental data are those obtained for Shuttle and CCV models at angles of attack of 25-40 deg over a wide range of ground test conditions. Comparisons with existing prediction methods are presented at ground test and free-flight conditions.

Experimental heat-transfer data for the windward ray of Shuttle models at 25, 30, 35, and 40 deg AOA are presented in Figs. 1-4, respectively. The experimental heating data are presented in terms of the heat-transfer coefficients normalized by a computed stagnation point heat-transfer coefficient for a scaled 0.3048 m spherical nose radius as a function of the model axial length ratioed to the overall length. Primarily the data were obtained on a Shuttle model in the AEDC Tunnel B facility.[1] The experimental data presented in Fig. 2c were obtained in the Ames Research Center 3.5 ft Hypersonic Wind Tunnel.[2] For the data presented in Figs. 1-4, the range of freestream Mach number is 7.0-8.0 and the range of freestream Reynolds number per meter is 1.7×10^6 to 2.3×10^7. The sharp decrease noted in almost all the heating data at about $X/L = 0.8$ is due to the presence of a body ramp where a flow expansion occurs. In Figs. 2 and 3, the heating data for angles of attack of 30 and 35 deg, respectively, are presented over a range of Reynolds numbers and illustrate that fully developed turbulent flow exists on the models at the highest Reynolds number for the AOA condition. However, note that the slope of the turbulent data for the 35 deg AOA condition (Fig. 3d) is much steeper than the turbulent data at 30 deg AOA (Fig. 2c). For the data presented in Fig. 3d, the flow becomes fully turbulent over the body ramp where the flow expansion affects the heating level and trend.

The laminar and turbulent heating results of the present engineering code are compared with the appropriate experimental data in Figs. 1-4. In general, discrepancies of less than 10% are noted for comparisons prior to the body ramp location. The results of existing laminar

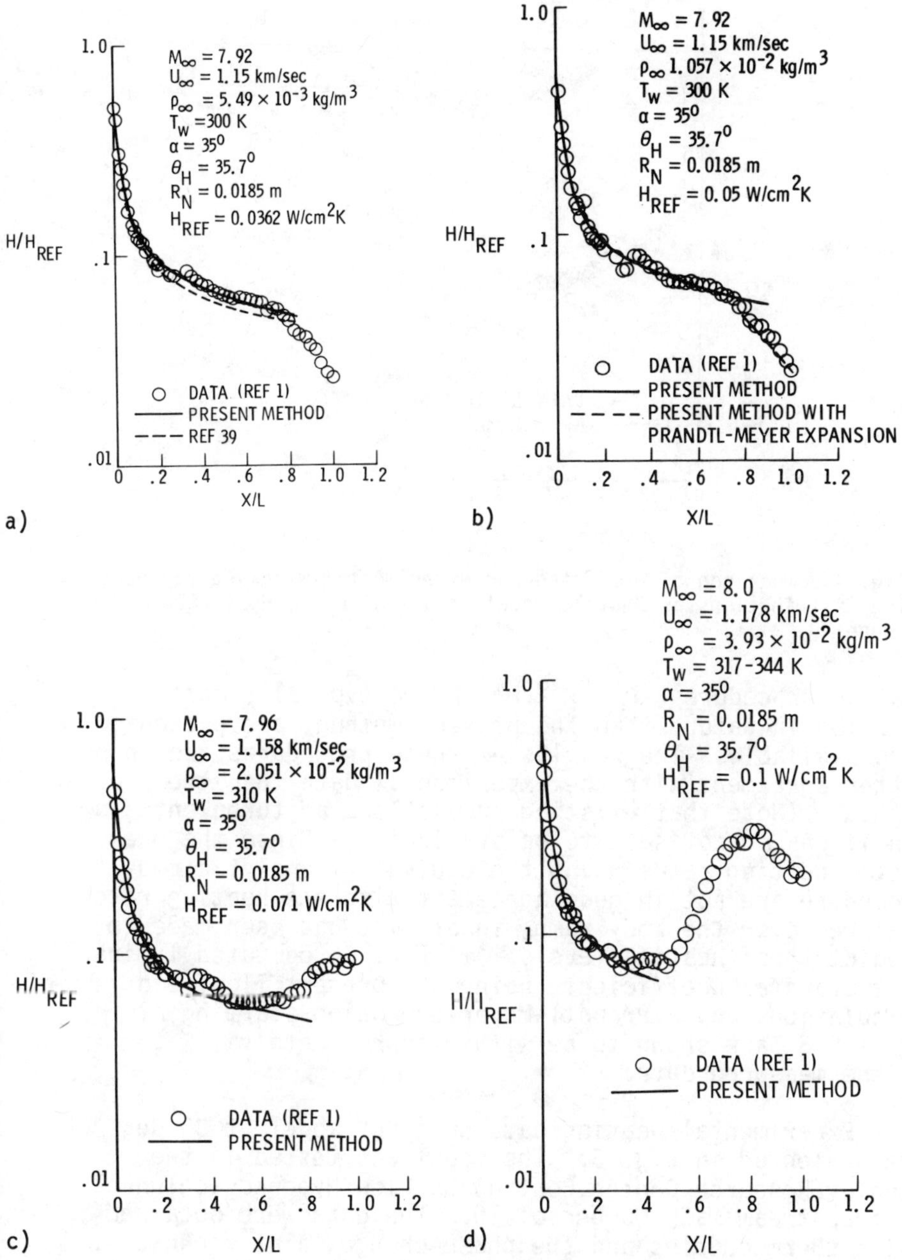

Fig. 3 Comparison of predicted and experimental windward-ray heating distributions on Shuttle model at α = 35 deg (OH4B, AEDC Tunnel B): a) $R_\infty/m = 1.73 \times 10^6$; b) $R_\infty/m = 3.35 \times 10^6$; c) $R_\infty/m = 6.49 \times 10^6$; d) $R_\infty/m = 1.29 \times 10^7$.

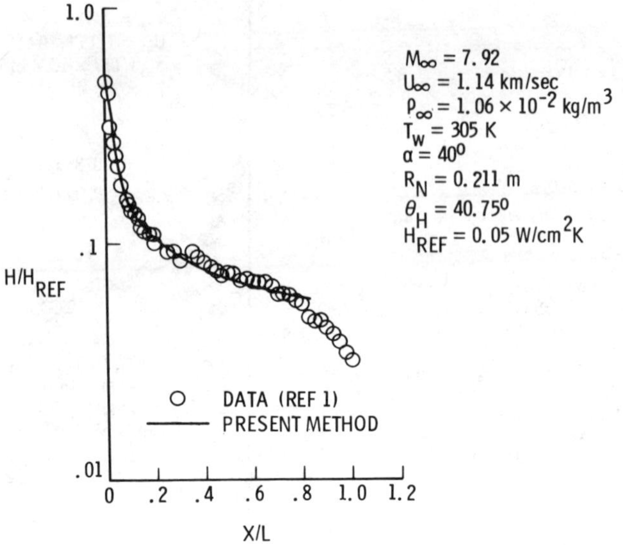

Fig. 4 Comparison of predicted and experimental windward-ray heating distributions on Shuttle model at α = 40 deg (OH4B, AEDC Tunnel B).

heating procedures,[18,24,39] which are typically more detailed in nature than the present method, are presented when available. The results of these procedures are in no better agreement with the experimental data than the present method. (Note that existing predictions at turbulent flow conditions do not seem to be available.) Since the predicted heating rates using the equivalent analytic body procedure are not in good agreement with the heating rates measured over the body ramp, an attempt has been made to predict these heating levels. In Fig. 3b computed laminar heat-transfer coefficients using the present flowfield calculations and a Prandtl-Meyer expansion starting at X/L = 0.8 are shown to be within approximately 15% of the measured data.

Experimental heating data on a CCV model at 36 deg AOA are presented in Fig. 5. The model was tested in the Langley Research Center Continuous Flow Hypersonic Tunnel at freestream Mach number of 10. The data were obtained using thermocouples and the phase-change paint technique. The experimental heat-transfer data and the present heating results are in good agreement. Heat-transfer predictions using the Miniver code[40] are not in good agreement with the experimental data from X/L of 0.2-0.8.

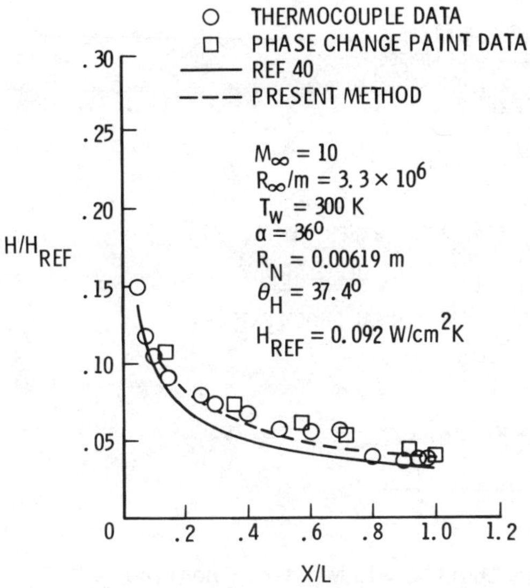

Fig. 5 Comparison of predicted and experimental windward-ray heating distributions on CCV model at α = 36 deg (LaRC Continuous Flow Tunnel).

The previous comparisons have been based on ground test data. The investigation of Refs. 19 and 20 demonstrated that the impact of variable entropy flow conditions on the surface heat transfer is much greater at flight conditions than at tunnel conditions. Thus, it is important to demonstrate the validity of the present method at flight conditions. Heating rates predicted[19,20] for normal shock and variable entropy conditions at a nominal design entry condition for the Shuttle are presented in Fig. 6. Appropriate ground test data have been extrapolated[19] to the flight conditions and are shown also in the figure. The present heating rate results are in very good agreement with the corresponding variable entropy results.[19,20] In Refs. 19 and 20, a three-dimensional inviscid flowfield calculation is employed and the heating calculations use the axisymmetric analog[21] technique. Thus, these analyses permit greater flexibility in defining the heating rates about a vehicle at angle of attack than does the present engineering code. However, the codes[19,20] require a rather large amount of computer storage and computational time and are limited presently to laminar flow conditions.

Since it is desirable to present herein comparisons of predicted laminar and turbulent heating rates at the nominal entry conditions for the first Space Shuttle flight, a VSL analysis[25] which includes a laminar and turbulent

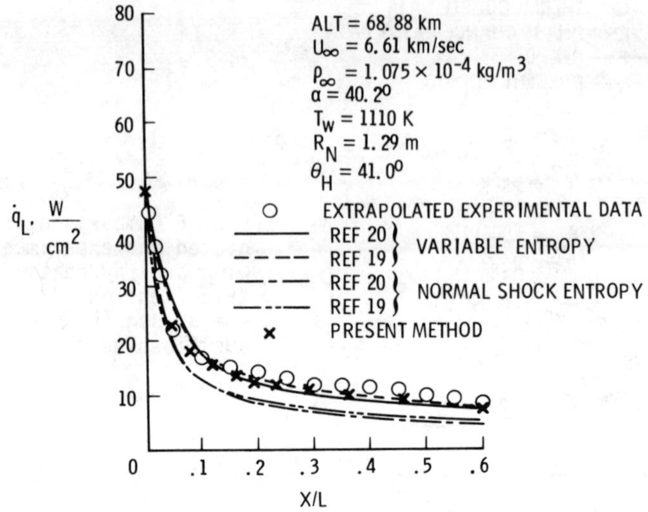

Fig. 6 Comparison of predicted Shuttle windward-ray heating distributions at entry conditions for maximum heating design trajectory.

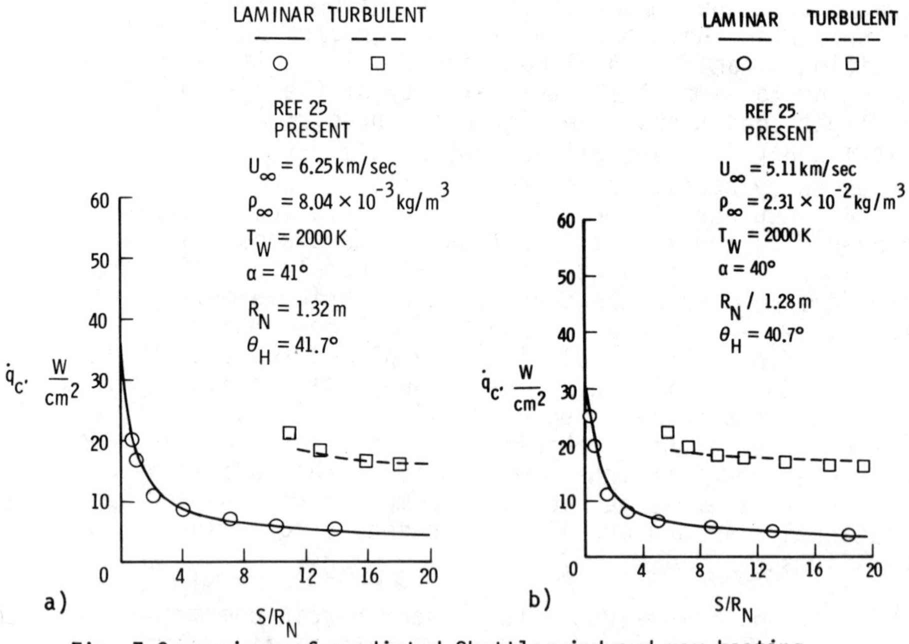

Fig. 7 Comparison of predicted Shuttle windward-ray heating distributions at entry conditions for nominal entry trajectory: a) 70.7 km altitude; b) 62.48 km altitude.

heating capability was used. Laminar and turbulent heating rates based on the present method and the VSL analysis are calculated for a 41 deg AOA condition at an altitude of 70.7 km and for a 40 deg AOA condition at an altitude of 62.48 km and are presented in Figs. 7a and 7b, respectively. Both analyses use an equivalent axisymmetric body to define the windward symmetry plane flowfield and instantaneous transition was assumed to occur at a local Reynolds number based on surface wetted length of 500,000. The results of the two methods are in very good agreement.

Note that the results presented herein for the engineering code require less than 50 s per case on the CDC 6600. Thus, as a result of the favorable comparisons demonstrated with the experimental data and solutions of existing prediction techniques, the present engineering code should be useful in parametric and preliminary design calculations of Shuttle-like bodies at large angles of attack.

Conclusion

An engineering method has been developed for computing the windward symmetry plane convective heat-transfer rates on Shuttle-like vehicles at large angles of attack. The engineering code includes an approximate inviscid flowfield technique, laminar and turbulent heating rate expressions, an approximation to account for the variable entropy effects on the surface heating, and the concept of an equivalent axisymmetric body. This concept presumes that the windward-ray flowfields of Shuttle-like vehicles at large angles of attack can be modeled by an axisymmetric body at zero angle of attack. For this investigation, hyperboloids of varying nose radii and asymptotic half angles were used to model the windward ray of such vehicles for angles of attack of 25-45 deg.

The laminar and turbulent heating results of the present technique were compared with ground test experimental heat-transfer data measured on both Shuttle and advanced transportation models at angles of attack of 25-40 deg, freestream Reynolds numbers per meter of 1.7×10^6 to 2.3×10^7, and a freestream Mach number of approximately 8. Also, the results of the engineering code and existing prediction techniques are compared at ground test and free-flight conditions. The results of the present method are shown to be in good agreement with the experimental data and the results of the existing prediction methods with discrepancies of less than 10% generally noted. Thus, since the present method has been demonstrated to provide rapid and reliable results, the engineering code should provide a useful technique for parametric or preliminary design calculations.

Acknowledgments

The author wishes to thank Kathryn Wurster and Pamela Bradley for providing the experimental heating measurements shown for the control configured vehicle (CCV).

References

[1]Martindale, W. R., "Test Results from NASA Rockwell International Space Shuttle Test (OH4B) Conducted in the AEDC-VKF Tunnel B," AEDC-DR-74-8, Jan. 1974.

[2]Dye, W. H., "Results of Heat Transfer Test of a 0.0175-Scale Space Shuttle Orbiter 104B Model (Modified 22-0) in the NASA/Ames Research Center 3.5-Foot Hypersonic Wind Tunnel (Test)H26)," NASA CR-151380, Oct. 1977.

[3]Harthun, M. H., "Orbiter Entry Heating Analysis-DCR Report," Rockwell International, Rept. SAS-AAT-79-048, Feb. 1979.

[4]"Space Transportation System Technology Symposium, Vol. I: Aerothermodynamics and Configurations," NASA TM X-52876, July 1970.

[5]Bertin, J. J., Musso, S. J., Barnette, D. W., and Goodrich, W. D., "Engineering Flowfields and Heating Rates for Highly Swept Wing Leading Edges," Journal of Spacecraft and Rockets, Vol. 13, Sept. 1976, pp. 540-546.

[6]Goodrich, W. D., Li, C. P., Houston, C. K., Meyers, R. M., and Olmedo, L., "Scaling of Orbiter Aerothermodynamic Data Through Numerical Flow Field Simulation, Aerodynamic Analysis Requiring Advanced Computer. Part II," NASA SP-347, 1975, pp. 1395-1411.

[7]Bertin, J. S. and Goodrich, W. D., "Effects of Surface Temperature and Reynolds Number on Leeward Shuttle Heating," Journal of Spacecraft and Rockets, Vol. 13, Aug. 1976, pp. 473-480.

[8]Lamb, J. P. and Murk, G. K., "Leeward Heat Transfer Experiments on the Shuttle Orbiter Fuselage," Journal of Spacecraft and Rockets, Vol. 13, Nov. 1976, pp. 701-703.

[9]Chaussee, D. S., Patterson, J. L., Kutler, P., Pulliam, T. H., and Steger, J. L., "A Numerical Simulation of Hypersonic Viscous Flows Over Arbitrary Geometries at High Angle of Attack," AIAA Paper 81-0050, Jan. 1981.

[10]Tannehill, J. C., Venkatopathy, E., and Rakich, J. V., "Numerical Solution of Supersonic Viscous Flow Over Blunt Delta Wings," AIAA Paper 81-0049, Jan. 1981.

[11]Helliwell, W. S., Dickinson, R. P., and Lubard, S. C., "Viscous Flow Over Arbitrary Geometries at High Angle of Attack," AIAA Paper 80-0064, July 1980.

[12]Lombard, C. K., Davy, W. C., and Green, M. J., "Forebody and Base Region Real-Gas Flow in Severe Planetary Entry by a Factored Implicit Numerical Method: Part I (Computational Fluid Dynamics)," AIAA Paper 80-0065, July 1980.

[13]Lubard, S. C. and Rakich, J. V., "Calculation of the Flow on a Blunted Cone at a High Angle of Attack," AIAA Paper 75-149, Jan. 1975.

[14]Szema, K. Y. and Lewis, C. H., "Three-Dimensional Hypersonic, Transitional and/or Turbulent Shock-Layer Flows," AIAA Paper 80-1457, July 1980.

[15]Hamilton, H. H., II, "Calculation of Laminar Heating Rates on Three-Dimensional Configurations Using the Axisymmetric Analogue," NASA TP 1698, Sept. 1980.

[16]Rakich, J. V. and Pegot, E. B., "Flow Field and Heating on the Windward Side of the Space Shuttle Orbiter, Aerodynamic Analysis Requiring Advanced Computers: Part II," NASA SP-347, 1975, pp. 1377-1394.

[17]DeJarnette, F., "Aerodynamic Heating on Complex Configurations," Technical Papers--Conference on Advanced Technology for Future Space Systems," May 1979, pp. 179-188 (also available as AIAA Paper 79-0891).

[18]Bertin, J. J., Idar, E. S., and Galanski, S. R., "Effects of Surface Cooling and of Roughness on the Heating (Including Transition) to the Windward Plane-of-Symmetry of the Shuttle Orbiter," University of Texas at Austin, Aerospace Engineering Rept. 77002, April 1977.

[19]Goodrich, W. D., Li, C. P., Houston, C. K., Chiu, P. B., and Olmedo, L., "Numerical Computations of Orbiter Flowfields and Laminar Heating Rates," Journal of Spacecraft and Rockets, Vol. 14, May 1977, pp. 257-264.

[20]Rakich, J. V. and Lanfranco, M. J., "Numerical Computation of Space Shuttle Laminar Heating and Surface Streamlines," Journal of Spacecraft and Rockets, Vol. 14, May 1977, pp. 565-572.

[21]Cooke, J. C., "An Axially Symmetric Analogue for General Three-Dimensional Boundary Layers," Ministry of Aviation, A.R.C. Tech. Rept. R&M No. 3200, June 1959

[22]Zoby, E. V. and Moss, J. N., "Thermodynamic Equilibrium-Air Correlations for Flowfield Applications," AIAA Paper 81-0280, Jan. 1981.

[23]Zoby, E. V., Moss, J. N., and Sutton, K., "Approximate Convective Heating Equations for Hypersonic Flows," Journal of Spacecraft and Rockets, Vol. 18, Jan. 1981, pp. 64-70.

[24]Adams, J. C., Jr., Martindale, W. R., and Maynes, A. W., Jr., "Real Gas Scale Effects on Hypersonic Laminar Boundary-Layer Parameters Including Effects of Entropy-Layer Swallowing," AEDC-TR-75-2, Dec. 1975.

[25]Moss, J. N., "Stagnation and Downstream Viscous Shock Layer Solutions with Radiation and Coupled Ablation Injection," AIAA Paper 74-73, Jan. 1974.

[26]Maslen, S. H., "Inviscid Hypersonic Flow Past Smooth Symmetric Bodies," AIAA Journal, Vol. 2, June 1964, pp. 1055-1061.

[27]Falanga, R. A. and Olstad, W. B., "An Approximate Inviscid Radiating Flow-Field Analysis for Sphere-Cone Venusian Entry Vehicles," AIAA Paper 74-758, July 1974.

[28]Zoby, E. V., Sutton, K., Olstad, W. B., and Moss, J. N., "An Approximate Inviscid Radiating Flow-Field Analysis for Outer Planet Entry Probes," Outer Planet Entry Heating and Thermal Protection, Progress in Astronautics and Aeronautics, Vol. 64, edited by R. Viskanta, AIAA, New York, 1979, pp. 42-64.

[29]Zoby, E. V. and Moss, J. N., "Preliminary Thermal Analysis for Saturn Entry," AIAA Paper 80-0359, Jan. 1980.

[30]Sutton, K. and Graves, R. A., Jr., "A General Stagnation-Point Convective-Heating Equation for Arbitrary Gas Mixtures," NASA TR R-376, 1971.

[31]Cohen, N. B., "Boundary-Layer Similar Solutions and Correlation Equations for Laminar Heat-Transfer Distribution in Equilibrium Air at Velocities Up to 41,000 Feet Per Second," NASA TR R-118, 1961.

[32]Eckert, E. R. G., "Survey on Heat Transfer at High Speeds," U.S. Air Force Rept. ARL 189, Dec. 1961.

[33] Phillips, R. L., "A Summary of Several Techniques Used in the Analysis of High Enthalpy Level, High Cooling Ratio Turbulent Boundary Layers on Blunt Bodies of Revolution," Ramo Wooldridge Corp. Rept. GM-TM-194, Sept. 1957.

[34] Vaglio-Laurin, R., "Turbulent Heat Transfer on Blunt Nosed Bodies in Two-Dimensional and General Three-Dimensional Hypersonic Flow," Journal of Aerospace Sciences, Vol. 27, Jan. 1960, pp. 27-36.

[35] Johnson, C. B. and Bushnell, D. M., "Power-Law Velocity-Profile-Exponent Variations with Reynolds Number, Wall Cooling, and Mach Number in a Turbulent Boundary Layer," NASA TN D-5753, April 1970.

[36] Kutateladze, S. S. and Leont' ev, A. I., Turbulent Boundary Layers in Compressible Gases, Academic Press, New York, 1964.

[37] DeJarnette, F. R. and Hamilton, H. H., "Inviscid Surface Streamlines and Heat Transfer on Shuttle-Type Configurations," Journal of Spacecraft and Rockets, Vol. 10, May 1973, pp. 314-321.

[38] Freeman, D. C., Jr. and Wilhite, A. W., "Effects of Related Static Longitudinal Stability on a Single-Stage-to-Orbit Vehicle Design," NASA TP 1594, Dec. 1979.

[39] Miner, E. W. and Lewis, C. H., "Viscous Shock-Layer Flows for the Space Shuttle Windward Plane of Symmetry," AIAA Paper 74-756, July 1974.

[40] "A Miniature Version of the JA70 Aerodynamic Heating Computer Program, H800 (MINIVER)," MDC 60642, revised Jan. 1972.

Catalytic Surface Effects Experiment on the Space Shuttle

David A. Stewart* and John V. Rakich*
NASA Ames Research Center, Moffett Field, Calif.
and
Martin J. Lanfranco†
Informatics, Inc., Palo Alto, Calif.

Abstract

A Space Shuttle experiment planned to measure the surface catalytic efficiency of the baseline high-temperature reusable surface insulation (HRSI) during Earth entry is described. A spray-on overcoat, with high catalytic efficiency, will be used as a comparative basis for determining the HRSI surface catalytic efficiency through surface temperature measurement. Catalytic efficiency, as well as aerothermal response of the overcoat, was evaluated, using various models made of HRSI material in arc-plasma flow environments. Agreement is obtained between the measured and computed heating rise of the coated surfaces. Computed predictions for the flight case are presented.

Nomenclature

H = total enthalpy
h_D = dissociation energy, MJ/kg
k = Boltzmann constant
k_w = surface reaction rate constant
L = Orbiter length (32.7 m, full-scale)
L_e = Lewis number
m_o = mass of atom or molecule
p = pressure, atm

Presented as Paper 81-1143 at the AIAA 16th Thermophysics Conference, Palo Alto, Calif., June 23-25, 1981.

*Research Scientist.
†Staff Engineer.

$\dot{q}$ = heat-transfer rate
R = radius
Re = Reynolds number
S = arc length from stagnation point
S_c = Schmidt number
St = local Stanton number
T = temperature
$\bar{t}$ = nominal structure thickness
U = velocity
X = axial distance
α = angle of attack
α_e = fraction of molecules dissociated
γ = energy recombination coefficient
λ = wavelength
μ = viscosity
ρ = mass density of mixture

Subscripts

b = body
CW = cold wall
e = boundary-layer edge
fc = fully catalytic
HW = hot wall
max = maximum
min = minimum
N = nitrogen
o = stagnation point
T = total
w = wall
∞ = freestream

Introduction

The catalytic efficiency of various material surfaces has been studied for many years in ground test facilities, using dissociated nitrogen or airstreams.[1-3] Also, the possibility of reduced entry heating over the Space Shuttle's lower surface thermal protection tiles due to noncatalytic surface effects has long been recognized and discussed.[4-5] These tiles have a reaction cured glass (RCG) surface that is noncatalytic at low temperatures similar to materials studied during early high-temperature arc-jet testing on possible Space Shuttle thermal protection systems.[4] However, whether this phenomenon occurs on a space vehicle heat shield during atmospheric entry has not been determined. Therefore, designers have not allowed for the possible substantial decrease in heating to a Space Shuttle thermal protection system (TPS) with noncatalytic surfaces. Direct side-by-side measurements of catalytic and noncatalytic surfaces on the Space Shuttle TPS are needed, because flight data from baseline tiles alone cannot establish firmly the exact cause-and-effect relationships between surface temperature and boundary-layer conditions.

A study was initiated some time ago to develop a spray-on catalytic overcoat that could be applied and removed (nondestructively) from selected areas of the Space Shuttle TPS. The Space Shuttle TPS consists of reinforced carbon-carbon composite (RCC) on the wing leading edges and on the nose cap, high-temperature reusable surface insulation (HRSI) on lower surfaces, and several low-temperature reusable surface insulations (LRSI) on upper fuselage and wing areas. This paper describes flight and ground test experiments to evaluate the thermal response of the RCC and HRSI with the catalytic overcoat, as well as to estimate the catalytic surface efficiency of each material for atom recombination. Qualitative agreement was obtained between predicted and side-by-side measurements from HRSI and RCC models with and without the catalytic overcoat in dissociated arc-plasma air flow. High-temperature surface properties obtained from these arc-jet data for the TPS glassy surfaces were used along with computational fluid dynamics codes[5] to define the flight experiment. Heating predictions are presented for a nominal design trajectory. The experiment, part of the Orbiter experimental program (OEX), is scheduled to start on the second Space Shuttle flight in late 1981.

Flight Experiment

In high-velocity dissociated flow, characteristic of Earth entry, it is predicted (from a reacting boundary-layer computation) that the Orbiter heat shield, whose surfaces are noncatalytic for atom recombination, will experience lower convective heating rates (about 30%) than one that promotes surface atom recombination (Fig. 1).

The catalytic surface effects Orbiter experiment (OEX) uses the baseline TPS and instrumentation without any effect on the Space Shuttle operations. A catalytic overcoat will be sprayed onto existing instrumented tiles. During Earth entry, the surface temperature of these tiles will be recorded for comparison with adjacent HRSI baseline

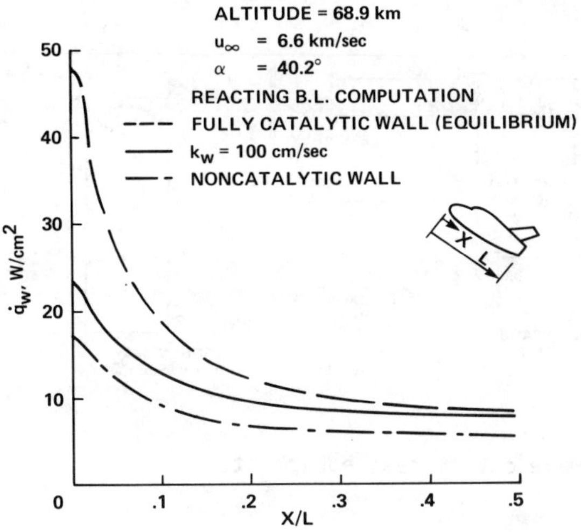

Fig. 1 Reduced Orbiter centerline heating, assuming a noncatalytic surface during flight.

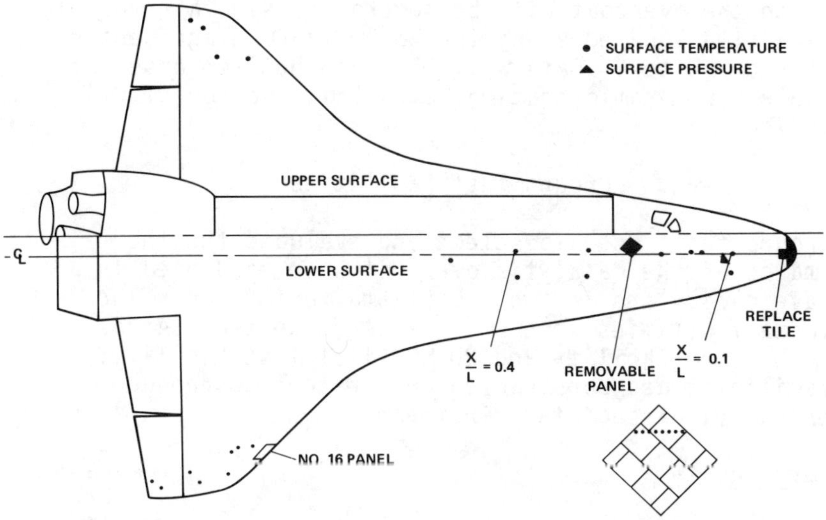

Fig. 2 Catalytic surface effects OEX experiment locations on Space Shuttle.

tiles. The experiment is planned for two phases:

1) Selected tiles along the midfuselage will be sprayed with the catalytic overcoat in early flights.

2) Areas in the stagnation region, wing leading edge (panel 16), and the upper wing surface will be sprayed in later flights (Fig. 2).

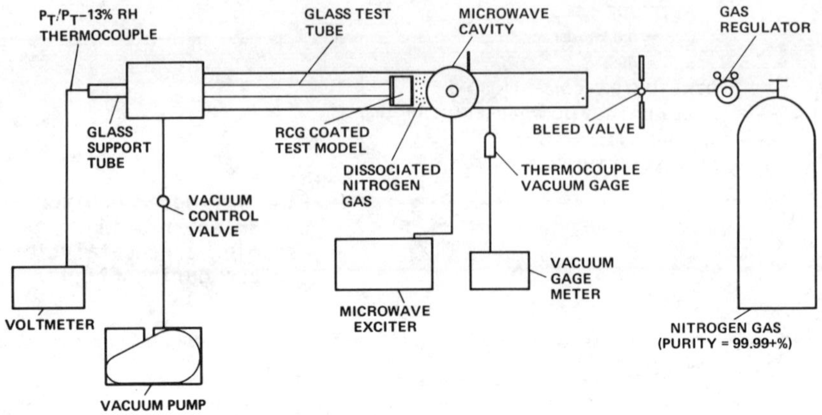

Fig. 3 Microwave cavity test equipment.

Also, during phase 2, temperature profiles across tiles with the overcoat will be recorded, using a removable panel and tile located along the lower midfuselage centerline of the Orbiter. Data will be taken between onset of measurable aerodynamic heating (at an altitude of 91 km) and landing.

Ground Test Facilities

Ground tests used to select and evaluate the thermal performance of the catalytic overcoat were conducted in a microwave cavity and in two arc-plasma facilities. The arc-plasma facilities were Ames Research Center's Aerodynamic Heating Facility and 20 MW Pilot Test Facility. Both facilities use constricted arc heaters to produce high-enthalpy, dissociated airstreams.

Microwave Cavity

A sketch of the microwave cavity test equipment is shown in Fig. 3. The equipment consists of a glass tube, 2.54 cm diam by 56 cm long, with a bleed valve and gas regulator at one end and vacuum pump and model support at the other. The Hg 198 microwave cavity is located roughly halfway along the test tube. Test models having various surface materials are installed just downstream of the microwave cavity. A surface thermocouple was installed at the center of each model. The thermocouple wires extend from the model surface through the support tube to a voltmeter. The support tube is sealed at the voltmeter end with RTV 560.

High-purity nitrogen gas is passed through the evacuated test tube at a constant flow rate. The nitrogen gas is dissociated as it passes through the microwave cavity (resonating at 24.5 MHz). During each test, a constant flow rate was maintained by monitoring the static pressure with a thermocouple vacuum gage. The reactive catalytic efficiency of the test material is measured by observing the rise in the model surface temperature at a constant microwave exciter power and test tube static pressure. The model surface thermocouple was referenced to room temperature.

Aerodynamic Heating Facility

This facility provides high-energy hypersonic flow (with total enthalpies 7.0 - 23 MJ/kg in air) by passing the gas between concentric toroidal electrodes and expanding it through a 30 deg total angle conical nozzle. During this investigation, the geometric area ratio of the nozzle was 144 to 961. The freestream Mach number was 4.7 - 6.0.

Blunt bodies were exposed to the high-enthalpy hypersonic airflow for test durations up to 600 s. The models were located 28 cm downstream of the conical nozzle on a water-cooled support. Thermocouple and pyrometers measurements were made during each exposure. Also, impact pressure and stagnation point heat-transfer rates were measured. The nozzle flow properties were estimated from the ratio of measured impact pressure to toal pressure, from calculated total enthalpy, and by assuming equilibrium flow from the reservoir to an arbitrary point where chemical reactions and molecular vibrations become frozen.[6]

20 MW Pilot Test Facility

High-enthalpy air from the constricted heater of the 20 MW Pilot Test Facility expands through a semielliptical nozzle with a throat aspect ratio of 4/1 and across a 48.3 × 50.8 cm flat-plate multiple HRSI tile configuration. A silica plate, 25.4 cm long and 40 cm wide, is located at the nozzle exit to insure a smooth hot-wall flow transition from the nozzle to the test article surface. Thermocouple and pyrometer measurements are recorded during each exposure. The total enthalpy of the stream can be determined within 15% from sonic flow measurements.[7] These measurements include total pressure, mass flow, and electrical power input. Exposures of about 300 s and a freestream Mach number of 4 were used during these tests.

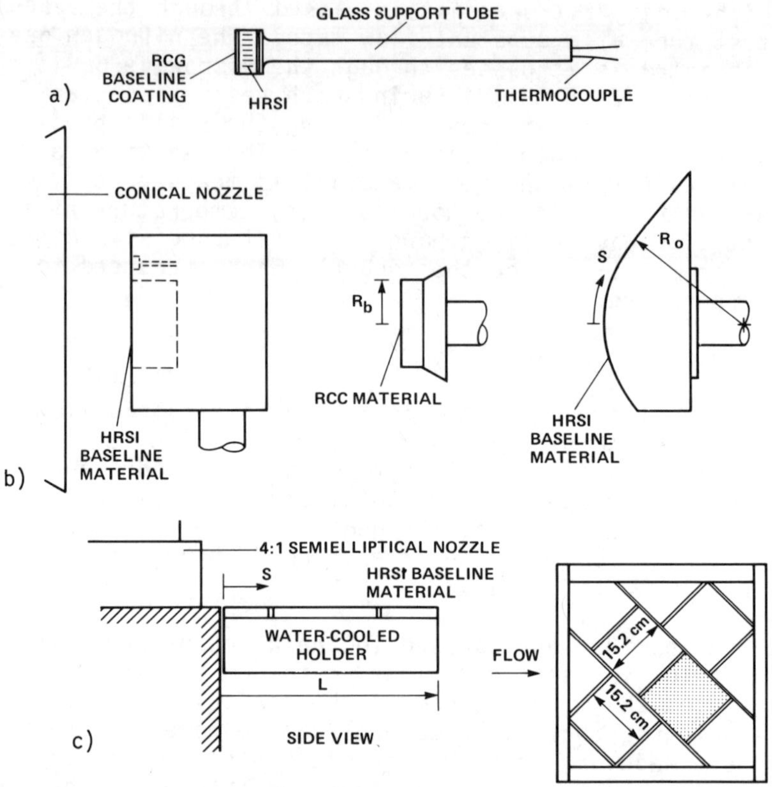

Fig. 4 Models for catalytic overcoat evaluation: a) microwave cavity test model; b) Aerodynamic Heating Facility models; c) 20 MW Pilot Test Facility model.

Test Models

The test models used in the various environments to select and evaluate the catalytic overcoat are shown in Fig. 4. The model used in the microwave cavity tests was a disk (1.91 cm in diameter and 1.27 cm thick) cut from a baseline high-temperature reusable surface insulation (HRSI) (Fig. 4a). A platinum/platinum-13% rhodium thermocouple was located at the center of the disk and in contact with the reaction-cured glass (RCG) coating.[8] The disk was mounted to a glass tube 0.98 cm in diameter and 36 cm long. The tube was sealed at both ends with RTV 560; the thermocouple wires extended 25 cm beyond to a voltmeter.

The three blunt-body configurations used to evaluate the catalytic overcoat in the Aerodynamic Heating Facility are shown in Fig. 4b. The first two configurations provided stagnation point surface temperature response data. With the smaller of the two disks (10 cm in diameter), a silicon-carbide holder was used to test a 7.11 cm diam by 0.66 cm thick reinforced carbon-carbon composite (RCC) specimen. Two types of RCC specimens were used in these tests: 1) baseline RCC with a tetraethyl-orthosilicate (TEOS) impregnation and 2) baseline RCC with TEOS impregnation and a type "A" glass coating. With the larger model (20 cm in diameter), a water-cooled copper cylinder holder was used to test a 10 cm diam by 5 cm thick disk made from HRSI. A pressure orifice and a calorimeter were located 90 deg apart and at a radius of 6.3 cm on the face of the holder. The third model, a truncated 40 deg blunted cone, also was made from an HRSI baseline material (RCG-coated, 3.52 kg/cm^3 silica, impregnated with silicon carbide). Thermocouples were installed in all three models at the stagnation point and in contact with the surface coating. Platinum/platinum-13% rhodium (0.013 cm in diameter) thermocouples were installed in the HRSI baseline models and tungsten-5% rhenium/tungsten-26% rhenium thermocouples were installed in the RCC baseline materials. In addition, five platinum/platinum-13% rhodium thermocouples were installed in a similar manner along the conical portion of the blunted cone at $S/R_0 \geq 1.15$.

A multiple-tile flat-plate configuration (48 × 50 cm) was tested in the 20 MW Pilot Test Facility (Fig. 4c). Each 15 × 15 cm tile had a platinum/platinum-13% rhodium thermocouple installed at the center and in contact with the RCG surface coating. The 3.8 cm thick tiles were bonded to a baseline strain isolation pad (SIP) which in turn was bonded to a 0.081 cm thick aluminum plate. The tiles were positioned so that their leading edges were swept 45 deg to the flow; there was a gap of about 0.15 cm between tiles. High-density silica strips were placed around the outer edge of the tile array to prevent damage to the edge tile by the effects of flow in the gaps.

Experiment

Overcoat Selection

The candidate catalytic overcoats were either vapor deposited or spray painted onto the RCG baseline coating of a 1.9 cm diam HRSI model for tests in the microwave cavity (Fig. 3). Low-velocity, high-purity nitrogen gas (99.99+%) flows through the cavity and around the model at a constant

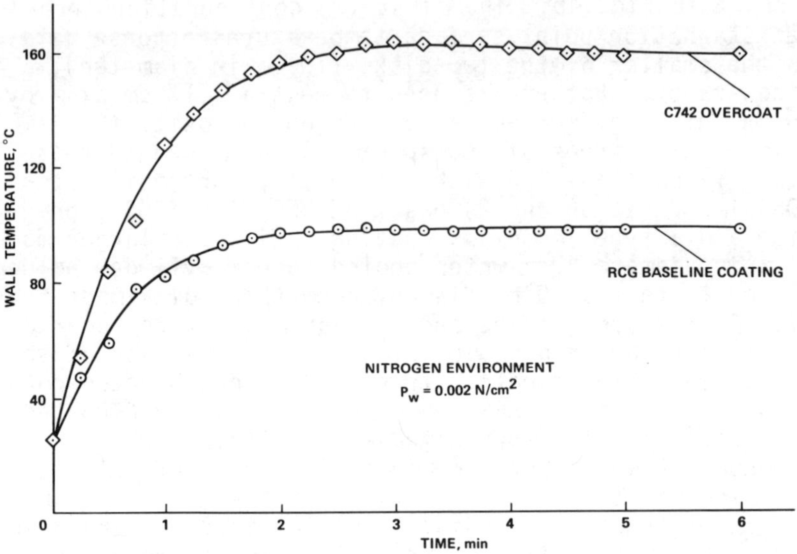

Fig. 5 Typical temperature response of HRSI surface with and without catalytic overcoat.

flow rate, the flow rate being controlled by maintaining a constant static pressure (p_W = 0.002 N/cm^2) in the tube during each test. Before each test, the glass tube was purged several times with the test gas. The coated model was exposed to dissociated nitrogen gas from the microwave cavity for 6 min. Surface temperature rise was measured to indicate the relative atom recombination at the surface of each material. Typical surface temperature histories for an RCG baseline coating with and without a catalytic overcoat are shown in Fig. 5. Steady-state conditions are reached within 3 min after the nitrogen gas is dissociated in the microwave cavity. Typically, a 50-60% increase in surface temperature resulted from the addition of any of the candidate high-catalytic efficiency materials to the RCG-coated model.

Catalytic Overcoat Evaluation

After selecting the catalytic overcoat formulation from microwave cavity tests, thermal response tests were performed with it in the convectively heated environments of the Aerodynamic Heating Facility and 20 MW Pilot Test Facility. First, flat-faced cylinders were used to obtain stagnation point surface temperature data from baseline HRSI and RCC materials with and without the catalytic overcoat. Dual models were used in the tests conducted in the Aerodynamic Heating Facility. The model without the catalytic overcoat was used to establish the desired surface temperature for the test. The model with the overcoat was then inserted into the arc-plasma stream and the surface temperature recorded. The HRSI baseline material in the water-cooled model holder is shown in the arc-plasma airstream in Fig. 6. Also, the stagnation point heat-transfer rates to the water-cooled 10 cm diam copper hemisphere and flat-faced cylinder were measured. These heating rates were used to determine the equivalent hemispherical nose radii for the flat-faced test models. These radii were used in the calculations involving the reaction rate constant and stagnation point heating rate discussed later.

Second, a truncated 40 deg blunted cone was used to measure the effect of the catalytic overcoat on heating rate distribution. The test conditions for this part of the investigation were established by measuring the stagnation point heat-transfer rate and surface pressure on a 10 cm diam hemisphere before and after each exposure.

Finally, a flat-plate model made up of multiple tiles was used to evaluate the catalytic overcoat on a full-scale tile. This test was conducted in the 20 MW Pilot Test Facility with the model mounted flush with the nozzle exit. Surface temperatures were recorded from two 15 × 15 cm

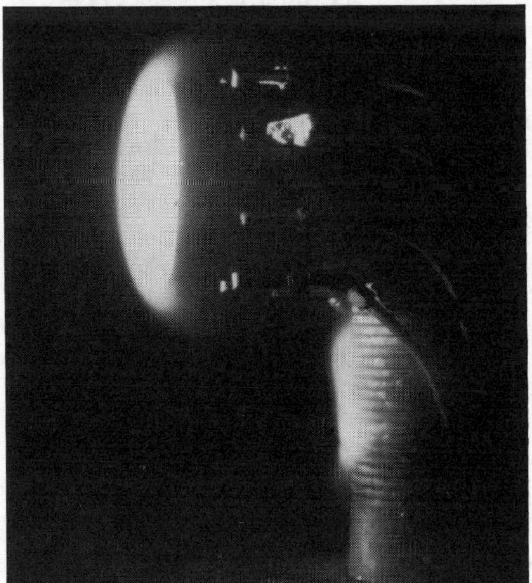

Fig. 6 HRSI baseline material (disk model) in Aerodynamic Heating Facility.

Table 1 Microwave cavity test results

Coating material	Application	Emittance, ε_{TH} T_w = 1366 K	$\frac{T_{w_i}}{T_w(RCG)}$	γ_w	k_w, cm/s
Cu	Vapor deposit	0.22	1.58	7×10^{-2}	964
Ni	Vapor deposit	0.32	1.52	10^{-1}	1352
Au-black	Vapor deposit	0.92	1.57	10^{-1}	1375
Al-black	Vapor deposit	0.95	1.48	10^{-1}	1390
Pt-black	Spray paint	0.90	1.61	5×10^{-2}	801
C742	Spray paint	0.88	1.54	10^{-1}	1360
RCG glass	Baseline	0.90	1.0	8×10^{-4}	10.0

tiles symmetric to the flow centerline. The model was exposed to the flow for 7 min. The catalytic overcoat was spray painted onto the HRSI tile. The tile was masked to prevent the catalytic overcoat from getting into the gaps between the tiles during spraying.

Results and Discussion

Several candidate catalytic overcoat materials were investigated in the microwave cavity. The test results are shown in Table 1. Surface temperature measurements from the various candidates were ratioed to the surface temperature of the RCG baseline coating. The manufacture of the RCG baseline coating[8] includes the mixing of a reactive high-silica borosilicate glass frit with 2.5% by weight tetraboron silicide (B_4Si). The mixture is sprayed onto the silica tile and glazed at 1473 K. When the coating is cooled rapidly to room temperature, a thin borosilicate glass surface layer (<0.001 cm) is formed which contains 6-7% B_2O_3. Since the surface character plays an important role in atom recombination, Vycor glass (3% B_2O_3) recombination coefficient data were chosen to approximate the RCG baseline coating.[9] Other low-temperature candidate catalytic material recombination coefficients, shown in Table 1, were obtained from the literature[10,11] for a nitrogen gas atmosphere. The reaction rate constants for the candidate catalytic overcoats were calculated assuming $\beta = 1$ from the well-known expression

$$k_w = \gamma_w (k\ T_w / 2\pi m_o)^{1/2} \tag{1}$$

All candidate materials appeared to have catalytic surfaces. However, with the first four materials listed in

Table 1 it would be necessary to replace the HRSI tiles on the Orbiter after each flight. These materials have to be vapor deposited on the RCG coating. The copper and nickel surfaces fail to meet the Space Shuttle design requirements for the total hemispherical emittance of $\varepsilon_{TH} \geq 0.8$. In addition, these four materials oxidize in air in a relatively short time.

A combustible turpentine carrier was required for spraying the platinum black overcoat onto the tile. The C742 overcoat was chosen as the only satisfactory candidate for the OEX experiment because of its water carrier and high total hemispherical emittance [$\varepsilon_{TH} = 0.88(T_W = 1366$ K)]. This overcoat consists of a black iron cobalt chromia spinal in a polyvinyl acetate binder. The polyvinyl acetate binder allows a hard, environmentally stable overcoat, 0.005 cm thick, to be applied easily to the RCG baseline coating. Tests in an environmental chamber showed the spinal catalytic overcoat was unaffected when exposed to a humidity greater than 95% for 72 h. Most of the binder burns out of the catalytic overcoat at a surface temperature of 622 K, as determined from temperature gas analysis (TGA) measurements at atmospheric pressure. This result suggests that the coating will be environmentally stable while on the ground and provide a catalytic surface during re-entry. Discussed next, are calculated reaction rate constants for both the baseline TPS and catalytic overcoat obtained from arc-jet data. Reaction rate constants for the catalytic overcoat were calculated at Orbiter flight temperatures in both laminar and turbulent boundary-layer flows.

Laminar Flow Prediction

Experimental reaction rate constants were calculated using the semiempirical relationship from Ref. 12 and measured stagnation point heat-transfer rates

$$k_W = \frac{(\rho_\infty U_\infty St_h)}{\rho_W} \cdot L_e^{2/3} \left(\frac{\dot{q}_o - \dot{q}_{min}}{\dot{q}_{max} - \dot{q}_o} \right) \quad (2)$$

Maximum stagnation point heating rate was calculated from the following equation from Ref. 13 for a cold-wall hemisphere

$$\dot{q}_{max} = \bar{c} \sqrt{\frac{P_o}{R_o}} \left(\frac{U_\infty}{10{,}000} \right)^{1.99} \left(1 - \frac{h_W}{H_{e_o}} \right) \quad (3)$$

The stagnation point heating rate $\dot{q}_o$ obtained from Goulard's theory[14] is

$$\frac{\dot{q}_o}{\dot{q}_{max}} = 1 - \frac{L_e^{2/3}\alpha_e h_D/H_e}{1 - (L_e^{2/3} - 1)\alpha_e h_D/H_e}(1 - \phi) \qquad (4)$$

where

$$\phi = 1 \Bigg/ \left\{ 1 + \frac{0.763}{k_w S_c^{2/3} \rho_w} \left(\frac{U_\infty \rho_\infty \mu_e}{R_o} \right)^{1/2} \left[\frac{\rho_\infty}{\rho_e} \left(2 - \frac{\rho_\infty}{\rho_e} \right) \right]^{1/4} \right\}$$

The minimum stagnation point heating rate $\dot{q}_{min}$ is obtained from Eq. (4) with $\phi = 0$.

Transport properties were calculated for an equilibrium-air mixture using cross-section data and equations developed by Hansen.[15] The corresponding hot-wall heat-transfer rates were then obtained from the well-known expression

$$\dot{q}_{HW} = \dot{q}_{CW}\,[1 - (h_w/H_{e_o})] \qquad (5)$$

Thermal response data from models tested in arc-plasma airstreams are plotted in Figs. 7-15. Data from the disk models are shown in Figs. 7 and 8. Figure 7 shows the correlation between experiment and theory for stagnation point heating rate as a function of reaction rate constant for the HRSI disk tested in the Aerodynamic Heating Facility.

The reaction rate constant obtained from Eq. (2) for the RCG baseline coating was $k_W = 100$ cm/s at $\dot{q}_o = 17.0$ W/cm^2; for the catalytic overcoat it was $k_W = 2900$ cm/s at $\dot{q}_o = 27.6$ W/cm^2.

In Fig. 7, these experimentally determined reaction rate constants are compared with Goulard's theory and a reacting boundary-layer computation. Stagnation point heat-transfer rates were calculated from Eq. (4) for various reaction rate constants, assuming a Mach freeze of 6. The maximum stagnation point heat-transfer rate calculated from Eq. (3) was assumed to be equivalent to an equilibrium value at $k_W = 10^5$ cm/s. Included in Fig. 7 is the reacting boundary-layer computation showing the variation in stagnation point heat-transfer rate with reaction rate constant for the HRSI disk. The computation utilized the

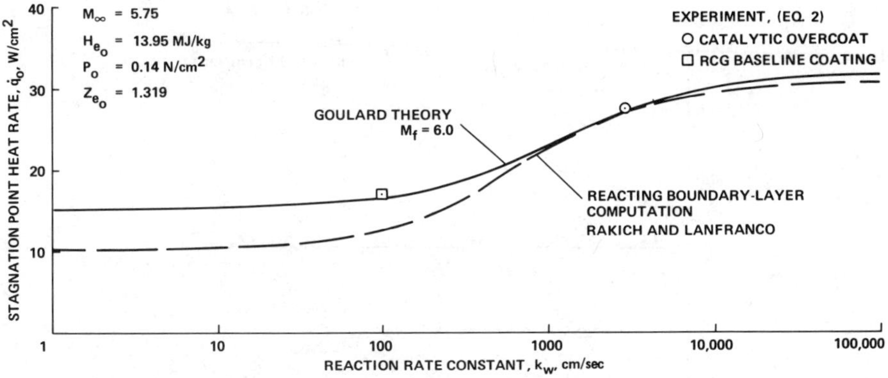

Fig. 7 Correlation between experiment and theory for flat-faced cylinder.

code described by Tong et al.,[16] which was modified to use a simplified gas model of five species (see Ref. 5 for details). There was good agreement between both theories and experiment above k_w = 1000 cm/s; at $k_w \leq 1000$ cm/s, the reacting boundary-layer computation predicted lower stagnation point heat-transfer rates than did Goulard's theory. Goulard's theory, basically for cool surfaces, assumes a frozen boundary layer. The reacting boundary-layer computation was for a hot wall and assumes equilibrium edge conditions with gas-phase recombination in the boundary layer. Total enthalpy, surface temperature, pressure, gas composition, and nose radius are input into the boundary-layer program. The reaction rate constants needed in the computation to predict the measured stagnation point heating rates were k_w = 350 cm/s at $\dot{q}_o$ = 17.0 W/cm^2 for RCG and k_w = 3400 cm/s at $\dot{q}_o$ = 27.6 W/cm^2 for the catalytic overcoat. Comparison with the values obtained from the simplified method suggests that the reaction rate constant for the RCG baseline coating at $\dot{q}_o$ = 17.0 W/cm^2 is between $100 \leq k_w \leq 350$ cm/s, and that for the catalytic overcoat at $\dot{q}_o$ = 27.6 W/cm^2 it is between $2900 \leq k_w \leq 3400$ cm/s.

Next, tests were conducted to determine what effect the catalytic overcoat would have on the RCC material in a convectively heated environment. Thermal response data from RCC disks with and without the catalytic overcoat are plotted in Fig. 8. Surface temperature response data from the two types of RCC without the catalytic overcoat are shown in Fig. 8a; data for the RCC with the catalytic overcoat are shown in Figs. 8b and 8c. Reaction rate constants

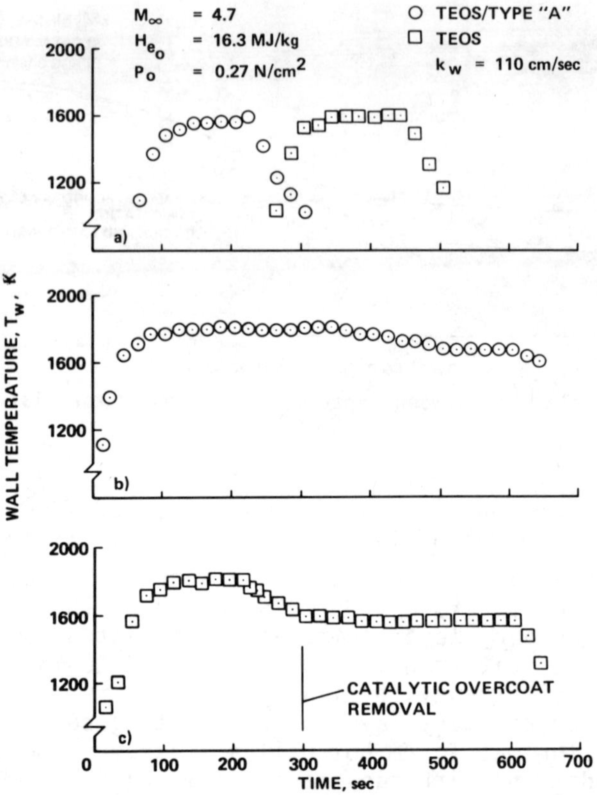

Fig. 8 Thermal response of RCC with and without catalytic overcoat: a) RCC materials; b) RCC/type A with catalytic overcoat; c) RCC with catalytic overcoat.

for the RCC materials with and without the catalytic overcoat were calculated using Eq. (2) and heating rates obtained from the measured surface temperatures. The heating rates were calculated using total hemispherical emittances for the RCC materials from Rockwell design values for total normal emittance[17] and for the catalytic overcoat from room-temperature measurements of spectral hemispherical reflectance [Willey 318 ($2.0 \leq \lambda \leq 15$ μm) and Beckman DKIA ($0.3 \leq \lambda \leq 2.3$ μm) spectrophotometers]. The reaction rate constants were $k_w = 110$ cm/s at $\dot{q}_0 = 29.0$ W/cm^2 for both RCC materials and $k_w = 6900$ cm/s at $\dot{q}_0 = 53.5$ W/cm^2 for the catalytic overcoat.

Looking at Figs. 8b and 8c we see by the drop in temperature that the catalytic overcoat was removed from the RCC with the added type A glass coating in 650 s, as com-

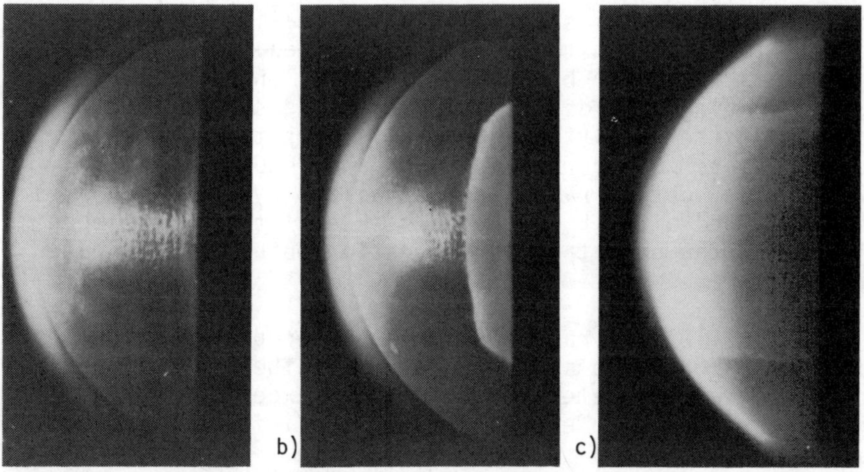

Fig. 9 Forty-degree blunted cone in arc-plasma airstream: a) RCC baseline; b) catalytic overcoat on cone trailing edge; c) catalytic overcoat strip.

pared with 300 s for a material without it, during arc-plasma exposures. The difference in time for the removal of the catalytic overcoat from the two RCC surfaces was probably due to softening of the type A glass (which is similar to Vycor glass) at $T_W = 1800$ K. The spinal species becomes mixed with the type A glass in one case, whereas it is directly exposed to the convectively heated environment in the other case. Upon examination of the disk surface under a magnification of 25X, it appeared that some of the type A glass was removed during the arc-plasma exposures. Further work is continuing in this area using these leading-edge materials.

The photographs in Fig. 9 show top views of the truncated 40 deg blunted cone without the catalytic overcoat and partially and fully coated with the catalytic overcoat. The brighter area along the cone surface indicates the location of the catalytic overcoat. The experimental heating rates from these tests, normalized by a calculated, fully catalytic wall stagnation point value, are shown in Figs. 10-12. In addition, the heating distribution along a cone with catalytic overcoat on the stagnation region and near the base is shown in Fig. 13. Heating rate distributions from the theory of Lees[18] and the reacting boundary-layer computations are compared with the data. The reacting boundary-layer computations were made by first varying the reaction rate constant until agreement was obtained with the experimental stagnation point heating rate. Thus, the reaction rate constants were determined independently of the theories described by Eqs. (2-4). In Figs. 10 and 11, the heating rate distribution over the sphere-cone with and without the catalytic overcoat is computed with the rate constants held constant at 460 and 7000 cm/s, respectively. The heating rates are normalized by the calculated value obtained with $k_W = 10^5$ cm/s. Lees' theory has been adjusted to the noncatalytic stagnation point heating rate (Fig. 10).

In Fig. 12, the rate constant was abruptly changed from 460 to 7000 cm/s at the point where the overcoat was applied, other parameters being fixed. Excellent agreement is obtained with experiment for the initial rise in heating on the coated section of the cone. The theory predicts a decreasing heat flux with distance, but no experimental data were obtained in that region. In Fig. 13, the computation was made with rate constant set at 7000 at the nose and near the base of the cone and at 460 in the remaining region. Excellent agreement is again obtained.

The theory of Lees, which approximates an equilibrium gas and constant wall properties, gives reasonably accurate results, except for the heating rise near the base of the cone (Figs. 12 and 13). This illustrates that, for a change from a noncatalytic to a catalytic material, the heating can overshoot the equilibrium value. This occurs because energy stored in the gas is suddenly released and is transferred to the body.

The reaction rate constants used in the reacting boundary-layer computations, as well as values from Eq. (2)

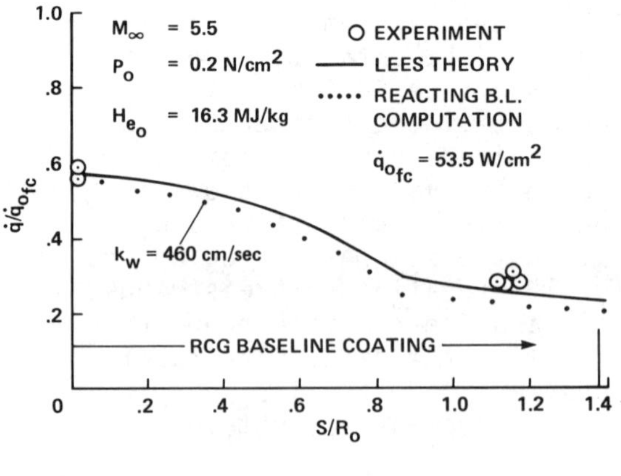

Fig. 10 Centerline surface heating for 40 deg blunted cone.

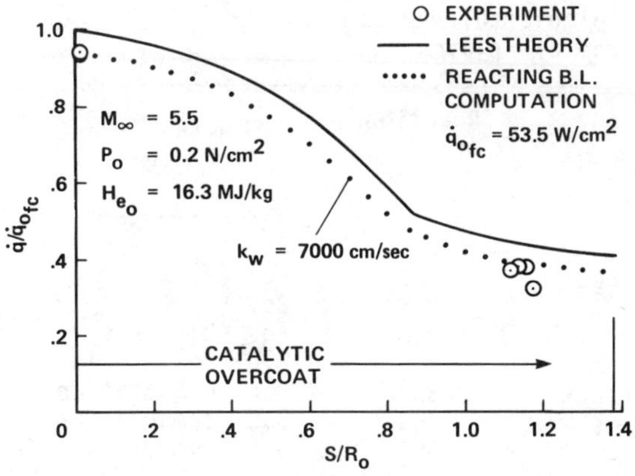

Fig. 11 Centerline surface heating for 40 deg blunted cone with partial catalytic overcoat.

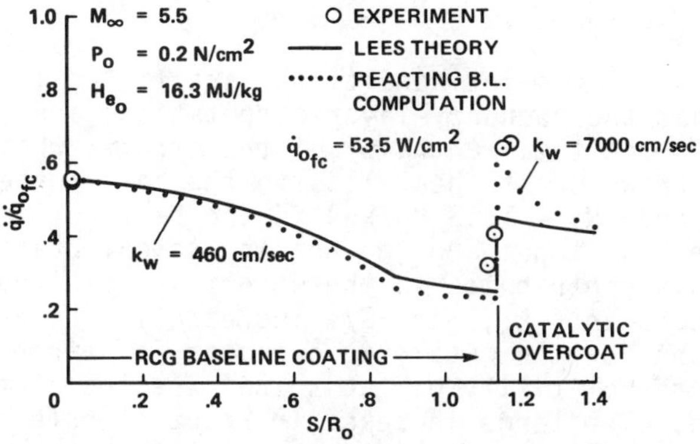

Fig. 12 Centerline surface heating for 40 deg blunted cone with partial catalytic overcoat.

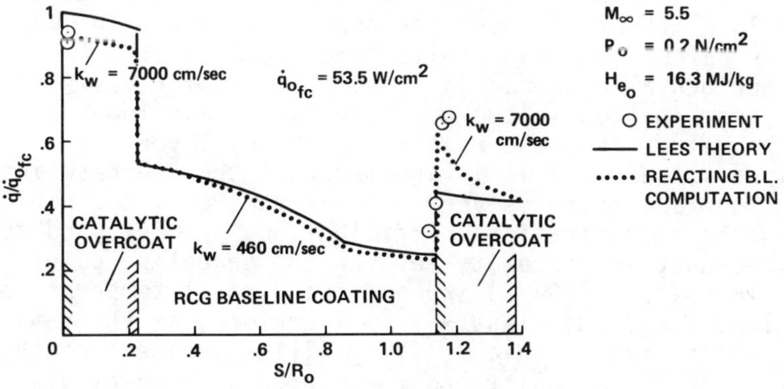

Fig. 13 Centerline surface heating for 40 deg blunted cone with partial catalytic overcoat

Table 2 Summary of calculated reaction rate constants obtained from arc-plasma data

Surface temperature, T_w, K	Material	Reaction rate constant, k_w, cm/s	Method determined
1344	RCG	100	Eq. (2)
1525	Catalytic overcoat	2900	Eq. (2)
1556	RCG	185	Eq. (2)
1800	Catalytic overcoat	6900	Eq. (2)
1344	RCG	350	B.L. computation
1525	Catalytic overcoat	2300	B.L. computation
1556	RCG	460	B.L. computation
1800	Catalytic overcoat	7000	B.L. computation

for the stagnation point conditions, are listed in Table 2. (We note that the boundary-layer computations are somewhat sensitive to the total enthalpy of the stream, which is not accurately known due to heat loss to the cold tunnel walls.) The value used (H_T = 16.3 MJ/kg) is the best estimate based on both the heat input and the heating to calibration models. This contributes to the uncertainty in the reaction rate constant, k_w. The 2/1 increase in the reaction rate constant for the catalytic overcoat is attributed to the increased volatility of the spinal species at the higher temperature. The large increase in reaction rate constant for the RCG baseline coating can be explained by examining the variation in the calculated recombination coefficients with surface temperature. In Fig. 14, Vycor data are plotted for surface temperatures between room temperature and 1100 K[19,20] and arc-plasma data using Eq. (1) between $1400 \leq T_w \leq 1650$ K. These data correlate well with coefficients calculated from similar arc-plasma tests.[21] The experimental data, which do not account for gas-phase recombination in the boundary layer, are lower than the values from the reacting boundary-layer computations. Again, if it is assumed that the RCG baseline coating surface layer is very similar to Vycor, the rapid rise in the recombination coefficient would correspond to a rapid decrease in viscosity between the annealing point temperature (T_w = 1183 K) and softening point temperature (T_w = 1650 K) of the glass.[22,23] Therefore, as the temperature increases, ionic mobility will increase because of the decrease in viscosity and the degree of thermal dissociation will increase. Both these factors could increase the catalytic efficiency of the RCG baseline coating at

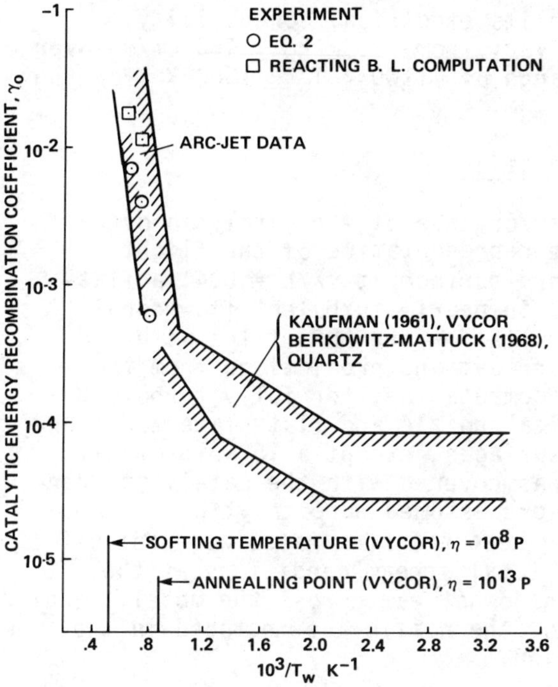

Fig. 14 Recombination coefficient for RCG as function of surface temperature.

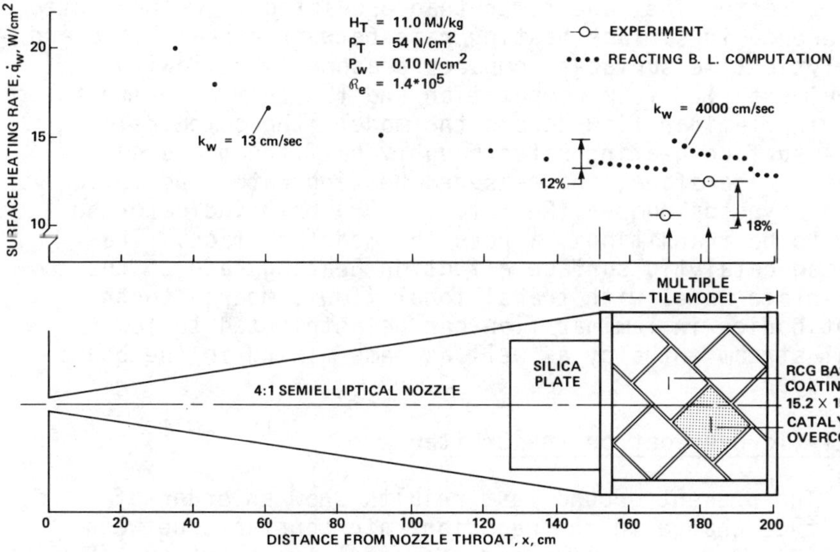

Fig. 15 Heating rate distribution over multiple-tile surface in 20 MW facility.

high temperature through its electrical conductivity. Reaction rate constants vary from $10 \leq k_w \leq 460$ cm/s over a surface temperature range of $1100 \leq T_w \leq 1800$ K for the RCG glass coating.

Near-Turbulent Flow Prediction

To determine the performance of the catalytic overcoat for conditions more representative of the flight experiment on the windward surface at $X/L = 0.4$, a flat-plate test was performed in nearly turbulent flow conditions. Experimental results for the HRSI tiles with and without the catalytic overcoat and predictions from the reacting boundary-layer computation, for fully turbulent flow along a semielliptical nozzle and flat-plate model are shown in Fig. 15. All surfaces, except a 15 × 15 cm tile in the flat plate that was covered with the catalytic overcoat ($k_w = 4000$ cm/s), were assumed noncatalytic ($k_w = 13$ cm/s). Surface heat-transfer rates were calculated starting with the total stream conditions at the nozzle throat and moving downstream across the model. Heat loss to the cold walls of the nozzle are included in the total enthalpy calculation.[7]

The experimental values, obtained from surface temperature data from tiles positioned symmetrically about the stream centerline, are lower than predicted. The predicted difference in surface heating rate between noncatalytic and catalytic tile surfaces compares reasonably well with experiment (12% from computation and 18% from experiment). Assuming laminar flow across the model, the computation gives surface heating rates roughly half the measured values. Therefore, the measured heating rates and calculated Reynolds number ($Re = 1.4 \times 10^5$) both indicate the flow to be transitional across the model surface. The reduced catalytic surface effect on heating rate to the flat-plate model with transitional flow compared to the blunt bodies in laminar flow can be attributed to lower total stream enthalpy as well as some mixing in the boundary layer.

Catalytic Overcoat on the Orbiter

The present ground test results show an order-of-magnitude change in the reaction rate constants between the RCG baseline coating and the catalytic overcoat. During Earth entry, the area of the midfuselage of the Orbiter in which the OEX experiment is to be conducted has a laminar boundary layer. The windward centerline heating rate distribution, obtained from the reacting boundary-layer computation, shows the catalytic surface effect along the midfuselage for a nominal design trajectory in Fig. 16. The peak surface heating rate to the tiles with the catalytic overcoat is about twice the heating rate on the (just upstream) RCG baseline coated tile. In addition, the surface heating rate to the tiles with the catalytic overcoat is about 40-50% higher than the equilibrium value, corresponding to a fully catalytic surface.

Figure 17 shows a more detailed profile of the surface heating rate across a tile with catalytic overcoat. Included in Fig. 17 is a cross-sectional sketch of the tile insulation. The 4.06 cm thick HRSI tile is attached to a strain isolation pad (SIP) which, in turn, is attached to the aluminum skin. An important concern in adding a catalytic overcoat to the tile surface is whether the aluminum skin of the Orbiter will overheat during Earth entry. Using an average surface heating rate, from calculated pulses at various times during a nominal Earth entry trajectory, the aluminum skin temperature rise was estimated from one- and three-dimensional models.[24] The one-dimensional model calculation — for a single tile at $X/L = 0.1$ with catalytic overcoat — showed the peak temperature of the aluminum structure to be well within the Rockwell design limit. When lateral conduction in the

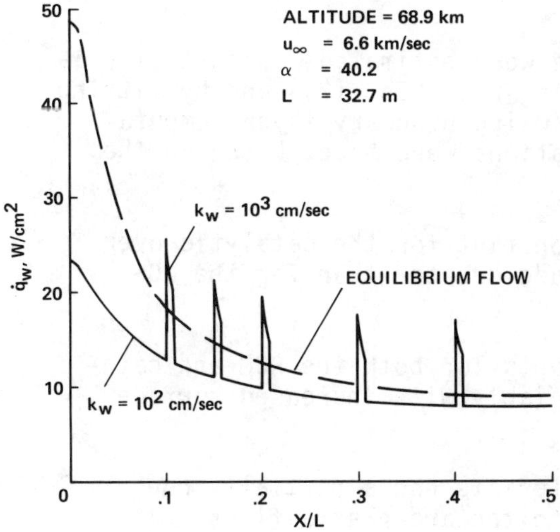

Fig. 16 Effect of catalytic overcoat on Orbiter centerline heating rate distribution.

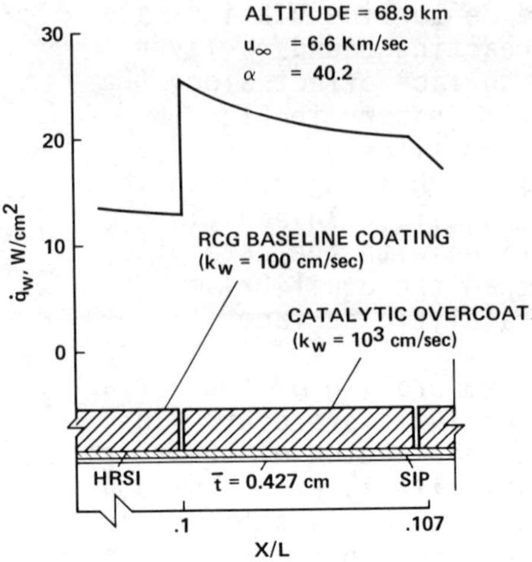

Fig. 17 Heating distribution across tile with catalytic overcoat.

aluminum structure is included, the increase in the peak aluminum temperature caused by the catalytic overcoat on a single tile at $X/L = 0.4$ is only a few degrees.

Conclusion

A spray-on overcoat, with high catalytic efficiency, has been developed for use in an OEX experiment to determine the HRSI surface catalytic efficiency through comparative suface temperature measurements. In addition, catalytic efficiency and aerothermal response of HRSI and RCC with and without the overcoat were evaluated using various models in arc-plasma flow.

Reaction rate constants were estimated, using a simplified relationship with heating rate [Eq. (2)] and by fitting heating rate data with a reacting boundary-layer computation. The following observations were made, based on the test data and computations:

1) The reaction rate constant for the catalytic overcoat was an order of magnitude greater than for the RCG baseline coating.

2) Reaction rate constants for both the RCG and catalytic overcoat varied appreciably with increased surface temperature.

3) Both RCC and RCG materials had a partially noncatalytic surface response to the arc-plasma flows. The reaction rate constants at 1344 K were about $k_w = 10^2$ cm/s.

4) The reacting boundary-layer computation accurately predicted the effect of the catalytic overcoat on hot-wall heating rate distribution along a 40 deg blunt truncated cone. The rise in surface heating rate in moving from a noncatalytic to catalytic surface on the conical portion of the blunted cone was greater than the predicted equilibrium value at that location. Similar predictions from the reacting boundary-layer computation were obtained for the Orbiter midfuselage during Earth entry, with the catalytic overcoat applied to selected tiles between $0.1 \leq X/L \leq 0.4$, and were used to design a flight experiment.

5) The reacting boundary-layer computation, assuming turbulent flow along the semielliptical nozzle and flat-plate model in the 20 MW Pilot Facility, overpredicted the surface heating rates on the tiles with and without the catalytic overcoat. However, a good estimate of the change in surface heating rate from the RCG to catalytic overcoat was obtained.

6) The effect of the catalytic overcoat on surface heating rate was much less in transitional flow (near turbulent) than laminar flow (15 and 35%, respectively).

References

[1]Goulard, R., "On Catalytic Recombination Rates in Hypersonic Stagnation Heat Transfer," Jet Propulsion, Vol. 28, Nov. 1958, pp. 737-745.

[2]Rosner, D. E., "Boundary Conditions for the Flow of a Multicomponent Gas," Jet Propulsion, Vol. 28, Aug. 1958, pp. 555-556.

[3]Sheldahl, R. E. and Winkler, E. L., "Effect of Discontinuities in Surface Catalytic Activity on Laminar Heat Transfer in Arc-Heated Nitrogen Streams," NASA TN-3615, 1966.

[4]Stewart, D. A., "Cyclic Arc Plasma Tests of RSI Materials Using a Preheater," NASA TMS-2720, 1973.

[5]Rakich, J. V. and Lanfranco, M. J., "Numerical Computation of Space Shuttle Heating and Surface Streamlines," Journal of Spacecraft and Rockets, Vol. 14, May 1977, pp. 265-272.

[6]Yoshikawa, K. K. and Katzen, E. D., "Charts for Air-Flow Properties in Equilibrium and Frozen Flows in Hypervelocity Nozzles," NASA TND-693, 1961.

[7]Winovich, W., "On the Equilibrium Sonic-Flow Method for Evaluating Electric Arc Heater Performance," NASA TND-2132, 1964.

[8]Goldstein, H. E., Leiser, D. G., and Katvala, V., "Reaction Cured Borosilicate Glass Coating for Low-Density Fibrous Silica Insulation, Borate Glasses, Plenum Corp., New York, 1978, pp. 623-634.

[9]Marshall, T. C., "Surface Recombination of Nitrogen Atoms Upon Quartz," Journal of Chemical Physics, Vol. 37, 1962, pp. 2501-2502.

[10]Wood, B. J. and Wise, H., "The Interaction of Atoms with Solid Surfaces," Stanford Research Institute Project Squid, Office of Naval Research, 1960.

[11]Anderson, L. A., "Effect of Surface Catalytic Activity on Stagnation Heat-Transfer Rates," AIAA Journal, Vol. 11, May 1973, pp. 649-656.

[12]Rosner, D. E., "Analysis of Air Arc-Tunnel Heat Transfer Data," AIAA Journal, Vol. 2, May 1965, pp. 945-948.

[13]Marvin, J. G. and Diewer, G. S., "Convective Heat Transfer in Planetary Gases," NASA TR R-224, 1965.

[14]Pope, R. B., "Stagnation-Point Convective Heat Transfer in Frozen Boundary Layers," AIAA Journal, Vol. 5, April 1968, pp. 619-626.

[15]Hansen, F. C., "Approximations for the Thermodynamic and Transport Properties of High-Temperature Air," NASA TR R-50, 1959.

[16]Tong, H., Buckingham, A. C., and Curry, D. M., "Computational Procedure for Evaluations of Space Shuttle TPS Requirements," AIAA Paper 74-518, 1974.

[17]Johansen, K. J., McGinnis, F. K., and Hill, C. M., "Shuttle Less Design Allowables Thermal Properties Test Program," Final Report, Vought Corp., 221RP00445, June 1976.

[18]Lees, Lester, "Laminar Heat Transfer Over Blunt-Nose Bodies at Hypersonic Flights Speeds," Jet Propulsion, Vol. 26, April 1956, pp. 259-269, 274.

[19]Kaufman, F., "Reactions of Oxygen Atoms," Progress in Reaction Kinetics, Vol. 1, 1961, pp. 1-39.

[20]Berkowitz-Mattuck, J. B., The Structure and Chemistry of Solid Surfaces, John Wiley & Sons, New York, 1969.

[21]Scott, C. D., "Catalytic Recombination of Nitrogen and Oxygen on High-Temperature Reusable Surface Insulation," AIAA Paper 80-1477, 1980.

[22]Harrison, D. E. and Moratis, C. J., "Ceramics, Glasses, and Micas," Handbook of Materials and Processes for Electronics, McGraw-Hill Book Co., New York, 1970, pp. 6-43.

[23]Van Vlack, J. L., "Thermal Properties and High Temperature Behavior," Physical Ceramics for Engineers, Addison-Wesley, Reading, Mass., 1964, pp. 156.

[24]Derry, S. M., Personal communication, 1981.

Space Shuttle Laminar Heating with Finite-Rate Catalytic Recombination

Carl D. Scott*
NASA Lyndon B. Johnson Space Center, Houston, Texas

Abstract

Temperature-dependent catalytic recombination coefficients are used to calculate the heat flux to the Space Shuttle Orbiter. With variable entropy boundary-layer edge properties obtained from a reacting inviscid flowfield calculation, the reacting boundary-layer equations are solved along the forward 50% of the windward centerline using the boundary-layer integral matrix procedure. The results are compared with fully catalytic and noncatalytic reacting solutions, with equilibrium solutions, and with the Orbiter design heating rates. The fully catalytic heat fluxes are about equal to the equilibrium predictions; whereas, the finite catalytic results are about 12-27% lower than the fully catalytic results (approaching the noncatalytic results downstream) and as much as 30-50% lower than design predictions away from the nose area.

Nomenclature

k = Boltzmann constant
k_f = forward (recombination) surface reaction rate
k_w = catalytic recombination speed
L = length of Orbiter
m = mass of atom
α = mass fraction
q = heat flux

Presented as Paper 81-1144 at the AIAA 16th Thermophysics Conference, Palo Alto, Calif., June 23-25, 1981.

*Research Engineer.

r_o = streamline metric
s = distance along surface
T = temperature
u = velocity in streamwise direction
x = axial distance from nosetip of Orbiter
y = normal distance from surface
β = chemical energy accommodation coefficient
γ' = energy-transfer catalytic recombination coefficient
γ = atom recombination coefficient
ε = emittance
η = normal coordinate defined by Eq. (3)
ξ = nondimensional streamwise coordinate defined by Eq. (4)
μ = viscosity
ρ = density

Subscripts

e = boundary-layer edge
N = nitrogen
O = oxygen
RI = design value
w = wall
$\gamma = 1$ = fully catalytic
$\gamma = 0$ = noncatalytic
$\gamma(T)$ = temperature-dependent finite catalytic recombination coefficient for high-temperature reusable surface insulation

Introduction

Measurements of nitrogen and oxygen energy-transfer catalytic recombination coefficients on high-temperature reusable surface insulation (HRSI)[1] have enabled more accurate predictions of the heating rate to the Space Shuttle Orbiter during re-entry. The method of calculation used herein resulted from a sequence of work that followed two parallel efforts - one for equilibrium flow and the other for chemically reacting flow. These approaches are described in papers appearing in the literature during the last few years in which predictions of heat fluxes to the Space Shuttle Orbiter are presented.[2-7] In these papers, the heating has been predicted using two-dimensional boundary-layer codes with edge conditions based on solutions to three-dimensional inviscid flow equations with an Orbiter-like geometry.[8,9] C. P. Li[10] developed a blunt-body time-dependent solution of the Euler equations and applied it to the nose of the Orbiter. His equilibrium version of the code was used to generate starting conditions for the inviscid three-dimensional shock-layer code developed by Kutler et al.[11,12] Subsequently, Goodrich et al.[3] used solutions obtained with these codes for the Orbiter geometry as variable entropy boundary-layer edge conditions to obtain equilibrium solutions using the nonsimilar boundary-layer code (BLIMP) of Tong, Buckingham, and Curry.[2]

The reacting version of this effort is based on flowfield solutions obtained by Rakich and Lanfranco[4] using the methods presented in Refs. 8, 13, and 14. They thus predicted heat fluxes to the Orbiter windward centerline with three-dimensional inviscid flowfield properties applied as boundary conditions for the reacting boundary-layer code (BLIMPK) of Bartlett and Kendall[15] and Tong, Buckingham, and Morse.[16] These calculations incorporated constant average catalytic recombination speeds k_w for oxygen and nitrogen. The values used by Rakich and Lanfranco were from Ref. 2 and were based on correlations of heat flux measurements on various thermal protection materials at various temperatures.[16] These values can be considered only approximate, but the heat fluxes obtained using them may indicate trends as the level of dissociation varies.

Until now, none of the calculations has included temperature-dependent energy-transfer recombination coefficients for nitrogen and oxygen recombination on the Orbiter high-temperature reusable surface insulation. These recombination coefficients were measured in an arc-jet flow and are reported in Ref. 1. As described, these coefficients were inferred from heat-transfer measurements at the stagnation point of flat-faced cylinders. The analysis of Goulard[17] was used to infer γ_N and γ_O from nitrogen and air experiments. Heat fluxes to a nickel surface having known recombination coefficients were measured on a geometrically similar model. These measurements were then used for reference to infer the energy-transfer catalytic recombination coefficients on HRSI.

This paper presents calculations using the temperature-dependent recombination coefficients as wall boundary conditions in boundary-layer solutions of the same type as in Ref. 4. As will be shown, the main significance of these coefficients to the present calculation is that the HRSI on the midfuselage lower surface of the Orbiter appears to be much less catalytic than previously assumed for design considerations. Also, the heat fluxes during the high heating regime of the entry trajectory are predicted to be 12-27%

lower than the corresponding fully catalytic predictions and as much as 50% lower than the design predictions on the midfuselage.

Analysis

The method of calculating Orbiter heat fluxes that account for finite-rate catalytic recombination is based on heat-transfer measurements on the glass coating of HRSI obtained in dissociated arc-jet flow from which the temperature-dependent energy-transfer recombination coefficients γ_O and γ_N were inferred. From heat-transfer measurements alone, differentiation between the fraction of atoms recombining γ' from the fraction of recombination energy absorbed per recombination β is not possible. Therefore, the assumption is made that the net recombination coefficient for energy transfer may be defined as $\gamma = \gamma'\beta$, thus henceforth no distinction will be made between γ and γ'. This assumption implies that γ_O and γ_N are suitable parameters in both the energy boundary equation as well as the species boundary equations. The assumption is satisfactory for heat-transfer predictions as long as the excited molecular species formed during a recombination event do not subsequently quench on the downstream wall or in the boundary layer near the wall. Further investigation in the laboratory and comparison of the heating rate predictions with flight measurements are necessary to completely justify this assumption. (This assumption and its consequences are examined in subsequent sections of this paper.)

The heat flux and other wall and boundary-layer flow properties were calculated by solving the nonsimilar boundary-layer equations using the reacting version of the integral matrix procedure (BLIMPK) developed by Bartlett and Kendall[15] and extended by Tong, Buckingham, and Morse[16] for axisymmetric analog geometries. The boundary-layer edge properties and streamline metric were obtained from Rakich,[18] who with Lanfranco published their method and heating results.[4] These variable entropy edge properties result from solutions to the reacting three-dimensional Euler equations for the Orbiter geometry. The wall boundary conditions are easily applied to the code in the form of temperatures along the body T_w and catalytic recombination rates. These first-order reaction rates were used in the form

$$k_f = \text{const} \times k_w = \text{const} \times \sqrt{kT/2\pi m}\,\gamma(T)$$

Table 1 - Parameters for nominal Shuttle flight (trajectory 14414)

Parameter	Case 1	2	3	4
Time, s	432	650	754	900
Velocity, km/s	7.62	6.614	5.944	5.121
Altitude, km	74.98	68.88	64.92	60.66
Angle of attack, deg	41.4	40.2	34.5	31.8
Mach number	27.5	22.6	19.6	16.4
Total enthalpy, MJ/kg	29.0	21.8	17.6	13.1
Freestream temperature, K	197	221	237	253
Freestream density, kg/m^3	3.795×10^{-5}	9.28×10^{-5}	1.55×10^{-4}	2.61×10^{-4}
Stagnation point pressure, N/m^2	2200	4100	5540	6930
Vehicle length, m	32.84	32.84	32.84	32.84

where the results from Ref. 1 are

$$\gamma_N = 0.071e^{-2219/T_w} \quad 1670\ K > T_w > 950\ K \tag{1}$$

$$\gamma_O = 16e^{-10271/T_w} \quad 1650\ K > T_w > 1400\ K \tag{2}$$

These energy-transfer catalytic recombination coefficients for nitrogen and oxygen recombination on HRSI were inferred from stagnation point heat flux measurements in arc-jet flow. These analytical expressions are curve fits of the inferred results. Iteration of the solutions of the boundary-layer code is necessary to converge to the correct wall temperature. The first wall temperature distribution down the vehicle was estimated. Each successive trial temperature distribution used radiation equilibrium values calculated from the heat flux results of the previous iteration. The wall temperatures converged ($\Delta T_w < 5$ K) within about three iterations. An emittance of 0.85 was assumed.

Results

Solutions to the boundary-layer equations were obtained for four points in the Orbiter design trajectory between 75 and 61 km altitude. The freestream properties for these four cases are given in Table 1. In each case, solutions were obtained for three wall recombination conditions.

1) Fully catalytic ($\gamma_O = \gamma_N = 1$).

2) Finite recombination rates that depend on the wall temperature expressed in Eqs. (1) and (2), ($\gamma(T)$).

3) Noncatalytic ($\gamma_O = \gamma_N = 0$).

These solutions extend along the windward centerline of the Orbiter from the stagnation point to about 50% of the Orbiter length. Since the Orbiter surface from the nose or the stagnation point to about $x/L = 0.02$ is reinforced carbon-carbon rather than HRSI, the results are only given for $x/L \geq 0.02$.

Most of the calculations incorporate 10 nodes normal to the body surface. In the three lower altitude cases with a noncatalytic wall, the number of nodes was increased to 14 to obtain more accurate species concentration profiles near the outer edge of the boundary layer.

Heating

The heat flux results are given in Fig. 1 as a function of distance down the Orbiter x/L. The reacting boundary-layer results with the three catalytic recombination coefficient conditions are presented and compared with the corresponding equilibrium cases calculated by Goodrich et al.[3] and the design predictions.[19] The design values are based on correlation factors obtained by comparing wind-tunnel measurements with simple one- and two-dimensional predictions for cones, wedges, etc. These factors are multiplied by heating rates predicted by corresponding simple flow and heating correlations for flight.

The results (Fig. 1) show that the fully catalytic heat fluxes are very nearly equal to the equilibrium values. This relationship is consistent with the correlation formulas for stagnation point solutions such as Fay and Riddell[20] and is also consistent with Fig. 3 of Ref. 4, in which the limiting case of a fully catalytic wall approaches the equilibrium prediction at the stagnation point as the catalytic recombination speed, k_w, is increased. This observation cannot be generalized too far because the dynamics of the flow far out of dissociative equilibrium may exhibit a different character; hence, the heating rate may differ from equilibrium predictions even though the wall is fully catalytic. This assertion is supported by the work of Rosner and Cibrian[21] and by

Goodrich et al.,[3] where their ideal-gas flowfield results in higher heat flux than their equilibrium calculations.

The design predictions of heat flux[20] q_{RI} differ qualitatively from the reacting fully catalytic and equilibrium solutions. As shown in Fig. 2, the fully catalytic results are as much as 20% greater than the design predictions for x/L less than about 0.1, but are as much as 35% lower than the design predictions for x/L > 0.1. This relationship obtains for all four trajectory points and indicates a basic difference in the present calculations and the design method.

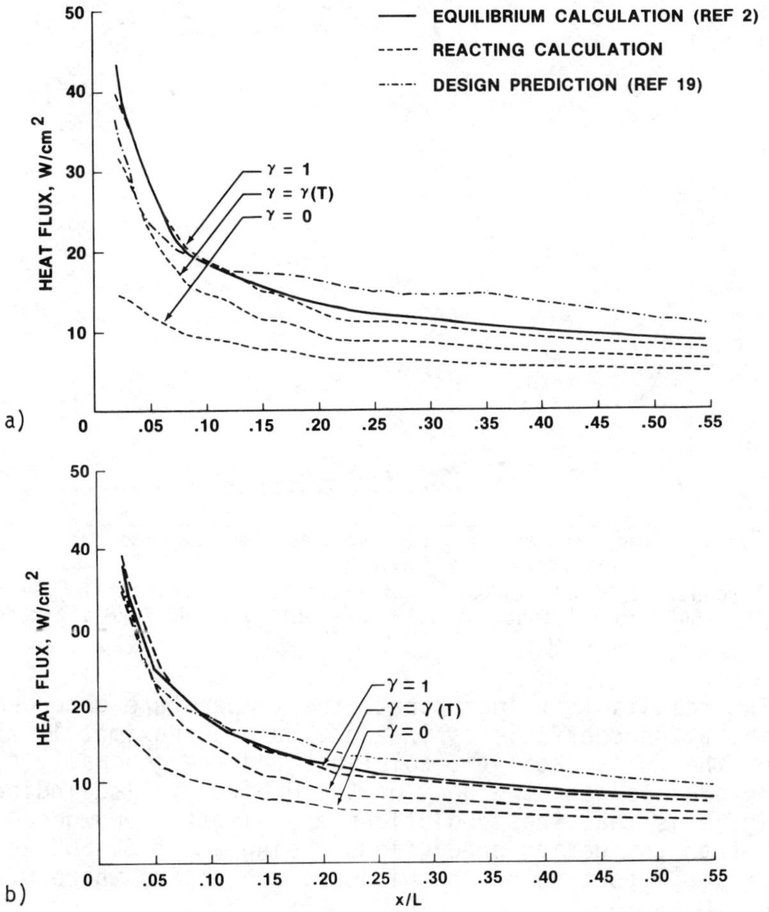

Fig. 1 Orbiter windward centerline heating rates: a) 7.62 km/s velocity, 75 km altitude; b) 6.61 km/s velocity, 68.9 km altitude.

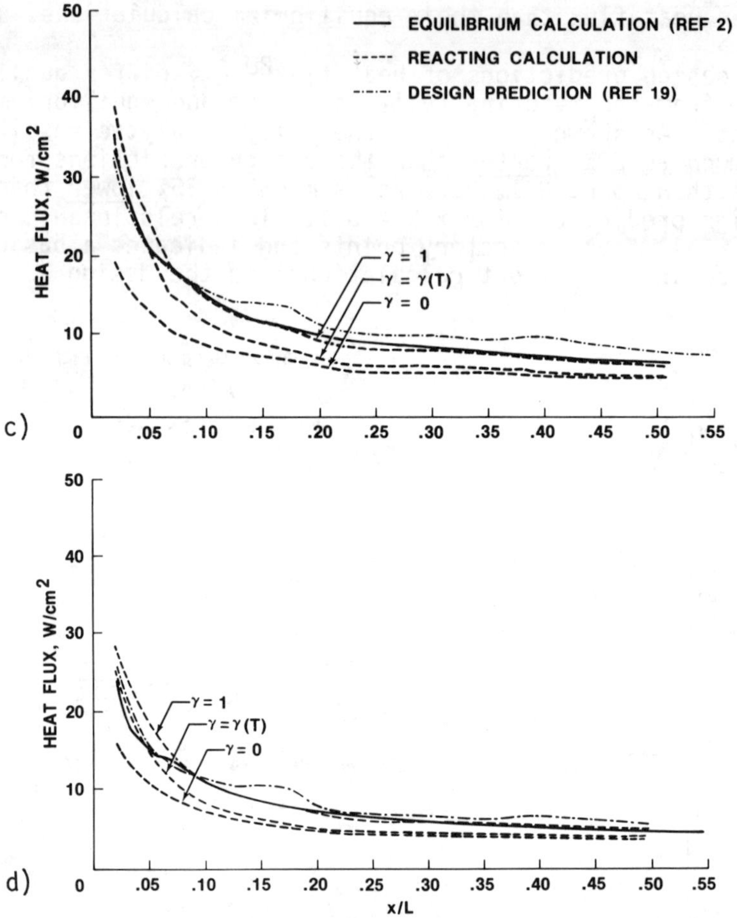

Fig. 1 (cont.) Orbiter windward centerline heating rates: c) 5.94 km/s velocity, 64.9 km altitude; d) 5.12 km/s velocity, 60.7 km altitude.

The results that incorporate the temperature-dependent recombination coefficients γ_N and γ_O are approximately 73-88% of the fully catalytic solutions and are generally similar in profile down the body of the Orbiter. Also indicated in Fig. 2 is that these solutions are almost everywhere lower than the design predictions, being about 30-50% lower over a large portion of the windward centerline where $0.15 < x/L < 0.55$.

Greater understanding of the nature of the reacting boundary layer and flowfield is obtained by examining the behavior of the noncatalytic heat flux $q_{\gamma=0}$ compared with

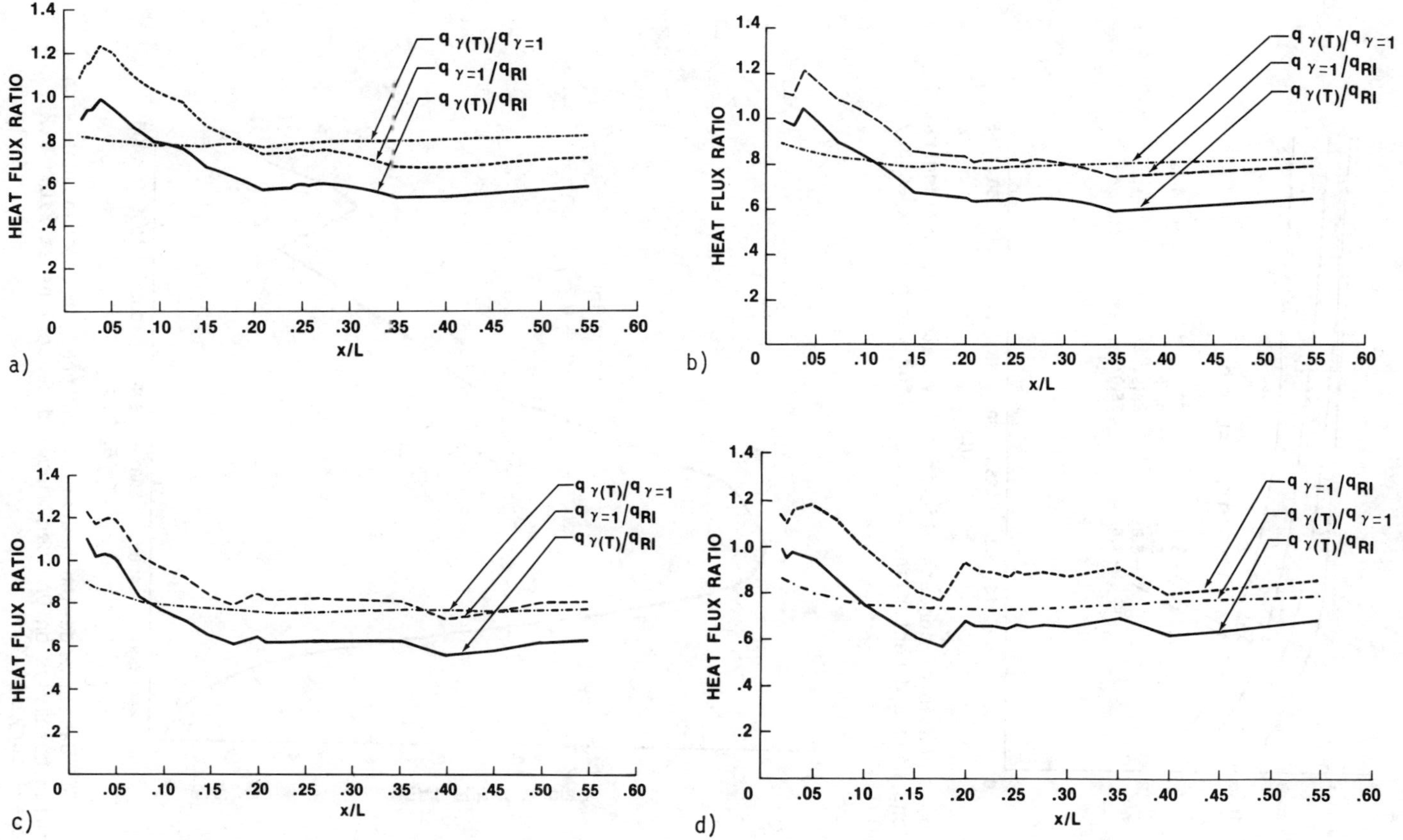

Fig. 2 Ratios of Orbiter windward centerline heating rates: a) 7.62 km/s velocity, 75 km altitude; b) 6.61 km/s velocity, 69 km altitude; c) 5.94 km/s velocity, 65 km altitude; d) 5.12 km/s velocity, 61 km altitude.

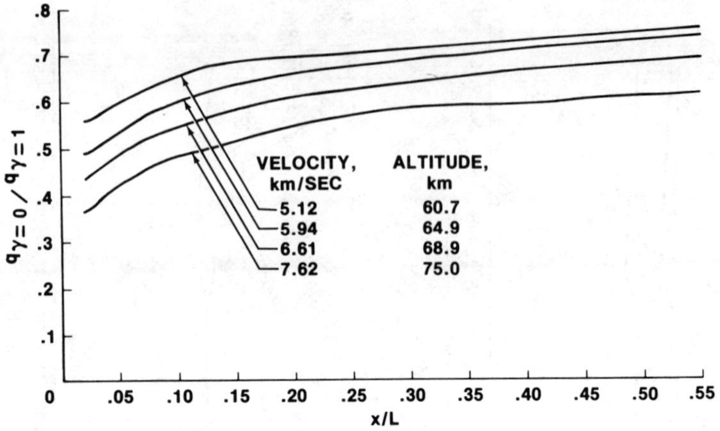

Fig. 3 Ratio of noncatalytic to fully catalytic heating rates along windward centerline of Orbiter for various trajectory points.

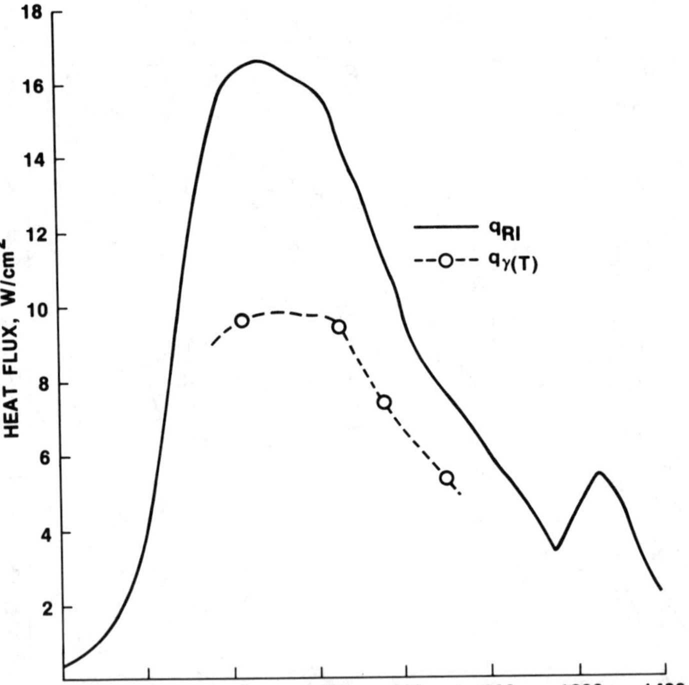

Fig. 4 Baseline design heat flux and present calculations with finite catalysis on Orbiter windward centerline at x/L = 0.2 (trajectory 14414).

the fully catalytic flux $q_{\gamma=1}$. In Fig. 3, the noncatalytic heat flux can be seen to remain significantly lower than $q_{\gamma=1}$, even well downstream on the Orbiter where the boundary-layer edge is influenced more by an oblique shock (rather than a normal shock) and where normal shock flow near the surface has had more time to approach equilibrium. These results imply that the reaction rates in the gas phase are slow.

The influence of the finite catalysis on the overall heat load is illustrated in Fig. 4 where the design heat flux for x/L = 0.2 is plotted against time. The heat fluxes for the four cases computed in the paper are also shown on the plot and can be seen to be about 35-46% lower than the design values during most of the high heating part of the trajectory. These results also mean the total heat load is lower by about the same factor.

Atom Mass Fractions

The behavior of the atom mass fractions at the edge of the boundary layer is shown in Fig. 5. These results of Rakich's calculations[18] show that the oxygen remains fully dissociated until late in the trajectory (altitude 61 km). Nitrogen atoms persist down to 65 km, decreasing with x/L down the vehicle.

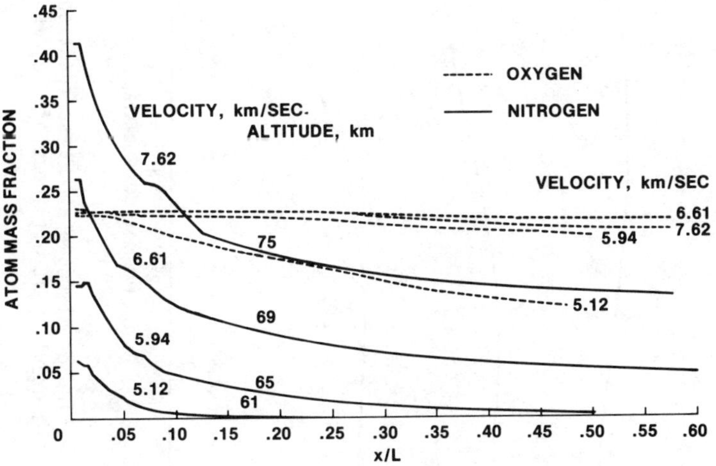

Fig. 5 Atom mass fractions at edge of boundary layer along Orbiter windward centerline (from Ref. 18).

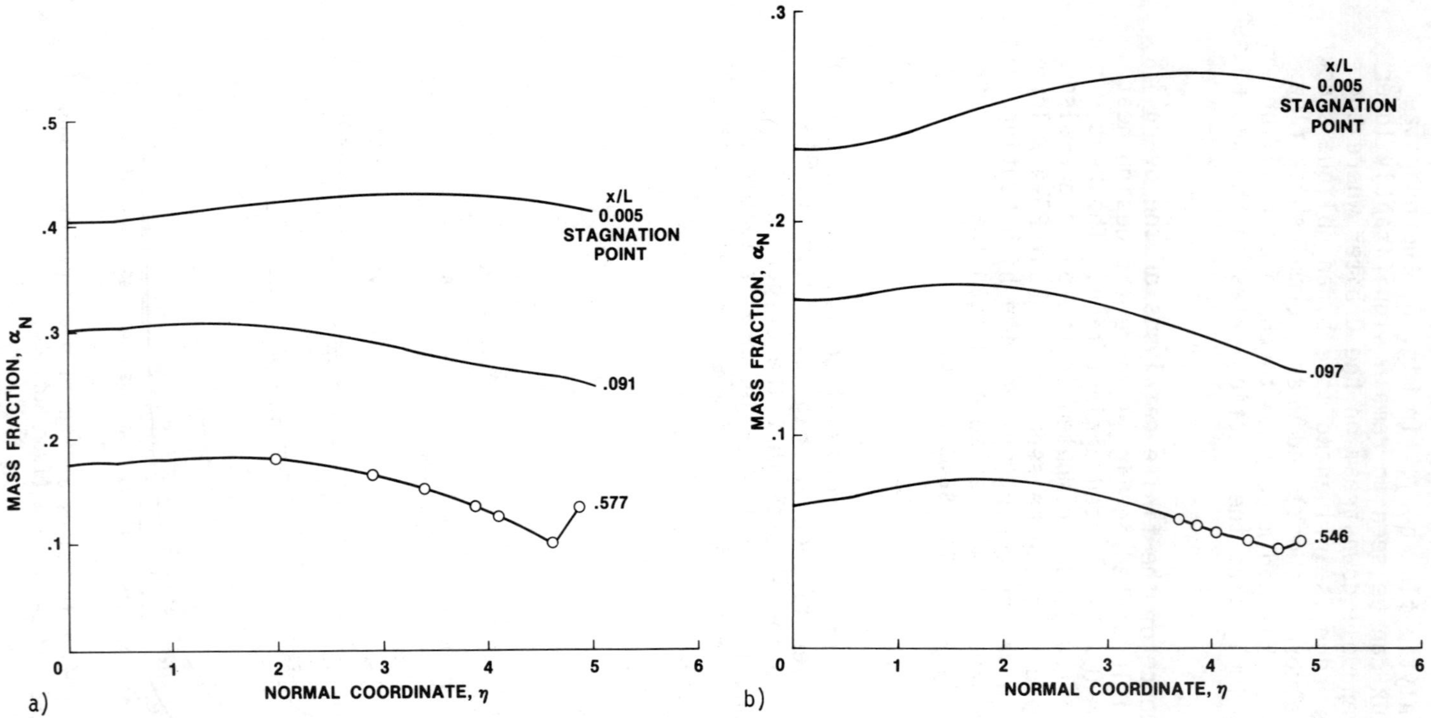

Fig. 6 Nitrogen atom mass fraction profiles normal to the surface at various x/L on the Orbiter centerline at altitudes of: a) 75 km and b) 69 km.

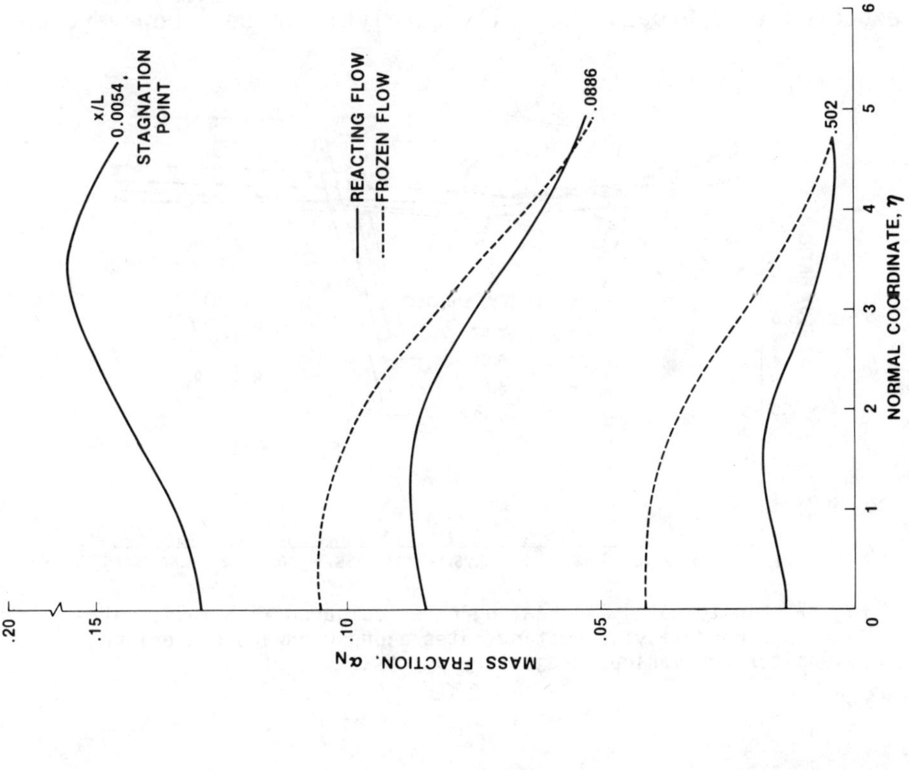

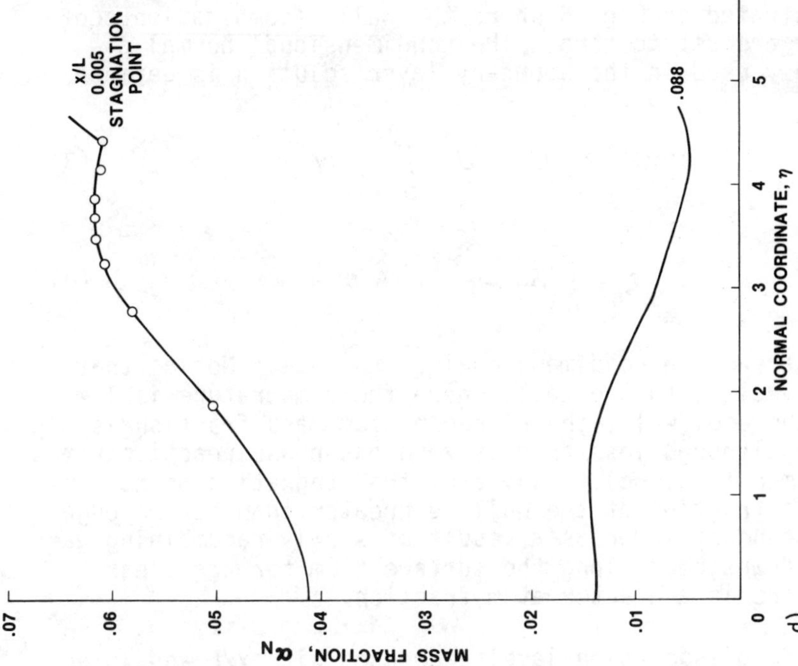

d)

Fig. 6 (cont.) Nitrogen atom mass fraction profiles normal to the surface at various x/L on the Orbiter centerline at altitudes of: c) 65 km and d) 61 km.

The slow gas-phase recombination in the boundary layer is illustrated in Fig. 6 where the wall recombination coefficients are set to zero. The nondimensional normal coordinate y used in the boundary-layer solution is defined by

$$\eta \equiv (r_o\ u_e)/\sqrt{2\xi}\ \int_0^y p\ dy \qquad (3)$$

where

$$\xi \equiv \int_0^s u_e\ \rho_e\ \mu_e\ r_0^2\ ds \qquad (4)$$

is the streamwise nondimensional coordinate. Notice that even very close to the wall, where the temperature is low due to the cool wall, the nitrogen atom mass fraction is significant although less than if zero gas-phase reaction rates are assumed (Fig. 6c). Away from the stagnation point, the atom mass fraction at the wall is greater than at the edge of the boundary layer as a result of slowly recombining gas flowing downstream along the surface from forward areas where there is a greater atom fraction.

As the dissociation level decreases with x/L and lower altitude, the finite catalytic rate heat flux $q_{\gamma(T)}$ would be expected to approach the fully catalytic value. However, be-

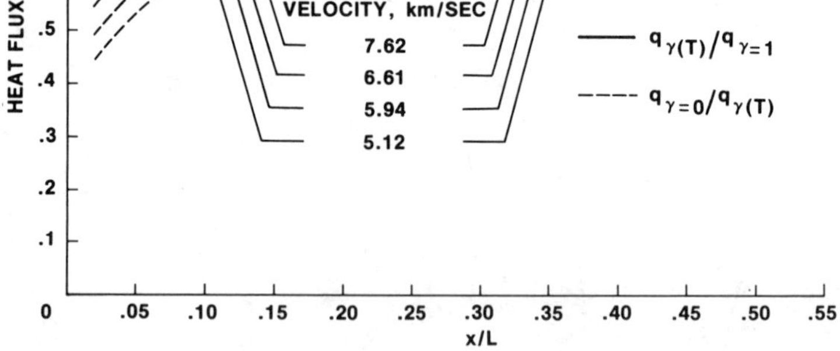

Fig. 7 Finite catalytic heating rates compared with fully catalytic and noncatalytic heating rates along windward centerline of Orbiter for various trajectory points.

cause the wall temperature also decreases with both x/L and low altitude, the relative heat flux remains fairly constant (0.73-0.88) as seen in Fig. 7. Indeed, $q_{\gamma(T)}$ approaches the noncatalytic value $q_{\gamma=0}$ at low altitudes and large x/L.

There is greater uncertainty in the low wall temperature results because of the necessity to extrapolate γ_0 to temperatures outside the range in which γ_0 was obtained. The minimum temperature in the measurements was about 1400 K, whereas the predicted temperatures were as low as 952 K.

Conclusions

This work presents the first systematic use of temperature-varying energy-transfer catalytic recombination coefficients in calculating the heat flux to the Space Shuttle Orbiter. These coefficients, applied as wall boundary conditions to reacting laminar boundary-layer calculations, have a significant influence on the heat flux distribution down the centerline of the Orbiter during the high heating regime of the Orbiter entry, which is represented by the four trajectory points for which calculations were made:

1) The heat flux to the high-temperature reusable surface insulation is 73-88% of the fully catalytic calculation over most of the forward 55% of the vehicle bottom centerline.

2) This heat flux prediction that includes finite rate recombination is also 30-50% lower than the baseline design predictions over a large part of the Orbiter ($0.15 \leq x/L \leq 0.55$).

3) Because of the lower temperature, the HRSI glass coating approaches noncatalytic behavior aft on the vehicle when the Orbiter reaches the lower altitudes but before transition to turbulence.

4) The boundary layer on the Orbiter is slowly reacting but is not completely frozen at the altitudes considered here.

5) The fully catalytic reacting heat fluxes are almost equal to equilibrium predictions obtained using the analogous flowfield and boundary-layer method.

Acknowledgments

The author is grateful to John V. Rakich of the NASA Ames Research Center for furnishing boundary-layer edge conditions for the calculations, to Dr. Chien Li of the NASA

Lyndon B. Johnson Space Center (JSC) for his assistance with the boundary-layer code, and to Dr. Li and Dr. Winston D. Goodrich of JSC for furnishing results of their equilibrium calculations.

References

[1]Scott, C. D., "Catalytic Recombination of Nitrogen and Oxygen on High-Temperature Reusable Surface Insulation," Aerothermodynamics and Planetary Entry, Progress in Astronautics and Aeronautics, Vol. 77, edited by A. L. Crosbie, AIAA, New York, 1981, pp. 192-212.

[2]Tong, H., Buckingham, A. C., and Curry, D. M., "Computational Procedure for Evaluation of Space Shuttle TPS Requirements," AIAA Paper 74-518, June 1974.

[3]Goodrich, W. D., Li, C. P., Houston, C. K., Chiu, P. B., and Olmedo, L., "Numerical Computations of Orbiter Flowfields and Laminar Heating Rates," Journal of Spacecraft and Rockets, Vol. 14, May 1977, pp. 257-264.

[4]Rakich, J. V. and Lanfranco, M. J., "Numerical Computation of Space Shuttle Laminar Heating and Surface Streamlines," Journal of Spacecraft and Rockets, Vol. 14, May 1977, pp. 265-272.

[5]Rakich, J. V. and Pegot, E. B., "Flow Field and Heating on the Windward Side of the Space Shuttle Orbiter," NASA SP-347, Pt. II, 1975, pp. 1377-1394.

[6]Goodrich, W. D., Li, C. P., Houston, C. K., Meyers, R. M., and Olmedo, L., "Scaling of Orbiter Aerothermodynamic Data Through Numerical Flow Field Simulations," NASA SP-347, Pt. II, 1975, pp. 1395-1410.

[7]Rakich, J. V. and Park, Chul, "Nonequilibrium Three-Dimensional Supersonic Flow Computations with Applications to the Space Shuttle Orbiter Design," Proceedings of the Symposium on Application of Computers to Fluid Dynamic Analysis and Design, Polytechnic Institute of Brooklyn Graduate Center, New York, Jan. 1973, pp. 205-212.

[8]Rakich, J. V. and Kutler, Paul, "Comparison of Characteristics and Shock Capturing Methods with Application to the Space Shuttle Vehicle," AIAA Paper 72-191, Jan. 1972.

[9]Rizzi, A. W. and Bailey, H. E., "Reacting Nonequilibrium Flow around the Space Shuttle Using a Time-Split Method," NASA SP-347, Pt. II, 1975, pp. 1327-1349.

[10]Li, C. P., "A Numerical Study of Laminar Flow Separation on Blunt Flared Cones at Angle of Attack," AIAA Paper 74-585, June 1974.

[11]Kutler, P., Reinhardt, W. A., and Warning, R. F., "Multishocked, Three-Dimensional Supersonic Flowfields with Real Gas Effects," AIAA Journal, Vol. 11, May 1973, pp. 657-664.

[12]Kutler, Paul, Warning, R. F., and Lomax, Harvard, "Computations of Space Shuttle Flowfields Using Noncentered Finite-Difference Schemes," AIAA Journal, Vol. 11, Feb. 1973, pp. 196-204.

[13]Rakich, J. V., Bailey, H. E., and Park, Chul, "Computation of Nonequilibrium Three-Dimensional Inviscid Flow over Blunt-Nosed Bodies Flying at Supersonic Speeds," AIAA Paper 75-835, June 1975.

[14]Rakich, J. V. and Mateer, G. G., "Calculation of Metric Coefficients for Streamline Coordinates," AIAA Journal, Vol. 10, Nov. 1972, pp. 1538-1540.

[15]Bartlett, E. P. and Kendall, R. M., "An Analysis of the Coupled Chemically Reacting Boundary Layer and Charring Ablator, Pt. III: Nonsimilar Solution of the Multicomponent Laminar Boundary Layer by an Integral Matrix Method," NASA CR-1062, June 1968.

[16]Tong, H., Buckingham, A. C., and Morse, H. L., "Nonequilibrium Chemistry Boundary Layer Integral Matrix Procedure," NASA CR-134039, July 1973.

[17]Goulard, R., "On Catalytic Recombination Rates in Hypersonic Stagnation Heat Transfer," Jet Propulsion, Vol. 28, Nov. 1958, pp. 737-745.

[18]Rakich, J. V., Private communication, May 1980.

[19]Haney, J. W. and Petrilla, C. T., "Space Shuttle Orbiter Entry Aerodynamic Heating Data Book," Rockwell International, SD73-SH-0184, Rev. C, Book I, Oct. 1978.

[20]Fay, J. A. and Riddell, F. R., "Theory of Stagnation Point Heat Transfer in Dissociated Air," Journal of Aeronautical Sciences, Vol. 25, Feb. 1958, pp. 73-85 and 121.

[21]Rosner, D. E. and Cibrian, Ramiro, "Non-equilibrium Stagnation Region Aerodynamic Heating of Hypersonic Glide Vehicles," AIAA Paper 74-755, July 1974.

Chapter IV Galileo Studies

Survey of the Supporting Research and Technology for the Thermal Protection of the Galileo Probe

J.T. Howe,* W.C. Pitts,† and J.H. Lundell†
NASA Ames Research Center, Moffett Field, Calif.

Abstract

The Galileo probe, which is scheduled to be launched in 1985 and to enter the hydrogen-helium atmosphere of Jupiter up to 1825 days later, presents thermal protection problems that are far more difficult than those experienced in previous planetary entry missions. The high entry speed of the Probe will cause forebody heating rates orders of magnitude greater than those encountered in the Apollo and Pioneer Venus missions, severe afterbody heating from base flow radiation, and thermochemical ablation rates for carbon phenolic that rival the freestream mass flux. This paper presents a comprehensive survey of the experimental work and computational research that provide technological support for the probe's heat-shield design effort. The survey includes atmospheric modeling; both approximate and first-principle computations of flowfields and heat-shield material response; base heating; turbulence modeling; new computational techniques; experimental heating and materials studies; code validation efforts; and a set of "consensus" first-principle flowfield solutions through the entry maneuver, with predictions of the corresponding thermal protection requirements.

Introduction

The Galileo probe is scheduled to be launched sometime in 1985 and to enter Jupiter's atmosphere as many as 1825 days later. Because this entry severely tests our thermal

Presented as Paper 81-1068 at the AIAA 16th Thermophysics Conference, Palo Alto, Calif., June 23-25, 1981.

*Head, Aerothermodynamics Section.
†Research Scientist, Thermal Protection Branch.

protection design capability, a significant research base—both experimental and analytical—has been formed in support of the probe heat-shield design effort. Since about July 1975[1] the supporting research and technology base has involved Ames Research Center, Langley Research Center, and several contractors.

The high-speed entry (up to 48.2 km/s relative to the hydrogen-helium atmosphere of Jupiter) is expected to lead to probe forebody heating rates that are hundreds of times greater than those of Apollo and 10 times the Pioneer Venus rates (Fig. 1). The corresponding thermochemical ablation rate of the carbon-phenolic heat shield is expected to rival the freestream mass flux because of the intense radiative and convective heating from the hot turbulent gas cap that will envelop the probe. Moreover, for the cool dense model of the Jovian atmosphere it is likely that radiative heating rates near the stagnation point will be so severe that the heat-shield material will undergo spallation as well as thermochemical ablation. It is calculated that the severe environment will cause the probe forebody to lose about one-third of its mass during the entry heating pulse. Not only is forebody heating a severe problem for the

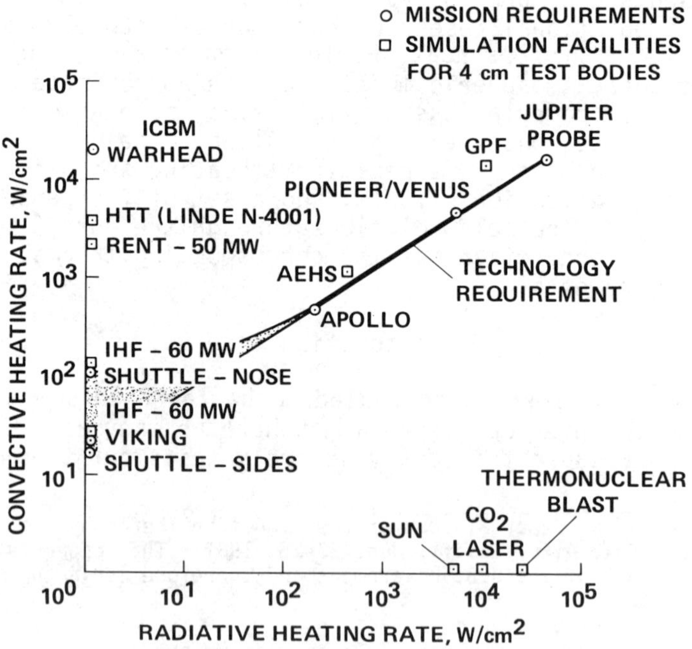

Fig. 1 Heating environments.

Jovian probe, but the afterbody will incur significant radiative heating from the wake region of the flow. As a result, it is necessary that the flowfield and the heating rates about the entire probe, including the wake, be understood as thoroughly as possible in order that the design of the thermal protection system be as effective as practical.

The flow environment and material response about the probe is beyond our previous capability to simulate experimentally or to compute realistically. In recent years, new experimental facilities have been devised, constructed, and used to simulate some but not all of the features of the entry environment; indeed they cannot simulate the flowfield in all respects. The Ames Giant Planet Facility can produce heating rates that simulate the heating levels on the flank of the probe, and the Ames Gas Dynamic Laser Facility can simulate heating rates to the stagnation region of the probe and produce heating levels that cause spallation of the heat-shield material. These facilities will be discussed subsequently. Both the forebody and afterbody flowfields and radiative transfer have been studied in shock-tube facilities, and by firing free-flight models into a noble gas mixture to simulate Mach number and Reynolds number in order to observe forebody and wake flowfield structure. Moreover, various computational codes have been developed to predict the effects of a great many coupled physical phenomena in the flowfield and in the heat-shield material.

A number of additional problems have been addressed to assess their effects on the probe heat shield. These include the composition and thermal structure of the atmosphere, probe shape, and targeting (initial entry) conditions. But some phenomena (such as spallation) that cannot readily be assessed by computational physical analysis must be investigated experimentally. The final objective of the comprehensive supporting research and technology effort is to provide an acceptable set of first-principle computational codes to be used as a standard for the more approximate design codes of others. To enhance confidence in the results of these first-principle or "benchmark" codes it is necessary to validate the codes experimentally, where possible, and to provide experimental corrections to account for phenomena that cannot be modeled in the codes.

Galileo Probe Science Objectives

The probe mission objectives are to determine by *in situ* measurements the atmospheric characteristics of

Jupiter to a depth corresponding to at least 10 bar pressure, to sense radio emissions from the planet, and to determine the radiation level trapped near the planet. Seven instruments will be used to make these measurements. Three of these instruments will start recording data prior to entry. At an altitude of 5 Jupiter radii the lightning and radio emissions detector will start to verify the existence of lightning on Jupiter and to measure its basic physical characteristics. The instrument consists of an rf antenna and a photodiode. The energetic particle instrument will also start to record data at an altitude of 5 Jupiter radii. From these data will be determined the fluxes, energies, and pitch angles of the inner magnetosphere electrons, protons, and alpha particles that are capable of penetrating the aft heat shield. The atmosphere structure instrument (ASI) will measure the thermodynamic state properties of the atmosphere from 1500 km above the 1 bar pressure level down to the end of the mission altitude. While the heat shields are attached, the ASI will measure the probe deceleration that is proportional to the atmosphere density. The pressure and temperature of the ambient atmosphere will be measured after the parachute is deployed and the heat shields are jettisoned. These pressure and temperature sensors as well as all other instruments will begin recording data at about the 0.08 bar level. The Nephelometer will determine the presence and extent of vertical clouds as well as the character of the particles within the clouds. The net solar flux and the net thermal flux from the planet will be measured by the net flux radiometer which alternately looks up and down. By using filters to isolate wavelength bands, water vapor and ammonia content can also be deduced from the data. The neutral mass spectrometer will measure the chemical and isotopic composition and the physical state of the Jovian atmosphere. The helium abundance detector will determine the He/H_2 mixing ratio to within 0.1% precision. This ratio is significant in the understanding of the evolution of our solar system. In addition to the data from the onboard experiments, the vertical shear of the horizontal wind speed in the east-west direction will be determined from the Doppler shift of the probe transmitter frequency.

Targeting, Atmospheric Modeling, and Trajectory Computations

Targeting for the entry probe has been at an altitude of 450 km at various latitudes near the Jovian equator. Effects of variations in the azimuthal angle (near equatorial), of entry angle of the probe with respect to the planet, and of probe mass and configuration on thermal protection requirements have been examined. That target altitude was fixed early in the study, because it corresponded to what was thought to be the "top" of the model atmosphere, according to Ref. 2. Subsequently, however, there were a number of other model atmospheres which differed both in thermal structure and in relative abundance of hydrogen and helium. The thermal structures of these different nominal atmospheres are shown in Fig. 2, which was adapted from an Acurex Corporation report. The original atmospheric model, which corresponded to that reported in Ref. 2, was composed of 85% hydrogen and 15% helium by volume. That was followed by the atmosphere shown by the solid line in Fig. 2; it was referred to as the Hunten model. That model had the thermal structure given by Ref. 3, but a composition of 89% hydrogen and 11% helium in accord with Ref. 4. This was followed by a nominal atmosphere model developed by G. Orton of the Jet Propulsion Laboratory which has the thermal structure shown in Fig. 2 and a composition of 89% hydrogen and 11% helium, as noted above.

The atmosphere was assumed to be in hydrostatic equilibrium. Hence, an ordinary differential equation could be solved for each model atmosphere composition and thermal structure to compute density altitude profiles for each model. A comparison of two sets of these atmospheres is shown in Fig. 3, in which the corresponding atmospheric composition is also shown. Thus for each atmosphere defined by a thermal structure there is a warm, nominal, and cool version, depending on the composition. The Orton cool and nominal model atmospheres are seen to be significantly

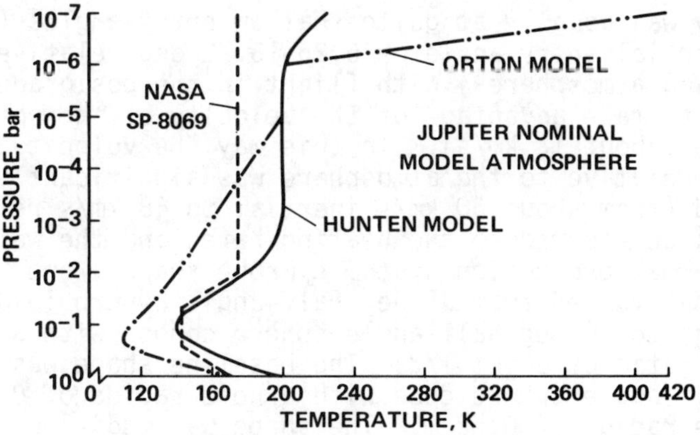

Fig. 2 Thermal structure for nominal model atmospheres of Jupiter.

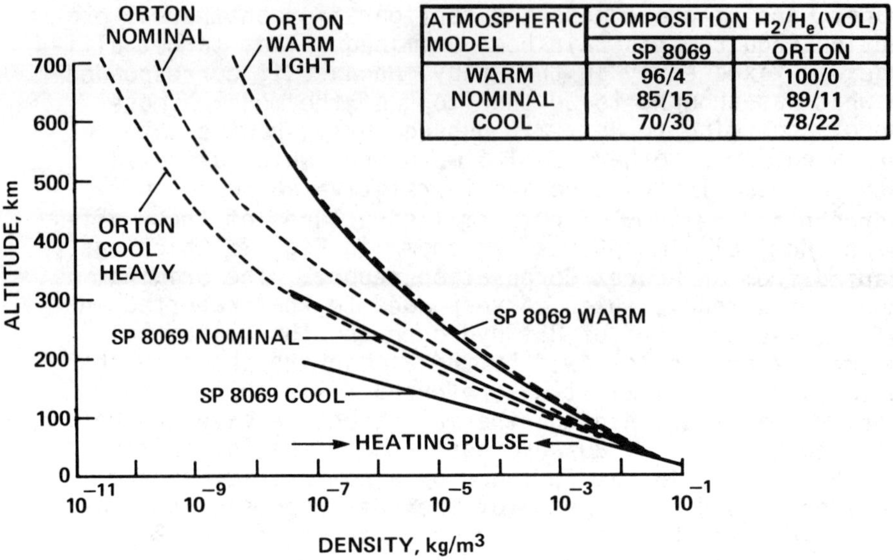

ATMOSPHERIC MODEL	COMPOSITION H_2/H_e (VOL)	
	SP 8069	ORTON
WARM	96/4	100/0
NOMINAL	85/15	89/11
COOL	70/30	78/22

Fig. 3 Atmospheric models of Jupiter.

higher than those of Ref. 2. [Subsequently, for the heavy (310 kg) probe study, a revised version of the Orton atmosphere was used.] Nevertheless, targeting was still specified at 450 km, and trajectories were calculated such that the probe would arrive at that altitude with the specified entry angle. Trajectories were calculated to include the effects of latitude, azimuthal angle, entry angle, and ballistic coefficient for each candidate probe shape, and to accommodate the fact that Jupiter is a rotating oblate spheroid.

Entry was usually at quite shallow entry angles (5-11 deg inertial entry angle or 6.25-13.72 deg relative to the rotating atmosphere[5]) with flight in the posigrade direction to take advantage of the velocity of the rotating atmosphere (about 12 km/s). In this way the velocity of the probe relative to the atmosphere was significantly diminished (from about 60 km/s inertial to 48 km/s relative) in order to reduce the heating rates and the weight of the thermal protection system. Probe shapes were studied that varied from 35 deg half-angle hyperboloid forebodies[6] to 60 deg half-angle sphere cones[7] with a bluntness ratio of about 1/2. The baseline shape was a 45 deg half-angle sphere cone with a nose radius of 0.31 m and a base radius of 0.62 m. The shape was modified slightly late in the study. The probe weight was increased during the course of the study as greater demands were placed upon the probe. Much of the early work was for 242 kg probe; later work was for 290 and 310 kg probes, as will be shown.

Flow about the Forebody and Heat-Shield Response

During the heating pulse the flow is determined over the forebody at a number of times through the trajectory. To provide a reference set of benchmark solutions, the physical phenomena illustrated in Fig. 4 (adapted from Ref. 8) must be included in the analysis. Thus the flow equations include the transport of mass, momentum, energy, and species. The gases are hot and reacting and emit thermal radiation which is incident on the heat shield. The radiative and convective heating cause the heat-shield surface to ablate at a rate that is sometimes comparable to the freestream mass flux. The ablated gases mix with, and react with the atmospheric gases and enter into the radiative transfer in the shock layer. The gas mixture makes a transition from laminar to turbulent flow as it is swept around the body. To obtain the benchmark flowfield solutions, a quasisteady-state ablation boundary condition is applied such that the ablation rate is simply related to the net heating to the wall. Subsequently, the heat shield is designed by applying the heating rates obtained from the flowfield to a transient material-response analysis of a

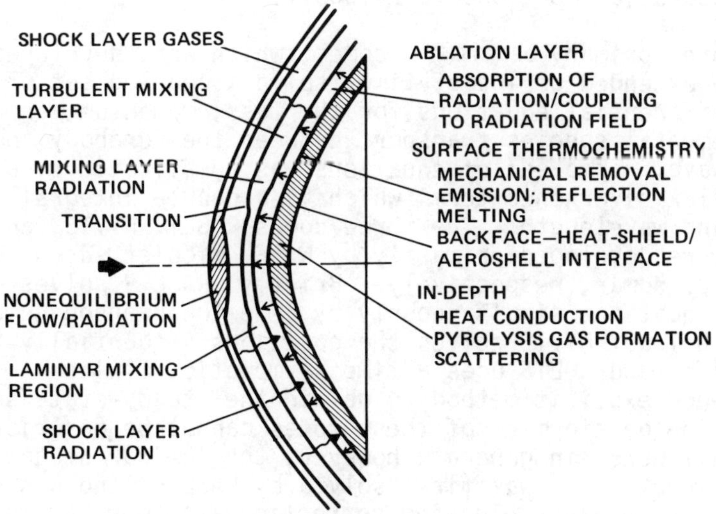

Fig. 4 Heat-shield and flow phenomenology.

material of finite thickness; the latter material is receding and losing mass by thermochemical ablation of the surface char and by the outgassing of the pyrolysis gases caused by the internal conduction of heat. The thickness of the heat shield is determined by a specified allowable temperature at its back surface.

A number of flowfield studies were performed early in the research effort, using either inviscid codes or various codes that employ correlations and physical estimates to assess the existing state-of-entry environment prediction methods[7]; and the extent and importance of atmospheric structures on radiative heating,[5] radiative heating distributions,[9] problems concerned with an atmospheric reconstruction experiment,[10] and configuration effects on radiative heating.[11]

Moreover, an extensive effort was made to acquire and refine the radiative properties of carbonaceous ablation gases both experimentally and theoretically.[12-22] The detailed effect of these carbonaceous species is illustrated in Fig. 5, which was adapted from Ref. 23. Figure 5a shows the spectral radiative flux incident on the stagnation region of the entry probe near peak heating, and Fig. 5b shows the spectral flux incident on the surface. The difference between the two spectra represents absorption by spectral bands of the ablation gases, as shown in the figure. The spectral property studies were used to provide input data primarily to the detailed benchmark flowfield codes, as mentioned previously.

Three primary benchmark codes, which were developed over an extended period of time, solve a coupled set of partial differential equations for the mass, momentum, energy, and elemental species transport between the forebody and shock wave. This set of equations was complicated by a radiative flux divergence term, which is a double integral over space and wavelength. The three codes—RASLE, HYVIS, and COLTS—were developed primarily by W. E. Nicolet, J. N. Moss, and A. J. Kumar, respectively. Briefly, RASLE[8] solves the set of equations simultaneously by a Newton-Raphson numerical technique; HYVIS solves the equations sequentially by iteration; and COLTS uses a time asymptotic, finite-difference explicit method to obtain the steady-state solution. The development of these codes cannot be described in detail here; in general, however, the laminar stagnation region shock layer was first solved by RASLE[24] and was used to provide ablation blocking corrections for more approximate codes used in a parametric study,[25] in which compari-

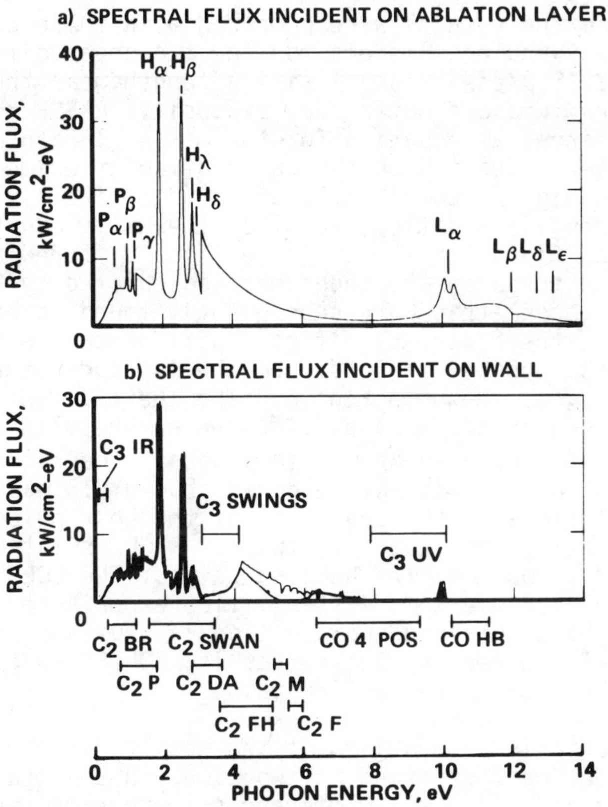

Fig. 5 Spectral flux attenuation by ablation species: shock temperature = 16,090 k, pressure = 5.1 atm, 89% H_2, 11% He atmosphere, carbon-phenolic heat shields.

sons with other parametric studies were made. These results were used in Ref. 26, to assess heat-shield requirements and were compared with the pioneering work of Tauber.[27]

The HYVIS code was a pioneering code in several respects. The code produced laminar solutions about hyperboloids of revolution. The solutions were over the entire carbon-phenolic forebody heat shield, and included coupled ablation as a quasisteady-state boundary layer.[28] Subsequently, the code also produced turbulent flow solutions about both sphere-cones and hyperboloids with coupled ablation for carbon-phenolic heat shields.[29]

The RASLE code also computed turbulent flow over a sphere-cone heat shield at various points in an entry trajectory,[30] and the results were compared with those of Ref. 29. The turbulence models for the two codes differed somewhat. HYVIS used a Prandtl mixing length near the wall and a Clauser-Klebanoff outer eddy viscosity; RASLE used an ordinary differential equation for the mixing length near the wall, selected partly on the basis of an experiment of Ref. 31 for a highly cooled wall, and a wake-like mixing length differential equation away from the wall. Surface radiative conditions for the two codes also differed and convergence criteria on the body shape calculated by RASLE were less stringent than were subsequently found to be necessary. Thus, there were significant differences in the integrated mass losses obtained from the two codes through the trajectory, as shown in Fig. 6. The inertial entry angle was -9 deg in the nominal atmosphere. Results from the COLTS code[32] are also shown; in general, they agree with the results of HYVIS. It may be noted that HYVIS was able to obtain solutions at the beginning of the heat pulse and after peak heating, but not in between. RASLE solutions were obtained throughout the heating pulse. The COLTS code seems to have overcome some instabilities experienced by HYVIS and to have obtained two additional solutions on the upward part of the heating pulse—that is, COLTS can provide solutions over the entire heating pulse.

A very detailed comparison of the results of these benchmark codes was performed at two times in the trajectory,[33] 110.2 and 111.3 s. Differences in physics, methods, boundary conditions, and "convergence" criteria were presented and the effects were assessed. An attempt was made to reconcile results of the codes by making adjustments to

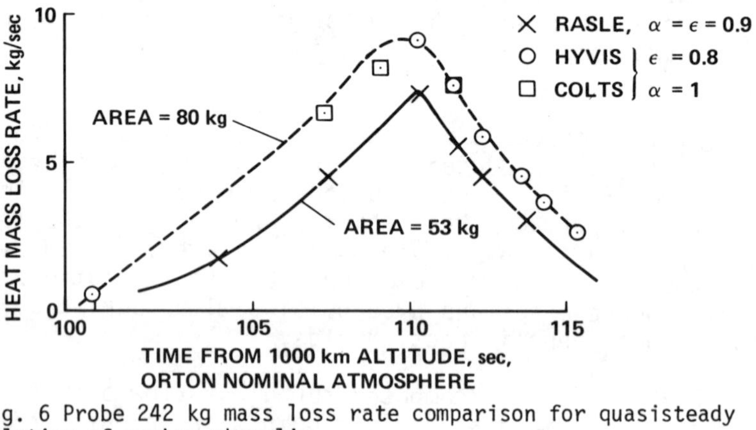

Fig. 6 Probe 242 kg mass loss rate comparison for quasisteady ablation of carbon phenolic.

correct for some differences. The work reported in Ref. 33 is a remarkable effort to sort out many coupled interrelated effects to determine those which are most significant. Global mass and energy conservation checks were made. Among three major contributors to differences in the results, the turbulence models are considered to be the most important because of their effect on radiative transfer to the wall. Although the various turbulence models employed do not affect state profiles near the wall significantly, profiles remote from the wall are affected. Thus, temperature profiles away from the wall differ because of the turbulence model. Correspondingly, species profiles differ as well. Thus the species that emit and absorb radiation are distributed differently in differing thermal environments by the different turbulence models, and the radiation emitted differs correspondingly. Because radiation is the most important phenomenon causing heat-shield ablation, the effect of turbulence modeling on radiative transfer is of primary importance. Results of this important paper[33] have affected subsequent benchmark solutions and contributed to a "consensus" set of benchmark solutions shown in part subsequently in this paper.

Turbulence modeling of a very hot, chemically reacting flow over a relative cool surface that is ablating at a rate comparable to the freestream mass flux is very poorly understood. A comparison of several candidate turbulent models (including those cited above) has been made[34] in the context of a given flowfield code to isolate the effect of the model alone and to rank the various models according to severity. It is clear that this is an important problem. Experiments must be devised that will make it possible to construct realistic turbulence models and to provide input data for those models.

As noted previously, the flowfield results shown in Fig. 6 are obtained by use of a quasisteady-state ablation condition. The heat shield designed by the CMA code—using the net heating output from the RASLE code as input to the in-depth materials-response code—is shown in Fig. 7. The CMA code computes the transient thermal response of a charring ablating material of finite variable thickness, and includes the effects of temperature-dependent properties and in-depth pyrolysis. The criteria that "design" the heat shield are the maximum allowable backface temperature of 700 K and the safety margins (shown in Fig. 7). These margins are percentages of the ablated material that are added for safety purposes. The figure shows the initial and final shapes of this probe designed for entry into the nominal atmosphere. The mass loss (47 kg) is less than that shown in Fig. 6 (53 kg). The difference is affected by the heat storage in the transient in-depth solution in only a minor way; this point will be discussed subsequently. When the heat shield with the safety margins defined by Fig. 7 is allowed to enter the cool dense atmosphere at an inertial entry angle of -11 deg, an additional thermal protection mass (22%) is required for survival, with no safety margin.[35] These preliminary results are for a 242 kg probe and are illustrative only. Subsequently, we shall cite research that reconciles some of the differences among the benchmark codes and present results for a more recent heavier probe (310 kg).

The codes that produced the results of Figs. 6 and 7 are based on first principles and are considered to be "benchmark" flowfield and material-response codes. They might even be labeled "first-generation" benchmark codes, because new computational techniques are being devised and applied to this problem, as will be shown subsequently. However, even these first-generation benchmark codes are very costly to operate and difficult and time consuming to run, especially for broad parametric studies. Thus for engineering purposes, more flexible, more approximate codes are useful for economic reasons. Zoby et al.[36] have developed an approximate code that simulates inviscid radiating flowfield analysis by the use of analytic shock shapes and

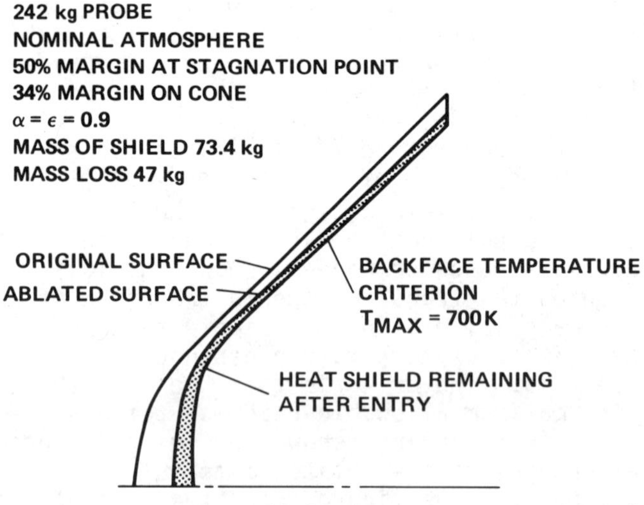

Fig. 7 Transient heat-shield response: carbon phenolic.

a 58-step radiative transfer model. Moreover a code has been developed to predict laminar and turbulent convective heating of reactive and nonreactive gases about blunt re-entry configurations for hypersonic flight.[37] These approximate codes produce results that are in good agreement with more complicated codes and are used extensively for parametric studies.

Flow about the Afterbody and Base Heating

Although the convective heating of the base regions is small,[38] radiative heating is not. The significance of the base radiative heating was noted by Stephenson[39] in 1965 by measurements made of radiation from the wake of an ablating blunt body launched into air; the measurements were made as the body flew past the radiation sensors.

Park[40] developed a base flowfield model and derived the base radiative heating from the model. Observations from both ablating and nonablating free-flight tests were used to devise the flowfield model shown in Fig. 8. The main fea-

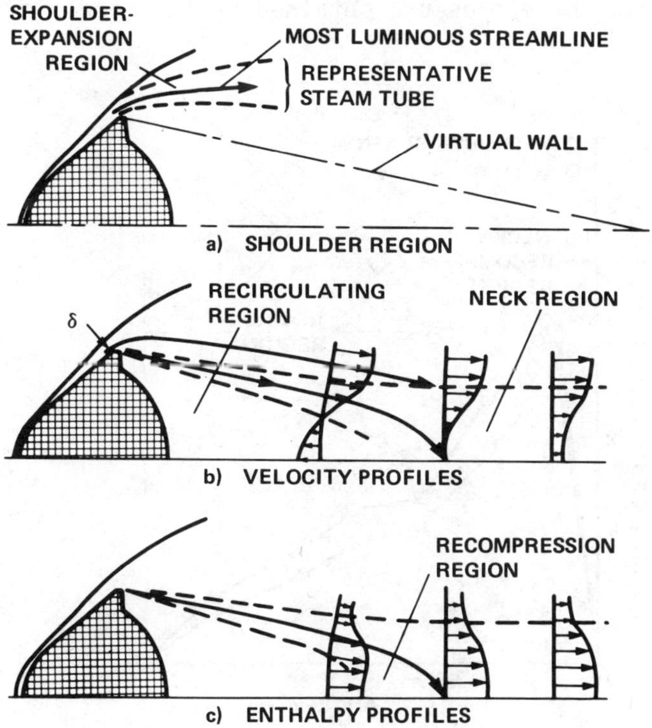

Fig. 8 Schematic of base region flowfield.

tures of the flowfield were an expansion about the shoulder; a recirculation region behind the body; and a wake which recompressed and formed a neck about 1.5 body diameters aft of the frustum, with a neck diameter between one-half and two-thirds of the body diameter. Method-of-characteristics solutions about the shoulder were matched to the recirculation region by a choice of base pressure that gave the observed turning angle. Physical reasoning led to a theoretical model of the recirculation, recompression, and neck region flows. In the ablation case of interest, chemical and radiative modeling predicted that the most severe radiation would occur immediately behind the frustum, because of radiation from hydrogen and carbon atoms that were not in chemical equilibrium in the expansion region. At the base stagnation point, radiation is received from both the neck and recompression region, and the recirculation region, as shown in Fig. 9. These regions are probably in chemical equilibrium, and detailed spectral radiation computations from the hydrogen-carbon mixture produce the upper line corresponding to the ablation case. (It is interesting that the nonablating case gives almost comparable results, probably because the temperature is higher.) It is believed that the high base pressure obtained by the model is the

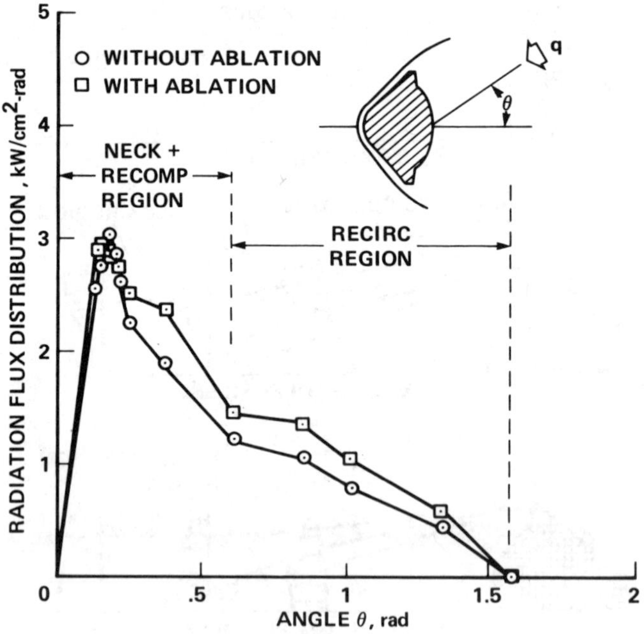

Fig. 9 Radiation at base stagnation point.

most significant physical property that leads to the high base radiative heating. The effect of base pressure was examined and the radiative heating was found to vary as the square of the ratio of base pressure to forward stagnation point pressure.

Experiments performed by instrumented nonablating models in shock tubes[41] showed that radiative heat fluxes are indeed significant, and the results were generally in agreement with those of Ref. 40. Comparison of the experimental and theoretical results was made with the early work of Ref. 39; agreement was generally good. It was determined that the ratio of base heat flux to front stagnation radiative heat fluxes and pressure ratios for a Jovian probe are significant, and are in agreement with the theory of Ref. 40.

A summary of the work of Refs. 40 and 41 is presented in Ref. 42. The work described in Refs. 40-42 is for laminar flow. Work in progress in free-flight facilities addresses the turbulent base flow problem.

Second-Generation Flowfield Codes

Although benchmark flowfield codes were being used extensively to produce forebody solutions with ever increasing complexity and physical detail, it was apparent that the method was being strained to the utmost and that accuracy and convergence problems were becoming more severe.[33] More important, perhaps, afterbody solutions have become necessary and they are intimately related to forebody events.[40] Thus efforts were initiated to develop a new method, one in which new coordinate systems and advanced numerical techniques would be used. A new code, CAGI,[43] that was "tailored" to capture the major flowfield features of both the forebody and the afterbody, was initiated; it is presently in an advanced state of development. The code solves the unsteady compressible Navier-Stokes equations for two-dimensional axisymmetric three-dimensional flow. A finite-volume formulation was devised, rather than the usual finite-difference approximations of the differential equations. The finite-volume formulation maintains global conservation of mass, momentum, and energy over the computational volume in accord with specified fluxes at the volume boundaries. A mesh is constructed over the entire probe configuration; the mesh is flow-aligned over four principal regions—forebody shock layer, base recirculation, outer inviscid wake, and inner inviscid wake.[44] There is a singu-

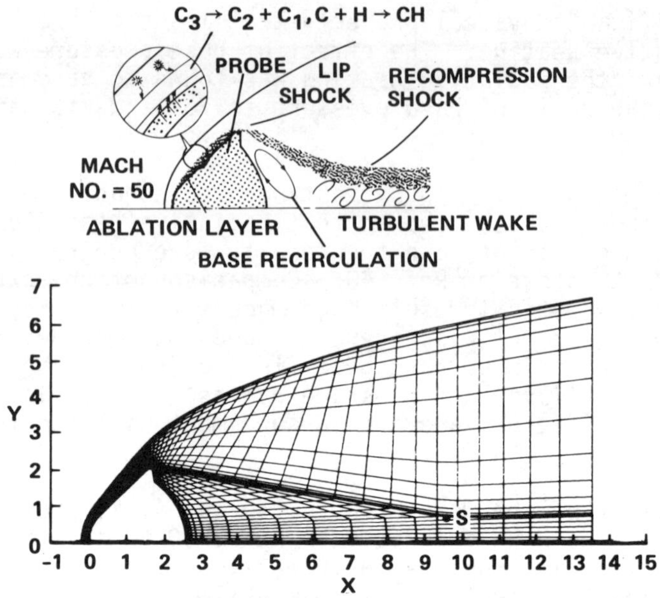

Fig. 10 Probe flowfield features and the computational grid.

lar coordinate mesh topology in the recompression region. The computational code employed is an outgrowth of the basic factored implicit algorithm of Refs. 45-49.

The first results of the code were presented in Ref. 43, where real-gas properties were modeled rigorously by a variable "gamma," which reduces to the ratio of specific heats for the case of a perfect gas. A "homogeneous sweep" algorithm was used in that study instead of the full-factored implicit algorithm. Results were presented in Ref. 43 for the case of flight at Mach 50; some results for mass addition along the forebody surface were included. Of particular interest is the flowfield structure about the entire bluff body and the corresponding computational mesh shown in Fig. 10 (actually adapted from Ref. 44). The computed pressure distribution is shown in Fig. 11, and a comparison of the computed velocity vector field with an experimental shadowgraph by C. Park is shown in Fig. 12.

A chemically reactive gas module of the code was introduced[50] and solutions were obtained. The pressure distribution and the temperature, enthalpy, compressibility, and species profiles were computed and compared with results of Ref. 6. There was generally good agreement between the results of the two studies.

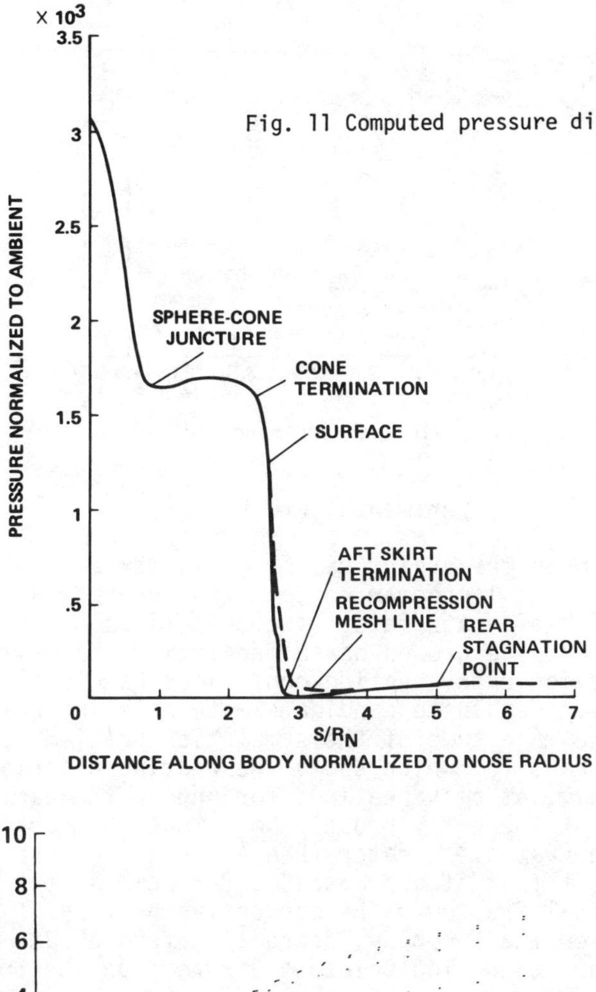

Fig. 11 Computed pressure distributions.

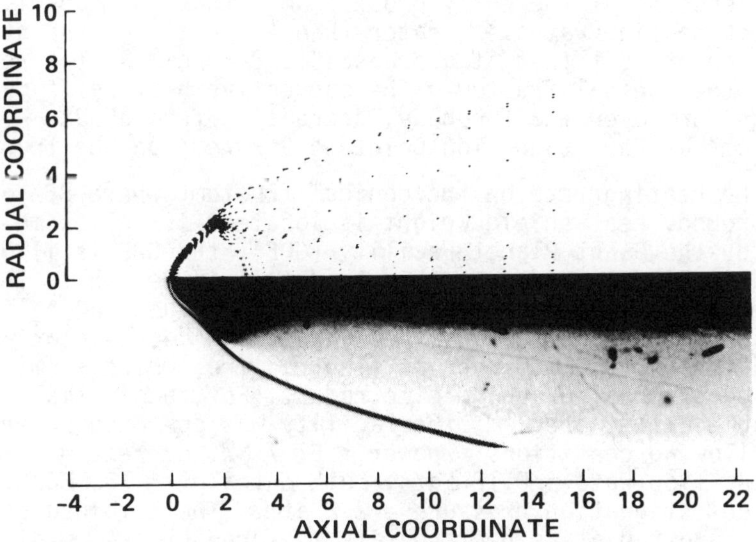

Fig. 12 Computed velocity field compared with free-flight shadowgraph.

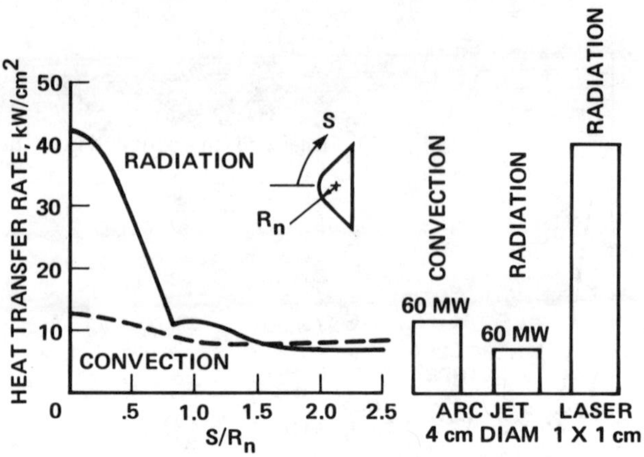

Fig. 13 Galileo peak-heating facilities.

Experimental Program

In addition to the analytical program, there is also an extensive experimental program on both the forebody and afterbody heat-shield materials. The objectives of the program are 1) to provide ground-based facilities that simulate the expected Jovian entry heating conditions as closely as possible and 2) to evaluate candidate materials in those facilities. The magnitude of the simulation problem is illustrated in Fig. 13, which shows the heating distribution over the forebody, at peak heating, for one of the early study versions of the entry probe. Note that the nonablating peak heating rate is greater than 40 kW/cm^2 at the stagnation point, but that it decreases to 8 kW/cm^2 at the rear end of the conical frustum. The convective heating is much more uniform over the forebody, decreasing from about 12 kW/cm^2 at the stagnation point to 9 kW/cm^2 on the frustum.

The heating rate on the conical frustum, where 85% of the forebody heat-shield weight is located, is well simulated by the Giant Planet Facility (GPF); the GPF is illustrated in Fig. 14 and described in detail in Ref. 51. In the arc heater, an equimolar mixture of hydrogen and helium is injected uniformly along the 4.32 m long constrictor and heated to about 15,000 K by an electric arc, which terminates on external graphite electrodes. For the initial heat-shield test program, the facility was operated under the following conditions: power = 57.7 MW, current = 5.33 kA, flow rate = 0.12 kg/s, bulk enthalpy = 185 MJ/kg, and model stagnation pressure = 2.2 atm. The combined convective and radiative heating rate measured by a 4 cm diam thin-shell tungsten calorimeter, with a radius of curvature

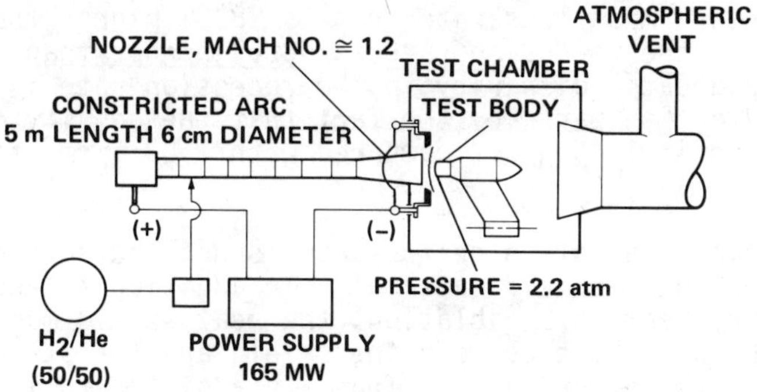

Fig. 14 Giant Planet Facility.

of 20 cm, was 14.6 kW/cm^2. A radiometer of similar geometry was used to measure the radiative heating component and the measured value was 6.0 kW/cm^2. By combining these two numbers and correcting for the reflectivity of the tungsten calorimeter cap, the applied convective heating rate was determined to be 12.0 kW/cm^2. These heating rates pertain to the model test position, which is 7 cm downstream of the 7 cm diam nozzle exit. The facility is equipped with four swing arms, so that a calorimeter, a pressure probe, and two ablation models can be inserted into the free jet stream during a given run. The ablation models are 4 cm diam flat-faced cylinders consisting of a concentrically wound carbon-phenolic shroud and a 2 cm diam core on which recession and weight measurements are made.

Heat-shield materials studied to date include six versions of the carbon-phenolic forebody materials, carbon-carbon composites, graphites, hyperpure silica reflective materials, and the nylon-phenolic afterbody material. Only the work on one version of carbon phenolic and graphite will be discussed. In all the test programs, ATJS graphite is used as a control material against which the other materials are compared. It also serves as a check on the arc-jet repeatability. Results of some of the earliest tests on carbon phenolic and graphite are compared with theoretical calculations in the next section. In later work, all materials were tested at four different exposure times so that the steady-state ablation rates could be established. A comparison of the steady-state rates for chop-molded carbon phenolic (the nose-cap material) with those of the control material is shown in Fig. 15. For ATJS graphite, the steady-state mass loss rate is 0.41 g/s and the recession rate is 0.070 cm/s. Corresponding values for carbon phenolic are 0.528 and 0.104 cm/s. Thus, the carbon-phenolic material has a mass loss rate that is 28.8% higher than that of ATJS, and the recession rate is 48.6% higher than that of ATJS. The greater disparity in the recession rate is accounted for, in part, by the fact that the density of carbon phenolic is 1.45 g/cm^3 compared with 1.83 g/cm^3 for the graphite.

Of course, ablation performance is not the sole criterion for a heat-shield material. In addition to accommodating entry heating by ablation, the heat shield must protect the bond line between the shield and the structure from excessive temperatures. Thus, in addition to its ablative properties, the insulative properties of a heat-shield material are important. In flight, the superior insulation properties of carbon phenolic will tend to compensate for its somewhat poorer ablation performance.

Since the GPF is not capable of producing radiative heating rates comparable to the peak values expected in flight, a gasdynamic laser (GDL) must be used to evaluate materials under these severe heating conditions. The

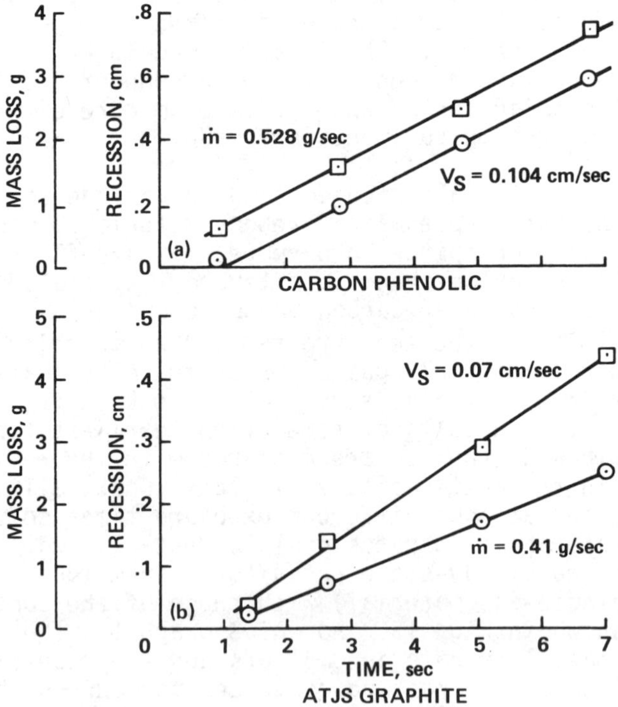

Fig. 15 Ablation results at 57.7 MW.

facility, which is described in detail in Ref. 52, is a conventional combustion-driven laser which produces radiation at 10.6 μm by burning CO to produce CO_2 as the lasing medium. Although the lasing wavelength is significantly different from the uv and visible radiation expected in flight, the difference is not expected to be important, because the heat-shield materials are graybody absorbers. When the internal mirrors are freshly polished and aligned, the laser will produce an output power of 33 kW when room-temperature nitrogen is injected along with the CO and up to 45 kW when the nitrogen is heated to 1400 K by means of a gas-fired heat exchanger. Thus, when the output beam is focused to a 1.0 cm^2 spot, the peak radiative heating rate expected in flight (Fig. 14) can be simulated.

Early in the laser testing,[53] it was discovered that carbon phenolic tends to spall (Fig. 16) when exposed to severe radiative heating. Although this phenomenon may be unique to laser testing, it must be assumed, to be conservative, that it will also occur during peak heating of the Galileo probe. In order to account for spallation in the heat-shield design, the phenomenon must be quantified; such tests are in progress.

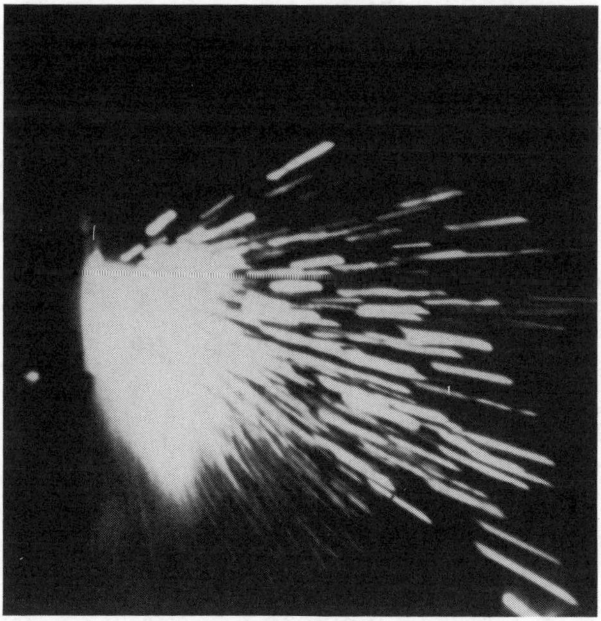

Fig. 16 Carbon-phenolic spallation.

The experimental setup for these tests is illustrated in Fig. 17 and described in Ref. 54. The output beam of the GDL is collimated into an 8 cm diam and then transformed into a spatially uniform 1.0 × 1.0 cm spot by means of a segmented mirror. This device consists of an eight-by-eight array of 1 × 1 cm polished molybdenum segments. Although the segments are optically flat, they are mounted on a curved water-cooled substrate which gives the device a 1.0 m focal length. The mirror divides the incident beam into many 1 × 1 cm beams and superimposes them at the focal point. The focal image is then reimaged one-to-one, by a 1.5 m focal length mirror, into a canister in which the test specimen is mounted. The canister (Fig. 18) is designed with suitable baffling so that the laser beam can enter but the spalled particles cannot escape. An air jet in front of the specimen deflects the laser plume upward and allows the beam to reach the surface without significant absorption in the plume.

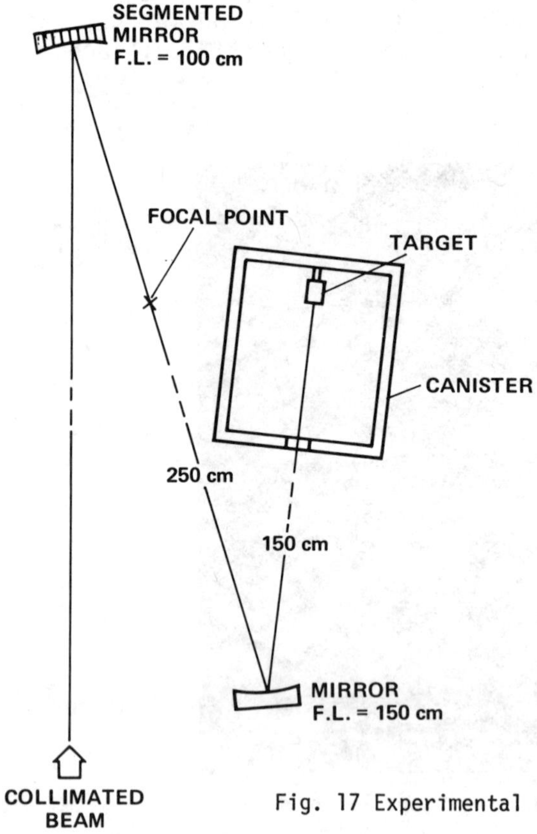

Fig. 17 Experimental setup for spallation tests.

Typical spallation results are shown in Fig. 19. The total mass loss is determined by weighing the sample (3 cm square × 5.1 cm long) before and after exposure, and the spallation mass loss is determined by collecting and weighing the particles in the bottom of the canister. At least three exposure times are used so that the steady-state total mass loss and spallation rates can be obtained. Note that for the data shown, 18.3% of the mass loss is by spallation at an intensity of 23 kW/cm^2. By repeating the test at a variety of intensities, a correlation of spallation fraction as a function of intensity will be developed. The calculated radiation history in flight will then be used in conjunction with the correlation to determine the total mass loss by spallation over the complete heating pulse and over the entire forebody heat shield.

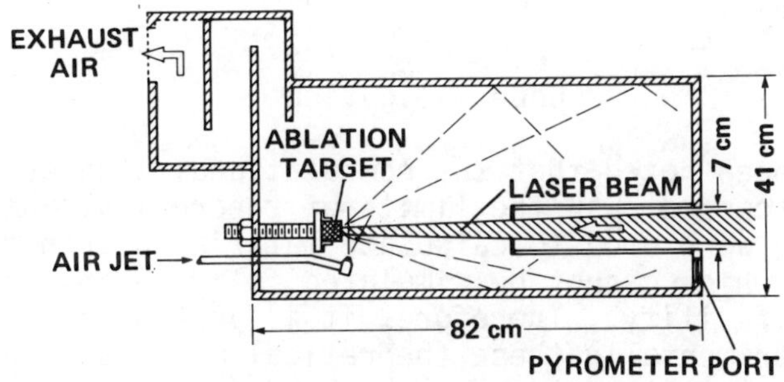

Fig. 18 Spallation canister.

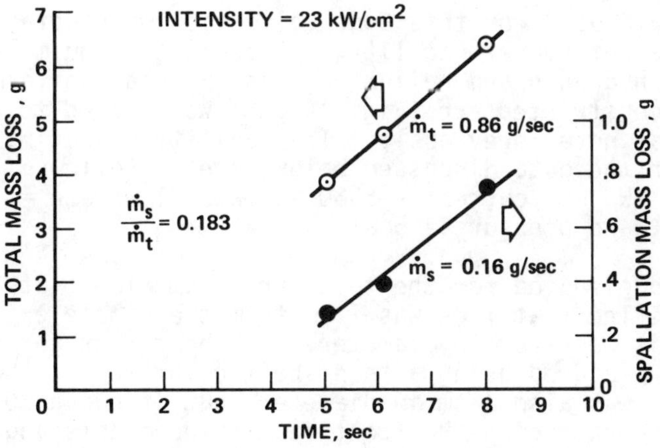

Fig. 19 Spallation results for carbon phenolic.

Because of the power limitations of the laser, the peak flight radiative rate cannot be simulated on a spot larger than 1 × 1 cm. Use of such a small spot size results in a hole being bored in the end of a square sample 3 × 3 cm in size and may produce results that are unique to the test setup. To check for scale effects, future work will involve tests with a square segmented mirror 2 cm on a side which will produce a square spot 2 cm on a side on the end of a square sample 2 cm on a side. Thus the test sample will be fully enveloped in the beam in a better simulation of the flight situation. Because of the power limitation, these tests will be limited to intensities of 10 kW/cm^2 or less. The results will be compared with the 10 kW/cm^2 results with the 1 cm segmented mirror to see if there are significant differences between the two test techniques. Note that the intensity of 10 kW/cm^2, which can be achieved with the 2 cm mirror, is a good simulation of the maximum radiative rate expected on the conical frustum of the entry probe.

Code Validation

It has been noted that the computational studies that have been discussed push the limits of theoretical analysis, and that they were used to calculate material responses to environments which cannot be simulated fully in any existing experimental facility. Therefore, it is very difficult to validate experimentally these theoretical methods. The best efforts to do this to date were reported in Refs. 55 and 56. We present here a very brief summary of some of the results presented in Ref. 56.

The data used for this code validation were obtained using the Giant Planet Facility,[51] wherein the equimolar mixture of hydrogen and helium gas was injected uniformly along the constricted arc (Fig. 14) and was heated to about 15,000 K, as noted previously. The facility operating conditions for the data discussed below were as follows: arc power = 55 MW, arc current = 5330 A, mass flow rate = 0.118 kg/s, and pressure = 5 atm.

The models used for the test were 4 cm diam flat-faced cylinders. The flat face was 7 cm from the nozzle exit, and the exit was 7 cm in diameter. The heat flux to the models was measured using a thin-shell calorimeter; the calorimeter was also 4 cm in diameter, but it had a 20 cm radius of curvature on the front face. The models and calorimeter were mounted on separate rotating arms so that they

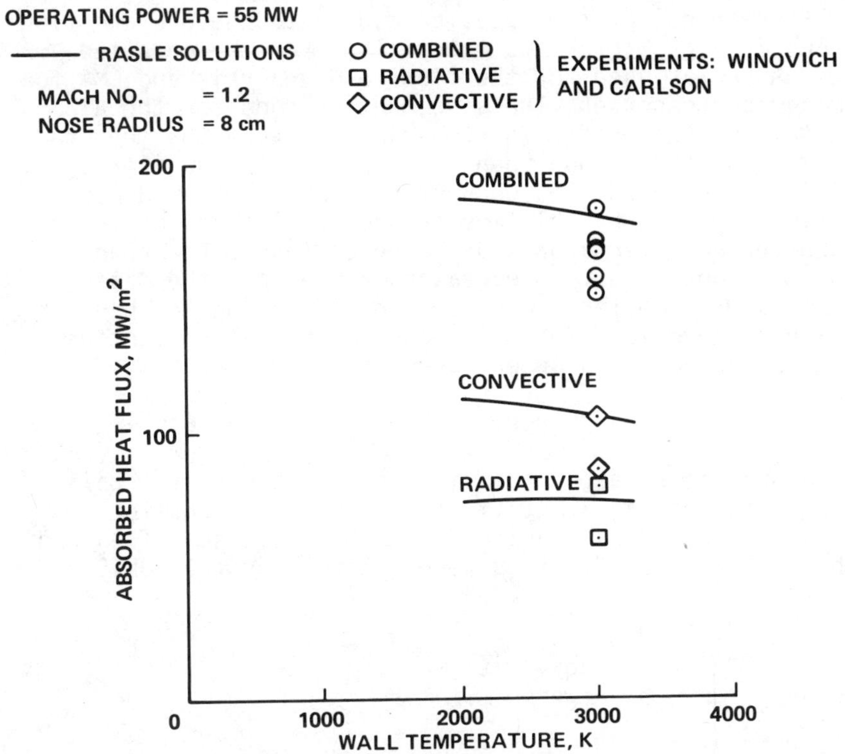

Fig. 20 Comparison of computed and measured absorbed calorimeter heating.

could be swung alternately to the same position of the arc stream, as described previously.

Absorbed heat flux data from this test are compared with heat fluxes computed by the RASLE code in Fig. 20. The calculated values generally fall within the estimated ± 10 MW/m^2 error bands of the calorimeter data. The RASLE solutions were obtained using a nonablating boundary condition for the tungsten surface of the calorimeter. The application of the RASLE code to these experimental conditions was not direct, primarily because the RASLE code is based on hypersonic approximations and the test Mach number was only 1.2. These approximations had to be modified for the subject analysis without significantly modifying the structure of the RASLE code. It was found convenient to obtain RASLE solutions for an equivalent sphere rather than the cylindrical models. From a parametric study,[56] it was found that a 4 cm diam cylinder is equivalent, in terms of stagnation point heat-transfer rate, to a spherical model with an 8 cm radius.

The response of the ablation models to the GPF arc flow is shown in Fig. 21 for two materials. For the computed curves RASLE was used for the flowfield analysis and CMA for the material response. The agreement is good for the ATJS graphite, but the predicted recession is twice the measured recession for the carbon phenolic. As for the Jupiter entry conditions, the predicted convective heating was reduced to a negligible value by ablation products. The reason for the much better agreement for ATJS graphite than for carbon phenolic is not clear. One reason may be that the material properties were better known for the ATJS graphite than for the carbon phenolic. The question of material properties for carbon phenolic is under investigation.

Consensus Benchmark Flowfield Solutions

To provide a set of solutions as a standard for calibrating other codes, a matrix of flight conditions for entry into the nominal Jupiter atmosphere was specified and forebody flowfield solutions were obtained by the COLTS and

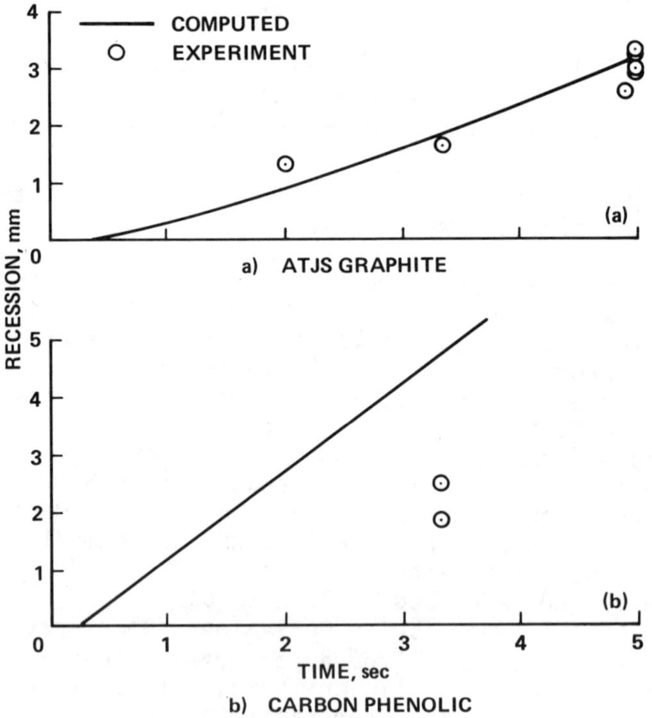

Fig. 21 Comparison of predicted and measured ablation tests in GPF.

RASLE codes. The probe was a 44.25 deg half-angle sphere-cone, which is described in Ref. 57. Two probe weights were studied: 290 and 310 kg. The results of the study are shown for the 310 kg probe in terms of mass loss rate at various times in the entry trajectory by the symbols in Fig. 22. It may be noted that there is consistently good agreement between COLTS and RASLE, except near the peak mass loss rates at 50.3 and 51.5 s. Examination of these results by J. N. Moss, W. E. Nicolet, A. Balakrishnan, W. C. Davy, M. J. Green, and J. T. Howe failed to resolve the differences near peak mass loss. Therefore, a "consensus" curve was agreed on; it is shown by the solid line in Fig. 22. The mass loss integrated over the trajectory is 101 kg. If the curve was drawn through the RASLE results (the square symbols), the result would be an additional 3 kg (approximate) of heat-shield mass loss. The integrated heat-shield mass loss is presented and discussed in Refs. 57 and 58 for both the nominal and cool dense atmospheres.

The distribution of mass loss about the body at 47, 50.3, and 54.1 s is shown by Fig. 23, which was adopted from Refs. 57 and 59. The mass loss distribution is shown non-dimensionally, where local mass loss rate was normalized by

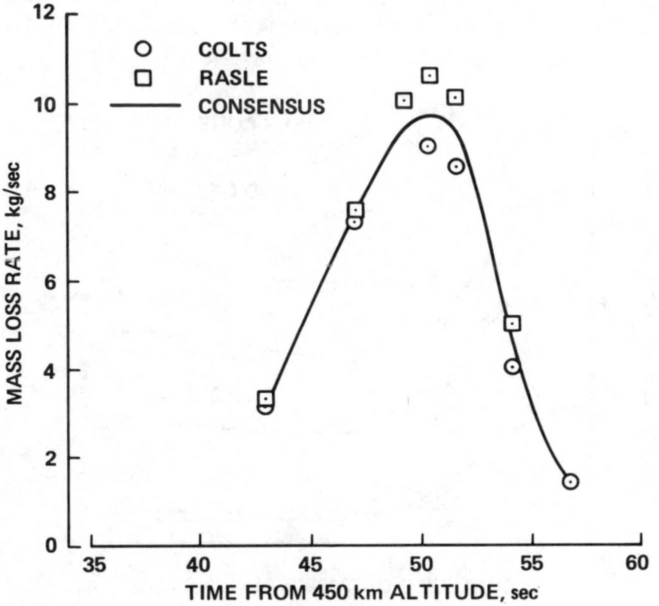

Fig. 22 Benchmark solution results of 310 kg probe (nominal Jovian atmosphere and nominal entry angle).

the product of instantaneous flight velocity and atmospheric density. At 47 s, both the integrated mass loss (Fig. 22) and the mass loss distribution about the body (Fig. 23) are in good agreement. At 50.3 and 54.1 s, both figures show a generally higher mass loss rate by RASLE especially on the conical frustum where most of the probe mass resides.

These consensus solutions had a quasisteady-state ablation boundary condition. Moreover, by prior agreement, the radiative boundary condition was that the surface absorbed all of the flux incident upon it—a very severe condition.

In Ref. 58, the more severe surface heating-rate history derived from the RASLE flowfield results[57] were input into the CMA code, which is a transient material-response code for a pyrolyzing ablating material of finite thickness wherein the material properties are temperature dependent. An allowable bond-line temperature between the aeroshell and the heat shield was specified, and the required heat-shield thickness distribution was determined. For the nominal atmosphere, safety margins were added to account for uncertainties (such as spallation effects). Thus a heat-shield "design" from a benchmark material code was produced (Fig. 24). Similar results were obtained for the cool dense atmosphere (without safety margins). The question of survivability of the nominal benchmark "design" in the cool

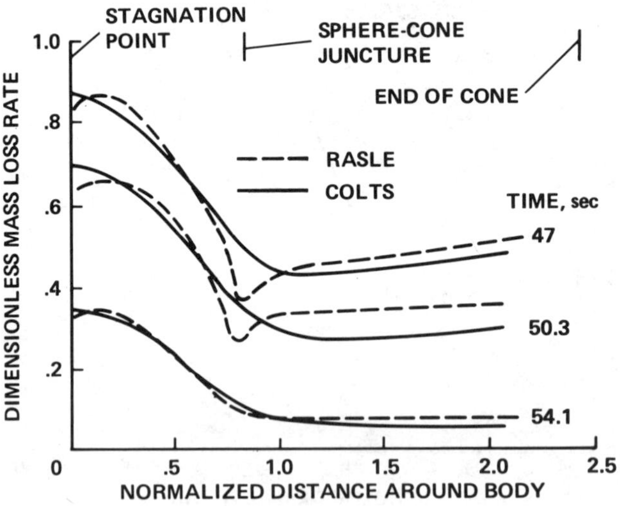

Fig. 23 Comparison of coupled carbon-phenolic mass loss distributions about 310 kg probe (nominal atmosphere).

dense entry environment was assessed in Ref. 58. Results of that paper suggested that for the Probe model considered, a thermochemically ablating heat shield with physical uncertainties that fall within the prescribed margin allowances, the 310 kg probe would survive the nominal Jovian atmosphere. Moreover, there appeared to be a reasonable probability that the probe would survive off-nominal atmospheres without margin.

Correlations were developed from these benchmark codes for heat-shield designs for other probe masses. The final probe mass was nearly 335 kg with a larger heat-shield thickness near the stagnation region than had been used in the analyses discussed herein. It was determined by the heat-shield designers, using these codes, that the final probe heat shield meets the design specifications: that the heat shield shall have a 32-45% margin of safety on ablation thickness for entry into the Jupiter nominal model

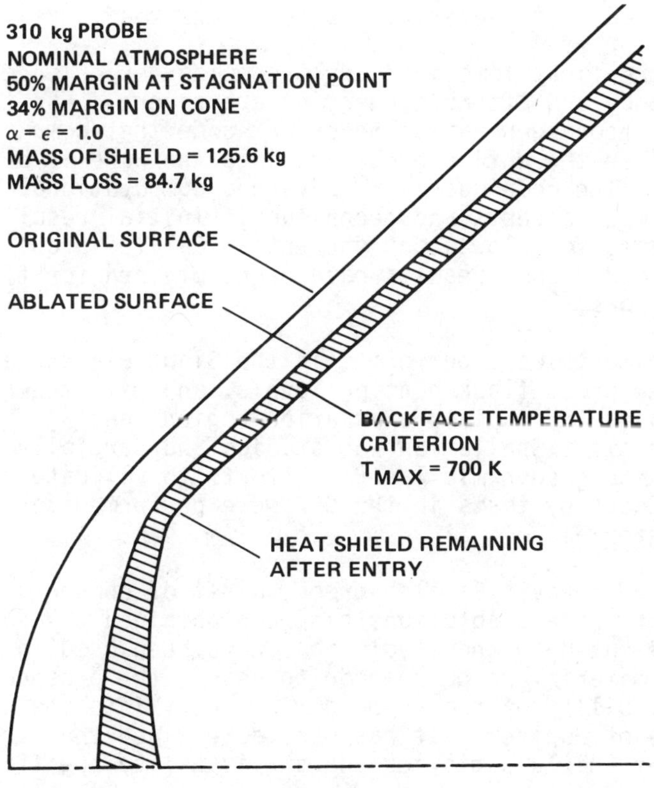

Fig. 24 Transient heat-shield response: carbon phenolic.

atmosphere and a capability of entering the most severe model atmosphere with zero margin.

Conclusions

A supporting research and technology effort related to the entry of the Galileo probe into the Jupiter atmosphere has been reported. Entry at various initial conditions into numerous model atmospheres by probes of various size and mass has been studied experimentally and computationally to determine both forebody and afterbody heating environments and forebody heat-shield requirements. Two kinds of codes were used: 1) codes that use engineering correlations and 2) benchmark codes that use basic modeled physical phenomena. Detailed comparisons of the benchmark flowfield codes have been performed, and they show that the turbulence modeling of a hot gas about a relatively cool probe that is massively ablating has important consequences and needs continued development. Ranking of various turbulence models according to severity has been performed.

For the first time in entry technology experience, both forebody and afterbody heating are severe. For that reason, a second-generation benchmark code that computes the flow over the entire probe and into the wake is being developed. The code uses very advanced computational fluid dynamic concepts and techniques. Initial results have been obtained for flow about the entire Galileo probe configuration, and some results have been obtained for flow of a reacting gas.

Experiments were performed in the Giant Planet Facility to simulate probe flank heating levels, and with the gasdynamic laser to simulate stagnation region heating for several materials. Spallation was studied and partially quantified in the gasdynamic laser. Efforts to validate the computer codes by tests in the GPF were performed for several materials.

For the "heavy" Galileo probe, a set of consensus benchmark flowfield solutions has been obtained. Results that contributed to these solutions have been used in a benchmark material-response code to assess the prospects for survivability of the probe during entry into the atmosphere of Jupiter. It has been determined that the flight heat shield satisfies the survivability specifications.

The overall effort has revealed areas for future entry technology research that will lead to understanding of phenomena that are currently uncertainly known and as a result are carried in the weight margin. These include turbulence modeling and mechanical spallation in particular.

References

[1]"Entry Probes in the Outer Solar System," Report of the Outer Planet Probe Science Study Group, D. M. Hunten, Chairman, NASA Ames Research Center, July 1975.

[2]"The Planet Jupiter," NASA SP-8069, 1971.

[3]Hunten, D. M., "Atmospheres and Ionospheres," Jupiter Studies of the Interior, Magnetosphere, and Satellites, edited by T. Gehrels, University of Arizona Press, Tucson, 1975, p. 23.

[4]Houck, J. R., Pollack, J. B., Schaak, D., Reed, R. A., and Summers, A., "Jupiter: Its Infrared Spectrum from 16 to 40 Micrometers," Science, Vol. 189, Aug. 1975, pp. 720-722.

[5]Sutton, K., Jones, J. J., and Powell, R. W., "Effects of Atmospheric Structure on Radiative Heating for Jupiter Entry Probe," Outer Planet Entry Heating and Thermal Protection, Progress in Astronautics and Aeronautics, Vol. 64, edited by R. Viskanta, AIAA, New York, 1979, pp. 3-21.

[6]Moss, J. N., "A Study of the Aerothermal Entry Environment for the Galileo Probe," AIAA Paper 79-1081, 1979.

[7]Sutton, K., Moss, J. N., Falanga, R. A., and Olstad, W. B., "Outer Planet Entry Probes Aerothermal Environment-Status of Prediction Methodology," AIAA Paper 75-1148, 1975.

[8]Nicolet, W. E., Waterland, L. R., and Kendall, R. M., "Methods for Predicting Radiation Coupled Flowfields about Planetary Entry Probes," Aerodynamic Heating and Thermal Protection Systems, Progress in Astronautics and Aeronautics, Vol. 59, edited by L. S. Fletcher, AIAA, New York, 1978, pp. 120-136.

[9]Sutton, K., "Radiative Heating about Outer Planet Entry Probes," AIAA Paper 75-183, 1975.

[10]Walberg, G. D., Jones, J. J., Olstad, W. B., Sutton, K., Moss, J. N., and Powell, R. W., "Mass Loss Shape Change and Real-Gas Aerodynamic Effects on a Jovian Atmospheric Probe," Acta Astronautica, Vol. 4, Pergamon Press, New York, 1977, pp. 555-575.

[11]Sutton, K., Jones, J. J., and Powell, R. W., "Effect of Probe Configuration on Radiative Heating During Jupiter Entry," Thermophysics of Spacecraft and Outer Planet Entry Probes, Progress in Astronautics and Aeronautics, Vol. 56, edited by A. M. Smith, AIAA, New York, 1977, pp. 355-376.

[12]Arnold, J. O., "A Shock Tube Determination of the Electronic Transition Moment of the C_2 (Swan) Bands," Journal of Quantitative Spectroscopy and Radiation Transfer, Vol. 8, Nov. 1968, pp. 1781-1794.

[13]Cooper, D. M. and Borucki, W. J., "Measurements of Hydrogen-Helium Radiation at Shock-Layer Temperatures Appropriate for Jupiter Entries," Journal of Quantitative Spectroscopy and Radiation Transfer, Vol. 13, Oct. 1973, 1047-1051.

[14]Cooper, D. M., "Absolute Measurements of the Electronic Transition Moments of Seven Band Systems of the C_2 Molecule," Ph.D. Thesis, York University, Toronto, Canada, 1974.

[15]Cooper, D. M. and Nicholls, R. W., "Measurements of the Electronic Transition Momenta of C_2 Band Systems," Journal of Quantitative Spectroscopy and Radiation Transfer, Vol. 15, Feb. 1975, pp. 139-150.

[16]Cooper, D. M. and Langhoff, S. R., "A Theoretical Study of Selected Singlet and Triplet States of the CO molecule," Journal of Chemical Physics, Vol. 74, Jan. 15, 1981, pp. 1200-1210.

[17]Arnold, J. O. and Langhoff, S. R., "A Theoretical Study of the Low-Lying Singlet States of C_3," Paper RD1 presented at Thirty-Third Symposium on Molecular Spectroscopy, Columbus, Ohio, June 12-16, 1978.

[18]Cooper, D. M., "A Theoretical Study of the Electronic Structure and Properties of Selected Triplet States of C_2," Paper presented at the American Conference on Theoretical Chemistry, Boulder, Colo., June 25-30, 1978.

[19]Cooper, D. M. and Jones, J. J., "An Experimental Determination of the Cross Section of the Swings Band System of C_3," Journal of Quantitative Spectroscopy and Radiation Transfer, Vol. 22, No. 2, 1979, pp. 201-208.

[20]Treffers, R. R. and Gilra, D. P., "The Vibrational Spectrum of C_3 in the 5-Micron Region," Astrophysics Journal, Vol. 202, No. 839, Dec. 15, 1975, pp. 839-843.

[21]Prakash, S. G. and Park, C., "Shock Tube Spectroscopy of C_3+C_2H Mixture in the 140-700 nm Range," AIAA Paper 79-0094, Jan. 1979.

[22]Cooper, D. M., "Theoretical Electronic Transition Moments for the Ballik-Ramsey, Fox-Herzberg, and Swan Systems of C_2," Journal of Quantitative Spectroscopy and Radiation Transfer, Vol. 26, Aug. 1981, pp. 113-118.

[23]Arnold, J. O., Cooper, D. M., Park, C., and Prakash, S. G., "Line-by-Line Transport Calculations for Jupiter Entry Probes," Entry Heating and Thermal Protection, Progress in Astronautics and Aeronautics, Vol. 69, edited by W. B. Olstad, AIAA, New York, 1980, pp. 52-82.

[24]Nicolet, W. E., "Radiative Heating Environments for Jovian Entry Conditions," Radiative Transfer and Thermal Control, Progress in Astronautics and Aeronautics, Vol. 49, edited by A. M. Smith, AIAA, New York, 1977, pp. 231-250.

[25]Nicolet, W. E., Morse, H. L., and Vojvodich, N. S., "Outer Planet Probe Thermal Protection, Part I: Aerothermodynamics Environment," AIAA Paper 74-700, 1974.

[26]Nicolet, W. E., Howe, J. T., and Mezines, S. A., "Outer Planet Probe Entry Thermal Protection, Part II: Heat Shielding Requirements," AIAA Paper 74-701, 1974.

[27]Tauber, M. E., "Heat Protection for Atmospheric Entry into Saturn, Uranus and Neptune," Paper AAS-7-145, 1971.

[28]Moss, J. N., Anderson, E. C., and Boltz, C. W., Jr., "Aerothermal Environment for Jupiter Entry Probes," Thermophysics of Spacecraft and Outer Planet Entry Probes, Progress in Astronautics and Aeronautics, Vol. 56, edited by A. M. Smith, AIAA, New York, 1977, pp. 333-354.

[29]Moss, J. N., Anderson, E. C., and Simmons, A. L., "The Impact of Turbulence on a Radiating Shock Layer with Coupled Ablation Injection," AIAA Paper 78-1186, 1978.

[30]Nicolet, W. E. and Balakrishnan, A., "Methods for Predicting Off-Stagnation Point Flow Fields for Planetary Entry Probes," Entry Heating and Thermal Protection, Progress in Astronautics and Aeronautics, Vol. 69, edited by W. B. Olstad, AIAA, New York, 1980, pp. 26-51.

[31]Hopkins, R. A. and Nerem, R. M., "An Experimental Investigation of Heat Transfer from a Highly Cooled Turbulent Boundary Layer," AIAA Journal, Vol. 6, Oct. 1968, p. 1912.

[32]Kumar, A., Graves, R. A., Weilmuenster, K. J., and Tiwari, S. N., "Laminar and Turbulent Flow Solutions with Radiation and Ablation Injection for Jovian Entry," AIAA Paper 80-0288, 1980.

[33]Menees, G. P., "An Evaluation of Computer Codes for Simulating the Galileo Probe Aerothermal Entry Environment," AIAA Paper 81-1069, 1981.

[34]Menees, G. P. and Lombard, C. K., "The Effect of Modeled Turbulence on a Hypersonic Shock Layer with Massive Ablation Injection," AIAA Paper 81-1071, 1981.

[35]Nicolet, W. E., Davy, W. C., and Wilson, J. F., "Galileo Probe Forebody Entry Thermal Protection: Aerothermal Environments and Heat Shielding Requirements," ASME Paper 80-ENAs-24, 1980.

[36]Zoby, E. V., Sutton, K., Olstad, W. B., and Moss, J. N., "An Approximate Inviscid Radiating Flowfield Analysis for Outer Planet Entry Probes," AIAA Paper 78-189, 1978.

[37]Zoby, E. V., Moss, J. N., and Sutton, K., "Approximate Convective Heating Equations for Hypersonic Flow," AIAA Paper 79-1078, 1979.

[38]Lockman, W. L., "Base Heating Measurements on Apollo Block 2 Command Module," Journal of Spacecraft and Rockets, Vol. 7, Jan. 1970, pp. 93-95.

[39]Stephenson, J. D., "Measurement of Optical Radiation from the Wake of Ablating Blunt Bodies in Flight at Speeds up to 10 km per Second," NASA TN D-2760, 1965.

[40]Park, C., "Modeling of Radiative Heating in Base Region of Jovian Entry Probe," Entry Heating and Thermal Protection, Progress in Astronautics and Aeronautics, Vol. 69, edited by W. B. Olstad, AIAA, New York, 1980, pp. 124-147.

[41]Shirai, H. and Park, C., "Experimental Studies of Radiative Base Heating of a Jovian Entry Model," Entry Heating and Thermal Protection, Progress in Astronautics and Aeronautics, Vol. 69, edited by W. B. Olstad, AIAA, New York, 1980, pp. 148-171.

[42]Park, C., "Problems of Radiative Base Heating," AIAA Paper 79-0919, 1979.

[43]Lombard, C. K., Davy, W. C., and Green, M. J., "Forebody and Base Region Real-Gas Flow in Severe Planetary Entry by a Factored Implicit Numerical Method, Part 1: Computational Fluid Dynamics," AIAA Paper 80-0065, 1980.

[44]Lombard, C. K., Lombard, M. P., Menees, G. P., and Yang, J. Y., "Some Aspects of Adapting Computational Mesh to Complex Flow Domains and Structures with Application to Blown Shock Layer and Base Flow," Paper presented at the Workshop on Numerical Grid Generation Techniques of Partial Differential Equations, NASA Langley Research Center, Oct. 1980.

[45]Beam, R. M. and Warming R. F., "An Implicit Finite-Difference Algorithm for Hyperbolic Systems in Conservation-Law Form," Journal of Computational Physics, Vol. 22, Sept. 1976, pp. 87-110.

[46]Beam, R. M. and Warming, R. F., "An Implicit Factored Scheme for the Compressible Navier-Stokes Equations," AIAA Paper 77-645, 1977.

[47]Warming, R. F. and Beam, R. M., "On Construction and Application of Implicit Factored Schemes for Conservation Laws," SIAM-AMS Proceedings, Vol. 11, 1977, pp. 85-129.

[48]Briley, W. R. and McDonald, H., "Solution of the Multidimensional Compressible Navier-Stokes Equations by a Generalized Implicit Method," Journal of Computational Physics, Vol. 24, Aug. 1977, pp. 372-397.

[49]Briley, W. E. and McDonald, H., "Solution of the Three-Dimensional Compressible Navier-Stokes Equations by an Implicit Technique," Proceedings of the Fourth International Conference on

Numerical Methods in Fluid Dynamics, Springer-Verlag, New York, 1975 (see also United Aircraft Research Laboratories Rept. M911363-6, 1973).

[50]Davy, W. C., Lombard, C. K., and Green, M. J., "Forebody and Base Region Real Gas Flow in Severe Planetary Entry by a Factored Implicit Numerical Method, Part II: Equilibrium Reactive Gas," AIAA Paper 81-0282, Jan. 1981.

[51]Winovich, W. and Carlson, W. C. A., "The Giant Planet Facility," Paper presented at the 25th Instrument Society of America Symposium, Anaheim, Calif., May 1979.

[52]Lundell, J. H., Otten, L. J., and Dickey, R. R., "The CO_2 Gas-dynamic Laser as a High-Intensity Radiation Facility," AIAA Paper 75-177, 1975.

[53]Lundell, J. H. and Dickey, R. R., "Response of Heat-Shield Materials to Intense Laser Radiation," Outer Planet Entry Heating and Thermal Protection, Progress in Astronautics and Aeronautics, Vol. 64, edited by R. Viskanta, AIAA, New York, 1979, pp. 193-209.

[54]Dickey, R. R. and Lundell, J. H., "A Technique for Evaluating the Jovian Entry-Probe Heat-Shield Material with a Gasdynamic Laser," Paper presented at the International Congress on Instrumentation in Aerospace Simulation Facilities, Monterey, Calif., Sept. 24-26, 1979 (IEEE Publication 79 CH 1500-8 AES).

[55]Davy, W. C., Menees, G. P., Lundell, J. H., and Dickey, R. R., "Hydrogen-Helium Ablation of Carbonaceous Material: Numerical Simulation and Experiment," Outer Planet Entry Heating and Thermal Protection, Progress in Astronautics and Aeronautics, Vol. 64, edited by R. Viskanta, AIAA, New York, 1979, pp. 228-244.

[56]Green, M. J. and Davy, W. C., "Numerical Simulation of Experiments in the Giant Planet Facility," Entry Heating and Thermal Protection, Progress in Astronautics and Aeronautics, Vol. 69, edited by W. B. Olstad, AIAA, New York, 1980, pp. 172-197.

[57]Balakrishnan, A. and Nicolet, W. E., "Galileo Probe Forebody Thermal Protection: Benchmark Heating Environment Calculations," AIAA Paper 81-1072, 1981.

[58]Green, M. J. and Davy, W. C., "Galileo Probe Forebody Thermal Protection," Thermophysics of Atmospheric Entry, Progress in Astronautics and Aeronautics, Vol. 82, edited by T. E. Horton, AIAA, New York, 1982, pp. 328-353.

[59]Moss, J. N. and Kumar, A., "Significance of Turbulence and Transition Location on Radiative Heating and Ablation Injection," Thermophysics of Atmospheric Entry, Progress in Astronautics and Aeronautics, Vol. 82, edited by T. E. Horton, AIAA, New York, 1982, pp. 354-381.

Galileo Probe Forebody Thermal Protection

Michael J. Green* and William C. Davy*
NASA Ames Research Center, Moffett Field, Calif.

Abstract

Material-response solutions for the forebody heat shield on the candidate 310 kg Galileo probe are presented. A charring material ablation analysis predicts thermochemical surface recession, insulation thickness, and total required heat-shield mass. Benchmark shock-layer solutions provide the imposed entry heating environments on the ablating surface. Heat-shield sizing results are given for a nominal entry into modeled nominal and cool heavy Jovian atmospheres and for two heat-shield property models. The nominally designed heat shield requires a mass of at least 126 kg and would require an additional 13 kg to survive entry into the less probable cool heavy atmosphere. The material-property model with a 30% surface reflectance reduces these mass requirements by as much as 16%.

Nomenclature

A = cross-sectional area
B = pre-exponential factor for decomposition reaction
C_{k_i} = number of atoms of element k in a molecule of species i
C_p = specific heat
E/R = activation energy for decomposition reaction
H_r = recovery enthalpy
h = enthalpy
h_F = enthalpy of formation of virgin material
$\bar{h}$ = density-weighted enthalpy difference between virgin material and char
I = total number of species
$\tilde{j}_{k_w}$ = diffusion flux of element k away from wall

---

Presented as Paper 81-1073 at the AIAA 16th Thermophysics Conference, Palo Alto, Calif., June 23-25, 1981.

*Research Scientist, Entry Technology Branch.

K = total number of elements in chemical system
$\tilde{K}_k$ = mass fraction of element k, regardless of molecular configuration
K_p = equilibrium coefficient
k = thermal conductivity; also element index
$\mathcal{M}$ = molecular mass
m = reaction order
$\dot{m}$ = mass flux
P = pressure
P_i = partial pressure of species i, equals $x_i P$
q_c = convective heat flux to solid surface by diffusional processes in the shock layer
q_{cond} = conductive energy flux
q_r = incident minus reflected surface radiative flux
R = probe radius
r = reflectivity of surface
S = distance from original location of receding surface to current surface location
$\dot{S}$ = surface recession rate
s = streamwise distance along probe surface
T = temperature
t = time
x = coordinate normal to ablating surface, fixed to receding surface
x_i = mole fraction of species i
y = $x + S$
Γ = volume fraction of phenolic resin in carbon phenolic, 0.35
ε = emissivity of surface
θ_c = cone half-angle
ρ = density
ρ_o = original density of a pyrolyzing component
ρ_r = final density of a pyrolyzing component
σ = Stefan-Boltzmann constant

Subscripts

b = base
c = char
g = pyrolysis gas
i = any identifiable species: atom, ion, molecule
k = chemical element
n = nose
w = wall

Introduction

The Galileo Orbiter-Probe Spacecraft is scheduled to be launched in 1985 and to arrive at Jupiter four years later. The Galileo entry probe will be instrumented so that as it descends through the Jovian atmosphere, the planet's atmospheric structure, composition, and radiation balance can be measured. These will be the first direct measurements of an outer-planetary atmosphere. Because of the large gravitational attractive force of the massive planet, the probe will enter Jupiter's atmosphere at a speed of 48 km/s (107,000 mph), a speed that must be greatly reduced before the measurements can be made. During deceleration in the upper atmosphere, a hypersonic shock layer will enshroud the probe, causing a hostile environment of combined radiative and convective heating. In protecting the scientific payload, the probe's forebody heat shield accommodates this heating by ablation. Since the heating is so intense (several times greater than for any previous mission) and, consequently, the ablative mass loss so great, almost half of the probe mass is allocated to the thermal protection subsystem. The heat-shield mass requirement is so large that a heat-shield design based on a conservative safety margin, although desirable, is not possible, except at the expense of other subsystems, because the total probe mass is constrained by launch requirement limits. Hence, an accurate heat-shield design is crucial to the success of the Galileo mission.

In the past decade, several studies of the heat-shield requirements for a Jupiter entry probe have been reported.[1-5] The requirements have been continually upgraded as mission parameters, such as atmospheric models, entry condition, and probe configuration, have been more accurately defined and as techniques for computing both the shock-layer environment and the material response have been improved. Within the last two years, detailed viscous shock-layer solutions, including turbulence and coupled ablation, have been generated throughout the entry heating pulse.[5-9] Recently, Brewer and Brandt[4] reported on a heat-shield design for a 290 kg probe based on an approximate engineering technique in which the effect of ablation on heating was determined by correlations developed from the benchmark results of Moss[6] for a 242 kg probe. Nicolet et al.[5] presented a heat-shield analysis for a 242 kg probe, using a different approach in which the ablative heating environment was defined directly from benchmark solutions, thus eliminating the need for correlations. Recently, Moss and Kumar[8] and Balakrishnan and Nicolet[9] have generated benchmark shock-layer results for a 310 kg probe and nominal atmosphere. Reference 9, in addition, gives results for a cool heavy atmosphere.

In this paper, using benchmark results[9] to define the aerothermal environment, we present forebody heat-shield requirements for the 310 kg probe. Nominal entry into modeled nominal and cool heavy atmospheres is considered. In addition, results are obtained for two different material-property models which have been used to characterize the carbon-phenolic heat-shield material, so that the effect of materials properties on the heat-shield design can be assessed. The results of this study, along with those for lighter probes, provide a data base which will be useful for making timely engineering judgments.

Method

The Galileo entry probe is illustrated in Fig. 1; the configuration corresponds to a probe mass of 310 kg. The main components are the descent module, which contains the scientific measurement package, the aeroshell structure, and the thermal protection subsystem, which is bonded to the aeroshell; it consists of two heat shields, one for the forebody and another for the afterbody. In the present study, the performance of only the forebody heat shield, composed of chopped-molded carbon phenolic in the spherical region and 30 deg tape-wrapped carbon phenolic on the frustum, is analyzed. An analysis of the complete thermal protection system including the afterbody heat shield for a 290 kg probe is described in Ref. 4.

Generally, a heat-shield material-response analysis can be divided into five components: 1) specifying the

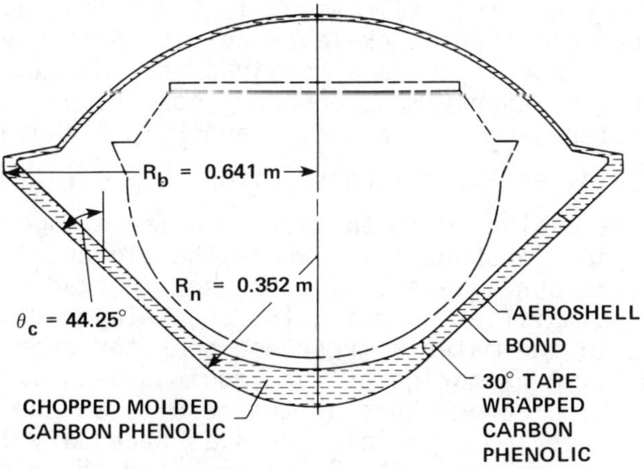

Fig. 1 Galileo entry probe (310 kg).

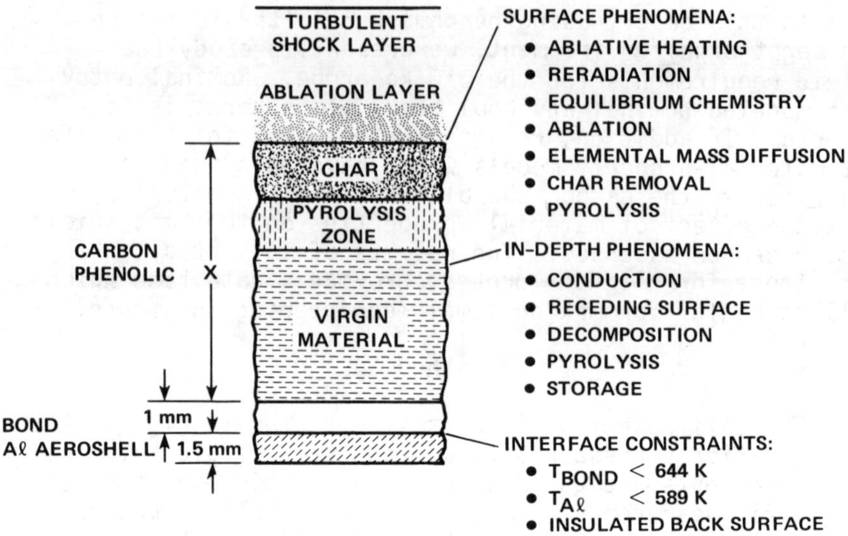

Fig. 2 Physical model and phenomena considered in material-response analysis.

entry heating environment, 2) modeling the in-depth phenomena, 3) modeling the surface phenomena, 4) specifying the back-surface requirements, and 5) imposing the constraints necessary to size the heat shield. In Fig. 2, the required components, as they apply specifically to the Galileo probe, are defined in terms of physical phenomena and physical constraints. In the remainder of this section, each component of the computational material-response analysis, as used in this study, will be described.

The entry heating environment is specified, using the recent RASLE benchmark shock-layer results presented in Ref. 9. The shock-layer data required for the material-response surface energy/mass boundary conditions, to be discussed later, are q_r, q_c, $\tilde{j}_{k_w}$, and P_w. Although $\tilde{j}_{k_w}$ is available directly, the correlation $\tilde{j}_{k_w} = (q_c/H_r)\tilde{K}_{k_w}$, where H_r is evaluated at the ablation layer edge (defined as the location in shock layer where the elemental mass fraction of carbon equals 0.01), is used instead. This correlation simplifies coupling the shock-layer mass transport events to the material-response computer program. Quasisteady time histories of the required data are produced by fitting curves through the results at the seven trajectory times. One set of curves is necessary for each of the nine streamwise probe locations studied in the material-response analysis.

One of the two carbon-phenolic material-property models to be investigated characterizes the surface as being gray, with $r = 0.3$ and $\varepsilon = 0.7$. Since shock-layer results are not available for this radiative boundary condition, an approximation is made: the heating environment for the gray surface is determined by reducing the radiative flux for the blackbody surface by 30%, but without modifying the other data. This approximation somewhat underpredicts both the radiative heating (because some of the reflected energy will be absorbed in the shock layer near the surface, increasing the incident flux) and the convective heating (because the net radiative heating will be lower, reducing the ablative mass flux). Based on previous shock-layer studies, however, the combined increase should be small.

With the heating environment defined, the second component, modeling of the in-depth phenomena, is now considered. When carbon phenolic is heated, the phenolic resin undergoes pyrolysis, forming gas and residual char. During ablation, a graphitic char is produced from both the residual resin char and the carbon in the carbon fabric. A distinct char layer is created between the pyrolysis zone and the shock layer (see Fig. 2). The charring material ablator (CMA) computer program[10] is used to analyze these phenomena. In fact, the CMA program controls all components in the present material-response analysis. It has been exercised in several previous heat-shield studies (for instance, Refs. 1-3, 5, 11-13). Basically, this program provides numerical solutions for the one-dimensional, transient conservation equations appropriate for a charring material. The basic equations will be summarized here; a comprehensive description of program can be found in Ref. 10.

Consider a point within the material a distance x from the surface that is receding and at a time t. The conservation of mass is written

$$\left.\frac{\partial\rho}{\partial t}\right)_x = \left.\frac{\partial\rho}{\partial t}\right)_y + \left.\frac{\partial\rho}{\partial x}\right)_t \dot{S} \tag{1}$$

where the phenolic decomposition rate is given by the Arrhenius expression

$$\left.\frac{\partial\rho}{\partial t}\right)_y = -B\,\exp(-E/RT)\rho_o\left(\frac{\rho - \rho_r}{\rho_o}\right)^m = \left.\frac{\partial\dot{m}_g}{\partial y}\right)_t$$

The conservation of energy is expressed as

$$\rho C_p \left.\frac{\partial T}{\partial t}\right)_x = \frac{1}{A}\frac{\partial}{\partial x}\left(kA\left.\frac{\partial T}{\partial x}\right)_t + (h_g - \bar{h})\left.\frac{\partial \rho}{\partial t}\right)_y + \dot{S}\rho C_p \left.\frac{\partial T}{\partial x}\right)_t + \dot{m}_g \left.\frac{\partial h_g}{\partial x}\right)_t \tag{2}$$

These terms are from left to right: energy storage, net energy transport by conduction, net energy convected by decomposition absorption, net energy convected as a consequence of the moving coordinate system, and net energy convected by the pyrolysis gas.

The third component, modeling the surface phenomena, is a major part of the analysis because it couples the shock-layer mass and energy transport events to the in-depth phenomena [Eqs. (1) and (2)]. The surface thermochemistry is determined from the Aerotherm chemical equilibrium (ACE) computer code,[14] which provides a solution to the conservation of chemical elements at the surface, assuming thermochemical equilibrium. The modeling equations are the following.

The chemical equilibrium relations for the chemical reactions are

$$\ln P_i - \sum_{k=1}^{K} C_{k_i} \ln P_k = \ln K_{p_i}(T) , \quad i = 1, \ldots, I - K \tag{3}$$

The conservation of chemical elements is written

$$K_{k_w} = \frac{\mathcal{M}_k}{P\mathcal{M}} \sum_{i=1}^{I} C_{k_i} P_i , \quad k = 1, \ldots, K \tag{4}$$

The definition of system pressure is

$$\sum_{i=1}^{I} P_i = P \tag{5}$$

The definition of molecular mass is

$$\mathcal{M} = \sum_{i=1}^{I} \frac{P_i}{P} \mathcal{M}_i \tag{6}$$

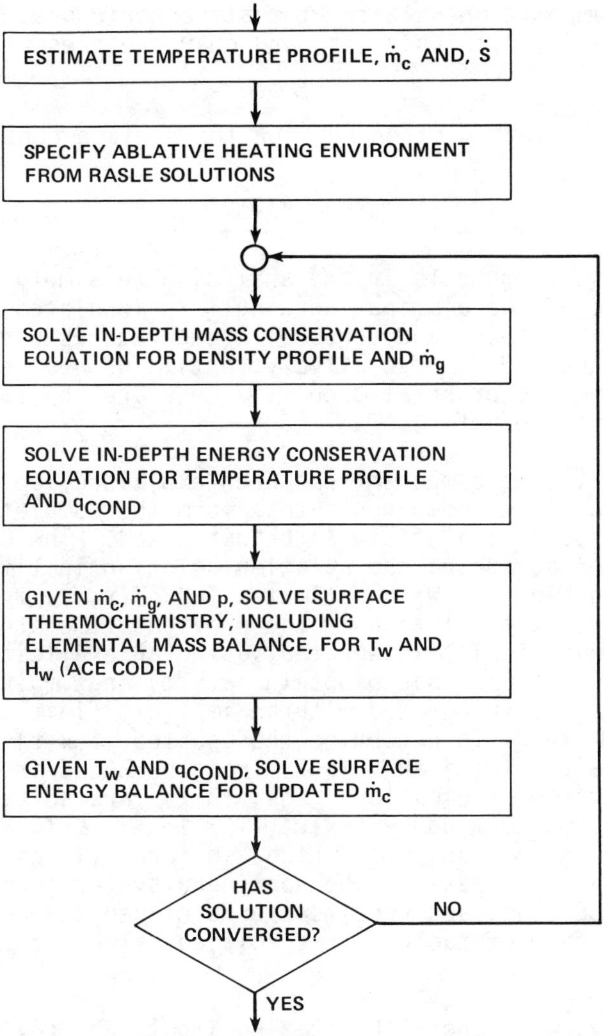

Fig. 3 CMA algorithm for advancing ablation solution.

Consider an infinitely thin control volume at the surface. The conservation of chemical elements is then written

$$\tilde{j}_{k_w} + \dot{m}\tilde{K}_{k_w} = \dot{m}_g\tilde{K}_{k_g} + \dot{m}_c\tilde{K}_{k_c}, \quad k = 1, \ldots, K \tag{7}$$

As is the case for the RASLE shock-layer solutions, the chemical species properties are obtained, with a few exceptions, from the JANAF thermochemical data.[15] Also, the elemental composition data for the atmospheric gas, virgin carbon phenolic, pyrolysis gas, and char are prescribed input.

The conservation of energy for the surface control volume is

$$(q_r + q_c) - (\dot{m}h_w - \dot{m}_g h_g - \dot{m}_c h_c) - \varepsilon\sigma T_w^4 = q_{cond} \quad (8)$$

The final two components in the analysis are simply that the back surface of the aluminum aeroshell is insulated, and that the maximum bond and aeroshell temperatures cannot exceed 644 and 589 K, respectively, until the heat shield is rejected from the probe at Mach 1. These are the same constraints given in Ref. 4.

Now that each component has been explained separately, the coupling of the components that form the complete material-response analysis will be discussed. The CMA algorithm for advancing the ablation solution in time is exhibited in Fig. 3. With estimates for $T(x,t)$, $m_c(t)$, and $\dot{S}(t)$ from the solution at the previous time step, Eq. (1) is numerically integrated, using an explicit finite-difference method to obtain $\rho(x,t)$ and $\dot{m}_g(x,t)$. Then, Eq. (2) is integrated with an implicit finite-difference method, to determine the updated $T(x,t)$ and $q_{cond}(t)$. Given $P(t)$ and $\tilde{j}_k(t)$ from the RASLE input, the nonlinear system of equations [Eqs. (3-7)] are solved by Newton's method to find $P_i(=x_iP)$, $i = 1, \ldots, I$; $\tilde{K}_k$, $k = 1, \ldots, K$; $\mathcal{M}$; and T_w, which, in turn, yields h_w. Incidentally, in practice, the nonlinear system is solved using the ACE code, and results are furnished to the CMA code in the form of tables for specified values of $\dot{m}_c$., $\dot{m}_g$, and P.

Finally, with the RASLE heating input, Eq. (8) can be solved to obtain an updated $\dot{m}_c$. The computational process is iterated until the solution converges. To obtain the heat-shield requirements at a particular probe body location, the initial required carbon-phenolic thickness is estimated. The initial temperature distribution for the thermal protection system is constant at 294 K. The CMA code is used to obtain the material response for the entire trajectory. The initial carbon-phenolic thickness is then adjusted and the material-response analysis is repeated. This iterative process is continued until the maximum allowable bond or aeroshell temperature is reached at the time the heat shield is ejected.

The thermochemical recession and insulation requirements are found for each of the nine body locations. For entry into the nominal atmosphere (but not the cool heavy atmosphere), a safety margin is added to the heat-shield thickness. The safety margin is 50% of the predicted recession at the stagnation point, 34% along the frustum, and a linear variation between the values along the spherical region. The required heat-shield mass and final shape are then computed.

Material Properties

The analysis of the performance of the thermal protection system requires knowledge of materials properties for the carbon-phenolic heat-shield material (including the char), the material that bonds the heat shield to the aluminum aeroshell, and the aeroshell itself. These properties include density, specific heat, and thermal conductivity. In addition, since carbon phenolic pyrolyzes and chars when heated, chemical composition data for the virgin material, pyrolysis gas, and char are needed, as are kinetic data for the decomposition of the phenolic resin. Carbon-phenolic surface properties, such as reflectance and emittance, are also required.

Two different models have been developed for characterizing carbon phenolic in planetary entry environments. The Underhill model, originally developed in 1967 and described in Ref. 16, has been used in studies associated with Jovian entry[4,5] and its experimental simulation.[11,12] It was evaluated, along with two other models for its ability to simulate material-response results for flight tests in convective heating environments. This model assumes a blackbody surface: no reflectance and unit emittance. (For the true Underhill model, $\varepsilon = 0.85$, but the value was modified in this study for consistency with the shock-layer radiative boundary conditions.)

Another carbon-phenolic property model, which will be denoted as the Wakefield model, has been developed in support of the Pioneer-Venus mission. The model's development, validation in combined radiative and convective heating environments, and application to the Pioneer-Venus heat-shield experiments are discussed in detail in Ref. 13. The emittance of 0.70 and surface reflectance of 0.30, which are the most significant differences between the two models, will be discussed later. The emittance value is consistent with those reported in Ref. 17 where emittance values between 0.60 and 0.71 were measured experimentally at heated surfaces (up to 2400 K) of decomposing, charring materials, including phenolic nylon. Each material-property model is employed in the material-response analysis. No distinction is made between the properties of the chopped-molded and tape-wrapped carbon phenolic.

Results and Discussion

The atmospheric models, probe characteristics, and entry conditions considered in Ref. 9 and pertinent to this study are listed in Table 1. Note that for the cool heavy entry, the solutions in Ref. 9 apply for a trajectory into the Orton cool heavy atmosphere ($0.83H_2 + 0.17He$) but with shock-layer results for a different atmospheric composition ($0.78H_2 + 0.22He$) which, because of the higher helium content, yields a more severe radiative heating environment. Therefore, the cool heavy solutions give a more conservative heat-shield design than needed for entry into the actual Orton cool heavy model.

Table 1 Probe configuration and entry parameters for RASLE shock-layer solutions

Atmosphere	
Orton models, revised Aug. 10, 1979	
Nominal ($0.89H_2 + 0.11He$, by volume)	
Cool-heavy ($0.83H_2 + 0.17He$)[a]	
Probe configuration	
Mass	310 kg
Nose radius	0.352 m
Base radius	0.641 m
Half cone angle	44.25 deg
Drag coefficient	1.05
Ballistic coefficient	228.72 kg/m^2
Entry condition	
Altitude	450 km
Relative velocity	48.2 km/s
Relative entry angle	-8.6 deg
Entry latitude	+3.4 deg
Relative azimuth	70.3 deg

[a]The solutions are for a trajectory into an Orton cool-heavy atmosphere, but with shock-layer heating for an atmospheric composition of $0.78H_2 + 0.22He$.

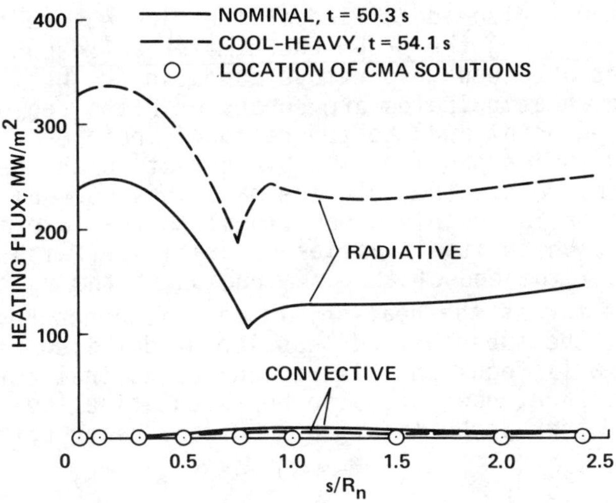

Fig. 4 RASLE peak heating distributions.

The probe's ballistic coefficient is assumed constant. According to Ref. 4, this assumption leads to a conservative heat-shield design because the ablative mass loss continually reduces the ballistic coefficient, leading to a trajectory with a less severe heating environment. The entry condition corresponds to one nominally selected for the Galileo mission. For the two entry trajectories of interest, the sonic condition occurs at 120.5 s for the nominal and at 117.5 s for the cool heavy; it corresponds to the deployment of the parachute when the heat shield will be jettisoned from the descent module and the maximum bond and aeroshell temperature constraints must be satisfied.

Typical results from the RASLE turbulent flow shock-layer solutions,[9] which define the ablative surface heating environment, are illustrated in Figs. 4 and 5. Steady-state ablation with no surface reflectance is assumed in these results. The heating flux distribution along the probe for the two entries at peak heating is shown in Fig. 4. For the massive ablation conditions, the convective heating is very small, as expected; the heating is clearly radiation dominated. The radiation flux for the cool heavy entry is 35% greater than the nominal at the stagnation point, and twice that at the end of cone ($s/R_n = 2.38$). In fact, the cool heavy radiative heating at the end of cone equals that of the nominal at the stagnation point. As was observed in Ref. 8, the radiative heating increases along the back half of the frustum. Also identified in the figure, are the nine locations along the probe where material-response solutions are obtained. To remove cusps in the heating distribution, heating rates are enhanced in the region of the stagnation point and the sphere-cone juncture (s/R_n = 0.80) which has a negligible effect on the heat-shield design. Since the ablative mass loss per unit area along the probe is roughly proportional to the lateral surface area, which itself varies quadratically with streamwise distance, the end of the cone region is the most important as far as the heat-shield mass is concerned. Accordingly, the subsequent discussion of detailed results (Figs. 5-9) will focus on this region for nominal entry. These results are, however, also representative (qualitatively) for other probe locations and cool heavy entry.

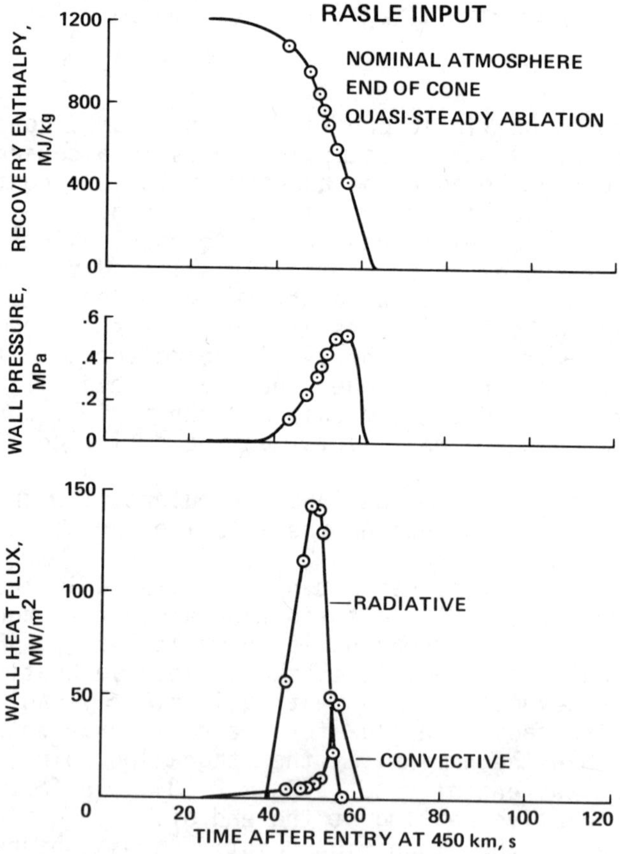

Fig. 5 Aerothermal environment at end of cone.

The RASLE heating data that specify the energy and mass transport from the turbulent shock layer to the surface, which is undergoing steady-state ablation, are given in Fig. 5. Since RASLE solutions were not available for the early and late times of the convective heating pulse, we used the length of the pulse given in Ref. 4 to extrapolate the wings of the pulse. The peak radiative flux is three times greater and occurs 7 s sooner than the peak convective flux, which occurs simultaneously with the maxi-

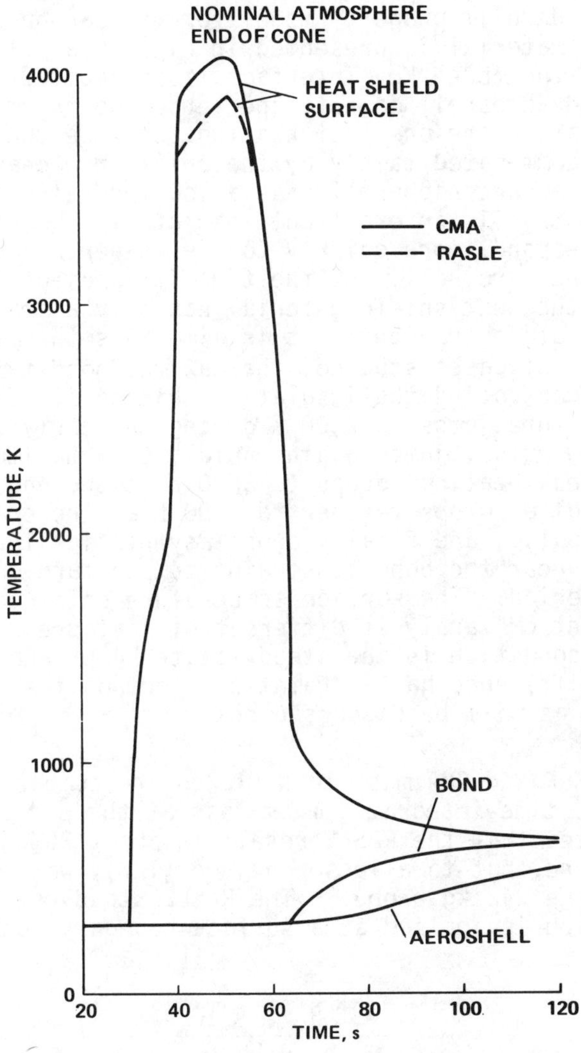

Fig. 6 Temperature history at interfaces.

mum surface pressure. The radiative pulse persists for 18 s, half as long as the convective. The elemental mass diffusion flux $\tilde{j}_{kw}$, when defined with a driving potential equal to q_c/H_r, agrees with actual values to within 15%. This level of accuracy is sufficient because (interestingly) at the surface the elemental diffusive flux is four orders of magnitude smaller than the ablative mass flux and hence has no effect on the material response.

The thermal protection system's transient response to the shock-layer heating as predicted by the CMA analysis using the Underhill property model for the carbon-phenolic heat-shield material is presented in Figs. 6-9. The response at the three key interfaces is exhibited in Fig. 6. The bond and aeroshell do not experience the thermal pulse until 30 s after the onset of heating, because the heating is being accommodated mainly by the ablation mechanism and, moreover, the carbon phenolic has a low thermal diffusivity $(k/\rho C_p)$ and thus is an excellent insulator. The bond interface responds more quickly to the in-depth heating than does the aeroshell. At the time it is ejected (120.5 s), the heat-shield material has only a small in-depth temperature gradient: it is almost isothermal. As is usual in the cases studied, the maximum bond-temperature constraint controlled the insulation thickness. The surface temperature jumps to 2000 K during the early convective heating pulse, jumps again rapidly to a maximum near 4100 K at peak heating, drops to 3500 K at the end of the radiative pulse, drops further to 1200 K at the end of the convective pulse, and finally drops asymptotically to a temperature near the bond-constraint temperature during the heat-soak period. The surface temperature obtained from the transient CMA analysis differs from that prescribed as a boundary condition in the steady-state RASLE analysis by 5%. This difference has a dramatic effect on the mass loss prediction, as will be discussed next.

The RASLE and CMA mass loss fluxes are compared in Fig. 7. The time-integrated mass loss at the end of the cone indicates that the RASLE result is about 20% higher. The same trend, but to a lesser degree (12%), was noted for a study of the 242 kg probe.[5] The RASLE steady-state energy balance, which is applied as a surface boundary condition, is

$$\dot{m} = \frac{q_r + q_c - \varepsilon\sigma T_w^4}{h_w - h_F} \qquad (9)$$

The lower RASLE surface temperature (Fig. 6) increases the mass loss flux in two ways: first, because the reradiation flux $\varepsilon\sigma T_w^4$ is decreased by 20%, the net surface heating is increased by about 3%; second, and more importantly, the wall enthalpy h_w is reduced by 16%. Since, in the cases studied, h_F is less than 5% of h_w, the difference in surface temperature explains almost entirely the difference in mass loss fluxes. The next problem, obviously, is to determine why the surface temperatures differ and which is correct. The sublimation temperature correlation given in Ref. 8 was used as the RASLE surface temperature boundary condition. The correlation was evaluated for an elemental composition corresponding to the RASLE solution at peak heating (t = 50.3 s, P = 0.033 MPa) and for a range of pressures, and is the curve labeled RASLE in Fig. 8. Using the same technique (the ACE computer code) that was used to generate the sublimation temperature, a new correlation was generated for the same condition and labeled ACE in the figure. Since these correlations should have been identical the temperature differences between them were puzzling.

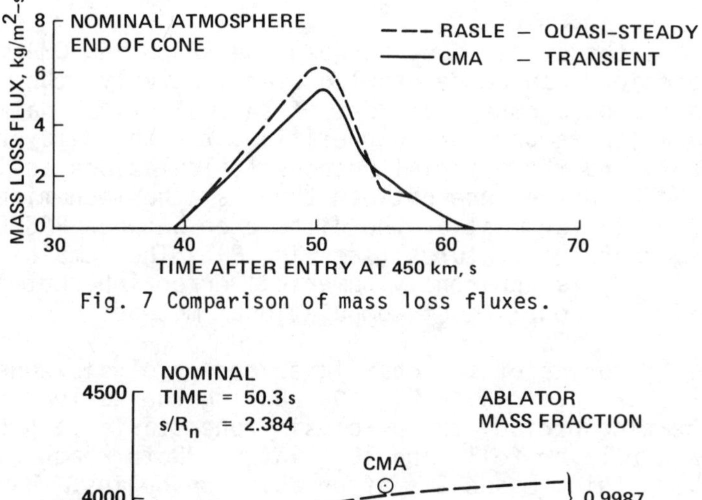

Fig. 7 Comparison of mass loss fluxes.

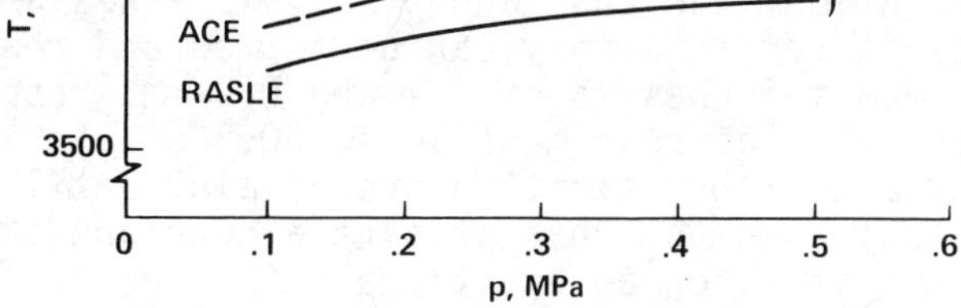

Fig. 8 Carbon-phenolic ablation temperature.

With the assistance of K. Sutton at NASA Langley, the two correlation techniques were compared in detail. We found that thermochemical data for the species C_3 was the principal difference in each technique. The correlation defined in Ref. 8 employed a new model,[18] whereas (as already discussed) the RASLE and ACE analyses used the JANAF model.[15] The effect of coupling different C_3 models is the following. For a given temperature, pressure, and elemental composition, the new C_3 model predicts a higher gas mixture enthalpy than the JANAF model. In the RASLE analysis, for a specified heating condition, the surface temperature would be prescribed as the sublimation temperature of the Ref. 8 correlation. However, when this surface temperature was used by RASLE to obtain the wall gas enthalpy, there resulted a lower enthalpy (because of the JANAF C_3 data) and, consequently, higher mass loss flux consistent with the heating condition. When the RASLE heating conditions were applied subsequently in the CMA-ACE analysis, the surface energy balance [Eq. (8)] converged to a higher enthalpy, comparable to the one corresponding to the new C_3 model at the sublimation temperature, but with a higher temperature corresponding to the JANAF thermochemistry.

So that the differences between the RASLE and CMA mass loss predictions can be reconciled quantitatively, the incorporation of a consistent set of thermochemical data in the sublimation temperature correlation, the shock-layer environment, and the material-response calculations is in progress. The difference between the C_3 thermochemical models accounts for most of the difference between RASLE and CMA surface temperatures (see Fig. 8). The remaining 1% error (40 K) is apparently numerical error inherent in the complicated coupling of computational methods.

Time histories of the char layer and pyrolysis zone thicknesses are plotted in Fig. 9. During the early convective heating period, the pyrolysis zone starts to grow, with the char layer following 10 s later. Both reach relative maximum thicknesses 2 s after ablation begins. With increasing ablation, the char and pyrolysis zones are stripped away faster than they can be formed and reach relative minimum thicknesses of 0.1 and 0.4 mm, respectively, at the time of peak mass loss, 50.3 s. As the ablation decreases, they start to grow again, until ablation terminates, when the char attains a constant thickness of 1.7 mm; it takes an additional 15 s for the pyrolysis zone to reach a constant 3.8 mm.

More general results are presented in Table 2 where heating loads and heat-shield accommodation are listed for both carbon-phenolic materials-property models, for both atmospheric models, and at the stagnation point and end of cone. The total heating load is the first term in Eq. (8) integrated in time over the complete trajectory. For both material-property models, the increase in heating resulting from entering the cool heavy atmosphere instead of the nominal atmosphere is proportionately greater at the end of cone than at the stagnation point. For the Wakefield model, the radiative heating is assumed to be 70% of the Underhill value, to approximately account for surface reflection. In all cases, the radiative heating dominates the convective by at least about three to one.

The modes, both at the surface and in-depth, with which the heat-shield material accommodates the heating is discussed next. At the surface, the net ablation energy leaving the surface, the time integral of the second term in Eq. (8), rejects the majority (57% on the average) of the total absorbed heating. The reradiated energy, the integral of the third term, emits 11%. The remaining 32%, not accommodated at the surface, is conducted into the material. The in-depth mode labeled "convection with solid" in Table 2 is the fourth term in Eq. (2), doubly integrated in space and time; it represents the stored energy removed by ablation as the solid material is convected toward the surface, because the coordinate system is fixed to the receding surface. This mode removes about 22%

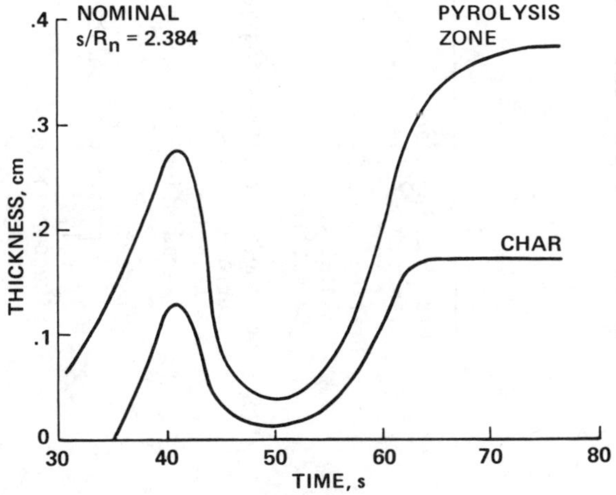

Fig. 9 Carbon-phenolic material response.

Table 2 Heating loads and heat-shield accommodation for 310 kg Galileo probe – carbon-phenolic properties for Underhill and Wakefield models

	Underhill model		Wakefield model	
	Nominal atmosphere[a]	Cool heavy atmosphere[a]	Nominal atmosphere[a]	Cool heavy atmosphere[a]
Total heating load (includes reflection), MJ/m^2	2829, 1694	3476, 2220	2000, 1283	2443, 1607
Heating load, %				
Radiative	98, 80	99, 92	97, 74	99, 89
Convective	2, 20	1, 8	3, 26	1, 11
Accommodation of heating load, %				
Surface				
Ablation	53, 54	58, 59	57, 56	59, 59
Reradiation	11, 16	7, 9	11, 15	8, 9
In-depth				
Convection with solid	24, 20	24, 21	23, 20	24, 22
Pyrolysis	9, 8	8, 8	7, 8	7, 7
Decomposition	2, 1	2, 2	1, 1	1, 2
Storage	1, 1	1, 1	1, 1	1, 1

[a] First value at stagnation point, second at end of cone.

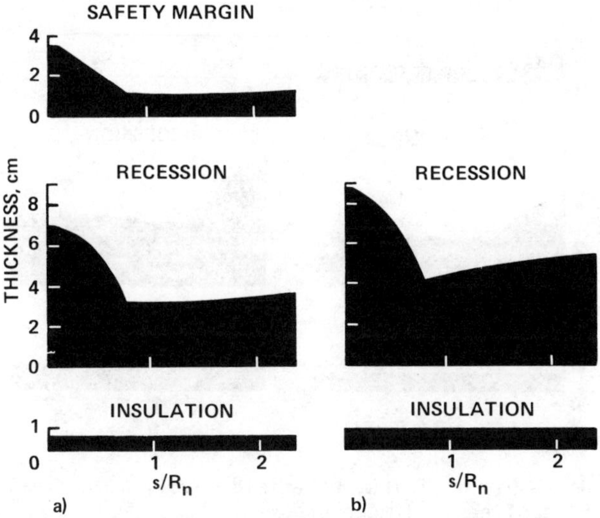

Fig. 10 Heat-shield sizing, Underhill model: a) nominal atmosphere; b) cool heavy atmosphere.

of the heating. Energy transported by the pyrolysis gases through the char, the double integral of the fifth term, accommodates 8%. The energy required to isothermally decompose the solid carbon phenolic to a gas, the double integral of the third term, absorbs only 1%. And, finally, energy stored in the material that remains at a given time, the double integral of the first term, also accommodates only 1%. Since the energy storage term is the principal difference between a transient and steady-state analysis, the steady-state energy balance [Eq. (9)] is an accurate approximation for Jovian entry problems. Note that the distributions of the accommodating modes for variations in material property models, atmospheric model, and probe location are very similar.

The remaining results are related to the heat-shield requirements for the Galileo probe. Heat-shield sizings for the Underhill property model are presented in Fig. 10. For the nominal entry (Fig. 10a), the insulation thickness is about 0.75 cm over the entire probe. The predicted thermochemical recession increases over the back half of the frustum in response to the radiative heating environment (Fig. 4). The safety margin distribution is also shown. The same type of results, except for the safety margin, are given in Fig. 10b for entry into the cool heavy atmosphere. The total heat-shield thickness is greater for the nominal case up to the body location

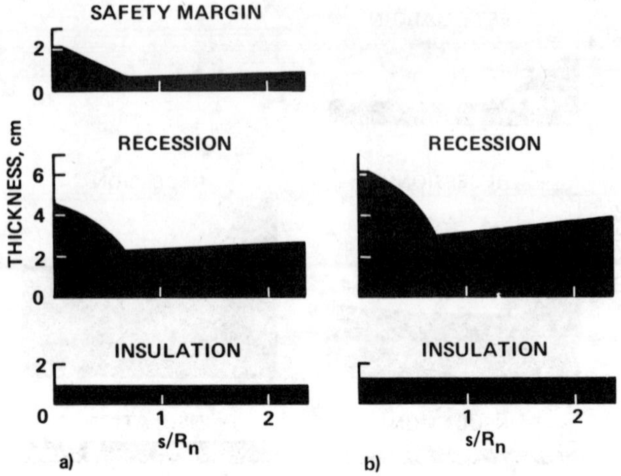

Fig. 11 Heat-shield sizing, Wakefield model: a) nominal atmosphere; b) cool heavy atmosphere.

s/R_n = 0.7, at which point thickness for the cool heavy case becomes greater.

The sizing for the Wakefield model is shown in Fig. 11. For the nominal entry (Fig. 11a), the insulation requirement is greater than predicted using Underhill's model by one-third because of the larger thermal conductivity. As expected in a radiation-dominated environment, the recession and, consequently, the safety margin are about 30% less than predicted using the Underhill model because of the difference in surface reflectance. Similar results are shown in Fig. 11b for the cool heavy entry.

The heat-shield shapes and mass requirement predicted by the Underhill model are illustrated in Fig. 12. Contours are shown for the initial shape, the final shape when the shield is ejected, and the shape of the inner surface that satisfies the maximum bond or aeroshell temperature constraints. For the nominal entry (Fig. 12a) the minimum mass required including safety margin is 126 kg, excluding the additional mass that may be needed to contour the heat shield's inner surface to the aeroshell. The predicted mass loss is 85 kg, which compares with 101 kg obtained from the steady-state RASLE solutions.[9] The reason for this difference was explained above. Results for the cool heavy entry, without safety margin, are presented in Fig. 12b. The minimum mass requirement is 139 kg, of which 120 kg is lost. The mass loss compares with 146 kg from the RASLE result. As mentioned previously, these results should be viewed as upper bounds for entry into the Orton cool heavy atmosphere.

The corresponding shapes and masses using the Wakefield model are given in Fig. 13. The nominal entry (Fig. 13a) requires a minimum mass of 110 kg and an ablated

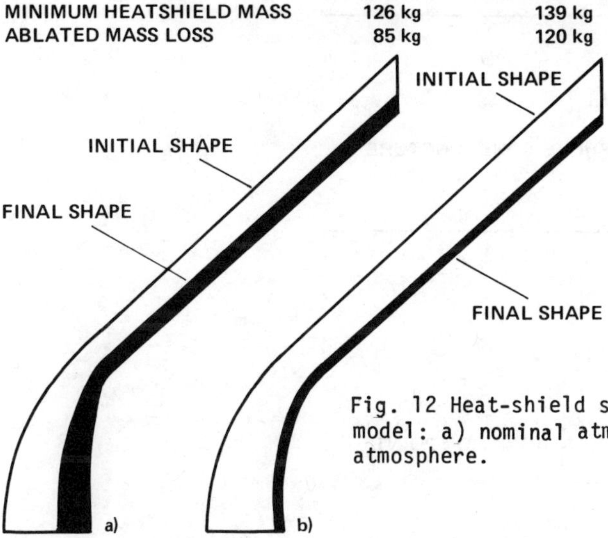

Fig. 12 Heat-shield shape and mass, Underhill model: a) nominal atmosphere; b) cool heavy atmosphere.

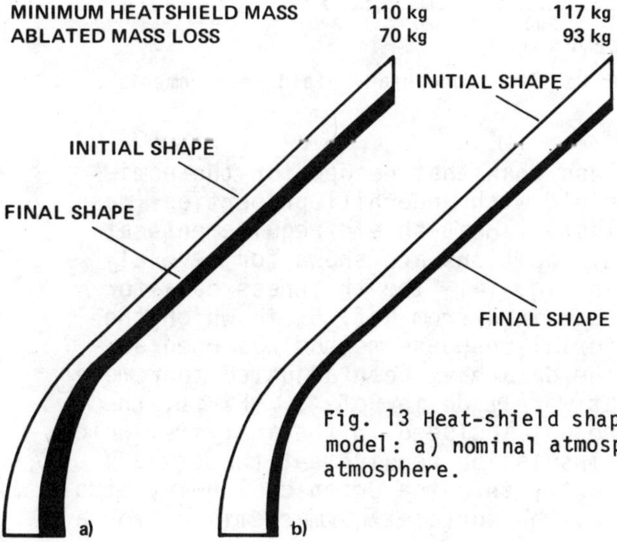

Fig. 13 Heat-shield shape and mass, Wakefield model: a) nominal atmosphere; b) cool heavy atmosphere.

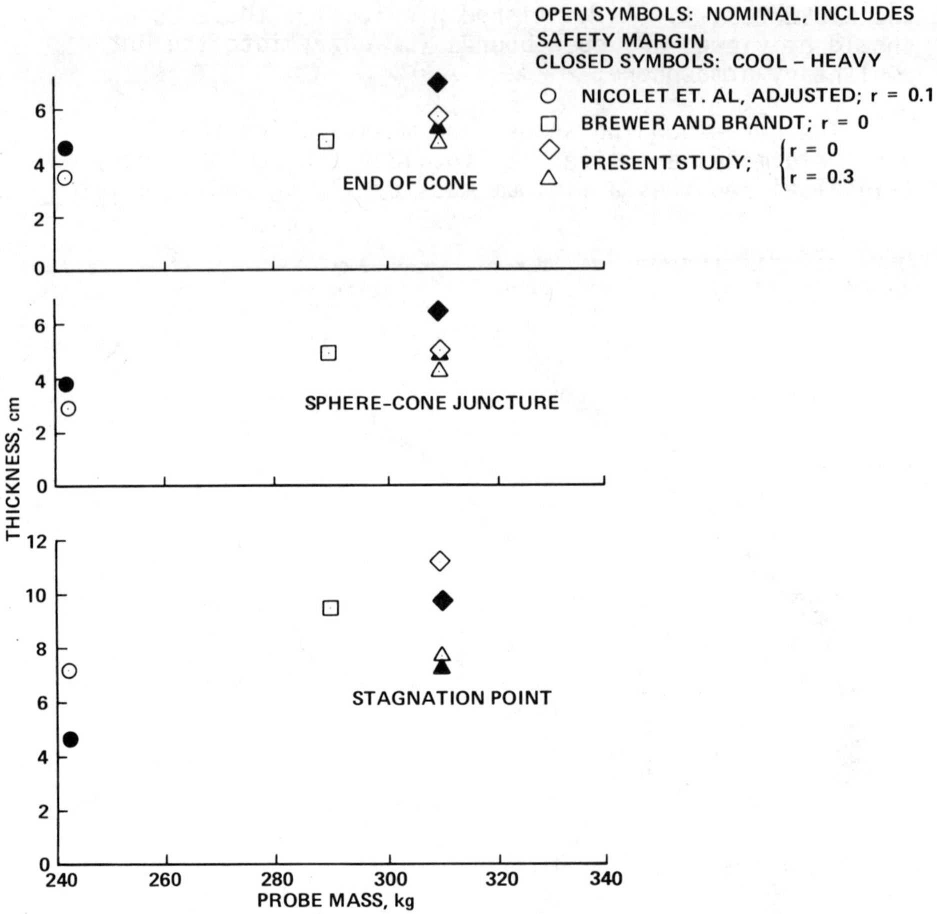

Fig. 14 Data base for Galileo probe heat-shield requirements.

mass loss of 70 kg - less than that needed for the nominally designed heat shield with Underhill properties (mass of 126 kg with 85 kg lost). Heat-shield requirements at the most important body locations are shown for several values of probe mass in Fig. 14. The thickness data for the 242 kg probe were obtained from Ref. 5, in which the same (essentially) material-response method was used as that employed here. The data have been adjusted to remove two approximations that were made in Ref. 5. First, the bond and aeroshell were not included in the analysis, which was found to yield an insulation requirement that is 33% too low. Second, for entry into the Orton cool heavy atmosphere ($0.78H_2 + 0.22He$) the surface thermochemistry for a different atmosphere ($0.89H_2 + 0.11He$) was used. This causes the recession to be underpredicted by about 6%. The surface reflectance was 10%.

The data for the 290 kg probe were obtained from Ref. 4; they were produced by a different computational technique. No reflectance was assumed in obtaining these results. The results for the 310 kg probe were from the sizing information given in Figs. 10 and 11. These data will be helpful, for example, in evaluating the heat-shield requirements for a probe having a mass greater than 310 kg.

Conclusion

Material-response solutions have been obtained for the forebody heat shield on a candidate 310 kg Galileo probe. Only thermochemical ablation has been considered; other phenomena which could affect the mass loss (e.g., spallation) have not been included in the analysis, except implicitly through the safety margin. The predicted mass loss from the RASLE shock-layer solutions, which define the aerothermal heating environment and are based on steady-state ablation boundary conditions, is greater by about 20% than that from CMA material-response solutions, which are based on a transient ablation analysis. This large difference is not caused by the difference between the steady-state and transient analyses, but rather by what at first appears to be an insignificant effect. The surface temperature used in obtaining the RASLE results was specified by a correlation based on thermochemical data that applied a different model for the C_3 chemical species than the one used in the RASLE and CMA analyses. Surprisingly, this difference in C_3 models led ultimately to most of the mass loss difference. The choice and accuracy of thermochemical data can have a significant effect on the heat-shield design. Work is under way to implement a consistent set of thermochemical data in all phases of the heat-shield analysis.

Two property models currently used to characterize the carbon-phenolic heat-shield material have been used to size the heat shield for a nominal entry into Orton's modeled nominal and cool heavy Jovian atmospheres. Based on the Underhill material-property model, the minimum required heat-shield mass is 41% of the total probe mass for the nominal atmosphere including safety margin, and 45% for the cool heavy atmosphere without safety margin. By adjusting the RASLE heating to approximately account for surface reflection, the corresponding results based on the Wakefield model are 35 and 38%, respectively. It should be noted that the preceding requirements given for the cool heavy atmosphere are conservative and should be viewed as upper bounds, because the applied heating environments were more severe than would be experienced by the probe during entry in the Orton cool heavy atmosphere ($0.83H_2 + 0.17He$). The principal difference between the two material-property models is the value of surface reflectance: 0 in the Underhill, 0.30 in the Wakefield.

The heat-shield requirements given in this study for a 310 kg probe, along with previous studies for lighter weight probes, provide a data base that heat-shield designers can use to make engineering judgments without doing a detailed, time-consuming analysis. For example, the heat-shield requirement for a still heavier probe, a different atmospheric model, or a different surface reflectance value can be assessed quickly in an approximate manner using the data base. Having this capability will be useful as the deadline for the Galileo probe heat-shield design approaches.

References

[1]Tauber, M. E. and Wakefield, R. M., "Heating Environment and Protection during Jupiter Entry," Journal of Spacecraft and Rockets, Vol. 8, June 1971, pp. 630-636.

[2]Nicolet, W. E., Balakrishnan, A., and Mezines, S. A., "Configuration Impact on Heatshield Requirements for Jupiter Entry," Thermophysics of Spacecraft and Outer Planet Entry Probes, Progress in Astronautics and Aeronautics, Vol. 56, edited by A. M. Smith, AIAA, New York, 1977, pp. 377-404.

[3]Nelson, H. F. and Mezines, S. A., "Analysis of Jupiter Probe Heat Shield Recession Uncertainties," Outer Planet Entry Heating and Thermal Protection, Progress in Astronautics and Aeronautics, Vol. 64, edited by R. Viskanta, AIAA, New York, 1979, pp. 293-320.

[4]Brewer, R. A. and Brant, D. N., "Thermal Protection Systems for the Galileo Mission Atmospheric Entry Probe," AIAA Paper 80-0358, Jan. 1980.

[5]Nicolet, W. E., Davy, W. C., and Wilson, J. F., "Galileo Probe Forebody Entry Thermal Protection: Aerothermal Environments and Heat Shielding Requirements," ASME Paper 80-ENAs-24, 1980.

[6]Moss, J. N., "A Study of the Aerothermal Entry Environment for the Galileo Probe," Entry Heating and Thermal Protection, Progress in Astronautics and Aeronautics, Vol. 69, edited by W. B. Olstad, AIAA, 1980, pp. 3-25.

[7]Kumar, A., Graves, R. A., Weilmuenster, K. J., and Tiwari, S. N., "Laminar and Turbulent Flow Solutions with Radiation and Ablation Injection for Jovian Entry," AIAA Paper 80-0288, 1980.

[8]Moss, J. N. and Kumar, A., "Significance of Turbulence and Transition Location on Radiative Heating and Ablation Injection," AIAA Paper 81-0281, 1981.

[9]Balakrishnan, A. and Nicolet, W. E., "Galileo Probe Forebody Thermal Protection: Benchmark Heating Environment Calculations," AIAA Paper 81-1072, June 1981.

[10]Moyer, C. B. and Rindall, R. A., "An Analysis of the Coupled Chemically Reacting Boundary Layer and Charring Ablator, Part II: Finite Difference Solution for the In-Depth Response of Charring Materials Considering Surface Chemical and Energy Balances," NASA CR-1061, June 1968.

[11]Davy, W. C., Menees, G. P., Lundell, J. H., and Dickey, R. R., "Hydrogen-Helium Ablation of Carbonaceous Material: Numerical Simulation and Experiment," Outer Planet Entry Heating and Thermal Protection, Progress in Astronautics and Aeronautics, Vol. 64, edited by R. Viskanta, AIAA, New York, 1979, pp. 228-244.

[12]Green, M. J. and Davy, W. C., "Numerical Simulation of Experiments in the Giant Planet Facility, Entry Heating and Thermal Protection, Progress in Astronautics and Aeronautics, Vol. 69, edited by W. B. Olstad, AIAA, New York, 1980, pp. 172-197.

[13]Wakefield, R. M. and Pitts, W. C., "Analysis of the Heat-Shield Experiment on the Pioneer-Venus Entry Probes," AIAA Paper 80-1494, July 1980.

[14]Powars, C. A. and Kendall, R. M., "User's Manual Aerotherm Chemical Equilibrium (ACE) Computer Program," Aerotherm Corp., Mountain View, Calif., Rept. UM 69-7, May 1969.

[15]Stull, D. R. and Prophet, H., "JANAF Thermochemical Tables," 2nd Ed., National Bureau of Standards, NBS-37, June 1971 (supplements through June 1977).

[16]Bueche, J. F., "Effects of Improvements and Uncertainties in Thermophysical Properties on Carbon Phenolic Heatshield Thermal Performance Predictions," AIAA Paper 77-787, June 1977.

[17]Pope, R. B., "Measurements of the Total Surface Emittance of Charring Ablators," AIAA Journal, Vol. 5, No. 12, Dec. 1967, pp. 2285-2287.

[18]Jones, J. J., "The Optical Absorption of Triatomic Carbon C_3 for the Wavelength Range 260 to 560 nm," NASA TP-1141, March 1978.

Significance of Turbulence and Transition Location on Radiative Heating with Ablation Injection

James N. Moss*
NASA Langley Research Center, Hampton, Va.
and
Ajay Kumar †
Old Dominion University, Norfolk, Va.

Abstract

This paper presents forebody flowfield solutions for Jupiter entry conditions where the ablation injection rate is coupled with the surface heating rate. The calculations are made with a time-dependent viscous-shock-layer analysis where the flow is assumed to be in chemical equilibrium. Both laminar and turbulent solutions are presented to describe the impact of turbulence on surface mass-loss rates for flow conditions where the heating is due primarily to radiation. Results are also presented where the transition location to turbulent flow is arbitrarily varied. Results show that the radiative heating rate values based on several downstream transition locations adjust quickly to the corresponding values based on transition near the stagnation point.

Nomenclature

$\bar{A}$ = surface area
C_A = ablator mass fraction
C_D = drag coefficient
C_i = mass fraction of species i, ρ_i/ρ
C_ℓ = mass fraction of element ℓ

Presented as Paper 81-0281 at the AIAA 19th Aerospace Sciences Meeting, St. Louis, Mo., Jan. 12-15, 1981. This paper is declared a work of the U.S. Government and therefore is in the public domain.

*Research Leader, Aerothermodynamics Branch, Space Systems Division.

†Research Associate Professor, Department of Mechanical Engineering and Mechanics. Presently, Aero-Space Technologist, Computational Methods Branch, High-Speed Aerodynamics Division.

H	=	total enthalpy, $\bar{H}/\bar{V}_\infty^2$
h	=	enthalpy of mixture, $\sum_{i=1}^{N} C_i h_i$
$\bar{h}_A$	=	enthalpy of undecomposed ablation material
h_i	=	enthalpy of species i, $\bar{h}_i/\bar{V}_\infty^2$
k	=	thermal conductivity of mixture, $\bar{k}/(\bar{\mu}_\infty \bar{V}_\infty^2/\bar{T}_\infty)$
Le	=	Lewis number
Le_T	=	turbulent Lewis number
$\bar{M}$	=	molecular weight of mixture
$\bar{M}_i$	=	molecular weight of species i
$\bar{M}_\ell$	=	molecular weight of element ℓ
M_∞	=	freestream Mach number
$\dot{m}$	=	mass injection rate, $\dot{\bar{m}}/\bar{\rho}_\infty \bar{V}_\infty$
N	=	number of species
n	=	coordinate normal to the body, $\bar{n}/\bar{R}_n$
p	=	pressure, $\bar{p}/\bar{\rho}_\infty \bar{V}_\infty^2$
q^R	=	net radiative heat flux in n-direction, $\bar{q}^R/\bar{\rho}_\infty \bar{V}_\infty^3$
q^{R+}	=	component of radiative flux toward the shock
q^{R-}	=	component of radiative flux toward the wall
$-q_w^C$	=	convective heat flux to the wall
R^*	=	universal gas constant, J/mole·K
$\bar{R}_n$	=	nose radius, m
Re	=	freestream Reynolds number, $\bar{\rho}_\infty \bar{V}_\infty \bar{R}_n/\bar{\mu}_\infty$
r	=	radius measured from axis of symmetry to point on body surface, $\bar{r}/\bar{R}_n$
s	=	coordinate measured along the body, $\bar{s}/\bar{R}_n$
Sc	=	Schmidt number
T	=	temperature, $\bar{T}/\bar{T}_\infty$
$\bar{T}_\infty$	=	freestream temperature, K
t	=	nondimensional time, $\bar{t}\bar{V}_\infty/\bar{R}_n$
u	=	tangential velocity, $\bar{u}/\bar{V}_\infty$
u_s	=	tangential velocity at shock, $\bar{u}_s/\bar{V}_\infty$
v	=	normal velocity, $\bar{v}/\bar{V}_\infty$
$\bar{V}_\infty$	=	freestream velocity, m/s
β	=	$r + n \cos\theta$
$\bar{\beta}$	=	ballistic coefficient, kg/m^2
$\beta_{i,j}$	=	coefficients for sublimation temperature, Eq. (5)
γ	=	streamwise transition intermittency factor
δ_s	=	shock standoff distance, $\bar{\delta}_s/\bar{R}_n$
$\delta_{i\ell}$	=	number of atoms of the ℓth element in species i
ε	=	emissivity
ε^+	=	normalized eddy viscosity, μ_T/μ
θ	=	body angle measured from the body axis

κ = local body curvature, $\bar{\kappa}\bar{R}_n$
λ = $1 + n\kappa$
μ = viscosity, $\bar{\mu}/\bar{\mu}_\infty$
μ_T = eddy viscosity, $\bar{\mu}_T/\bar{\mu}_\infty$
$\bar{\mu}_\infty$ = freestream viscosity, $N \cdot s/m^2$
ρ = density, $\bar{\rho}/\bar{\rho}_\infty$
$\bar{\rho}_\infty$ = freestream density, kg/m^3
σ = Prandtl number
σ_T = turbulent Prandtl number
$\bar{\sigma}$ = Stefan-Boltzmann constant

Superscript

$(\bar{\ })$ = dimensional quantities

Subscripts

i = ith species
ℓ = ℓth element
s = shock values
sub = sublimation
w = wall values
∞ = freestream values
$-$ = values for the solid ablation material at the surface
o = stagnation point

Introduction

The Galileo Mission to Jupiter is scheduled for launch in early 1984. This mission will require a dual launch of an orbiter and a probe. The probe will experience a unique flow environment as it encounters Jupiter's atmosphere. During this encounter, most of the kinetic energy of the probe, entering at an inertial velocity of approximately 60 km/s, will be depleted, thereby producing a high energy flowfield that envelops the probe. For about 20 s the probe's forebody heat shield will experience very intense heating rates which will be dominated by radiative heating. Consequently, the probe's forebody heat shield will experience large mass-loss rates.

The massive ablation injection rates, driven by radiative heating from the high energy, shock-layer gases, create the unique flow environment for the Galileo probe. Characterization of the flowfield environment has relied heavily on numerical computations because of the difficulty of duplicating the entry environment in ground experimental facilities. Three computer codes (HYVIS,[1] COLTS,[2] and RASLE[3]) have been developed that provide fully coupled flowfield solutions; that is, solutions where the ablation injection rate, radiation transport, and the flow conservation equations are fully coupled. The detailed solutions that have been obtained show good qualitative agreement and current studies are in progress to isolate any quantitative differences that exist.

The present study was initiated to focus on the significance of turbulence as it influences radiative heating and ablative mass-loss rates and to provide solutions for entry conditions that reflect the current probe mass which is about 28% greater than that considered in previous studies.[1-3] The previous coupled flow solutions have shown that turbulence produces a situation where the injected ablation species are much less effective in absorbing or blocking the radiation than would be the case if the flow was laminar. These calculations assumed that the transition from laminar to turbulent flow occurred immediately downstream from the stagnation point. However, the actual Galileo probe heat shield design studies[4] have been made assuming that instantaneous transition from laminar to turbulent flow occurs at a wetted length Reynolds number (boundary-layer edge properties) of 10^5. With this transition criterion, the flow is laminar initially; then the transition location moves forward early in the heating pulse until it settles on the nose of the blunted cone prior to peak heating and remains in this general location for the remainder of the heating pulse. Since the radiation absorption by the ablation species is substantially different for laminar flow and turbulent flow, the question arises about the nature of the radiative heating as the transition front moves forward. This question is addressed in this study by calculating fully coupled solutions and arbitrarily varying the location of transition. The coupled solutions are made with the time dependent viscous-shock-layer analysis of Ref. 2 and the flow is assumed to be in chemical equilibrium.

Analysis

Governing Equations

The flowfield solutions presented were obtained by solving the viscous-shock-layer equations of motion for a reacting gas mixture in chemical equilibrium. The equations account for surface mass injection and are appropriate for both laminar and turbulent flow. In the present study, the solutions without ablation injection were obtained with the computer program HYVIS which solves the steady-state[1] viscous-shock-layer equations. A recently developed COLTS computer code[5] that provides improved solution capability for coupled solutions was used to generate the coupled solutions. This code solves the time-dependent viscous-shock-layer equations which are described herein. These equations when represented in the body-oriented coordinate system shown in Fig. 1 and for flow at zero angle of attack are expressed as

$$\partial U/\partial t + \partial M/\partial s + \partial \hat{N}/\partial n + Q = 0 \qquad (1)$$

where

$$U = \lambda \begin{bmatrix} \rho \\ \rho u \\ \rho v \\ \rho H - p \\ \rho C_\ell \end{bmatrix} \qquad M = \begin{bmatrix} \rho u \\ p + \rho u^2 \\ \rho uv \\ \rho uH \\ \rho uC_\ell \end{bmatrix}$$

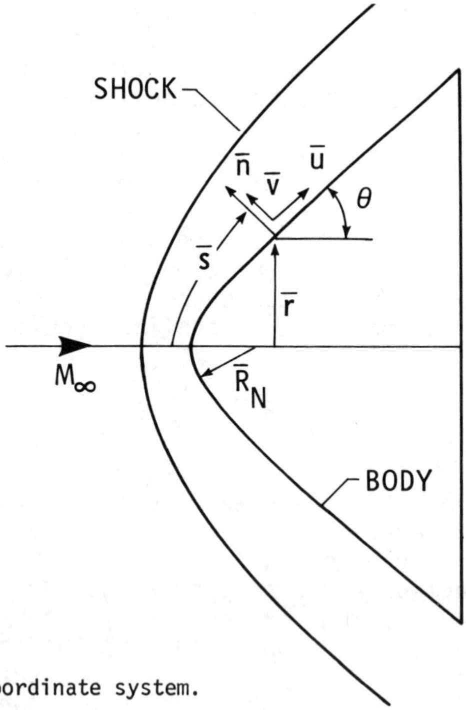

Fig. 1 Coordinate system.

$$\hat{N} = \lambda \begin{bmatrix} \rho v \\ \rho u v - \tau \\ p + \rho v^2 \\ \{\rho v H - (\mu/\sigma Re)(1 + \varepsilon^+(\sigma/\sigma_T)(\partial h/\partial n) - u\tau \\ - (\mu/\sigma Re)[Le - 1 + \varepsilon^+(\sigma/\sigma_T)(Le_T - 1)]\sum_{i=1}^{N} h_i(\partial C_i/\partial n) \\ + q^R\} \\ \rho v C_\ell - (\mu/\sigma Re)(Le + \varepsilon^+(\sigma/\sigma_T)Le_T)\partial C_\ell/\partial n) \end{bmatrix}$$

$$Q = (\lambda \sin\theta/\beta)M + (\cos\theta/\beta)\hat{N} + (1/\beta)\begin{bmatrix} 0 \\ \beta\kappa(\rho u v - \tau) - p\lambda \sin\theta \\ -\beta\kappa\rho u^2 - p(\lambda\cos\theta + \beta\kappa) \\ 0 \\ 0 \end{bmatrix}$$

Here

$$C_\ell = \sum_{i=1}^{N} \delta_{i\ell}(\overline{M}_\ell/\overline{M}_i)C_i$$

and

$$\tau = \mu(1 + \varepsilon^+)/Re(\partial u/\partial n - u\kappa/\lambda)$$

The above equations are not valid at the axis of symmetry where $\beta = 0$. A limiting form of the governing equations is obtained at $s = 0$ by differentiating Eq. (1) with respect to s and taking a limit as $s \to 0$. The following equations are obtained at the axis of symmetry:

$$(\partial U_0/\partial t) + (\partial M_0/\partial s) + (\partial\hat{N}_0/\partial n) + Q_0 = 0 \qquad (2)$$

where

$$U_o = \lambda \begin{bmatrix} \rho \\ \rho u \\ \rho v \\ \rho H - p \\ \rho C_\ell \end{bmatrix} \qquad M_o = \begin{bmatrix} 2\rho u \\ p + 2\rho u^2 \\ 2\rho uv \\ 2\rho uH \\ 2\rho uC_\ell \end{bmatrix}$$

$$\hat{N}_o = \lambda \begin{bmatrix} \rho v \\ \rho uv - \tau \\ p + \rho v^2 \\ \{\rho vH - (\mu/\sigma Re)[1 + \varepsilon^+(\sigma/\sigma_T)](\partial h/\partial n) - u\tau - (\mu/\sigma Re)[Le \\ - 1 + \varepsilon^+(\sigma/\sigma_T)(Le_T - 1)]\sum_{i=1}^{N} h_i(\partial C_i/\partial n) + q^R\} \\ \rho vC_\ell - (\mu/\sigma Re)(Le + \varepsilon^+(\sigma/\sigma_T)Le_T)(\partial C_\ell/\partial n) \end{bmatrix}$$

$$Q_o = \kappa \begin{bmatrix} \rho v \\ 2(\rho uv - \tau) \\ [-p + \rho(v^2 - u^2)] \\ \{\rho vH - (\mu/\sigma Re)[1 + \varepsilon^+(\sigma/\sigma_T)](\partial h/\partial n) - u\tau - (\mu/\sigma Re)[Le \\ - 1 + \varepsilon^+(\sigma/\sigma_T)(Le_T - 1)]\sum_{i=1}^{N} h_i(\partial C_i/\partial n) + q^R\} \\ \rho vC_\ell - (\mu/\sigma Re)[Le + \varepsilon^+(\sigma/\sigma_T)Le_T](\partial C_\ell/\partial n) \end{bmatrix}$$

The equation of state is given by

$$p = (R^*\bar{T}_\infty/\bar{M}\ \bar{V}_\infty^2)\ T\ \rho \qquad (3)$$

Two independent variable transformations are applied to the governing equations. The first transformation maps the computational domain into a rectangular region in which both the shock and the body are made boundary mesh lines. The second transformation further maps the computational region into another plane to allow higher resolution near the body surface without any significant increase in the number of mesh points in the normal direction. The details of these transformations are given in Ref. 6.

Boundary Conditions

No-slip boundary conditions are used at the surface. The wall temperature and mass injection rate are either specified or calculated. The boundary conditions at the shock are calculated by using the shock relations.

For the calculated mass injection conditions, the ablation process is assumed to be quasi-steady and the wall temperature is the sublimation temperature of the ablator surface. The coupled mass injection rate is then given by

$$\dot{m} = \left(-\bar{q}_w^C - \bar{q}_w^R\right) / \left[\sum_{i=1}^{N} (C_i \bar{h}_i)_w - \bar{h}_A\right] / \bar{\rho}_\infty \bar{V}_\infty \tag{4}$$

The sublimation temperature for the carbon-phenolic ablator is given by

$$\bar{T}_{sub} = \sum_{j=1}^{5} \beta_{T,j} C_A^{j-1} + \log \bar{p}_w \sum_{j=1}^{5} \beta_{2,j} C_A^{j-1} + (\log \bar{p}_w)^2 \sum_{j=1}^{5} \beta_{3,j} C_A^{j-1} \tag{5}$$

where $\bar{p}_w$ is the wall pressure in atmospheres and C_A is the ablator mass fraction at the wall. The values of $\beta_{i,j}$ are given in Table 1. These are applicable for a freestream gas composition of 89% H_2, 11% He by volume and for ablator mass fractions of 0.4 to 1.0.

For ablation injection, the elemental concentrations at the wall are governed by convection and diffusion as given by the equation

$$(\partial C_\ell / \partial n)_w - (1/Re)(\dot{m}\, Sc/\mu)_w [(C_\ell)_w - (C_\ell)_-] = 0 \tag{6}$$

Table 1 Sublimation temperature coefficients

Coefficients	j: 1	2	3	4	5
$\beta_{1,j}$	5552	-20184	53058	-57933	23243
$\beta_{2,j}$	1798	-12049	30145	-32045	12457
$\beta_{3,j}$	322	-2208	5270	-5450	2092

The net radiative flux, q^R, can be represented as

$$q^R = q^{R+} - q^{R-} \tag{7}$$

At the surface

$$\bar{q}_w^{R+} = \varepsilon\bar{\sigma}\bar{T}_w^4 \tag{8}$$

The surface is treated as a gray surface with the absorptivity and emissivity both equal to one. The heat transferred to the wall due to conduction and diffusion is

$$-q_w^C = (1/Re)\left[k(\partial T/\partial n) + (\mu/Sc)\sum_{i=1}^{N} h_i(\partial C_i/\partial n)\right]_w \tag{9}$$

Radiative Transport

The radiative flux, q^R, is calculated with the radiative transport code RAD[7,8] which has been incorporated into the present computer code.

The RAD accounts for the effects of nongray self absorption. Molecular band, continuum, and atomic line transitions are included. A detailed frequency dependence of the absorption coefficient is used for integrating over the radiation frequency spectrum and the tangent slab approximation is used for integrating over physical space. The chemical species considered in the present study for determining the radiative transport are H, H_2, H^+, e^-, C, C_2, C_3, C^+, C^-, CO, O, O_2, O^+, and O^-. The radiation properties used in this study for C_2 and C_3 are those reported in Refs. 9 and 10, respectively.

Thermodynamic and Transport Properties

The equilibrium composition is determined by a free energy minimization calculation as developed in Ref. 11 and written for the present code in Ref. 12. Thermodynamic properties for specific heat, enthalpy, and free energy and transport properties for viscosity and thermal conductivity are required for each species considered. Values for the thermodynamic and transport properties are obtained by using polynomial curve fits. Mixture viscosity is obtained by using the semiempirical formula of Wilke[13] while the Lewis number and Prandtl number of the mixture are set equal to 1.1 and 0.64, respectively. For the solutions without injection, the mixture transport properties for viscosity and Prandtl number are calculated using the correlations of Zoby[14] and the Lewis number is set equal to 1.1.

Six chemical species are used to describe the hydrogen-helium gas mixture: e^-, H, H_2, H^+, He, and He^+. With carbon-phenolic injection, 13 additional species are used which are C, C_2, C_3, C^+, C_2H, C_3H, C_4H, C_2H_2, O, O_2, O^+, CO, and CO_2.

Eddy-Viscosity Approximations

A two-layer eddy-viscosity model consisting of an inner law based upon Prandtl's mixing length concept and the Clauser-Klebanoff expression (based on Refs. 15 and 16) for the outer law is used in the present investigation. This model, introduced by Cebeci,[17] assumes that the inner law is applicable for the flow from the wall out to the location where the eddy viscosity given by the inner law is equal to that of the outer law. The outer law is then assumed applicable for the remainder of the viscous layer. It is noted that the eddy viscosity degenerates to approximately zero in the inviscid portion of the shock layer. The degeneracy is expressed in terms of the normal intermittency factor given by Klebanoff.[16] Reference 1 gives a more detailed description of the turbulence model and various expressions for it. The only difference between the models used in the present calculations and that used in Ref. 1 is in the boundary-layer edge definition. The present analysis uses the definition given in Ref. 18 which is based on an index of diffusion, conduction, and dissipation.

The turbulent Prandtl number and Lewis number are assumed to be 0.9 and 1.0, respectively.

Method of Solution

A time-asymptotic two-step finite-difference method due to MacCormack[19] is used to solve the governing equations. The details of the method are given in Ref. 6, and a description of the computer code COLTS, which is written in Control Data CYBER-203 FORTRAN language, is given in Ref. 5.

The calculations of equilibrium chemistry, radiative heat flux, and eddy viscosity requires a significant amount of computing time and it is not feasible to perform these calculations in each time step. In the present analysis, the eddy viscosity is calculated after every 25 time steps, the equilibrium chemistry after every 200 time steps, and the radiative heat flux after every 1000 or 2000 time steps.

Results and Discussion

The primary motivation of this study was to examine numerically the impact of transition location on surface heating and ablation mass loss downstream of the transition location. In addition, solutions were obtained at selected times along nominal entry trajectories for two similar probes so as to provide information concerning the impact of turbulence on heating and the probe forebody heat shield mass-loss rate.

The coupled solutions were obtained using a computational mesh of 101×14 grid points. The hundred and one grid points normal to the body were located using a variable grid spacing whereas those along the body were evenly spaced at Δs values of 0.1597. The COLTS program is currently dimensioned for 14 body stations (no restriction on grid dimensions), and this combined with the Δs step size (see Ref. 5 for step-size options) used in this study provided forebody calculations to an s value of 2.08 rather than the end of probe ($s = 2.38$). The forebody mass-loss rates, which are discussed later, were obtained by extrapolating the values to the end of probe.

For all turbulent solutions unless stated otherwise, the transition to turbulent flow was instantaneous and located at the first body grid downstream of the stagnation point. For the COLTS and HYVIS solutions, this was at s values of 0.1597 and 0.1, respectively.

For the coupled solutions, the wall temperature is a variable and is given by Eq. (5). The wall temperature for the no-injection solutions is a function of the entry conditions [equal to the sublimation temperature given by Eq. (5)

for an ablative mass fraction of one and for the stagnation-point pressure], yet constant along the body surface for a given entry condition.

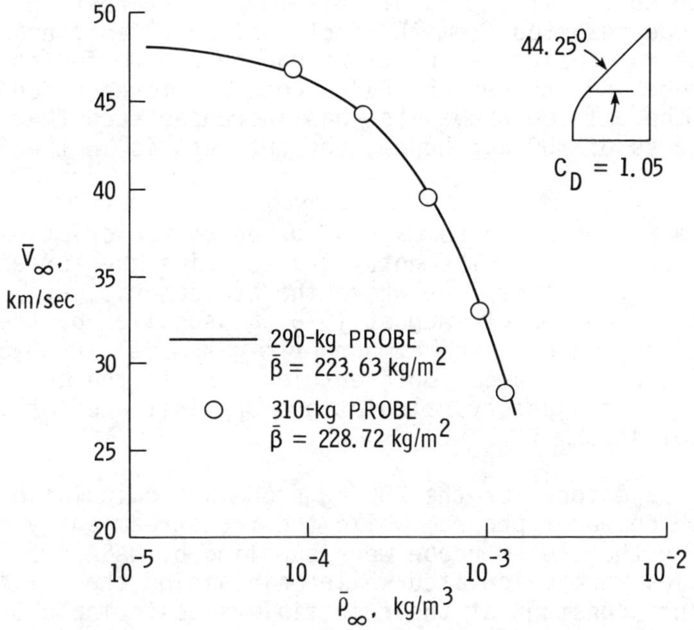

Fig. 2 Nominal entry trajectories for Jupiter entry.

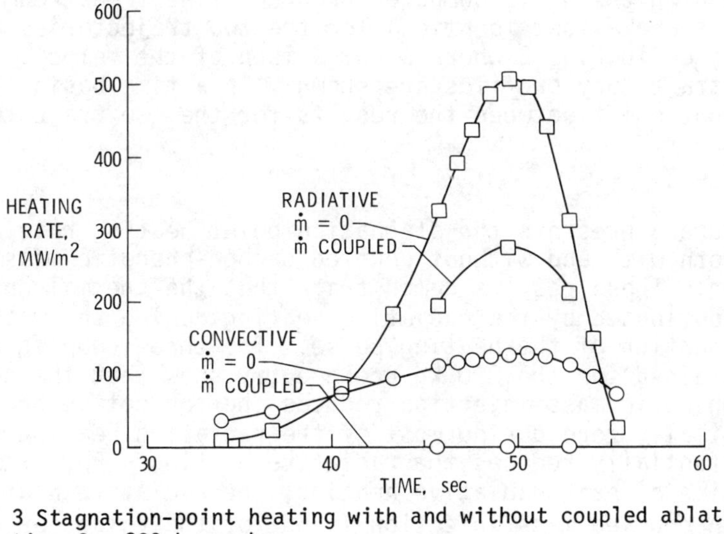

Fig. 3 Stagnation-point heating with and without coupled ablation injection for 290 kg probe.

Entry Conditions

Solutions are presented for two probes: a 290 kg probe and a 310 kg probe, both of which are spherically blunted cones with cone half angles of 44.25 deg. Interest in the 290 kg probe resulted from the fact that an extensive amount of aerothermal design calculations had been made for the 290 kg probe and, therefore, had become a reference condition. However, the Galileo probe mass has increased such that it is now in excess of 300 kg; hence, the interest is in the 310 kg probe.

Table 2 presents a summary of probe characteristics and entry conditions. The entry trajectories are for an entry into the Jupiter atmosphere where the atmospheric gas model is the Orton nominal model (August 1979) as supplied by the NASA Ames Research Center. The atmospheric gas model is 89.5% H_2 + 10.5% He by volume. Both entries are off the equatorial plane and the trajectory calculations are initiated at an altitude of 450 km.

The trajectory for the 290 kg probe was calculated with the POST[20] computer program while the freestream entry conditions for the 310 kg probe were supplied by NASA Ames. Both trajectories were calculated while maintaining the ballistic coefficients constant at their initial values. Table 3 presents freestream conditions and stagnation results for both probes. These results are for no mass injection and were obtained with the HYVIS computer program. The freestream conditions are almost identical for the two trajectories as demonstrated in Fig. 2 where a comparison of the velocity-density trajectory profiles are shown. On a time basis, there is a slight shift between the results for the two trajectories.

290 kg Probe

Figure 3 presents the stagnation-point heating rate pulses both with and without coupled carbon-phenolic mass injection. These results demonstrate that the thermal environment is dominated by the radiative heating during the most intense portion of the heating pulse. The three coupled solutions obtained for the 290 kg probe entry show that the coupled carbon-phenolic mass injection reduces the convective heating to essentially zero during most of the radiative heating pulse and substantially reduces the radiative heating. For example, at the time of peak radiative heating, the radiative heating rate with coupled mass injection is 55% of the value for no injection.

Table 2 Probe characteristics and entry conditions

Parameters	Conditions	
	290 kg Probe	310 kg Probe
Body half angle, deg	44.25	44.25
Nose radius, m	.344	.352
Base to nose radius ratio	1.823	1.821
Drag coefficient	1.05	1.05
Mass, kg	290	310
Ballistic coefficient	223.63	228.72
Atmosphere[a] (by volume)	0.895 H_2 + 0.105 He	0.895 H_2 + 0.105 He
Entry altitude, km	450	450
Inertial (relative) entry velocity, km/s	60.56 (48.2)	60.11 (48.2)
Inertial (relative) entry angle, deg	-6.835 (-8.6)	-6.886 (-8.6)
Entry latitude, deg	-5.5	+3.4
Azimuth angle of inertial velocity vector, deg	82	70.3

[a]Orton nominal (Aug. 1979)

Table 3 Freestream conditions and stagnation-point results without ablation injection[a]

a) 290 kg probe									
Time, s	Altitude, km	$\bar{V}_\infty$, km/s	$\bar{\rho}_\infty$, kg/m^3	$\bar{p}_s$, atm	$\bar{\rho}_s/\bar{\rho}_\infty$	$\bar{T}_s$, K	$\bar{\delta}_s$, cm	$-\bar{q}_w^C$, MW/m^2	$\bar{q}_w^{R-}$, MW/m^2
34.00	216.70	48.02	1.28×10^{-5}	0.27	13.70	14502	1.87	37	10
36.75	199.24	47.77	2.42×10^{-5}	0.50	13.21	15075	1.91	50	24
40.50	175.96	47.05	5.86×10^{-5}	1.18	12.45	15833	1.98	74	82
43.25	159.43	45.96	1.12×10^{-4}	2.15	11.93	16260	2.04	94	182
45.75	144.99	44.22	2.03×10^{-4}	3.57	11.39	16430	2.13	111	327
46.75	139.43	43.25	2.56×10^{-4}	4.30	11.13	16403	2.17	117	392
47.50	135.36	42.39	3.04×10^{-4}	4.90	10.93	16323	2.21	121	437
48.50	130.08	41.07	3.18×10^{-4}	5.74	10.65	16149	2.28	124	486
49.50	125.00	39.53	4.74×10^{-4}	6.61	10.33	15858	2.36	126	507
50.50	120.14	37.76	5.87×10^{-4}	7.43	9.98	15453	2.46	130	495
51.50	115.52	35.80	7.20×10^{-4}	8.16	9.60	14900	2.57	125	440
52.75	110.12	33.10	9.17×10^{-4}	8.84	9.12	13952	2.73	113	313
54.0	105.17	30.22	1.15×10^{-3}	9.18	8.73	12586	2.85	97	150
55.25	100.68	27.28	1.41×10^{-3}	9.22	8.84	10457	2.78	74	27

(Table 3 continued on next page)

Table 3 (cont.) Freestream conditions and stagnation-point results without ablation injection[a]

Time, s	Altitude, km	$\bar{V}_\infty$, km/s	$\bar{\rho}_\infty$, kg/m^3	$\bar{p}_s$, atm	$\bar{\rho}_s/\bar{\rho}_\infty$	$\bar{T}_s$, K	$\bar{\delta}_s$, cm	$-\bar{q}_w^C$, MW/m^2	$-\bar{q}_w^R$, MW/m^2
43.0	167.8	46.58	8.059×10^{-5}	1.58	12.21	16059	2.05	85	123
47.0	147.0	44.47	1.869×10^{-4}	3.33	11.46	16411	2.16	108	306
49.2	135.3	42.30	3.046×10^{-4}	4.89	10.90	16293	2.27	119	437
50.3	129.7	40.37	3.861×10^{-4}	5.77	10.59	16097	2.34	123	485
51.5	123.9	39.04	4.966×10^{-4}	6.74	10.23	15738	2.44	123	503
54.1	112.5	34.12	8.262×10^{-4}	8.48	9.30	14323	2.73	112	361
56.7	102.7	28.37	1.290×10^{-3}	9.08	8.68	11377	2.91	81	62

[a]HYVIS stagnation results.

Included in Table 4a is a listing of stagnation-point results for the coupled carbon-phenolic injection cases. Note that the stagnation-point results (Table 4) are different depending on whether the flow is laminar or turbulent downstream of the stagnation point. For the coupled turbulent solutions, the COLTS program predicts a temperature overshoot occurring at the outer edge of the turbulent mixing layer. This instability, which occurs only for the coupled turbulent flow, is caused by the sudden large increase in the magnitude of the viscosity and is confined mainly in the region of tran-

Table 4 Stagnation-point results with coupled carbon-phenolic injection and forebody ablative mass loss rates[a]

a) 290 kg probe

Time, s	$\bar{q}_w^{R-}$, MW/m^2	$-\bar{q}_w^{C}$, MW/m^2	$\bar{\delta}_s$, cm	$\dot{m}$	$\int\bar{\rho}_w\bar{v}_w d\bar{A}$, kg/s
45.75	175.8	0.20	0.283	0.758	2.86
	197.1	0.33	0.344	0.857	7.41
49.50	253.3	0.38	0.301	0.540	3.41
	277.8	0.14	0.341	0.595	8.47
52.75	194.0	0.79	0.303	0.252	1.07
	213.4	2.56	0.331	0.282	3.22

b) 310 kg probe

Time, s	$\bar{q}_w^{R-}$, MW/m^2	$-\bar{q}_w^{C}$, MW/m^2	$\bar{\delta}_s$, cm	$\dot{m}$	$\int\bar{\rho}_w\bar{v}_w d\bar{A}$, kg/s
43.0	85.2	0.36	2.72	0.785	1.12
	95.9	1.13	3.33	0.898	3.12
47.0	166.0	0.24	2.88	0.767	2.75
	187.2	0.30	3.59	0.873	7.30
50.3	239.8	0.46	2.99	0.605	3.64
	274.0	0.08	3.50	0.695	9.03
51.5	247.3	0.37	3.08	0.510	3.37
	284.0	0.46	3.47	0.590	8.48
54.1	188.1	0.51	3.15	0.262	1.55
	242.2	1.56	3.41	0.346	4.00
56.7	63.3	7.43	3.00	0.067	0.56
	65.8	11.08	3.14	0.074	1.41

[a]COLTS results. for a given time, the first line denotes results for a laminar solution while the second line denotes results for a turbulent solution (transition at s = 0.1597).

sition. As an example, the overshoot in temperature was about 7% with respect to the laminar value and the extent of the overshoot was about 9% of the shock-layer thickness for the solution at 49.5 s. The fact that the coupled stagnation-point heating was greater when the downstream flow was turbulent than when it was laminar is believed to be due in part to this instability. Also, with the COLTS program the effects of instantaneous transition diffuse about two grid points upstream and downstream of the disturbance due to the artificial damping required in the program. Consequently, the effects on the coupled stagnation-point results of the artificially imposed instantaneous transition just downstream of the stagnation point is more an artifact of the solution and not a modeled phenomena. If this is the case, then the stagnation-point results (downstream flow turbulent) are conservative, unquestionably so if the actual transition location is far downstream of the stagnation region.

There is an increasing amount of experimental data[21,22] that suggests that once the transition front moves forward into the subsonic portion of the flow, then turbulence in the presence of mass injection influences the stagnation-point flow conditions and convective heating. In light of these results, the present stagnation-point results are, even though fortuitously, probably not conservative when the transition front is near the stagnation point.

Figure 4 shows the impact of both coupled ablation injection and turbulence on the probe forebody surface heating. The trends evident in the present results are consistent with previous calculations[1-3] in that both radiative (Fig. 4a) and convective (Fig. 4b) heating experience substantial reductions due to injection over the entire probe forebody for laminar flow. In fact, the convective heating with coupled injection is negligible and never exceeds 10 MW/m^2 even when the flow is turbulent. The radiative heating for turbulent flow is reduced appreciably with respect to the no-injection results on the blunted nose portion of the probe; however, on most of the conical portion of the probe the coupled radiative heating values exceed the corresponding results for no injection.

References 1 and 10 discuss in detail why turbulence influences coupled radiative heating adversely, yet has no impact on radiative heating for no mass injection. Briefly, turbulence alters the chemical composition (higher temperatures) within the ablation layer such that the effectiveness of the relatively cool ablation layer to absorb or block the radiation that originates in the high temperature region of the shock layer is greatly reduced. A potential source of the radiation enhancement with mass injection has been attributed[1] to the increased number of atomic hydrogen radiators (thicker shock layer), because the number density of atomic hydrogen remains about the same throughout the shock layers with or without injection.

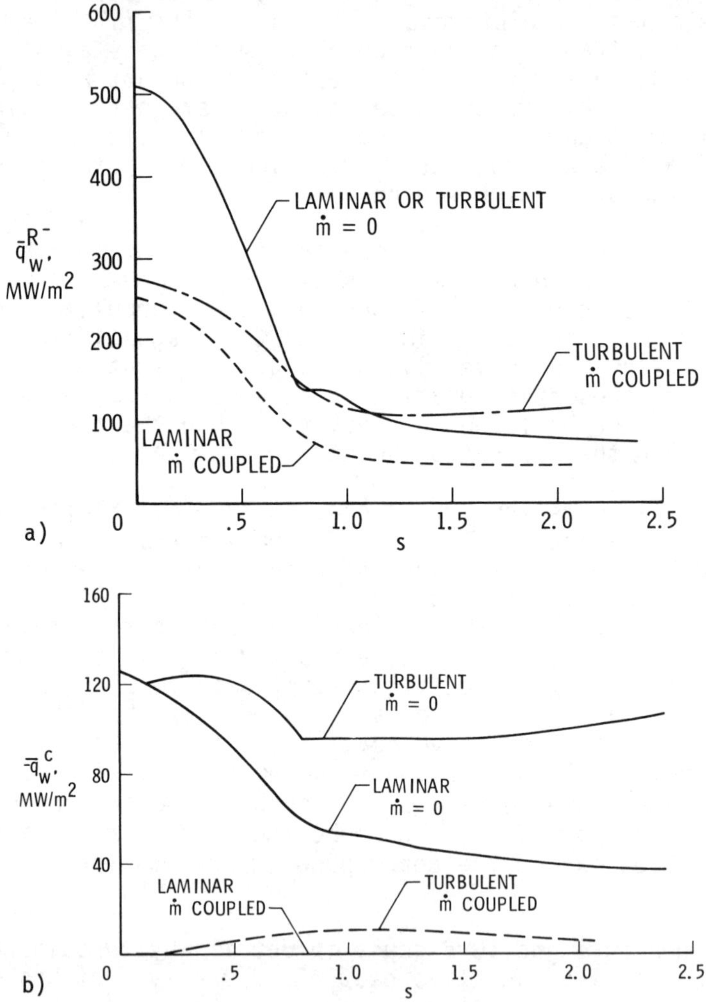

Fig. 4 Effect of coupled carbon-phenolic injection and turbulence on surface heating (290 kg probe, time = 49.5 s): a) radiative heating; b) convective heating

Figure 5 presents the coupled carbon-phenolic mass-injection distributions for both laminar and turbulent flow. The significance of turbulence on the mass-loss rate is very dramatic as is clearly demonstrated in Table 4 where the forebody surface mass-loss rate is tabulated for both laminar and turbulent flow. At 49.5 s, the mass-loss rate for turbulent flow (8.47 kg/s) is 2.5 times that for laminar flow. The increased mass addition to the shock layer due to the effect of turbulence produces a much thicker shock layer as evidenced in Fig. 6.

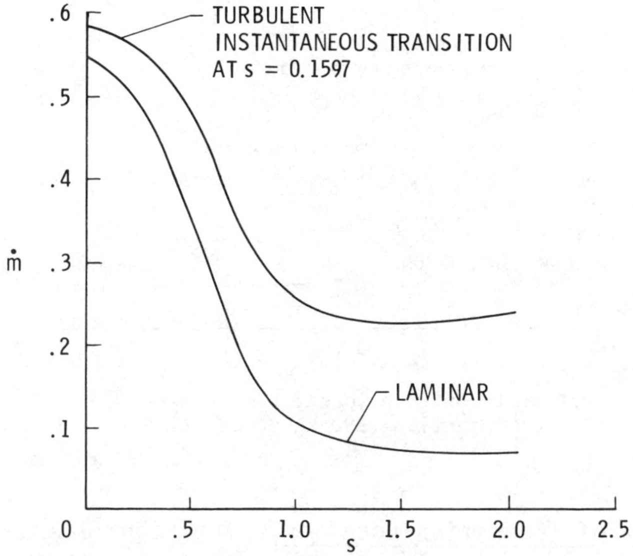

Fig. 5 Effect of turbulence on the coupled carbon-phenolic mass-injection distribution.

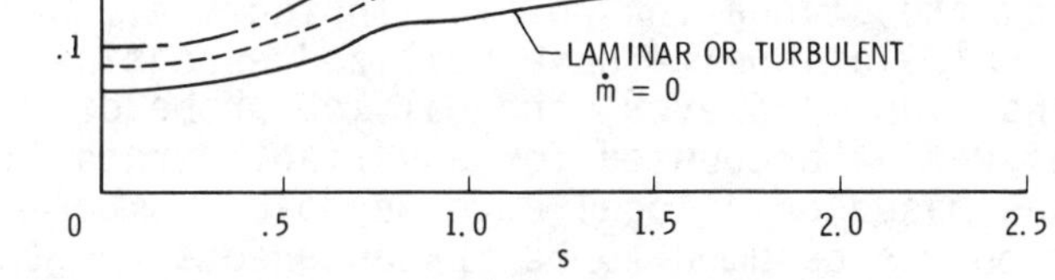

Fig. 6 Effect of coupled carbon-phenolic injection on shock-layer thickness distribution (290 kg probe, time = 49.5 s).

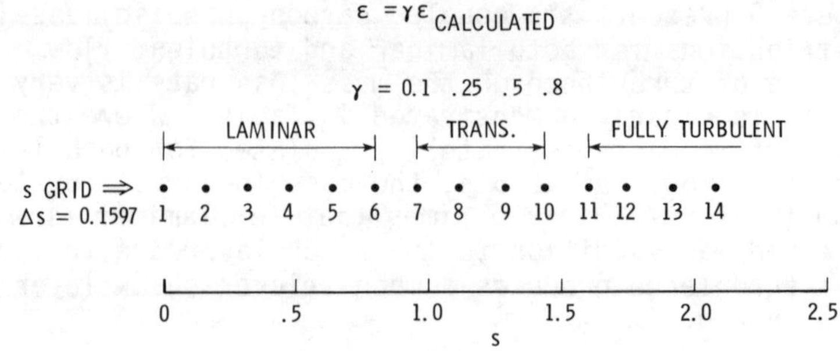

Fig. 7 Example of transition condition used.

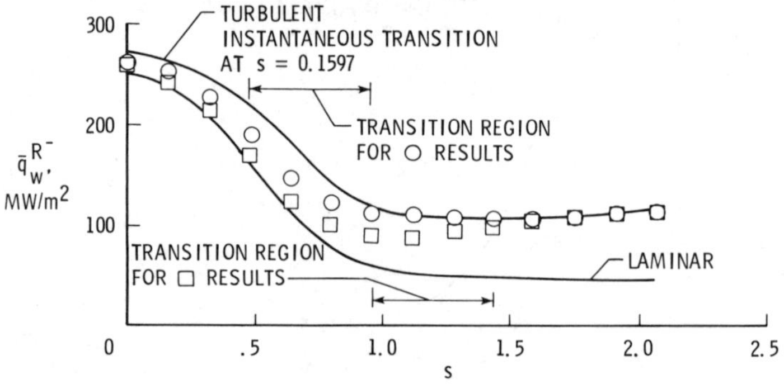

Fig. 8 Effect of transition location on radiative heating with coupled ablation injection (290 kg probe, time = 49.5 s).

Effect of Transition Location. The significance of transition location as it influences the radiation that penetrates the ablation layer was accomplished by making flowfield calculations with different transition locations. The solutions provide information on the radiative heating and ablative mass loss downstream of the transition location. The calculations were made for the 290 kg probe at a time of 49.5 s.

The previous discussion has concentrated on the results where instantaneous transition was imposed one body grid downstream of the stagnation point. In fact, all previous "benchmark" calculations[1-3] have utilized essentially this type of transition. However, the Galileo probe design calculations of Ref. 4 accounted for a variable transition location that was based on a local edge Reynolds number of 10^5. The transition was assumed to be instantaneous and the heating environment was determined by multiplying the calculated no-injection values by a correction factor (obtained from the "benchmark" analysis) to account for the effects of ablation injection. From the results presented herein, it is clear that there are large differences in correction factors depending on whether the flow is laminar or turbulent. Thus, the question arises as to what is the nature of the radiative heating (correction factor) downstream of the transition location as the transition front moves forward during entry.

Since an instantaneous transition required special treatment for the coupled solutions when the transition occurred downstream of the stagnation region (larger gradients in viscosity), the calculations were made assuming that the transition occurred over a specific spatial region (Fig. 7). That is, only a fraction of the calculated eddy viscosity is used but the fraction increases over a spatial distance of five grid points until fully turbulent flow is achieved.

Figures 8 and 9 present results that demonstrate the impact of transition location on radiative heating and carbon-phenolic mass-loss rate, respectively. Results are shown for four separate flowfield solutions: laminar, turbulent with instantaneous transition at the first body station aft of the stagnation point (s = 0.1597), turbulent with the onset of

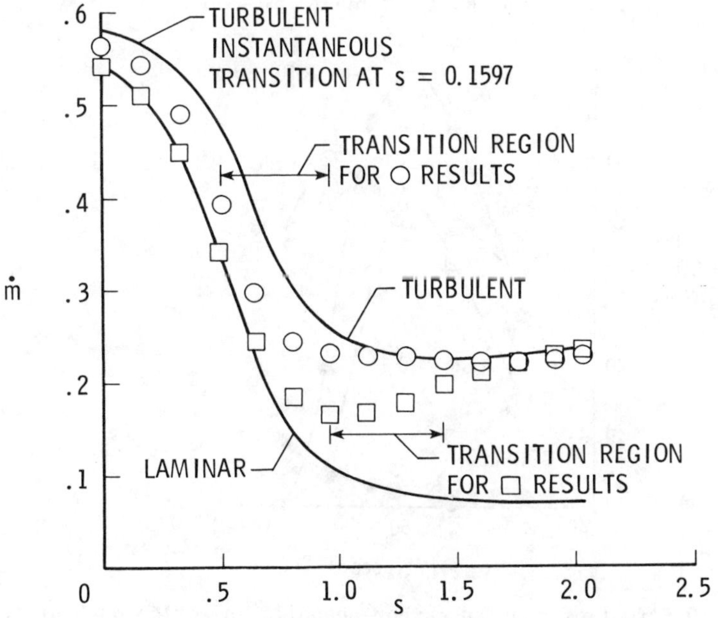

Fig. 9 Coupled carbon-phenolic mass-injection distributions for various transition conditions (290 kg probe, time = 49.5 s).

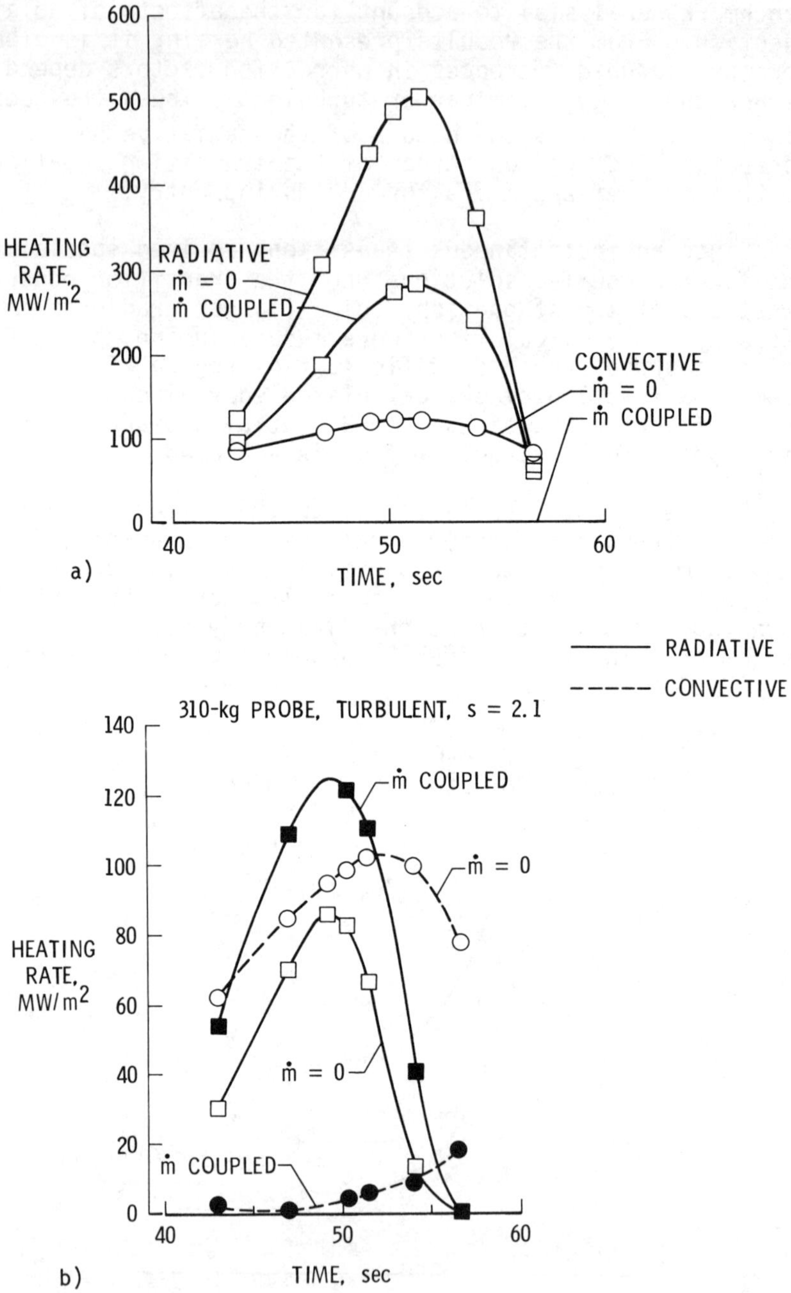

Fig. 10 Effect of coupled carbon-phenolic injection on heating (310 kg probe): a) s = 0.0; b) s = 2.1 and turbulent.

transition at s = 0.479, and turbulent with the onset of transition at s = 0.958. The last two cases had an arbitrarily defined transition region in which the flow made the transition from laminar to fully turbulent flow. The significance of these results is that regardless of where transition occurs, the radiative heating (mass-loss rate) quickly adjusts to the values of the coupled solution with transition located one station downstream of the stagnation point. Therefore, once knowledge of the impact of coupled injection on radiative heating is known for both laminar and turbulent flow (blockage correlations developed), approximate heat-shield design studies should be able to adequately model the effect of a variable transition location.

Additional research, particularly experimental, must be carried out to provide results even if qualitative to determine the reliability of current turbulence models, the location of transition, and the extent of the transition region for problems such as considered herein.

310 kg Probe

For the 310 kg probe, both laminar and turbulent solutions were obtained at six times along the entry trajectory. These

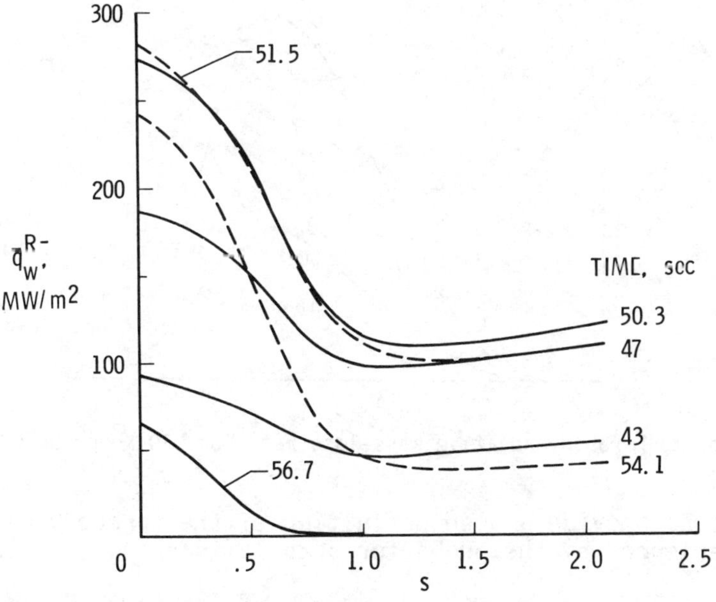

Fig. 11 Radiative heating-rate distributions with coupled carbon-phenolic injection for 310 kg probe.

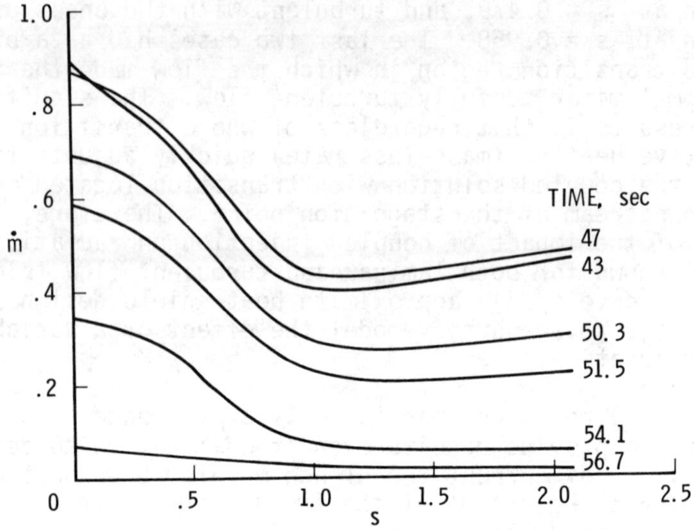

Fig. 12 Coupled carbon-phenolic mass-injection distributions for 310 kg probe.

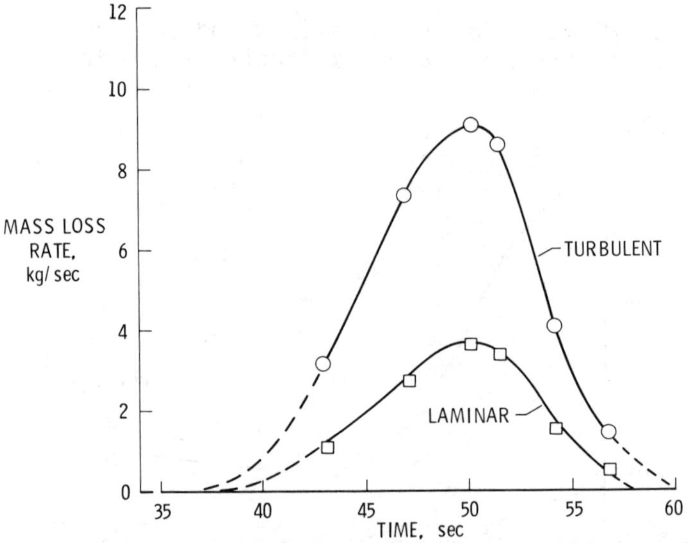

Fig. 13 Forebody ablative mass-loss rate for 310 kg probe.

calculations provide a good definition of the forebody mass loss experienced by the probe for both laminar and turbulent flow.

The important results of these calculations are briefly summarized since the results are very similar to those described for the 290 kg probe. Freestream conditions and stagnation-point results without ablation injection are presented in Table 3b while the stagnation-point results with coupled ablation injection are included in Table 4b. The heating-rate histories both with and without injection are presented for the stagnation point and a value of $s = 2.1$ in in Figs. 10a and 10b, respectively. The coupled radiative heating-rate distributions for turbulent flow are shown in Fig. 11 while the corresponding mass injection rates are presented in Fig. 12. Figure 13 shows a comparison of the mass-loss rates (see Table 4b) for laminar and turbulent flow. When the total probe mass loss for laminar flow is 35 kg while that for turbulent flow is 2.57 times the laminar value or 90 kg. These results clearly underscore the adverse effect that turbulence has on radiative heating with ablation injection.

Conclusion

Coupled flowfield solutions are presented for the forebody of a probe entering Jupiter's atmosphere. The calculations are for entry trajectories where the initial probe mass is approximately that of the current Galileo probe. Results of the present calculations are in good agreement with previous studies concerning the effect of turbulence on radiative heating with ablation injection. The present results are the first "benchmark" solutions to show the effect of transition location as it influences the heating and ablator mass loss downstream of transition for massively blown flows, and the results show that the heating-rate values adjust quickly to the values of the coupled solution with transition located one station downstream of the stagnation point. The calculated forebody ablative mass loss for a 310 kg probe was 90 kg for turbulent flow and only 35 kg for laminar flow. These results clearly demonstrate the large penalty encountered due to the adverse effect that turbulence has on the ability of the injected ablation species to block or absorb radiation.

Acknowledgment

The authors would like to thank Ann L. Simmonds and James W. Fenbert for their help in obtaining the computer solutions and in the data reduction.

References

[1]Moss, J. N., "A Study of the Aerothermal Entry Environment for the Galileo Probe," AIAA Paper 79-1081, June 1979.

[2]Kumar, A., Graves, R. A. Jr., Weilmuenster, K. J.,and Tiwari, S. N., "Laminar and Turbulent Flow Solutions with Radiation and Ablation Injection for Jovian Entry," AIAA Paper 80-0288, Jan. 1980.

[3]Nicolet, W. E. and Balakrishnan, A., "Methods for Predicting Off-Stagnation Point Flowfields for Planetary Entry Probes," AIAA Paper 79-1083, June 1979.

[4]Brewer, R. A. and Brant, D. N., "Thermal Protection System for the Galileo Mission Atmospheric Entry Probe," AIAA Paper 80-0358, Jan. 1980.

[5]Kumar, A. and Graves, R. A. Jr., "User's Guide for the Computer Code 'COLTS' for Calculating the Coupled Laminar and Turbulent Flows Over a Jovian Entry Probe," NASA TN 81848, 1980.

[6]Kumar, A. and Graves, R. A. Jr., "Numerical Solution of the Viscous Hypersonic Flow Past Blunted Cones at Angle of Attack," AIAA Journal, Vol. 15, Aug. 1977, pp. 1061-1062.

[7]Nicolet, W. E., "Advanced Methods for Calculating Radiation Transport in Ablation-Product Contaminated Boundary Layers," NASA CR-1656, 1970.

[8]Nicolet, W. E., "User's Manual for the Generalized Radiation Transfer Code (Rad/EQUIL)," NASA CR-116353, 1969.

[9]Sutton, K. and Moss, J. N., "Radiation Absorption by the C_2 Band Systems for Jupiter Entry Conditions," AIAA Paper 79-0033, Jan. 1979.

[10]Moss, J. N., Jones, J. J., and Simmonds, A. L., "Radiative Flux Penetration Through a Blown Shock Layer for Jupiter Entry," Outer Planet Entry Heating and Thermal Protection, Progress in Astronautics and Aeronautics, Vol. 64, edited by R. Viskanta, AIAA, New York, 1979, pp. 22-41.

[11]Stroud, C. W. and Brinkley, K. L., "Chemical Equilibrium of Ablation Materials Including Condensed Species," NASA TN D-5391, 1969.

[12]Kumar, A., Graves, R. A. Jr., and Weilmuenster, K. J., "User's Guide for the Vectorized Code 'EQUIL' for Calculating Equilibrium Chemistry on CDC-STAR-100 Computer," NASA TN 80193, 1980.

[13]Wilke, C. R., "A Viscosity Equation for Gas Mixtures," Journal of Chemical Physics, Vol. 18, April 1950, pp. 517-519.

[14]Zoby, E. V., Graves, R. A. Jr., Moss, J. N., Kumar, A., and Simmonds, A., "Correlations for Determining Transport Properties of Hydrogen-Helium Gas Mixtures at Temperatures from 1000 to 25 000 K," AIAA Journal, Vol. 18, April 1980, pp. 463-470.

[15]Clauser, F. H., The Turbulent Boundary Layer, Vol. IV of Advances in Applied Mathematics, edited by H. L. Dryden and Th. Von Karman, Academic Press, New York, 1956, pp. 1-51.

[16]Klebanoff, P. S., "Characteristics of Turbulence in a Boundary Layer with Zero Pressure Gradient," NACA Rept. 1247, 1955 (supersedes NACA TN 3178).

[17]Cebeci, T., "Behavior of Turbulent Flow Near a Porous Wall With Pressure Gradient," AIAA Journal, Vol. 8, Dec. 1970, pp. 2152-2156.

[18]Anderson, E. C. and Wilcox, D. C., "Vorticity Interaction Effects on Blunt Bodies," NASA CR-2778, 1977.

[19]MacCormack, R. W., "The Effect of Viscosity in Hypervelocity Impact Cratering," AIAA Paper 69-354, April 1969.

[20]Brauer, G. L., Cornick, D. E., and Stevenson, R., "Capabilities and Applications of the Program to Optimize Simulated Trajectories (POST)," NASA CR-2770, 1977.

[21]Feldhuhn, R. H., "Heat Transfer from a Turbulent Boundary Layer on a Porous Hemisphere," AIAA Paper 76-119, Jan. 1976.

[22]Holden, M., "Studies of Transpiration Cooling, Surface Roughness and Entropy Swallowing in Transitional and Turbulent Boundary Layers Over Nose Tips," IAF-79-F42, Sept. 1979.

An Experimental Simulation of Massive Blowing from a Nosetip During Jovian Entry

Michael S. Holden*
Calspan Corporation, Buffalo, N.Y.

Abstract

A research program has been conducted to examine the structure and stability of the shock layer and the detailed distribution of heating to a highly blowing blunt body in an environment simulating entry into the Jovian atmosphere. The experimental studies were conducted in the 96-inch Shock Tunnel at Calspan in 80% H_2-20% He gas flow at Mach 12. Measurements were made at blowing rates from $0<\dot{m}/\rho_\infty u_\infty<0.7$ using CF_4 as the principal injectant; selected measurements were also made with N_2, CO_2, and SF_6 as injectants. High-speed photography was used to examine the structure and stability of the shock layer. The experimental studies demonstrated that for low blowing rates, the heating rates in the stagnation region can be enhanced as blowing promotes boundary-layer transition. While increasing the blowing rate decreased the surface heating for $0.3<\dot{m}/\rho_\infty u_\infty<0.5$, further increases in blowing rates did not significantly lower the heating levels below $C_H/C_{H_0} = 0.1$. For injection ratios below 0.1, the shock layer was relatively stable; however, for blowing rates of $0.1< \dot{m}/\rho_\infty u_\infty< 0.5$, high-speed photography suggests that the turbulent viscous layer above the body is unsteady and a description of this flow in terms of conventional boundary-layer theory may be inaccurate. For injection ratios of over 0.5, the viscous layer can become grossly unstable. For a given injection rate, shock-layer stability can be increased by increasing the molecular weight of the injectant; however, the thermal protection decreases with increased molecular weight. It is recommended that the theoretical modeling of the viscous layer for high mass addition rates be re-examined in the light of these studies.

Presented as Paper 81-1070 at AIAA 16th Thermophysics Conference, Palo Alto, Calif., June 23-25, 1981.

*Prinicipal Aeronautical Engineer, Aerodynamic Research Department.

Introduction

Ablative heat shields have been used successfully for many years to achieve the thermal protection of spacecraft and missiles during re-entry into the Earth's atmosphere. Under most practical re-entry conditions, the mass of gas ablated into the boundary layer is a small fraction of the mass flow rate in the inviscid flow (i.e., $\dot{m}/\rho_\infty u_\infty < 0.05$). Unlike the conventional re-entry into the Earth's atmosphere, the high-energy entry into the He-H_2 atmosphere of Jupiter generates extremely large radiative heating rates in the presence of a relatively light convective heating load. Under these conditions, the mass addition from the surface blowing will be a significant fraction of the unit mass flow rate in the inviscid flow. Analyses[1,2] have suggested that these large mass addition rates block a significant portion of the radiative flux from reaching the surface and also reduce the convective heat transfer to nearly zero for laminar boundary-layer flows. At lower altitudes, where the boundary layer is turbulent, the convective heating may exceed the radiative flux.[1] In an earlier experimental investigation conducted at Calspan by Holden,[3] high-speed Schlieren photography and heat-transfer measurements were made with a transpiration-cooled large hemispherical nosetip to study the stability of the shock layer for large blowing rates ($0.03 < \dot{m}/\rho_\infty u_\infty < 0.5$). These studies demonstrated that the structure of the shock layer is significantly modified by the injectant (Fig. 1) and modeling these flows using conventional boundary-layer theory is open to serious question. For high blowing rates ($\dot{m}/\rho_\infty u_\infty > 0.3$), the shock layer became unstable (as shown in Fig. 2), with the flow exhibiting a pulsating oscillation similar to that exhibited by spiked blunt bodies. In these studies, air was used as the freestream gas and mixtures of helium and nitrogen (used to simulate the molecular weight and specific heat of steam) were injected through the porous surface of the model. Thus, a relatively light gas with a high volumetric flow rate was used in these studies. This contrasts with the relatively heavy injectants which must be used to simulate the Jovian entry.

The major objective of earlier studies of transpiration cooling techniques was to evaluate how the blockage heat transfer C_H/C_{H_0} varied with the Mach number, Reynolds number, and the physical and chemical properties of the freestream and injectant. The lack of definitive techniques to predict the effectiveness of transpiration/film cooling techniques reflects the lack of fundamental understanding of turbulent mixing in the presence of mass injection and surface roughness. There is a dearth of data at hypersonic speeds where transpiration cooling is of considerable interest. While experimental studies have been conducted in

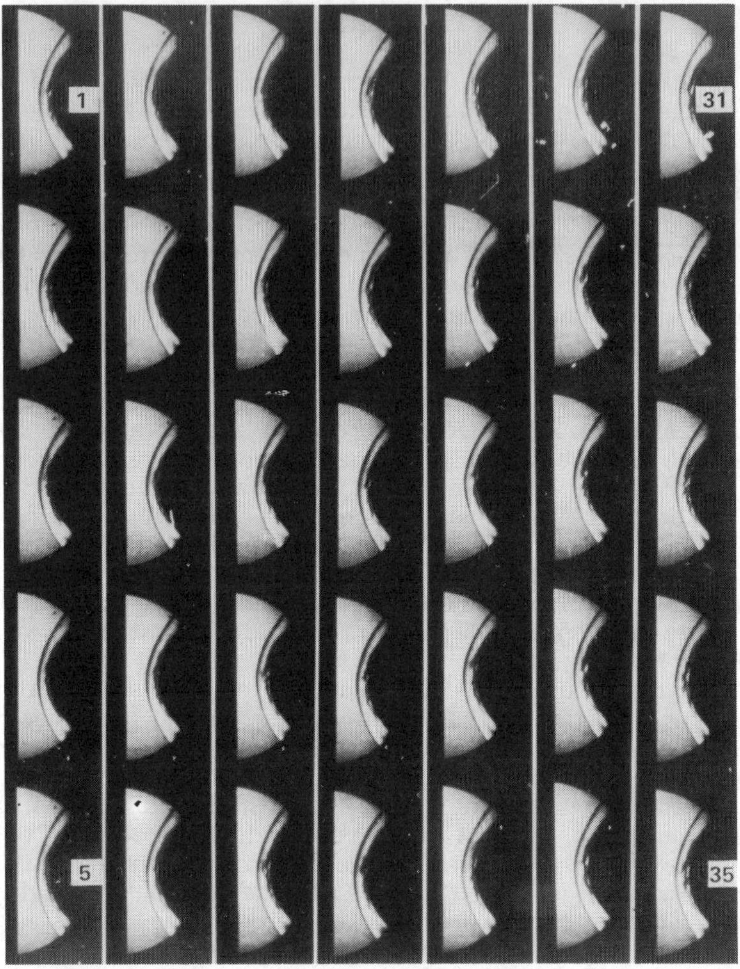

Fig. 1 High-speed Schlieren photographs (5000 frames/s) of the turbulent mixing of the injectant N_2 in the blunt-body shock layer for $\dot{m}/\rho_\infty u_\infty = 0.16$ in a Mach 11 airflow.

supersonic flow with flat plates[4-6] and cones,[7-10] there has been little work on the transpiration cooling of blunt nosetips. In a recent study by Holden,[3] a survey of the existing experimental data was made and the correlation shown in Fig. 3 was obtained. The correlation indicates that for large blowing rates ($\dot{m}/\rho_\infty u_\infty C_{H_0} > 20$), increased blowing does not significantly improve thermal protection. This may well result from a decrease in the stability of the mixing layer and an increase in the scale of turbulence with increasing blowing. However, Holden's measurements on a spherical nosetip, shown in Fig. 4, suggests

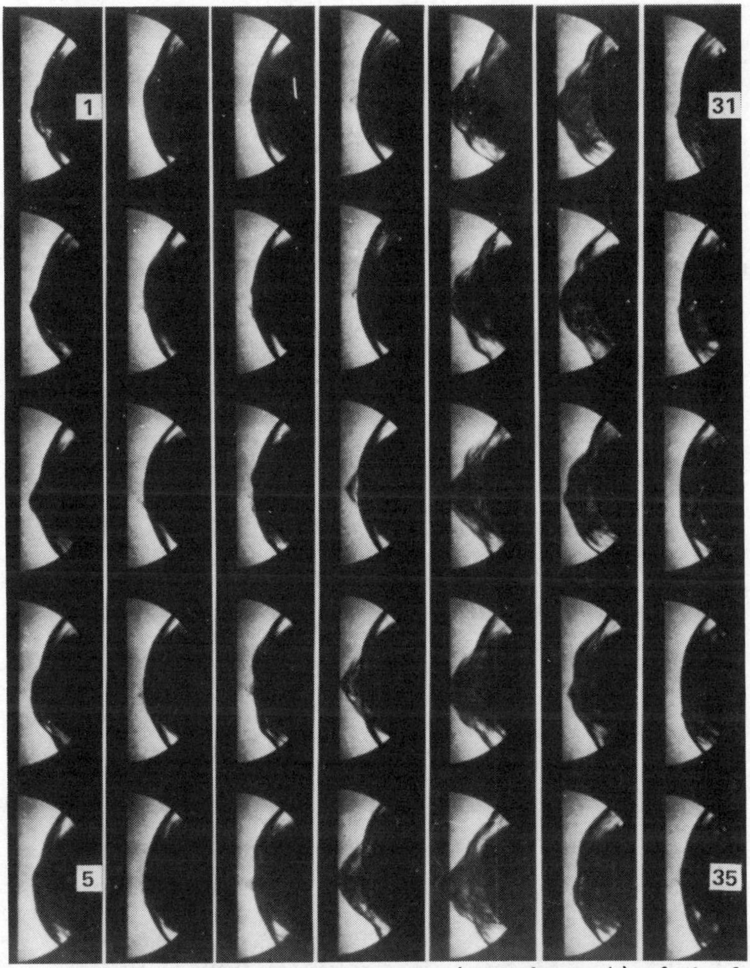

Fig. 2 High-speed Schlieren photographs (5000 frames/s) of the large-scale shock-layer instabilities resulting from nitrogen injection rates ($\dot{m}/\rho_\infty u_\infty$) of 0.4 in Mach 11 airflow.

that heating levels significantly lower than those found on flat plates and cones were obtained for the higher blowing rates. These latter measurements could be correlated in the form

$$(C_{H_0} - C_H)/C_{H_0} = \frac{1}{3}\left(\frac{\rho_\infty u_\infty}{\rho_e u_e}\right)^{1/3}\left(\frac{\dot{m}}{\rho_\infty u_\infty C_{H_0}}\right)^{1/3}$$

The object of the present program was to examine the fluid mechanical structure and stability of the boundary layer and shock layer ahead of a highly blowing heat shield in a H_2-H_e freestream. While these early experiments do not provide a complete simulation

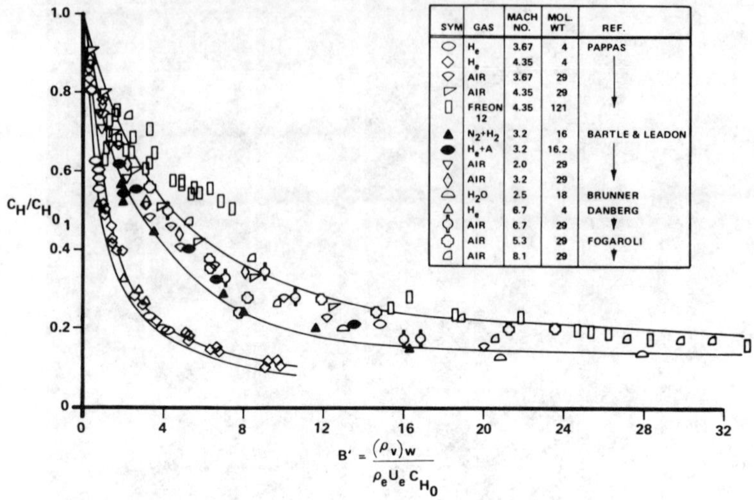

Fig. 3 C_H/C_{H_0} vs B' for flat plates and cones.

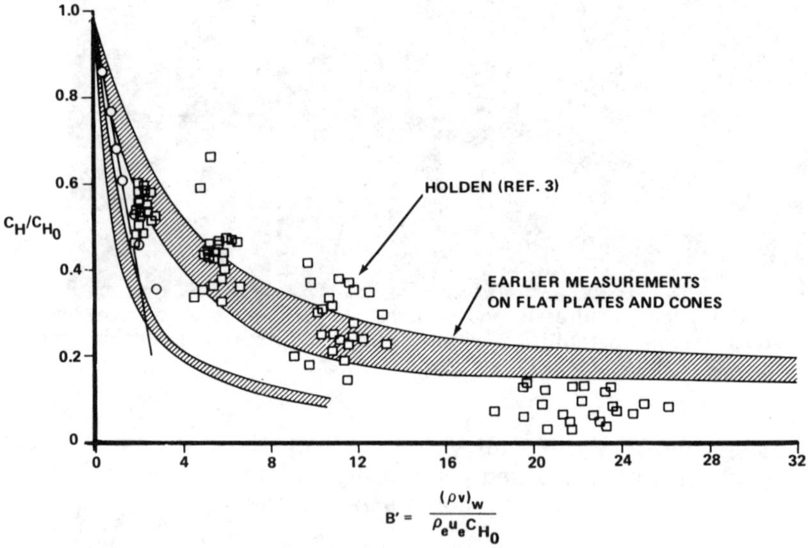

Fig. 4 C_H/C_{H_0} vs B' for N_2 injectant—blunt nose.

of Jovian entry because of the absence of radiative heat transfer, they do provide an insight into the basic fluid dynamic stability and turbulence processes that occur with massive blowing and the associated changes in the convective heat transfer. In these studies, which were conducted in Calspan's 96-inch Shock Tunnel, injectants with molecular weights of 44 ⟶ 120 were used to examine the effect of the volumetric flow rate on boundary-layer structure and shock-layer stability over a spherical nose shape. High-speed Schlieren photographs were used to visualize the flowfield. Detailed heat-transfer and pressure measurements were made on the surface of the porous model. The experiments were conducted with an 80% H_2-20% He mixture for the freestream to simulate the anticipated environment on Jupiter at a Mach number of 12 and freestream Reynolds numbers from 1.2×10^7 to 1.9×10^7 per meter.

In the following section, the model design, the instrumentation, and the modifications to the experimental facilities required for operation with the H_2-H_e environment are described. The measurements made in the experimental studies are presented and discussed together with the results from earlier studies. We conclude with a discussion of the implications of these measurements to the modeling of the shock-layer flow over ablating bodies and the thermal protection of a transpiration-cooled body with a gaseous injectant.

Model Design, Instrumentation, and Experimental Facilities

Model and Instrumentation

In this study, we used a model (Fig. 5) which employed a gaseous injection technique patterned after that used by the Aerojet transpiration-cooled nosetip (TCNT). In this design, which employs some of the better features of both the transpiration and film cooling techniques, the coolant is injected through a large number of slots distributed uniformly over the hemispherical surface of the model. The slots, from which the injectant flows at a low subsonic speed, are fed from tiny sonic orifices machined at the base of each individual slot. These orifices are fed from eight concentric zones in the model which are in turn fed through eight fast-acting "Valcor" valves from eight reservoirs mounted in a cylindrical body at the base of the nosetip. By controlling the initial pressure in each reservoir, the distribution of the injectant around the sphere can be specified. In the present studies, where an 80% H_2-20% He mixture was employed as the test medium, CF_4 was used as the principal injectant; however, CO_2, N_2, and SF_6 were also used in studies to evaluate the influence of molecular weight on the character of the flowfield. The porous nosetip, the

Fig. 5 Porous nosetip model with fast-acting valve and reservoir assembly.

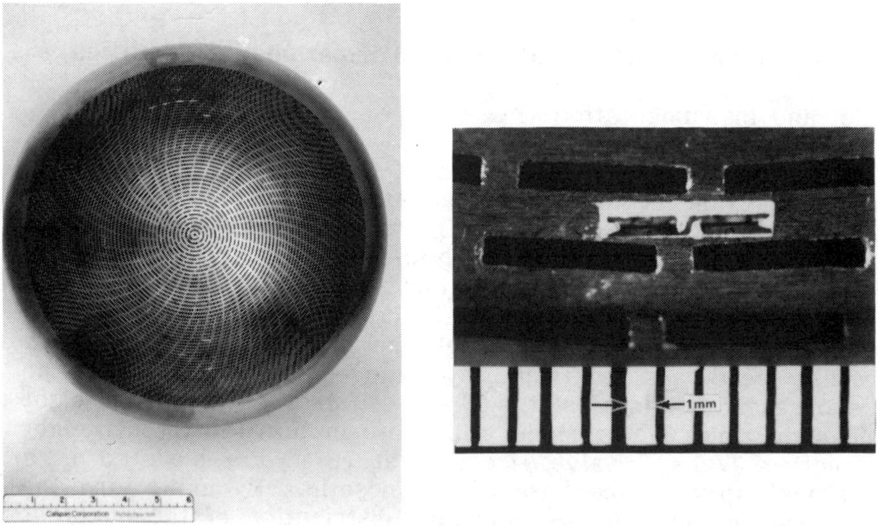

Fig. 6 Instrumented porous nosetip showing typical heat-transfer gage.

fast-acting valve assemblies, and the reservoirs are shown mounted in the 96-inch Shock Tunnel in Fig. 5. The geometry of the slots and their relative positions on the nosetip are shown in Fig. 6. The 54 rows of slots are staggered in a 35 deg swirl extending in an arc back to 50 deg from the stagnation point. The mass flow rate from the slots in each of the eight annular zones on the model were calibrated with each of the gaseous injectants used in the experimental studies. Miniature heat-transfer gages were developed specifically for this study so that detailed measurements could be made in the "bands" between the slots (see inset, Fig. 6). The instrumentation was concentrated in a 25 deg segment of the nosetip and positioned in such a manner that three-dimensional flow structures can be examined. Pairs of heat-transfer gages were positioned to obtain measurements behind the center of the slot and between the slots at the same radial location. The model was also instrumented with pressure gages to measure the surface pressure and the pressure through the internal porting of the models. The second model used in these studies was a smooth-wall replica of the TCNT shown in Fig. 7. The model was highly instrumented along orthogonal rays with thin-film heat-transfer and pressure gages. The heat-transfer distributions obtained with this nonporous model were used to nondimensionalize the measurements made with the blowing nosetip.

Experimental Facility

The experimental program was conducted in the Calspan 96-inch Shock Tunnel.[11] This facility has an advantage in producing and safely handling the 80% H_2-20% He environment required in these studies; however, there are a number of features associated with this problem which present significant technical difficulties which are common to any experimental facility. The extremely large mass injection rates ($\dot{m}/\rho_\infty u_\infty$ as large as 0.7) required for this study can cause an elevation in the test section pressure which leads to flow breakdown if the facility is too small. Although the dump tank in the 96-inch Shock Tunnel is very large, for the largest injection rates, maintaining a test section low enough to permit a smooth start requires that the steady flow through the tunnel must be established within 30 ms of the establishment of the steady flow of injectant through the model. Since the flow of injectant had to be initiated with the same switch used to fire the tunnel, this mandated that the tunnel be fired rapidly with exact repetition. To achieve these objectives, it was necessary to redesign the the double-diaphragm rig venting system to achieve a venting rate of 100,000 SCFM. Venting hydrogen this rapidly directly to the atmosphere is clearly hazardous. This problem was resolved by fabricating a secondary high-pressure reservoir into

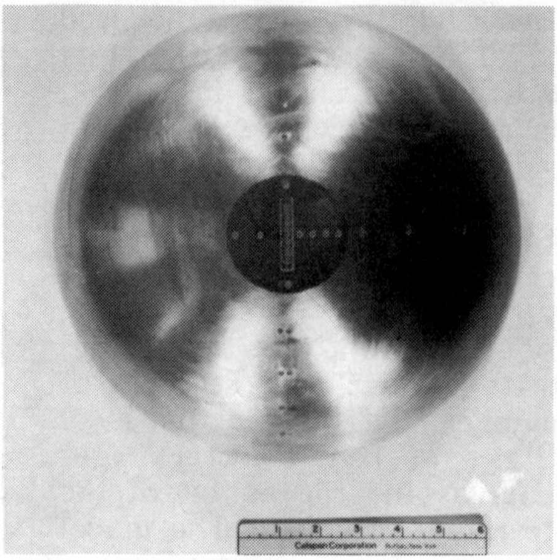

Fig. 7 Smooth wall hemispherical nosetip.

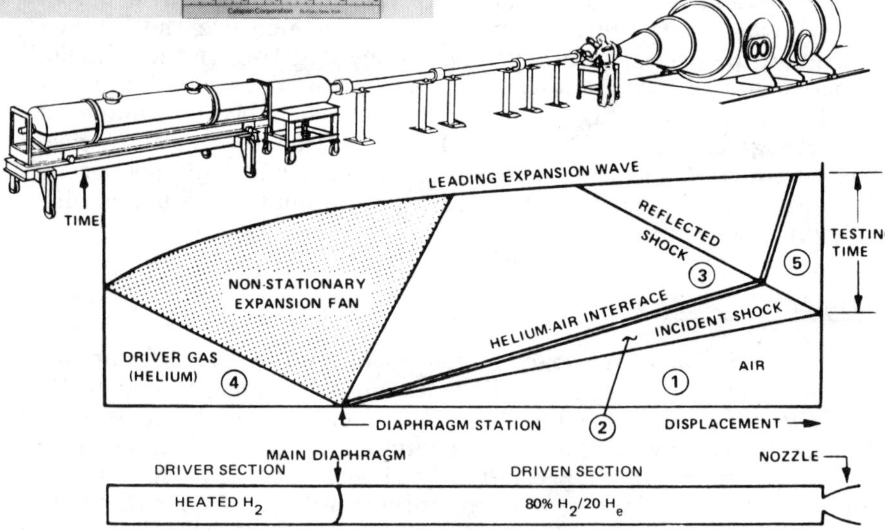

Fig. 8 Wave diagram for tailored-interface shock tube.

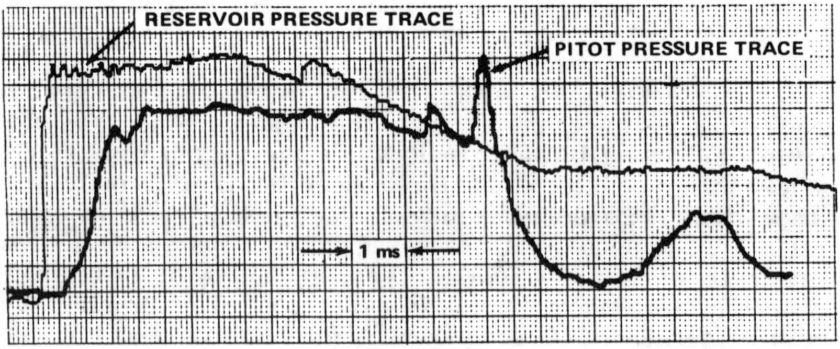

Fig. 9 Reservoir and pitot pressure time histories.

which hot hydrogen from the tunnel could be dumped and then bled slowly to the atmosphere. After making these modifications, we achieved a repeatability of better than 25 ms in the time between closing the switch to fire and the establishment of steady flow through the test section, making this test program feasible.

A significant effort was devoted to generating test conditions that not only simulated the fluid-dynamic environment of the Jovian atmosphere but also provided an ample test period. The high Mach numbers required in both the test medium (80% H_2-20% He) and the driver gas (heated H_2) made the generation of long run times in a shock tunnel inherently difficult. The earlier studies[3] had suggested that the shock layer over the TCNT model was fully established within 2.0 ms from the data rise; we sought to obtain at least an additional 4 ms of steady flow for this study. A

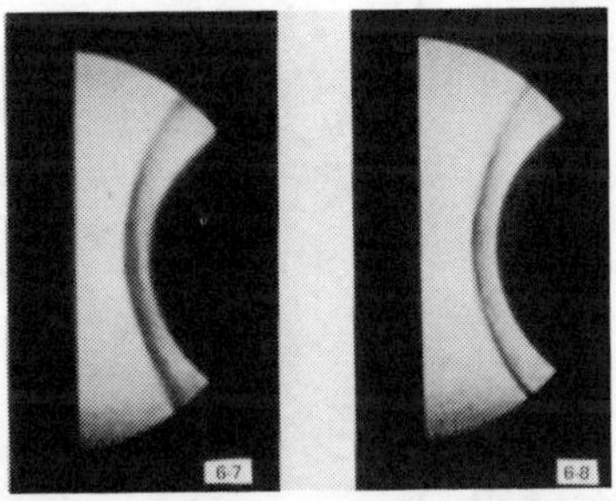

Fig. 10 Schlieren photographs with no mass flow (run 6).

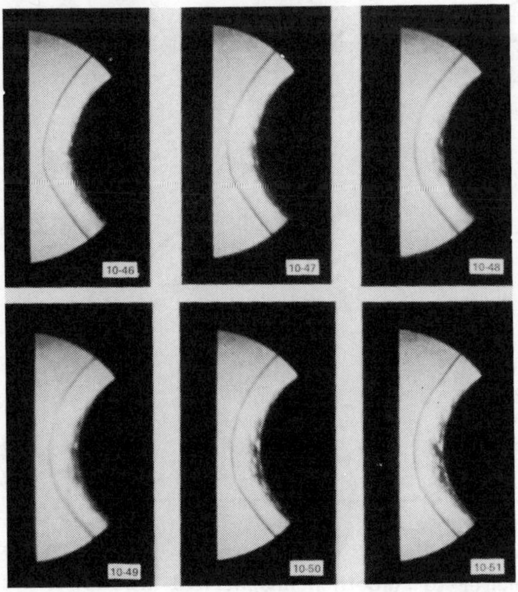

Fig. 11 Schlieren photographs of flow with CF_4 injection, $\dot{m}/\rho_\infty u_\infty = 0.3$.

typical wave diagram for the driver and driven section of the shock tunnel is shown in Fig. 8. Because hydrogen has such a high sound speed, the wave trajectory is relatively flat and the associated transit times are inherently short. To maximize run times, we operated with a low incident-shock Mach number and we tailored the prerun conditions to slow the contact surface after it interacts with the bow shock (i.e., we operated in a tailored-interface mode). It can be shown that to operate under tailored conditions, the speeds of sound ahead and behind the contact surfaces a_2 and a_3, respectively, must be related by the equation

$$\frac{a_3}{a_2} = \left[\frac{\gamma_4(\gamma_4 - 1)\left[1 + \left(\frac{\gamma_4 + 1}{\gamma_4 - 1}\right) P_{52}\right]}{\gamma_1(\gamma_1 - 1)\left[1 + \left(\frac{\gamma_1 + 1}{\gamma_1 - 1}\right) P_{52}\right]} \right]^{0.5} \tag{1}$$

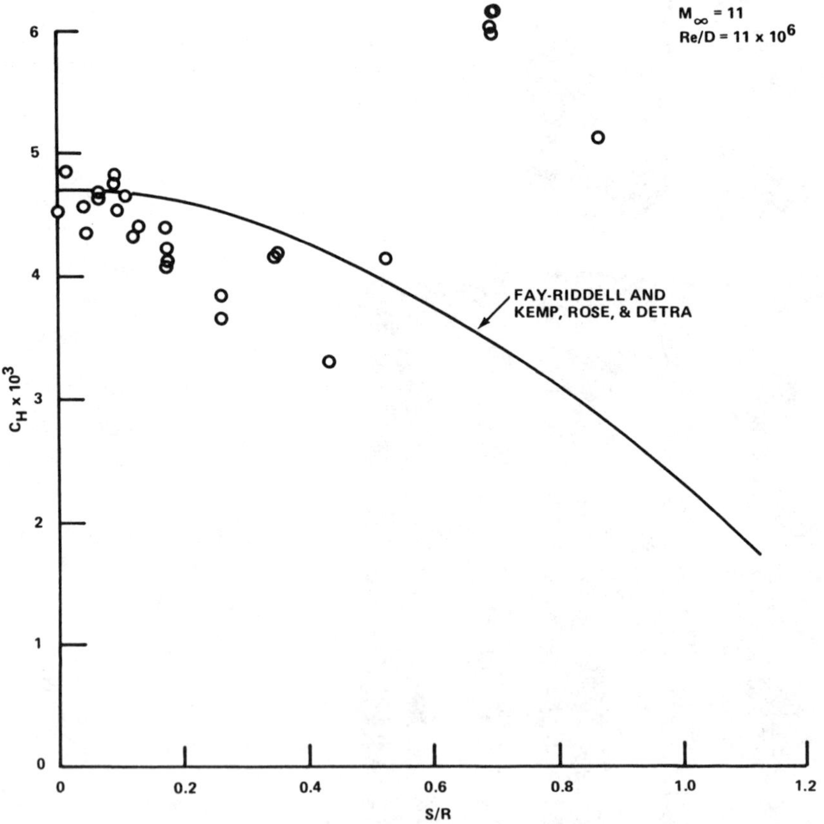

Fig. 12 Distribution of heat transfer to smooth hemispherical nosetip.

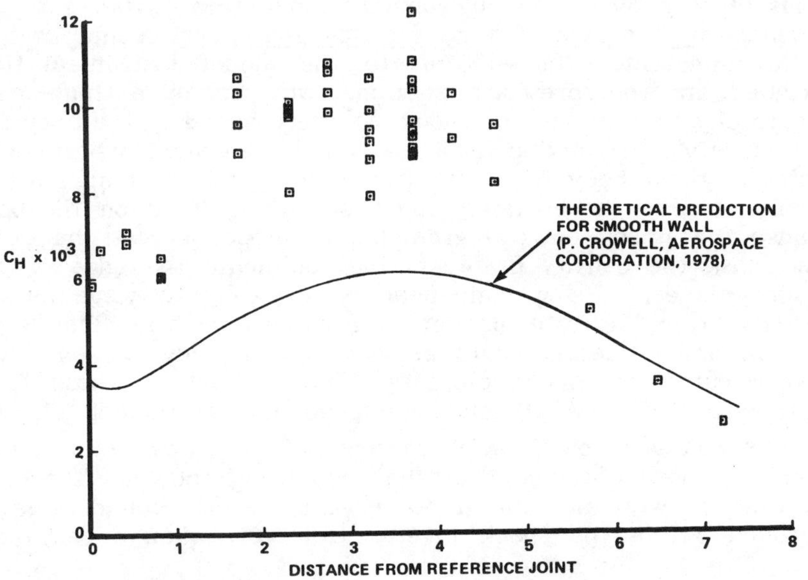

Fig. 13 Streamwise distribution of heat-transfer coefficient to the porous nosetip for $\rho_w v_w/\rho_\infty u_\infty$ = 0 (M_∞ = 11.3, Re_D = 11.1 x 10^7).

To obtain tailored-interface operation, it is necessary to control the temperature of the driver gas and the properties of gases in the driver and driven tubes to achieve a shock Mach number, M_S, at which the pressure behind the reflected shock remains constant until the expansion fan from the driver interacts with the reflected shock. Using Eq. (1) as a guide and with further experimentation, we were able to obtain excellent H_2/H_2-He tailoring, as can be seen from the reservoir and pitot pressure traces shown in Fig. 9. It can be seen that even operating with hydrogen in the driver and driven tubes, test times of over 8 ms could be obtained. With the increased accuracy of the firing sequence, we were able to initiate the injectant flow so that it was fully established between 10 and 18 ms prior to flow establishment through the tunnel. Since the flow of injectant was controlled by sonic orifices, its flow rate was invariable with tunnel conditions, and we found that the shock-layer flow was fully established within 2.0 ms from initiation of flow.

Experimental Results and Discussion

Flow Visualization

The principal objective of this study was to examine how the structure and stability of the hypersonic shock layer over a porous blunt body, in an environment simulating Jovian entry, is influenced by the rate of mass addition and the molecular weight of the injectant. In addition to the model instrumentation, described in the previous section, we employed high-speed photography to examine the shock structure ahead of the body and the boundary layer/mixing layer beneath it. In most experimental studies of blunt body flows at hypersonic speed, it is difficult to visually observe the boundary layer or mixing layer on the body because the strong entropy gradients developed across the shock layer mask the density gradients that delineate the edge of the boundary layer. Using high-speed photography to visualize the boundary layer over a blunt body depends on correctly illuminating the flow and processing light beam. In this study, we employed circular cutoff with a single-pass Schlieren system whose focal length was 10 ft. A Hycam high-speed camera running at 7000 frames/s was used to obtain photographs of the flow. Photographs typical of those obtained over the nonblowing body are shown in Fig. 10. We were not able to observe a distinct boundary layer embedded within the shock layer. However, when a relatively

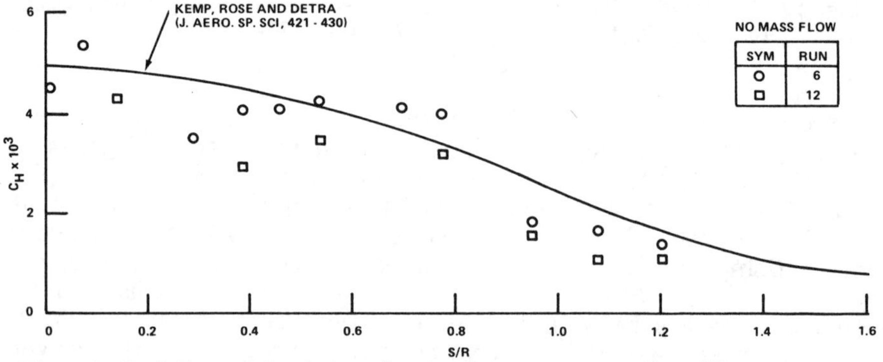

Fig. 14 Variation of heat transfer to the porous nosetip with zero mass flow rate.

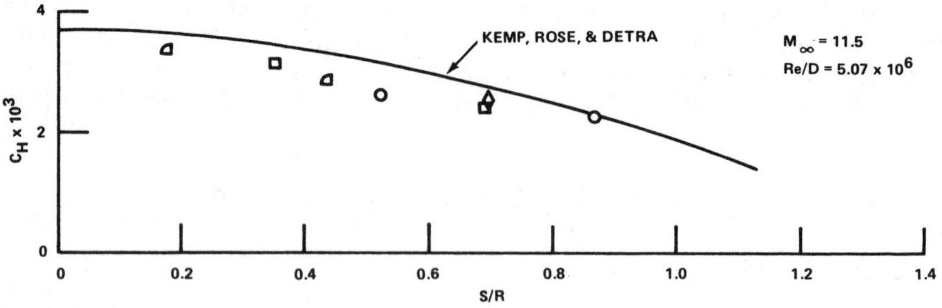

Fig. 15 Heat-transfer distribution to the smooth replica of the transpiration-cooled nosetip.

Table 1 Test conditions

Flow properties	Run 1	2	3	4	5	6	7	8
Mi	1.850	1.850	1.850	1.840	1.860	1.820	1.880	1.890
P_0, N/m^2	4.266(7)	5.418(7)	3.223(7)	3.030(7)	3.058(7)	2.865(7)	3.177(7)	3.305(7)
H_0, m^2/sec^2	7.801(6)	7.801(6)	7.801(6)	7714(6)	7.889(6)	7.563(6)	8.002(6)	8.040(6)
T_0, ^{o}K	691.1	691.1	691.1	683.3	698.9	670.0	708.9	711.7
M_∞	12.09	13.22	11.62	11.61	11.71	11.52	11.35	12.55
u_∞, m/s	3.888(3)	3.897(3)	3.883(3)	3.861(3)	3.906(3)	3.822(3)	3.929(3)	3.951(3)
T_∞, ^{o}K	20.84	17.52	22.51	22.31	22.42	22.19	24.16	19.98
P_∞, N/m^2	450.4	324.0	437.5	414.6	395.3	410.8	500.6	275.2
Q_∞, N/m^2	4.736(4)	4.073(4)	4.249(4)	4.017(4)	3.900(4)	3.922(4)	4.638(4)	3.118(4)
ρ_∞, kg/m^2	6.267(-3)	5.360(-3)	5.638(-3)	5.391(-3)	5.113(-3)	5.370(-3)	6.009(-3)	3.995(-3)
μ_∞, N/m s	1.324(-6)	1.119(-6)	1.426(-6)	1.410(-6)	1.414(-6)	1.407(-6)	1.528(-6)	1.266(-6)
Re/m	1.841(7)	1.868(7)	1.535(7)	1.476(7)	1.412(7)	1.459(7)	1.545(7)	1.247(7)
Flow properties	**9**	**10**	**11**	**12**	**13**	**14**	**15**	**16**
Mi	1.890	1.900	1.900	1.900	1.900	1.890	1.900	1.931
P_0, N/m^2	3.287(7)	3.267(7)	3.309(7)	3.284(7)	3.232(7)	3.278(7)	3.308(7)	8.733(7)
H_0, m^2/sec^2	8.109(6)	8.096(6)	8.140(6)	8.077(6)	8.159(6)	8.071(6)	8.159(6)	8.356(6)
T_0, ^{o}K	718.3	717.2	721.1	715.6	722.8	715.0	722.8	740.0
M_∞	12.24	12.36	12.57	11.72	11.73	11.55	11.96	11.48
u_∞, m/s	3.965(3)	3.963(3)	3.976(3)	3.952(3)	3.972(3)	3.949(3)	3.975(3)	4.016(3)
T_∞, ^{o}K	21.15	20.72	20.17	22.92	23.11	23.58	22.26	24.67
P_∞, N/m^2	321.0	299.9	273.2	422.3	413.4	463.6	374.1	547.2
Q_∞, N/m^2	3.459(4)	3.295(4)	3.103(4)	4.172(4)	4.091(4)	4.445(4)	3.849(4)	5.187(4)
ρ_∞, kg/m^2	4.400(-3)	4.196(-3)	3.926(-3)	5.344(-3)	5.185(-3)	5.700(-3)	4.873(-3)	6.432(-3)
μ_∞, N/m s	1.344(-6)	1.312(-6)	1.282(-6)	1.450(-6)	1.456(-6)	1.491(-6)	1.409(-6)	1.554(-6)
Re/m	1.298(7)	1.268(7)	1.218(7)	1.456(7)	1.415(7)	1.510(7)	1.375(7)	1.662(7)

cold, high-density injectant is introduced at the base of the boundary layer, the turbulent mixing region close to the body can be seen clearly (see Fig. 11). The exposure time of the film was sufficient to "stop" the motion of large-scale turbulent structure.

Measurements on the Smooth and Nonblowing Models

As part of the initial phase of this study, we examined flow over the porous nonblowing nosetip (see Fig. 5) and the smooth replica of this configuration shown in Fig. 7. We sought to determine whether the surface roughness would trip the boundary layer and cause roughness heating augmentation. In an earlier program,[3] measurements were made on both smooth and porous models in air at Mach 11 for a Reynolds number of 10×10^6. We observed boundary-layer transition on the smooth configuration close to the sonic point, as shown in Fig. 12. On the porous model, transition was tripped by the roughness close to the stagnation region, and the roughness apparently enhanced the heating beneath the turbulent boundary layer because heat-transfer rates were at least 50% greater than predicted theoretically (Fig. 13). The measurements made with smooth and porous models at Mach 12 in the 80% H_2-20% He gas flow at Reynolds numbers of 2×10^6 are shown in Figs. 14 and 15. Here it can be seen that the boundary layer remains laminar on the smooth configuration and that the rough surface of the porous model does not trip transition or induce higher than laminar heating rates. Thus, when we observe heating levels on the blowing model that are higher

Table 2 Test matrix

Run	Test gas, % He/% H_2	Model configuration	$\frac{\dot{m}}{\rho_\infty u_\infty}$	Injectant	Mode
1	20/80	Porous	0.1	CF_4	Uniform
2	20/80	Porous	0.3	CF_4	Uniform
3	20/80	Porous	0.5	CF_4	Uniform
4	20/80	Porous	0.5	CO_2	Uniform
5	20/80	Porous	0.5	CO_2	Peaked
6	20/80	Porous	0	—	—
7	20/80	Porous	0.5	CF_4	Peaked
8	20/80	Porous	0.7	CF_4	Peaked
9	20/80	Porous	0.5	CF_4	Uniform
10	20/80	Porous	0.3	CF_4	Uniform
11	20/80	Porous	0.5	SF_6	Uniform
12	20/80	Porous	0	—	—
13	20/80	Porous	0.5	N_2	Uniform
14	20/80	Porous	0.3	CF_4	Uniform
15	20/80	Porous	0	—	—
16	20/80	Nonporous	0	—	—

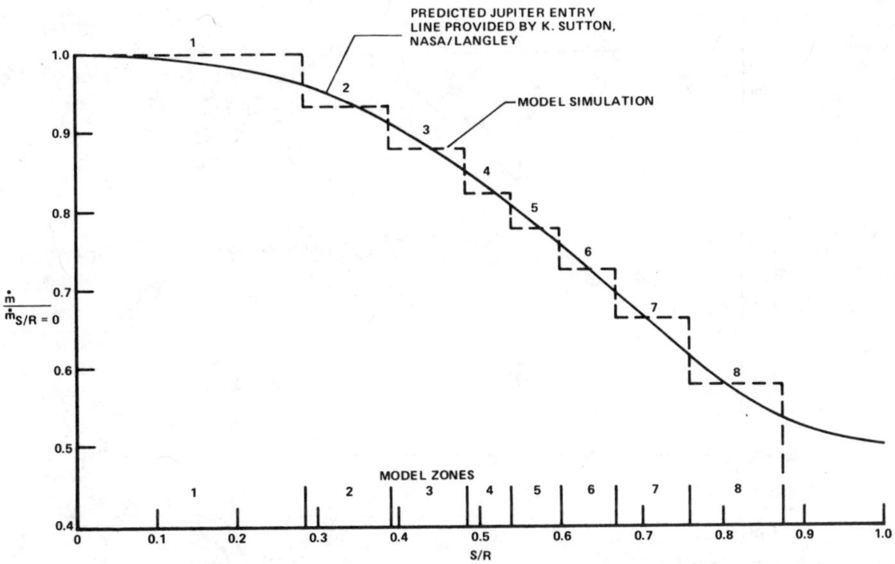

Fig. 16 "Peaked" mass injection distribution.

than those with zero mass injection, we must conclude that blowing has tripped the boundary layer, inducing turbulent heating.

Photographic Studies of Shock and Mixing-Layer Structure and Stability

Experimental studies were conducted in a 80% H_2-20% He gas flow at a Mach number of 12, with the freestream conditions as listed in Table 1. Table 2 lists the matrix of model configurations and injection conditions that were used for the runs specified in Table 1. CF_4 was used as the principal injectant in these studies to closely match the molecular weight ratio encountered during Jovian entry. Further studies were performed with N_2, CO_2, and SF_6 as injectants to explore the effect of the molecular weight ratio on the structure and stability of the flowfield. While the major segment of the program was conducted with a uniform distribution of blowing around the model, the "theoretical" blowing distribution, as shown in Fig. 16, was used to pattern the "peaked" distribution in the experimental program.

The photographs of the flow over the smooth and nonblowing porous model shown in Fig. 10 do not, as mentioned earlier, provide evidence of the size or condition of the boundary layer over the body. However, we know the boundary layer to be laminar from the surface heat-transfer measurements discussed previously. The

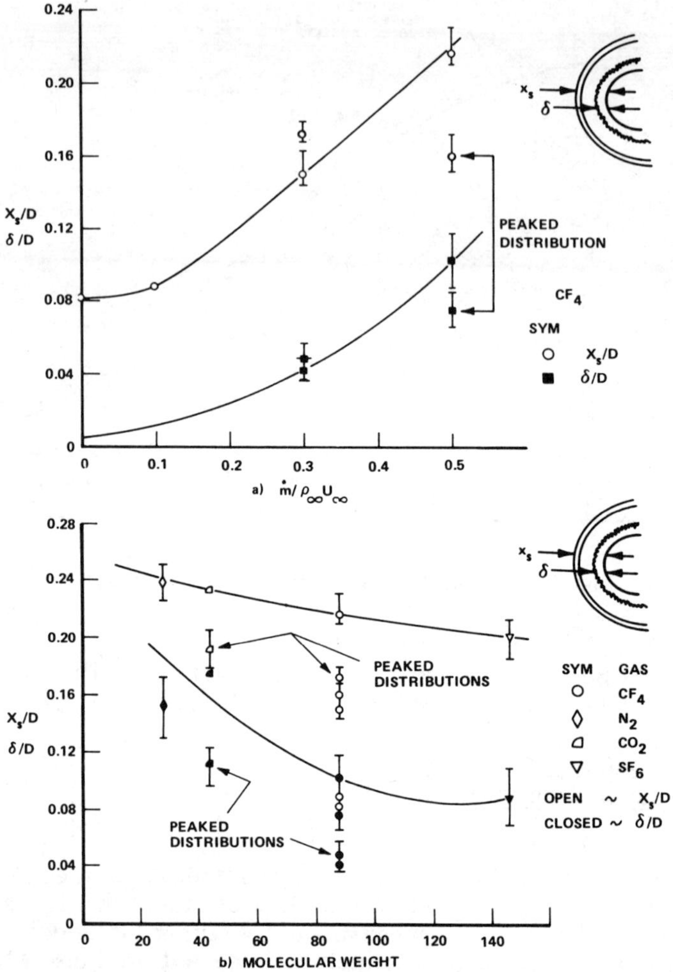

Fig. 17 Variation of shock standoff distance and mixing-layer thickness with mass injection rate and molecular weight of injectant.

measured shock standoff distance is in good agreement with theoretical predictions for an 80% H_2-20% He environment (see Fig. 17). The first photographic evidence of the size of structure of the boundary layer and its influence on the shock layer occurs for $\dot{m}/\rho_\infty\ u_\infty = 0.1$. The photographs shown in Fig. 18 indicate that the shock standoff distance has been increased and the boundary layer, which is just visible, is transitional and turbulent. Increasing the nondimensional blowing rate to 0.3 causes significant increase in the shock standoff distance, and at this condition the boundary layer/mixing layer is highly visible. Figure 19 was taken at a slightly different cutoff and shows that the bow shock has been displaced forward by 0.1 body diameters and the viscous layer now occupies one-third of the shock layer. It is clear from the photographs shown in Figs. 19 and 20 that the boundary layer is turbulent in the stagnation region and the scale of the turbulence layer is approximately equal to the thickness of the mixing layer. While the stability and shape of the bow shock does not appear to be strongly influenced by the turbulence over the body, it is clear that a major part of the subsonic inviscid flow in the shock layer is. When $\dot{m}/\rho_\infty u_\infty$ is increased to 0.5, the high-speed photographs (shown in Fig. 21) clearly demonstrate that both the stability and shape of the bow shock and the structure of the shock layer are strongly influenced by the turbulent mixing layer. The shock-layer flow is not grossly unstable; however, for CF_4 flows with $\dot{m}/\rho_\infty u_\infty > 0.3$, it is clear that a conventional boundary-layer approach could not be used in a theoretical model of these flows.

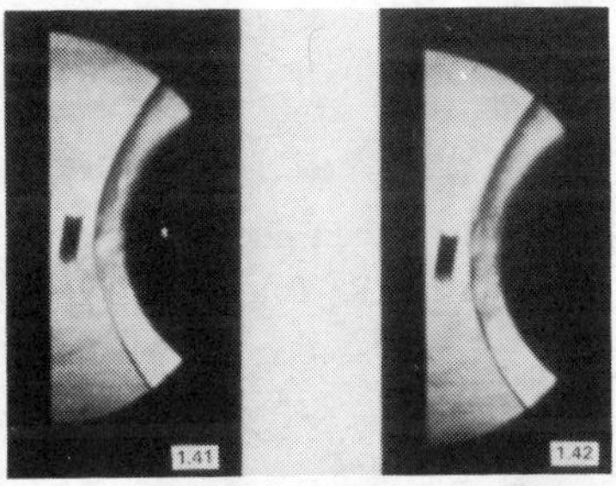

Fig. 18 Schlieren photographs of flow with CF_4 injection, $\dot{m}/\rho_\infty u_\infty = 0.1$.

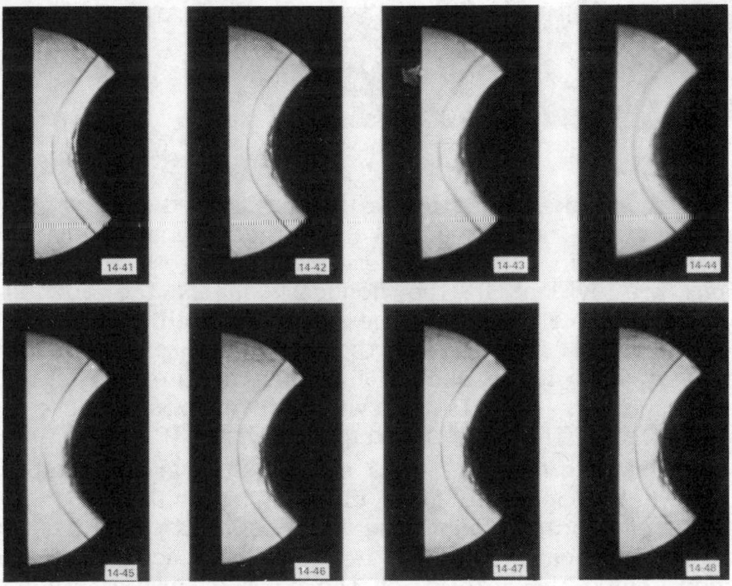

Fig. 19 Schlieren photographs of flow with CF_4 injection, $\dot{m}/\rho_\infty u_\infty = 0.3$.

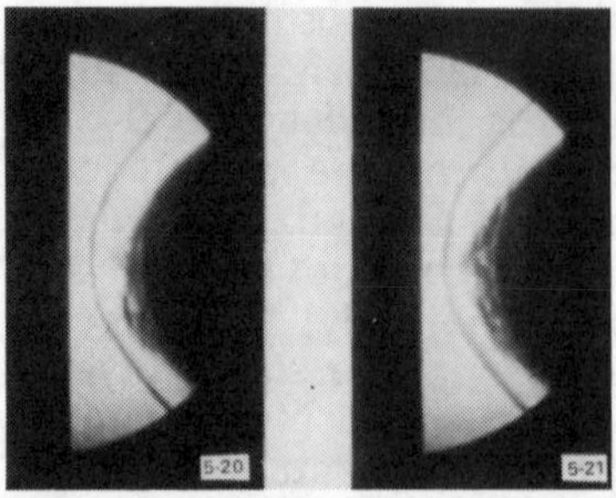

Fig. 20 Schlieren photographs of flow with CO_2 (peaked) injection, $\dot{m}/\rho_\infty u_\infty = 0.5$.

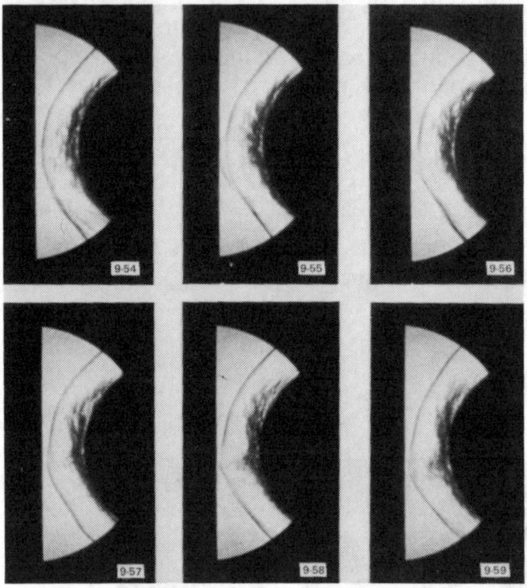

Fig. 21 Schlieren photographs of flow with CF_4 injection, $\dot{m}/\rho_\infty u_\infty = 0.5$.

Using the peaked injection profile prescribed in Fig. 16 with a maximum blowing rate of 0.5, we obtained the somewhat more stable flow shown in Fig. 22. Here, while large-scale turbulent fluctuations are evident in the boundary layer, the bow shock appears to be relatively stable. The relative stability of this latter flow encouraged us to perform the remaining experiments for $\dot{m}/\rho_\infty u_\infty = 0.7$ with a peaked injection profile rather than a uniform distribution. The latter, we believe, would cause gross flow instabilities. The high-speed photographs (Fig. 23) for the peaked injection with $\dot{m}/\rho_\infty u_\infty = 0.7$ clearly show gross instabilities exist, with what appears to be a circular instability mode. The aerodynamics of a nosetip exhibiting this instability could seriously jeopardize the mission on which it is flown. If a conical afterbody were placed behind the nosetip, on the basis of earlier studies by Holden,[12] we would expect that such a phenomenon would cause large and destabilizing effects on the cone stability.

Figure 17a shows the variation of shock standoff distance and the mean thickness of the boundary layer with the mass injection rate of CF_4 for both uniform and "peaked" injection patterns. With zero mass addition, the boundary layer is thin and, as mentioned earlier, the shock standoff distance is close to the theoretical "inviscid" value for a 80% H_2-20% He freestream. Both the mixing layer thickness and the shock standoff distance increase with increased blowing rate. When a blowing rate of 0.5 is obtained, the mixing layer occupies approximately one-half the thickness of the shear layer. These measurements demonstrate that despite the gross unsteadiness of the mixing layer, the bow shock is effectively displaced ahead of the body by the thickness of the mixing layer. Decreasing the molecular weight of the injectant at a fixed value of $\dot{m}/\rho_\infty u_\infty$ increases the volume of gas introduced into the shock layer and the forward displacement of the bow shock. Figure 17b shows the variation of shock- and mixing-layer thickness with molecular weight of the injectant. This figure shows that the lowest molecular weight fills over half the shock-layer thickness and is highly unsteady. When SF_6 was used as an injectant, the flow was steadier and the mixing layer occupied over one-third the thickness of the shock layer.

Rather than explore the effects of the molecular weight for $\dot{m}/\rho_\infty u_\infty = 0.7$ where the flow was grossly unstable, we chose to perform these studies for $\dot{m}/\rho_\infty u_\infty = 0.5$, where the flow was believed to be incipiently unstable. N_2, CO_2, and SF_6 were

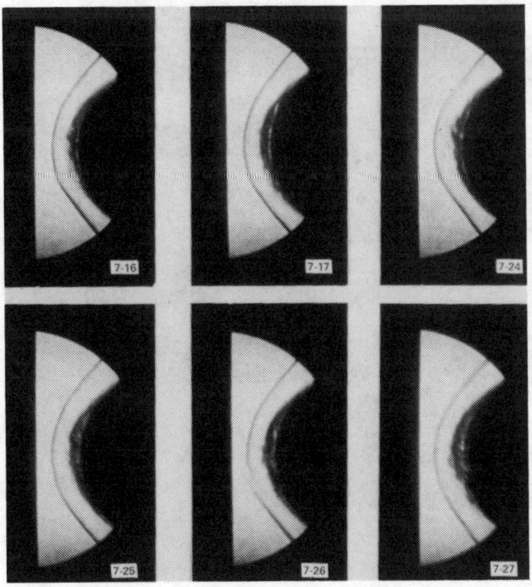

Fig. 22 Schlieren photographs of flow with CF_4 (peaked) injection, $\dot{m}/\rho_\infty u_\infty = 0.5$.

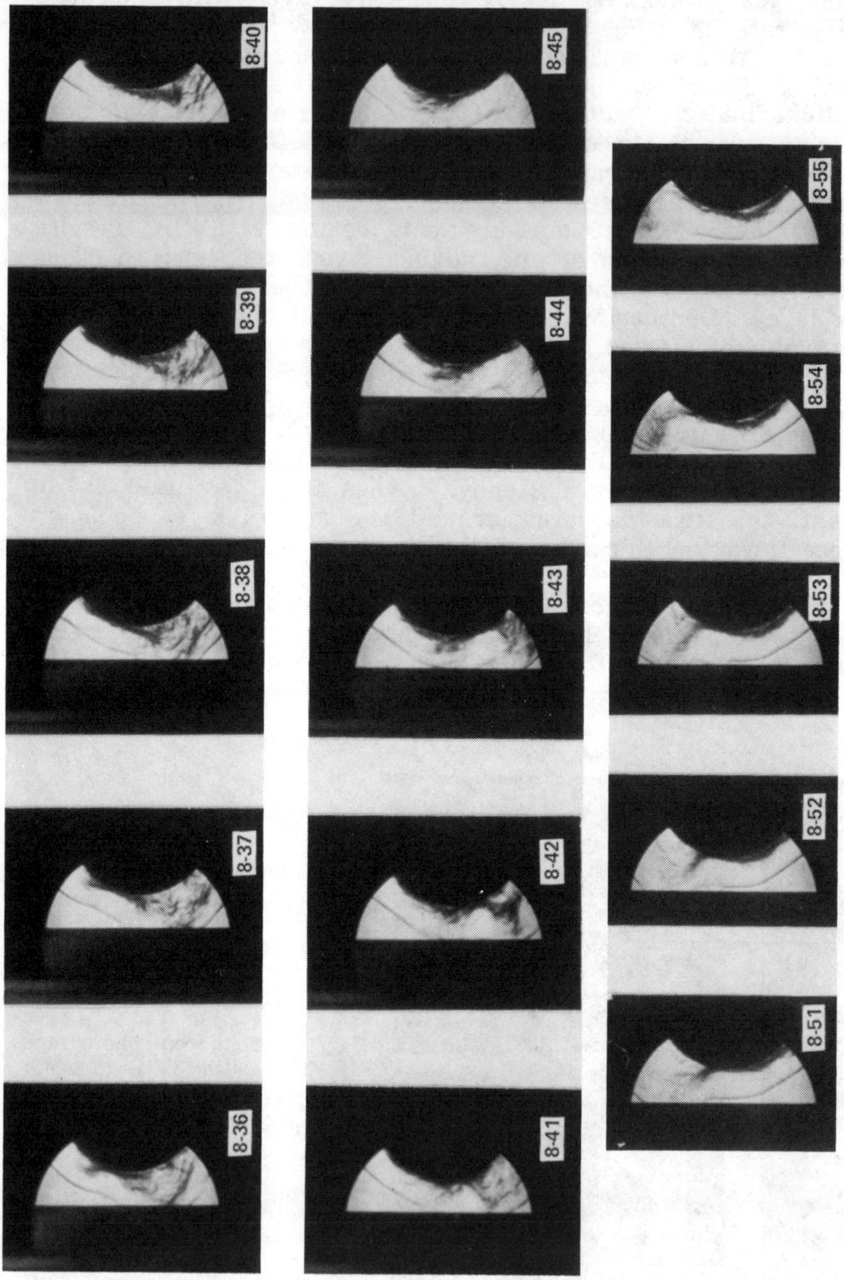

Fig. 23 Schlieren photographs of flow with CF_4 (peaked) injection, $\dot{m}/\rho_\infty u_\infty = 0.7$.

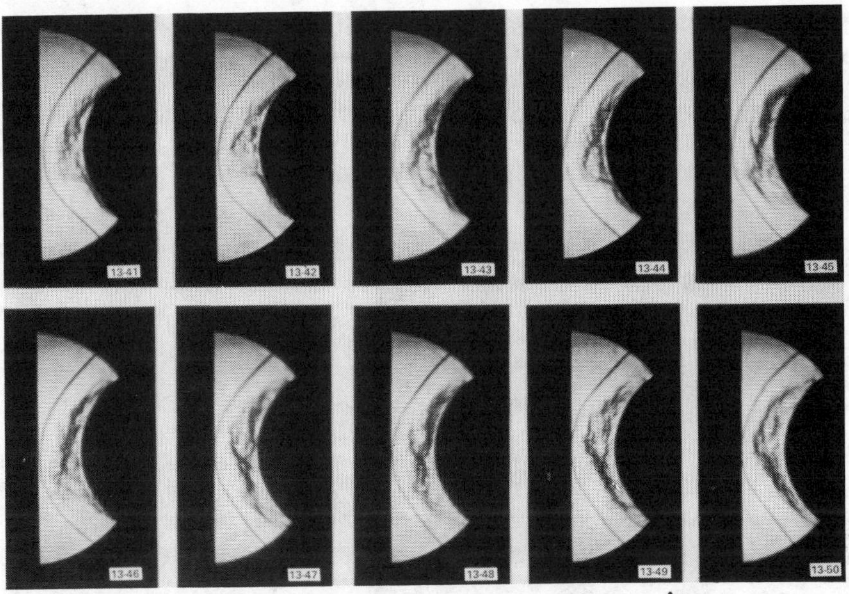

Fig. 24 Schlieren photographs of flow with N_2 injection, $\dot{m}/\rho_\infty u_\infty = 0.5$.

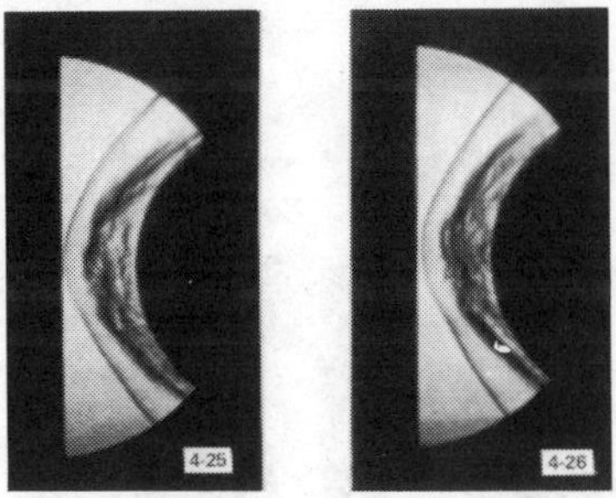

Fig. 25 Schlieren photographs of flow with CO_2 (uniform) injection, $\dot{m}/\rho_\infty u_\infty = 0.5$.

employed as injectants with a uniform injection distribution. In addition, the flow for a "peaked" CO_2 distribution was examined. Comparing the high-speed photographs of the flow for N_2, CO_2, and SF_6 injectants, shown in Figs. 24, 25, and 26, respectively, with those for CF_4 in Fig. 21, it can be seen that both the structure of the bow shock and the shear layer become increasingly stable as the molecular weight of the injectant increases. We believe that this results directly from the inverse relationship between the volume of gas injected through the nosetip and its molecular weight for a given $\dot{m}/\rho_\infty u_\infty$. Clearly, a much greater volume of N_2 than SF_6 is being injected for the same value of $\dot{m}/\rho_\infty u_\infty$. Again, as shown in Fig. 27 for CO_2, we found the flowfields over the models with the "peaked" distribution were significantly more stable than those with uniform blowing. It is clear from these studies that for flows with $\dot{m}/\rho_\infty u_\infty > 0.3$, any theoretical modeling must acknowledge the first-order relationships between the shock structure and the structure of the turbulent mixing layer. The interaction between the bow shock and mixing layer, combined with the highly turbulent nature of the mixing

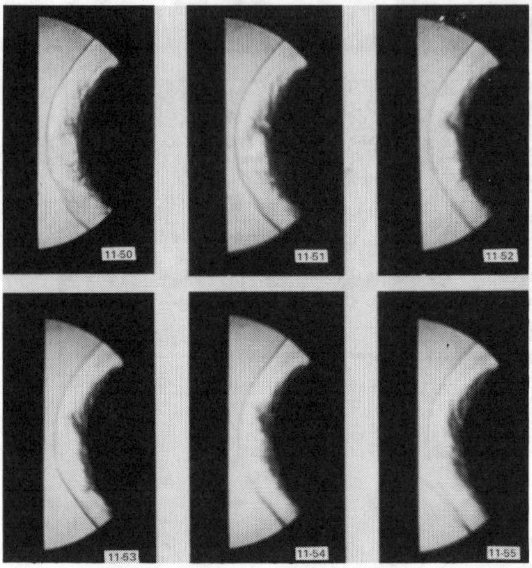

Fig. 26 Schlieren photographs of flow with SF_6 injection, $\dot{m}/\rho_\infty u_\infty = 0.5$.

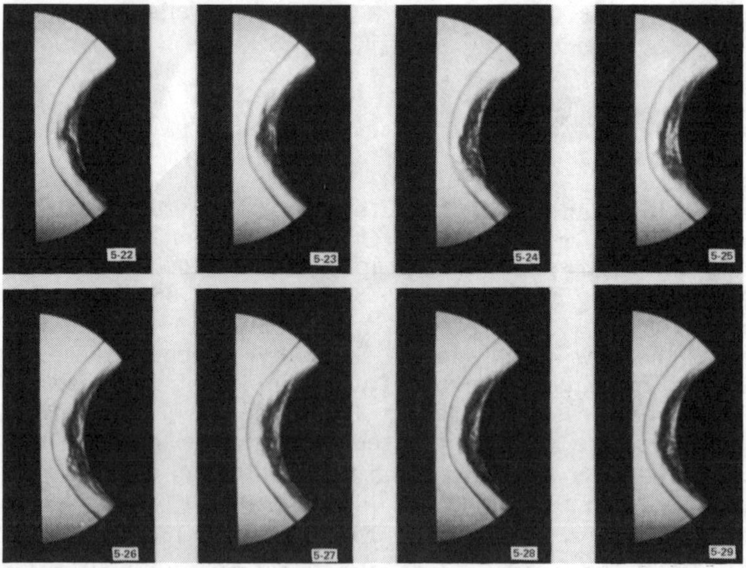

Fig. 27 Schlieren photographs of flow with CO_2 (peaked) injection, $\dot{m}/\rho_\infty u_\infty = 0.5$.

layer, is a major reason for the relatively poor thermal protection of the body by the massive amounts of injectant and will be described in the following section.

Distribution of Heating in the Presence of Massive Blowing

The earlier studies of the transpiration cooling of flat plates and cones and, most recently, the studies of spherical nosetips used gaseous injectants whose molecular weights were significantly smaller than those of the freestream, because such gases make the most efficient coolants. However, it was also recognized in these studies that flow instabilities can occur at relatively low blowing rates. In the present study, we used gases which were very heavy compared with the freestream, to obtain the correct Jovian simulation. While these transpiration-cooled flows should be intrinsically more stable, they also might be expected to be less efficient.

The heat-transfer measurements on the smooth and nonblowing models demonstrated, as described earlier, that the flows over both models were fully laminar. Introducing mass into the boundary layer perpendicular to the surface is destabilizing, particularly in the stagnation region, and, for the injection rates

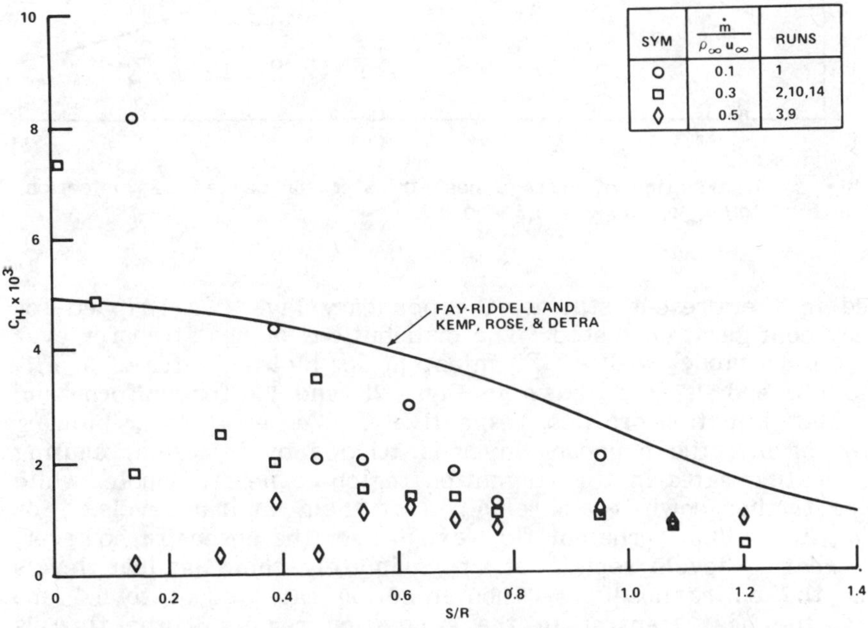

Fig. 28 Variation of heat-transfer distribution with blowing rate for uniform CF_4 injection.

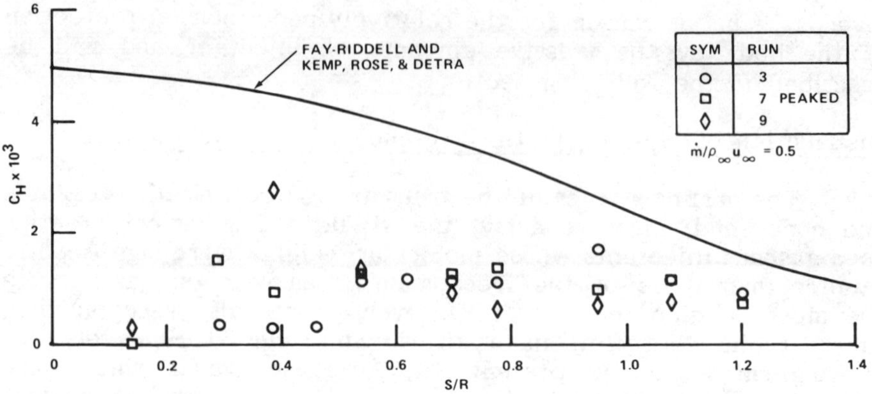

Fig. 29 Distribution of heating with uniform and peaked injection with CF_4 for $(\dot{m}/\rho_\infty u_\infty)_{MAX}$ = 0.5.

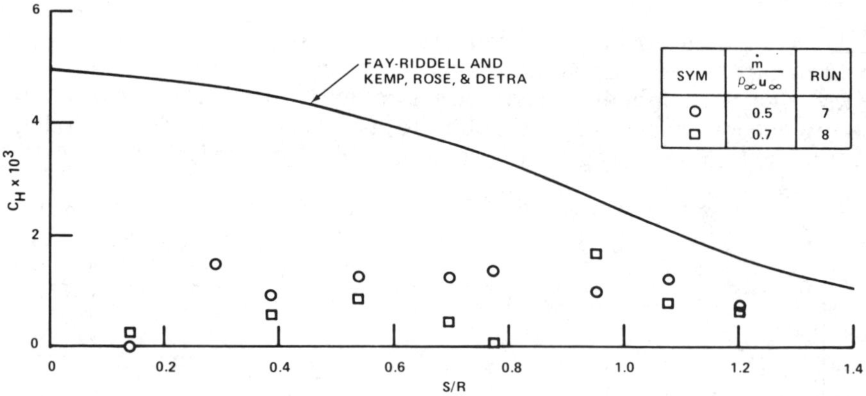

Fig. 30 Distribution of average heat transfer for peaked CF_4 injection rates of $(\dot{m}/\rho_\infty u_\infty)_{MAX}$ = 0.5 and 0.7.

used in the present studies, the boundary layer was tripped for every configuration tested. The distributions of heat transfer over the porous model with a CF_4 injectant for blowing rates of 0, 0.1, 0.3, 0.5, and 0.7 are shown in Figs. 28 and 29 for uniform and "peaked" injection profiles, respectively. We see that for blowing rates of 0.1, the boundary layer is tripped by injection, causing the heating rates in the stagnation region to nearly double, while those further downstream remained at their laminar levels. For $\dot{m}/\rho_\infty u_\infty$ = 0.3, turbulent flow exists over the nosecaps; however, the heating levels remain at approximately their laminar levels over the entire nosetip. When injection rates equal to 0.5 are used, the heat transfer to the stagnation region of the flow is greatly reduced for both uniform and "peaked" injection profiles

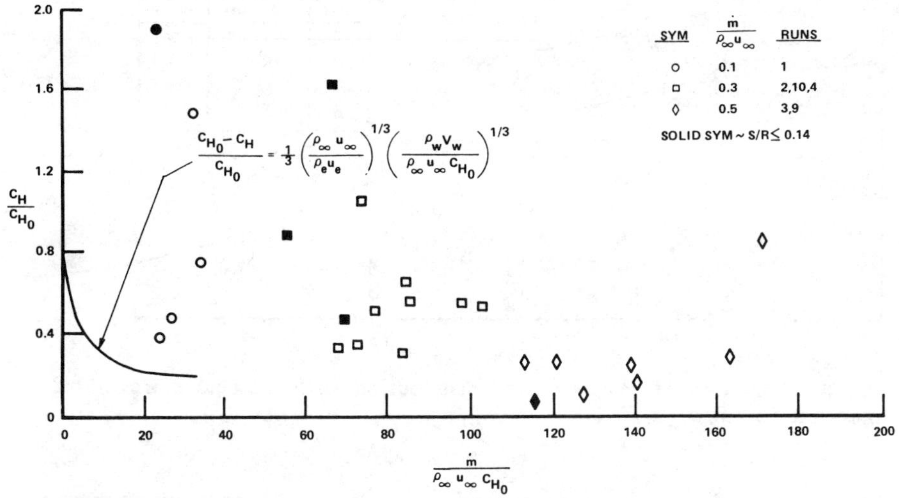

Fig. 31 Variation of normalized heating rate with blowing parameter for uniform CF_4 injection.

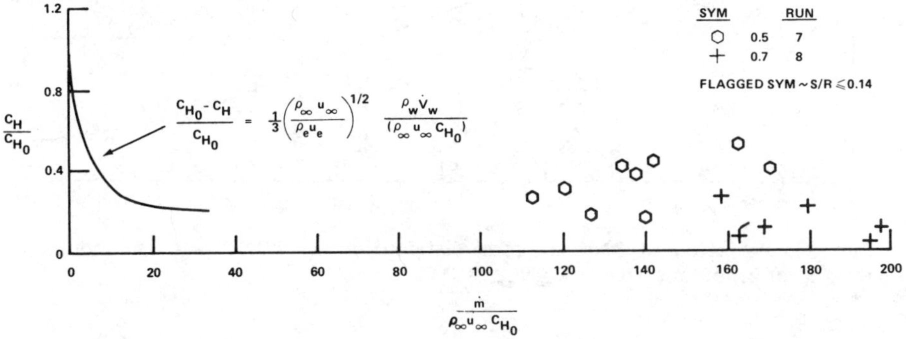

Fig. 32 Variation of normalized heating rate with blowing parameter for peaked CF_4 injection.

(see Figs. 29 and 30); however, further downstream, transpiration cooling proved less effective.

These measurements, normalized by the laminar nonblowing heating, are also shown in Figs. 31 and 32. It is readily apparent that to achieve reductions in heating of a magnitude similar to the "light gas studies" (represented by the correlation formula), significantly larger nondimensional blowing rates are required and that at these large blowing rates, the flows are again highly unstable. The measurements with the peaked distributions at the high injection rates (shown in Fig. 32) also demonstrate that it is

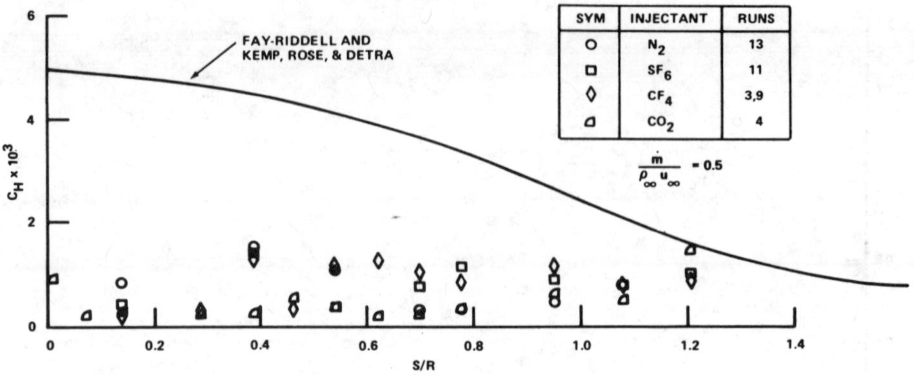

Fig. 33 Variation of heat-transfer distribution with molecular weight of injectant for $\dot{m}/\rho_\infty u_\infty = 0.5$.

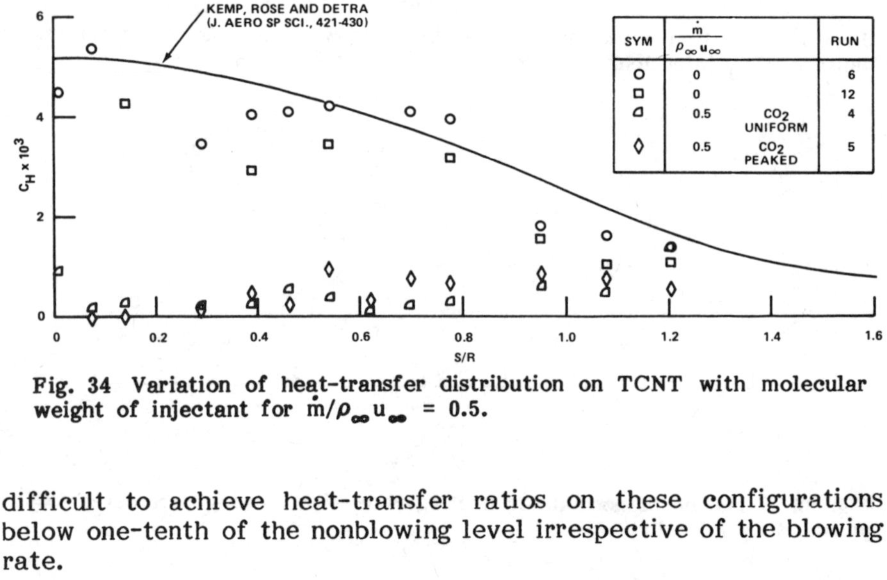

Fig. 34 Variation of heat-transfer distribution on TCNT with molecular weight of injectant for $\dot{m}/\rho_\infty u_\infty = 0.5$.

difficult to achieve heat-transfer ratios on these configurations below one-tenth of the nonblowing level irrespective of the blowing rate.

A comparison between the heat-transfer distributions for the gaseous injectants N_2, CO_2, CF_4, and SF_6 with a common blowing rate of 0.5 is shown in Fig. 33. We find that the gases with the lower molecular weight are more effective as coolants, despite the greater instability of the shock layer for these lighter gases. The heat-transfer measurements for the uniform and peaked CO_2 distributions are shown in Figs. 34 and 35. Both sets of data show a 90% reduction in heating over the nosetip, which is in agreement with the earlier measurements. As one might have anticipated from the earlier tests, the uniform distribution is more effective in reducing heat transfer than the peaked distribution;

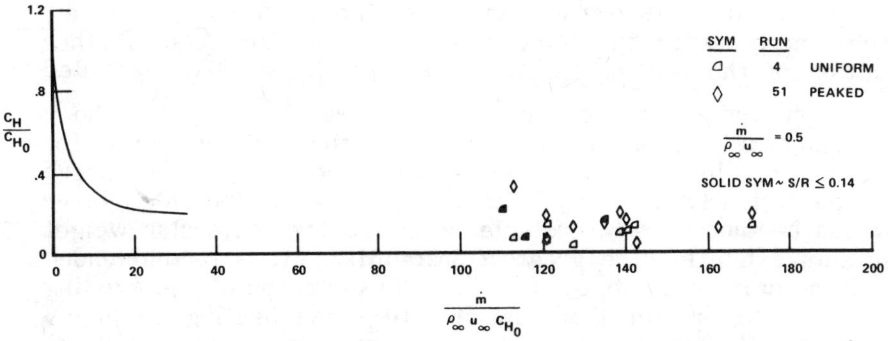

Fig. 35 Variation of nondimensional heating with blowing parameter for uniform and peaked CO_2 injection.

however, since significantly less gas is involved in cooling the nosetip when a peaked distribution is employed, this latter method of introducing the gas must be considered the most efficient. Considering the unstable nature of the flows depicted in the high-speed photographs (see Figs. 16-26) with $\dot{m}/\rho_\infty u_\infty > 0.5$, it is perhaps surprising that 90% reductions in heating can occur. However, at the large injection rates, a large amount of cool gas is being deposited in front of the body.

Conclusions

An experimental study has been conducted to examine the structure and stability of the shock layer and the distribution of heating to a highly blowing blunt body in an environment simulating the fluid dynamics of Jovian entry. High-speed Schlieren photography was employed to examine the structure and stability of the shock layer, heat transfer, and pressure measurements made on a porous blunt nosetip, for blowing rates $0 < \dot{m}/\rho_\infty u_\infty < 0.7$ in a 80% H_2-20% He gas flow. In this investigation, which was conducted in the Calspan 96-inch Shock Tunnel at Mach 12, CF_4 was used as the principal injectant with N_2, CO_2, and SF_6 used in additional selected experiments. The initial studies demonstrated that it was possible to generate an environment in the shock tunnel that simulated the fluid dynamic aspects of the Jovian environment. Flow durations of close to 8 ms were generated for the H_2-He test gas used in the shock tunnel; this is four times that required to establish flow over the highly blowing models. Heat-transfer measurements both on a smooth replica of the porous nosetip and on the nonblowing porous model indicated that the flow over both these configurations was fully laminar; however, we observed that for low blowing rates, the heat transfer was enhanced when fluid injection tripped the boundary layer. For blowing rates between

0.1 and 0.5, the heat transfer to the surface decreased; however, further increases in the blowing rate did not result in further reduction of the heating levels below C_H/C_{H_0} = 0.1. Detailed photography indicated that for $\dot{m}/\rho_\infty u_\infty$ less than 0.3, the shock layer remains stable, although a description of the flows for $\dot{m}/\rho_\infty u_\infty > 0.1$ indicates that conventional boundary-layer theory could be seriously in error. For $\dot{m}/\rho_\infty u_\infty > 0.5$, the shock-layer flow can become grossly unstable, with the low molecular weight injectants exhibiting the greatest instability. It is recommended that these measurements be used in an examination of the existing codes used to predict the flow structure and heating on highly blowing nosetips.

References

[1]Moss, J.N., Anderson, E.C., and Bolz, C.W., Jr., "Aerothermal Environment for Jupiter Entry Probes," Progress in Astronautics and Aeronautics: Thermophysics of Spacecraft and Outer Planet Entry Probes, Vol. 56, edited by A.M. Smith, AIAA, New York, 1977, pp. 333-354.

[2]Moss, J.N., Jones, J.J., and Simmonds, A.L., "Radiative Flux Penetration Through a Blown Shock Layer for Jupiter Entry Conditions," Progress in Astronautics and Aeronautics: Outer Planet Entry Heating and Thermal Protection, Vol. 64, edited by R. Viskanta, AIAA, New York, 1979, pp. 22-41.

[3]Holden, M.S., "Studies of Transpiration Cooling, Surface Roughness and Entropy Swallowing in Transitional and Turbulent Boundary Layers over Nose Tips," Paper presented at IAF-79-F-42, XXX Congress International Astronautical Federation, Sept. 17-22, 1979.

[4]Brunner, M.J., "Transpiration Cooling Tests at a Sharp Sphere-Cone in a Rocket Exhaust Using N_2, H_e and H_2O Coolants," Paper 64-WA/Ht-50 presented at ASME Winter Annual Meeting, New York, 1964.

[5]Pappas, C.C. and Okuno, A.F., "Measurements of Heat Transfer and Recovery Factor of a Compressible Turbulent Boundary-Layer on a Sharp Cone with Foreign Gas Injection," NASA TN D-2230, 1964.

[6]Fogaroli, R.P., "Measurements of Turbulent Heat Transfer and Skin Friction Reduction on a Porous Cone with Air Injection at High Mach Numbers," General Electric Rep. 64SD5291, 1964.

[7]Bartle, E.R. and Leadon, B.M., "The Effectiveness as a Universal Measure of Mass-Transfer Cooling for a Turbulent Boundary-Layer," Proceedings of 1962 Heat Transfer and Fluid Mechanics Institute, Stanford University Press, Stanford, Calif., 1962.

[8]Danberg, J.E., "Characteristics of the Turbulent Boundary-Layer with Heat and Mass Transfer at Mach Number 6.7," Proceedings 5th U.S. Navy Symposium on Aeroballistics, White Oak, Md., 1961.

[9]Spalding, D.B. and Chi, S.W., "The Drag of a Compressible Turbulent Boundary Layer on a Smooth Flat Plate with and without Heat Transfer," Journal of Fluid Mechanics, Vol. 18, 1964, pp. 117-143.

[10]Van Driest, E.R., "Problem of Aerodynamic Heating," Aeronautical Engineering Review, Vol. 15, No. 10, 1956.

[11]"Calspan Hypersonic Shock Tunnel, Description and Capabilities Brochure," 1975.

[12]Holden, M.S., "Accurate Vehicle Experimental Dynamics Program — Studies of Aero-Thermodynamic Phenomena Influencing the Performance of Hypersonic Re-Entry Vehicles," Calspan Rep. AB-6072-A-2, SAMSO-TR-79-47, Aug. 1979.

Chapter V Future Planetary Missions

Trends in Unmanned Planetary Entry

J.R. French*
Jet Propulsion Laboratory,
California Institute of Technology, Pasadena, Calif.

Abstract

Entry systems used in unmanned planetary exploration have followed a steady trend toward higher velocity and greater complexity. Simple unguided probes still have a major place in future plans, but the demands of exploring the giant planets are driving these vehicles into ever more demanding entry environments. Concurrently, the increased interest in the use of aerodynamic capture into orbit (aerocapture) results in applications for high lift-to-drag ratio guided vehicles. Controlled orbit decay (aerobraking) requires improved understanding of rarefied flow environments.

Introduction

To no one's surprise, the trends in unmanned planetary entry systems can best be capsulized as "faster and more complicated." Figure 1 shows the trend from the first planetary entry systems for Mars through the currently planned Galileo and postulated outer planet probes. Concurrently, the trend has been from pure ballistic, through unguided lifting, to possible application of highly maneuverable, actively guided vehicles. In this paper we consider requirements for probes, entry-to-landing systems, and orbit capture and modification systems.

Probes

Atmospheric probes in general tend to be pure ballistic vehicles since there is no requirement for

---

Presented as Paper 81-1125 at the AIAA 16th Thermophysics Conference, Palo Alto, Calif., June 23-25, 1981.

*Manager, Advanced Planetary Studies.

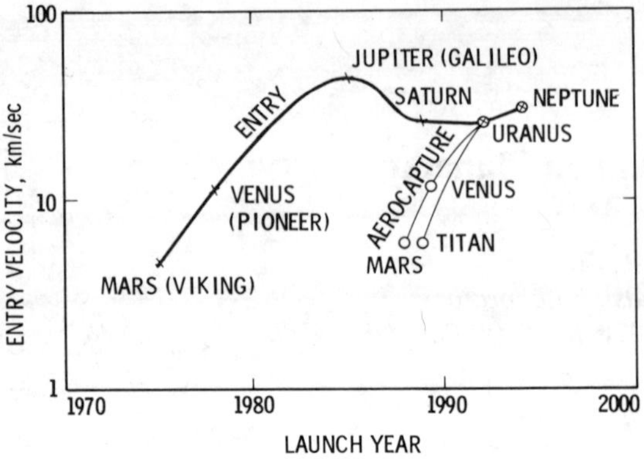

Fig. 1 U.S. entry system trends.

pinpoint accuracy and the high-gravity loads are not usually a problem for the vehicles. Steep entries and high deceleration rates generally characterize such vehicles. An atmospheric probe is designed to obtain a vertical profile of the atmosphere in a relatively short time. Survival of surface contact is a secondary goal at best. Atmospheric probes tend to be simple vehicles in terms of guidance and control, with orientation and targeting accomplished by the carrier vehicle. The configuration of choice for such vehicles has been the high-angle blunted cone.

Although atmospheric probes were considered early in the planetary program, especially for Mars, it was not until 1978 that the United States launched its first such vehicle. The Pioneer Venus mission featured four probes, one large and three small, and in fact the carrier vehicle functioned as an upper atmosphere probe before burning up in the denser regions of the Venus atmosphere.

Atmospheric probes figure prominently in plans for future exploration of the outer planets. The presently on-going Galileo project involves a Jupiter orbiter and a probe of the Jovian atmosphere. This mission in some ways presents the most demanding entry requirements yet encountered. Entry will take place at a velocity of 48 km/s and an angle of -7 to -10 deg into an atmosphere composed largely of hydrogen. Only slightly less demanding are the other gas giant planets, Saturn, Uranus, and Neptune. Each of these presents its own set of peculiarities. By contrast the other major target of atmospheric probes, Titan, the major satellite of Saturn, may be the least demanding target in the entire solar system.

Because of many factors, mostly launch vehicle changes and budgetary problems, the Galileo mission has undergone numerous changes. Originally a 1982 launch of a single orbiter carrying a probe, Galileo became a 1984 launch of an orbiter and a separately launched probe/probe carrier; now it has evolved to a 1985 launch, again of a combined vehicle. Despite the changes, the probe mission remains essentially the same. The spin-stabilized probe is separated from its carrier vehicle some 50 days prior to closest approach. The probe will enter the atmosphere of Jupiter at full hyperbolic velocity. Table 1 summarizes the entry environment for the Jupiter probe. This probe will send back data, via a relay through its carrier, until it reaches the 20 bar (20×10^5 Pa) pressure level. The necessity to provide this relay function places substantial geometric constraints on trajectory and spacecraft design, especially for the combined launch where the carrier is also an orbiter and must be properly positioned and oriented for an orbit insertion burn just after the relay period.

The other outer planets present similar problems. None present the challenge of Jupiter in terms of entry velocity since, although approach velocities tend to be higher, the less massive planets contribute less to the final velocity. Entry conditions for Saturn, Uranus, and Neptune are summarized in Table 1. The major problems that arise in probing these planets involve the avoidance of known or suspected rings and the need for the probe to

Table 1 Outer planet probe summary

Planet	Entry velocity, km/s	Entry flight path angle, deg	Approx. communication range, km	Probe communication aspect angle, deg
Jupiter	48	-7 to -10	250,000	15
Saturn	27	-30	105,000	18
Uranus	22	-50	55,000	29
Neptune	24.5	-40	50,000[a]	50[a]
			200,000[b]	30[b]
Titan				
Direct	7	-30	<50,000[c]	TBD[c]
Saturn orbit	3.5	-3.0	<30,000[c]	TBD[c]

[a]Prograde orbiter.
[b]Retrograde orbiter.
[c]Entry time 0.5-2 h.

reach deeper (high-pressure) levels in order to be sure of reaching the water cloud level. Other constraints such as satellite flybys and the peculiar nature of each system further complicate the design. Figures 2 and 3 show typical geometries for outer planet probe missions.

Paradoxically, the easiest entry environment also lies in the outer solar system. Titan, the large moon of Saturn, has an extensive, dense atmosphere composed mostly of nitrogen. Because Titan itself is small, it contributes little to the velocity at entry and, because Titan is a substantial distance from Saturn, the primary does not contribute greatly either. As a result the entry velocity is a little higher than the hyperbolic approach velocity. If the probe is delivered from elliptical orbit about Saturn, the entry velocity becomes still more benign. Table 1 presents the data for Titan. The most difficult aspect of the Titan probe is guidance. For the case of direct hyperbolic entry, onboard optical guidance will almost certainly be required because of the small size of the target and ephemeris uncertainty.

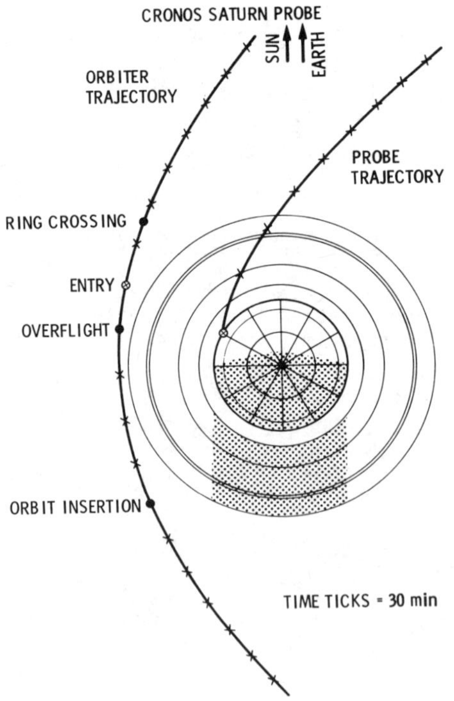

Fig. 2 Orbiter-probe overflight, view from Saturn north pole.

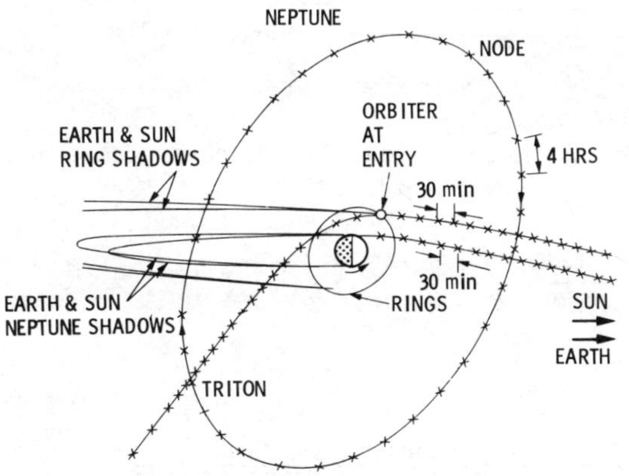

Fig. 3 Neptune 14 year mission prograde probe/flyby, view from north trajectory pole.

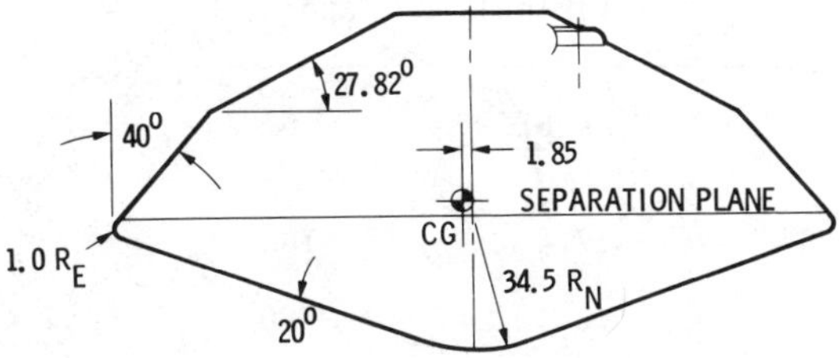

Fig. 4 Viking entry system.

Entry to Landing

The single example of this class in the U.S. planetary program is the Viking entry system. The configuration in Viking was a high-angle blunted cone (Fig. 4). This vehicle was trimmed via c.g. location to achieve a modest lift-to-drag ratio (L/D) of about 0.18. No effort was made to modulate the lift, entry being effected with a fixed attitude. Entry was initiated from an elliptical orbit about Mars with entry conditions as defined in Table 2. It can be seen that Mars entry is relatively benign either from orbit or directly from a hyperbolic flyby as is also shown in Table 2. The challenge in Mars entry to landing for the future will lie in improved accuracy. For the Viking mission, the landing ellipse length was on the order of 300 km. This was mostly due to orbit uncertainties, the accumulation of errors after separation from the orbiter, and the fact that no effort was made to modify the trajectory by lift modulation. Studies in support of the Mars Rover and Mars Sample Return missions have indicated that a Viking-type aeroshell employing lift modulation via roll

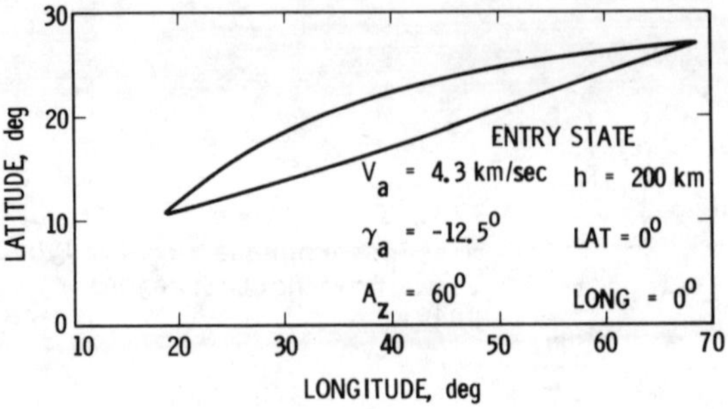

Fig. 5 Maneuver footprint for L/D = 0.3, W/C_DA = 350 kg/m^2.

Table 2 Mars entry conditions

Parameter	Elliptical orbit (Viking)	Hyperbolic approach
Entry velocity, km/s	4.62	6
Entry flight path angle, deg	-17.4	TBD
Trim L/D	0.18	TBD
Max dynamic pressure, Pa	6890	TBD

could reduce the landing ellipse to the order of 25 x 40 km. (It can be reduced to about 60 x 80 km by improved tracking and procedures.) The cross-range maneuverability of such vehicles is quite limited, restricting landing sites to those essentially in the orbit plane (Fig. 5).

Application of modern entry system and guidance technology to Mars landing systems offers substantial improvement. Moderate L/D (1 - 2) systems derived from modern military system technology render the entire planet accessible from easily attainable orbits (Fig. 6) or direct entry (Fig. 7) with landing ellipses of 19 x 18 km for the orbital case and 21 x 17 km for the direct. The maneuverability of these systems plus map matching to terrain features for guidance initialization can reduce error to the point that map error (say, 5 km) is the dominant error source. Such accuracy allows precise selection of sample sites or rendezvous with a rover. Application for manned missions, especially permanent bases, is obvious.

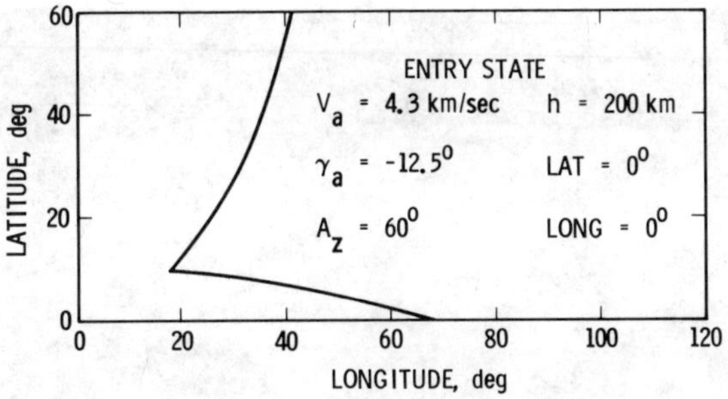

Fig. 6 Maneuver footprint for L/D = 1.4, W/C_DA = 500 km/m^2.

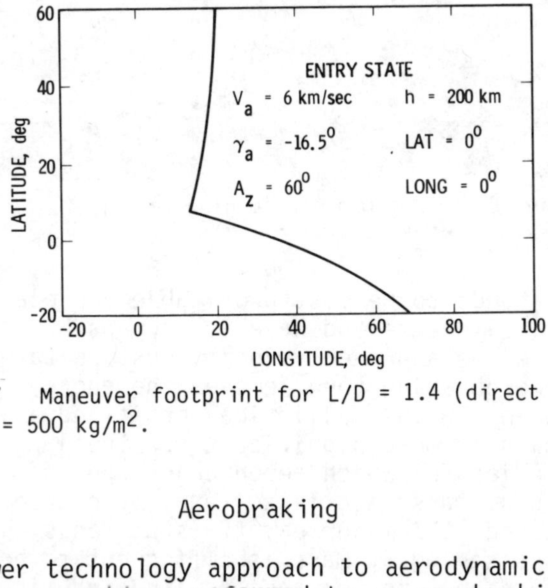

Fig. 7 Maneuver footprint for L/D = 1.4 (direct entry), W/C_DA = 500 kg/m^2.

Aerobraking

The lower technology approach to aerodynamic orbit modification, sometimes referred to as aerobraking, involves relatively slow circularization of an initial elliptical orbit by repeated passes through the upper atmosphere. Keeping periapsis in the free molecular-to-continuum transition region allows relatively simple drag devices of high-temperature metal foil and minimum impact on the spacecraft configuration. On the other hand, it is time consuming and does require expenditure of propellant (typical ΔV = 1 km/s) to achieve an initial elliptical orbit. The large area required to achieve reasonable circularization times will typically require deployable brake structures. This concept is under active consideration for the Venus Orbiter imaging radar mission (Fig. 8).

Fig. 8 Venus Orbiter imaging radar.

Aerobraking tends to be most profitable for planets which are relatively large and have low approach velocities, since the approach velocity must be taken out propulsively. Aerobraking then removes the energy difference between a highly elliptical orbit and a low circular one which is more significant for the larger planets. The bodies for which aerobraking appears feasible are Venus, Mars, and Titan (and, of course, Earth). As expected, the major payoff is at Venus, although the mass saving at Mars is sufficient to be attractive for a highly performance-constrained mission. It is of doubtful validity for Titan. Table 3 illustrates operating environments at the various targets.

The major attractions of aerobraking are the modest technology needs, small impact on spacecraft design, and flexibility. Just as much aerobraking can be done as is required to meet launch mass constraints. This can range from aerobraking to circular orbit from a barely closed ellipse all the way down to very modest changes in orbit. This flexibility essentially eliminates constraints on arrival velocity, thus allowing the launch and interplanet trajectory to be optimized for maximum launch performance. The basic feasibility of the operation has been demonstrated by the Atmospheric Explorer and Pioneer Venus Orbiter. The major weakness of the concept is the difficulty of accurately analyzing flowfields in the transition regime.

Table 3 Aerobraking environments

	Target			
Parameter	Venus	Earth	Mars	Titan
Peak stagnation pressure, first pass, Pa	2.90	3.39	0.67	0.15
Peak heating rate, first pass, W/cm^2	1.43	1.80	0.16	0.016
Time-integrated heating, first pass, J/cm^2	93.4	139.1	20.6	12.0
Time-integrated heating, last pass, J/cm^2	183	272	34.2	13.2
Fall in apoapsis altitude, last pass, km	70	100	33	70
Fall in periapsis altitude, first pass, km	0.0044	0.0087	0.0042	0.134
Fall in periapsis altitude, last pass, km	0.42	0.90	0.30	0.80

[a]All data for initial elliptical orbits of 24 h period; density at periapsis altitude 3×10^{-8} kg/m^3.

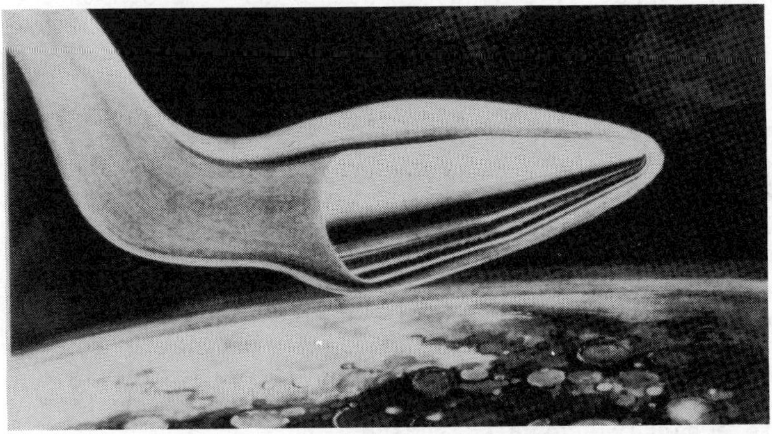

Fig. 9 Aerocapture vehicle.

Aerocapture

A more technologically demanding but higher payoff concept is that of aerocapture. This approach makes use of moderate L/D entry vehicles ($1 < L/D < 2$) which, using controlled flight paths through the atmosphere, can go directly from a hyperbolic flyby trajectory to low circular orbit. Aerodynamic shapes based upon modern missile entry system technology offer excellent potential for these applications (Fig. 9). Approach guidance requirements, while more stringent than for propulsive insertion, seem well within current capability. The same type of vehicle shows excellent potential for precision landing as well.

Aerocapture at Mars is well within current technology, the environment being in fact more benign than entry from Earth orbit. Venus is more demanding but still not beyond current technology. Somewhat surprisingly, insertion into Saturn orbit via aerocapture is not highly demanding. The secret is to use the atmosphere of Titan rather than Saturn. Aerocapture using the atmosphere of the giant planets is probably beyond current capability, although Uranus and Neptune offer some promise.

The most extensive studies of aerocapture in recent years have involved Mars. The concept has been around for many years and was studied in conjunction with manned Mars missions. The most recent resurgence of interest was generated in JPL studies of unmanned Mars Sample Return missions. These missions required more mass than could be delivered by a single Shuttle/IUS. Aerocapture was suggested as a means of eliminating most of the propellant required for the Mars orbit capture. Because of the benign entry conditions (never in excess of 6 km/s), aerocapture at Mars was clearly well within the capability of lightweight thermal protection systems. In order to achieve highly accurate orbit insertion, higher L/D configurations than the blunt Viking/Apollo shapes are required. The biconic shapes ($L/D = 1 - 2$) developed for maneuvering warheads fit the desired criteria very well. These shapes lend themselves to the large cross range and accurate targeting for landing discussed earlier. This leads to the concept of a double-duty vehicle which first inserts into orbit aerodynamically and then, after a landing site survey, performs the final entry to a highly accurate landing.

At the same time interest developed in aerocapture at other bodies. Venus is an obvious target and has been studied to some degree although, at the moment, aerobraking generates the most interest for Venus since it is so well adapted. The most interesting and surprising result involves the Saturn/Titan system. Achieving orbit about Saturn with adequate payloads is difficult, especially with short flight times. Orbiting Titan is essentially impossible with present propulsion technology. Aerocapture makes both achievements possible, not by using Saturn's atmosphere (possible but technically very demanding) but by using that of Titan. Titan's nitrogen atmosphere is dense but, because of the low gravity, quite extended. Thus it is ideally suited to aerocapture. Depending upon flight times, planetary positioning, and Titan's location, entry conditions for aerocapture range from being comparable to Mars to only slightly worse than Venus. The ΔV can be selected to achieve a variety of Saturn orbit characteristics or an orbit about Titan itself, all for essentially the same mass penalty.

A somewhat more speculative application of the aerocapture concept is in the atmospheres of the giant planets. Jupiter probably exceeds our technical capability, and Saturn is at least marginal (and unnecessary because of Titan). Uranus and Neptune may be possible targets but are clearly far more demanding than the terrestrial planets or Titan. Figure 10 compares the

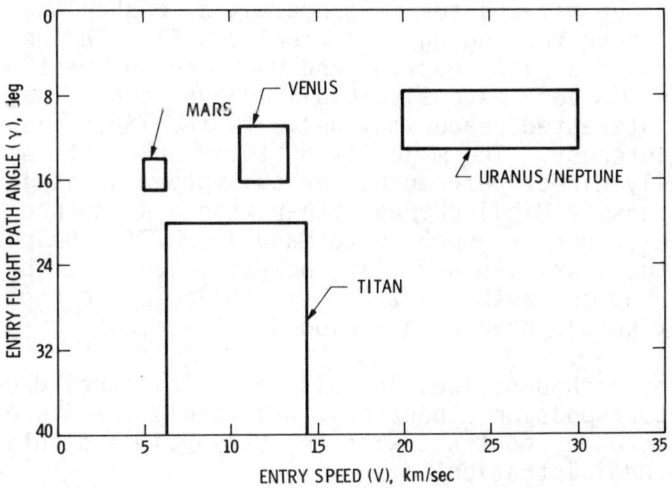

Fig. 10 Air relative planetary entry V-γ requirements for aerocapture.

entry angles and velocities for the major aerocapture targets.

Because the focus of this paper is based upon entry requirements, there is no attempt to present performance data to justify the need for aerocapture or aerobraking. Suffice to say, use of these techniques can reduce the mass penalty for achieving desired orbits from the 50-90% range (propulsive) to perhaps 15% of the approach mass. The impact of this ranges from improving margins to enabling otherwise impossible missions.

Conclusions

The field of planetary entry systems continues to provide interesting and exciting challenges. Probes of the outer planets will push heat-shield design to the state-of-the-art limits. Demands for increased payload and limitations on launch vehicle capability will generate a demand for the substantial performance improvements offered by aerobraking and aerocapture. Advanced exploration of the Martian surface, whether manned or unmanned, will demand the accurate landing offered by highly maneuverable entry systems. If the United States is to maintain more than a minimal planetary program, these advancements in entry systems will be mandatory.

Note

This paper was written deliberately as a shallow overview rather than going into great detail. The goal was to provide an introduction and whet the interest of the reader. A very extensive Bibliography is provided so that the interested reader may delve further into his areas of interest. The majority of these documents are not actually direct references for this paper so it is in the true sense a Bibliography rather than a Reference list. The author is indebted to Manuel Cruz for helping assemble the list. Such a list, however, can never be complete, and the author extends his apologies to those whose work should have been included but was not.

The research described in this paper was carried out by the Jet Propulsion Laboratory, California Institute of Technology, under contract with the National Aeronautics and Space Administration.

Bibliography

Armento, R., "Mars Aerocapture Vehicle Definition Study Final Report," G.E. Rept. 79SDR2258, Aug. 17, 1979.

Barbieri, R. W. and Cooley, J. L., "Atmospheric Explorer (AE-C) Adaptive Mission Design and Trajectory Performance," presented to AAS/AIAA Astrodynamics Conference, Nassau, Bahamas, July 1975.

Brewer, R. A. and Brant, D. N., "Thermal Protection System for the Galileo Mission Atmospheric Entry Probe," Aerothermodynamics and Planetary Entry, Progress in Astronautics and Aeronautics, Vol. 77, edited by A. L. Crosbie, AIAA, New York, 1981, pp. 309-334.

Butts, A. J., "Study of Advanced Atmospheric Entry Systems for Mars," Document MCR-78-582, Martin Marietta Corp., Denver Div., July 1978.

Cruz, M. I., "Aerocapture Atmospheric Flight Trajectory Guidance," JPL EM 312/79-78, Dec. 6, 1979 (JPL internal document).

Cruz, M. I., "The Aerocapture Vehicle Mission Design Concept," AIAA Paper 79-0893, May 8, 1979.

Cruz, M. I., "Aerocapture Vehicle Mission Design Concepts for the Inner and Outer Planets," AIAA Paper 79-115, June 25, 1979.

Cruz, M. I., "Aerodynamics Mission Concepts for the Mars Sample Return Mission," AIAA Paper 81-1859, Aug. 19, 1981.

Cruz, M. I., "Design of an Aerocapture Mission to Uranus," JPL IOM 312/81.1-3-MIC, Feb. 13, 1981 (JPL internal document).

Cruz, M. I., "Generic Aerocapture Technology Interchange Meeting Minutes, June 26, 1981, at Ames Research Center," JPL IOM 312/81.8-568, July 14, 1981.

Cruz, M. I., Generic Aerocapture Technology Requirements Review," JPL internal document 725-26, June 18, 1980.

Cruz, M.I., "Mars Sample Return Aerocapture and Aeromaneuver Mission and System Design Concepts," JPL Report 312/79-93, Sept. 19, 1979 (JPL internal document).

Cruz, M. I., "Maximum Mars Orbital Payload Capability for an Aerocapture Vehicle," JPL IOM 312/89.8-365, Feb. 11, 1980 (JPL internal document).

Cruz, M. I., "Performance and Accuracy of an Entry Guidance Option at Mars," JPL EM 312/78-71, Aug. 3, 1978 (JPL internal document).

Cruz, M. I., "System Design Data Book for Aerocapture and Aeromaneuvering Entry Vehicles," JPL Reference Document for Contract 955344 to G.E., Jan. 1979.

Cruz, M. I., "Technology Requirements for a Generic Aerocapture System," Aerothermodynamics and Planetary Entry, Progress in Astronautics and Aeronautics, Vol. 77, edited by A. L. Crosbie, AIAA, New York, 1981, pp. 412-439.

Cruz, M. I., "A Venus Aerocapture Concept," JPL EM 312-79-82, Feb. 7, 1979 (JPL internal document).

Cruz, M. I., Armento, R. F., and Giles, W. H., "Aerocapture - A System Design for Planetary Exploration," Paper 79-160, 30th International Astronautical Federation Congress, Munich, F.R.G., Sept. 16, 1979.

Dallas, S. S., "Venus Orbiting Imaging Radar, Aerobraking Mission, Vol. I: Mission Analysis and Navigation," JPL internal document 630-11, Oct. 1980.

Dallas, S. S. and Kwok, J. H., "The Venus Orbiting Imaging Radar Mission," AIAA Paper 81-183, Aug. 3, 1981.

Ellis, J. and Laing, P., "Navigation for Titan Aerocapture and Titan Probe Delivery," JPL IOM 314.5-498, April 30, 1980 (JPL internal document).

Etkin, Dynamics of Atmospheric Flight, John Wiley & Sons, Inc., New York, 1972.

Florence, D. E., "Aerothermodynamic Design Feasibility of a Generic Planetary Aerocapture/Aeromaneuver Vehicle," Thermophysics of Atmospheric Entry, Progress in Astronautics and Aeronautics, Vol. 82, edited by T. E. Horton, AIAA,New York, 1982, pp.

Florence, D. E., "Aerothermodynamic Design Feasibility of a Mars Aerocapture/Aeromaneuver Vehicle," AIAA Paper 81-0350, Jan. 12, 1981.

Florence D. E., "Generic Aerocapture Atmospheric Entry Study Final Report," G.E. Rept. 80SDR2226, Oct. 24, 1980.

French, J. R. and McRonald, A., "Thermophysical and Systems Integration Considerations in Aerobraking Design," Thermophysics of Atmospheric Entry, Progress in Astronautics and Aeronautics, Vol. 77, edited by A. L. Crosbie, AIAA, New York, 1981, pp. 396-411.

French, J. R. and Uphoff, C. W., "Aerobraking for Planetary Missions," Paper 79-286, AAS 1979 Annual Meeting, Space Shuttle: Dawn of an Era, Oct. 29, 1979.

Harpold, J., "Shuttle Program Analytic Drag Control Entry Guidance System Revision I," JSC Internal Note 74-FM-25, Jan. 21, 1975.

Loh, W. H. T., Re-entry and Planetary Entry Physics and Technology II/Advanced Concepts, Experiments, Guidance-Control and Technology, Springer-Verlag, Inc., New York, 1968.

Macomber, G. and Fernandez, M., Inertial Guidance Engineering, Prentice-Hall, Inc., Englewood Cliffs, N. J., 1962.

"Mars '84 Landing System Definition," Martin Marietta Report, April 1977.

McRonald, A., "Comparison of Aerobraking at Venus, Mars and Earth," JPL IOM 312/79.8-289, Sept. 13, 1979 (JPL internal document).

McRonald, A., "Use of Atmospheric Drag to Circularize an Elliptic Spacecraft Orbit at Venus," JPL internal document 760-163, March 1, 1977.

McRonald, A. D. and Nock, K. T., "Atmospheric Braking to Circularize an Elliptical Venus Orbit," presented to AAS/AIAA Astrodynamics Conference, Jackson, Wyo, Sept. 1977.

Miller, L., "Navigation Delivery Accuracy of Titan and Saturn Probes," JPL IOM 314.2-45, July 15, 1981.

Morth, R., "Re-entry Guidance for Apollo," MIT Instrumentation Laboratory Document R-532, Vol. I, Jan. 1966.

Noon, E. E., "Aerocapture Report," JPL IOM 3526/81/DEN/22, May 11, 1981.

Pivirotto, D. L. S., "Saturn Orbiter With Dual Probes Mission: A Progress Report," AIAA Paper 81-0310, Jan. 1981.

Rinderle, E. A., "ATMOS User's Guide and Program Description," JPL EM 312/78-79, Dec. 29, 1978 (JPL internal document).

Skerbetz, F. G. and Cruz, M. I., "Low-Speed Re-entry Guidance Dispersion Evaluation," TRW Note 69-FMT-764, July 29, 1969.

Tiwari, S. N., Chow, H., and Moss, J. N., "Analysis of Aerothermodynamic Environment of a Titan Aerocapture Vehicle," Thermophysics of Atmospheric Entry, Progress in Astronautics and Aeronautics, Vol. 82, edited by T. E. Horton, AIAA, New York, pp. 430-454.

Vinh, N. X., Optimal Trajectories in Atmospheric Flight, Elsevier Scientific Publishing Co., Amsterdam, Netherlands, 1981.

Vinh, N. X., Buseman, A., and Culp, R. D., Hypersonic and Planetary Entry Flight Mechanics, The University of Michigan Press, Ann Arbor, Mich., 1980.

Wallace, R. A. et al., "Missions to the Far Outer Planets in the 1990s," AIAA Paper 81-0311, Jan. 1981.

Zoby, E. V. and Moss, J. N., "Preliminary Aerothermal Analysis for Saturn Entry," Aerothermodynamics and Planetary Entry, Progress in Astronautics and Aeronautics, Vol. 77, edited by A. L. Crosbie, AIAA, New York, 1981, pp. 374-395.

Analysis of Aerothermodynamic Environment of a Titan Aerocapture Vehicle

S.N. Tiwari* and H. Chow†
Old Dominion University, Norfolk, Va.
and
J.N. Moss‡
NASA Langley Research Center, Hampton, Va.

Abstract

The extent of convective and radiative heating for a Titan entry vehicle is investigated. The flow in the shock layer is assumed to be axisymmetric, steady, laminar, viscous, and in chemical and local thermodynamic equilibrium. The implicit finite-difference technique is used to solve the viscous shock-layer equations for an equivalent body which is a 45 degree sphere cone at zero angle of attack. Different compositions for the Titan's N_2-CH_4 atmosphere were assumed and results were obtained for the entry conditions specified by the Jet Propulsion Laboratory. The results show that the heating rates, in general, increase with increasing N_2 concentration. Both convective and radiative heating increase with increasing initial entry velocity. The radiative heating increases but the convective heating decreases with increasing body nose radius. The amount of CN concentration in the shock-layer gas determines the extent of radiative heating to the body. Radiative heating will be important for freestream gas compositions with N_2 concentration between 50 and 90%. For the atmospheric composition of 99.5% N_2+0.5% CH_4, the radiative heating is insignificant in comparison to the convective heating.

Presented as Paper 81-1128 at the AIAA 16th Thermophysics Conference, Palo Alto, Calif., June 23-25, 1981.

*Eminent Professor, Department of Mechanical Engineering and Mechanics.

†Graduate Research Assistant, Department of Mechanical Engineering and Mechanics.

‡Aerospace Research Scientist.

Nomenclature

C_i	=	mass fraction of species i in the shock layer, ρ_i/ρ
C_p	=	equilibrium specific heat of mixture, $\Sigma\, C_i\, C_{p,i}$
$C_{p,i}$	=	specific heat of specific i, $C^*_{p,i}/C^*_{p,\infty}$
D_{ij}	=	binary diffusion coefficient
V^{*2}_∞	=	specific enthalphy, h^*/V^{*2}_∞
J_i	=	mass diffusion flux of species i, $J^*_i\, R_N/\mu^*_{ref}$
k	=	thermal conductivity of mixture, $k^*/\mu^*_{ref}\, C^*_{p,\infty}$
L/D	=	lift/drag ratio
Le	=	Lewis number, $\rho^*\, D^*_{ij}\, C^*_p/k^*$
M^*_i	=	molecular weight of species i
$\overline{M}^*$	=	molecular weight of mixture
n	=	coordinate normal to the body, n^*/R^*_N
p	=	pressure, $p^*/(\rho^*_\infty\, V^*_\infty)$
Pr	=	Prandtl number, $\mu^*\, C^*_p/k^*$
q_r	=	net radiant heat flux, $q^*_r/(\rho^*_\infty\, V^{*3}_\infty)$
r	=	radius measured from axis of symmetry to a point on the body surface, r^*/R^*_N
R^*	=	universal gas constant
R^*_b	=	radius of the body
Re_∞	=	Reynolds number, $\rho^*_\infty\, V^*_\infty\, R^*_N/\mu^*_\infty$
R^*_N	=	body nose radius
s	=	coordinate along the body surface, s^*/R^*_N
T	=	temperature, T^*/T^*_{ref}
T^*_{ref}	=	reference temperature, $V^{*2}_\infty/C^*_{p,\infty}$
u	=	velocity tangent to body surface, u^*/V^*_∞
U_E	=	initial entry velocity
v	=	velocity normal to body surface, v^*/V^*_∞
V^*_∞	=	freestream velocity
α	=	shock angle defined in Fig. 1
β	=	ballistic coefficient, $W/(C_D A)$
γ	=	inertial entry angle
ε	=	Reynolds number parameter
θ	=	body angle defined in Fig. 1
κ	=	body survature, $\kappa^* R^*_N$
μ	=	viscosity of mixture, μ^*/μ^*_{ref}
μ^*_{ref}	=	reference viscosity, $\mu^*(T^*_{ref})$
ρ	=	density of mixture, ρ^*/ρ^*_∞

<u>Subscripts</u>

i	=	ith species
s	=	shock value
w	=	wall value
∞	=	freestream condition

Introduction

The feasibility of aerocapture vehicle mission has been emphasized recently for inner and outer planetary missions.[1-6] The aerocapture vehicle essentially is a contoured heat shield that encloses the planetary orbital spacecraft and its subsystems. The vehicle has a communication system that reaches the Earth through interface with spacecraft electronics. Originally, the aerocapture study was undertaken for a Mars sample return mission.[3,6] The aerocapture missions under present consideration are the Mars Surface Sample Return (MSSR), Saturn Orbiter Dual Probe (SO2P), and Titan Orbiter (TO) missions.

The aerocapture mission is possible for any planet (or satellite) which has an atmosphere. The aerocapture technique begins with the interplanetary cruise vehicle (aerocapture vehicle) approaching the reference body (planet or satellite) on a hyperbolic trajectory and removing its excess energy in a single pass through the atmosphere to achieve the initial required orbit about the celestial body. As such, an aerocapture is a system concept that utilizes aerodynamic drag to acquire the velocity depletion necessary to obtain a closed planetary orbit from a hyperbolic flyby trajectory. It is accomplished through an aerodynamically controlled atmospheric entry during which the vehicle's in-plane lift-to-drag ratio is varied to maintain a constant drag. The aerocapture offers significant gains in payload and choice of orbits and eliminates fuel-costly retropropulsion module (RPM) for planetary orbiter missions.[4]

For missions to outer planets, use of the aerocapture concept in a convenient atmosphere-bearing satellite of the target planet has been emphasized. Thus, it has been proposed to use the atmosphere of Titan for braking into a Saturn orbit. The use of Titan's atmosphere would minimize the entry speed requirement for aerocapture and this, in turn, would minimize the thermal protection requirements of the aerocapture vehicle. The Titan's aerocapture concept (for Saturn orbital mission) is expected to cut the interplanetary cruise travel time to Saturn from 8 to 3.5 years. A Titan orbiter mission using anything other than aerocapture is presently impossible.[4] For Titan's aerocapture mission, the need for high-performance entry vehicle geometrics and high-performance thermal protection systems has been stressed.[4,5] In partial support of this need, it is essential to provide a complete analysis of aerothermodynamic environment of the Titan aerocapture vehicle.

The main objective of this study, therefore, is to determine the extent of convective and radiative heating to the aerocapture vehicle under different entry conditions. This essentially can be accomplished by assessing the heating rates in the stagnation and windward regions of an equivalent body. The equivalent body configuration considered for this study is a 45 deg sphere cone at zero angle of attack. Different compositions for the Titan's N_2-CH_4 atmosphere have been assumed and results have been obtained for various entry trajectories suggested by the Jet Propulsion Laboratory (JPL).

Basic Formulation

The physical model and coordinate system considered for the equivalent body are shown in Fig. 1. The hypersonic flow of radiating and reacting gas mixture in the shock layer is assumed to be axisymmetric, steady, viscous, and in chemical and local thermodynamic equilibrium. Basic governing equations for this physical model are available in the literature.[7-9] The nondimensional form of the viscous shock-layer equations that are applicable in the

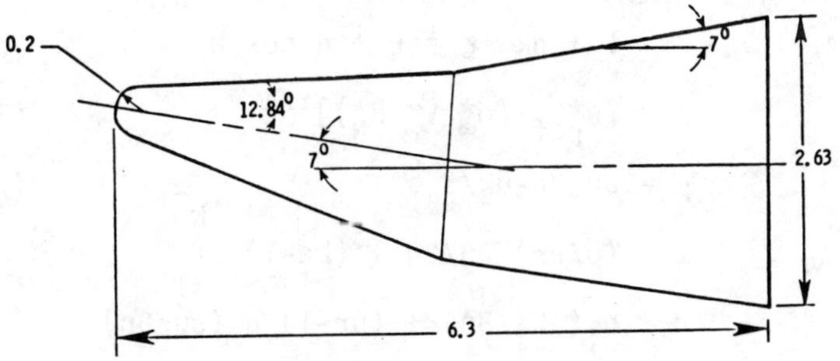

Fig. 1a Titan aerocapture vehicle configuration.

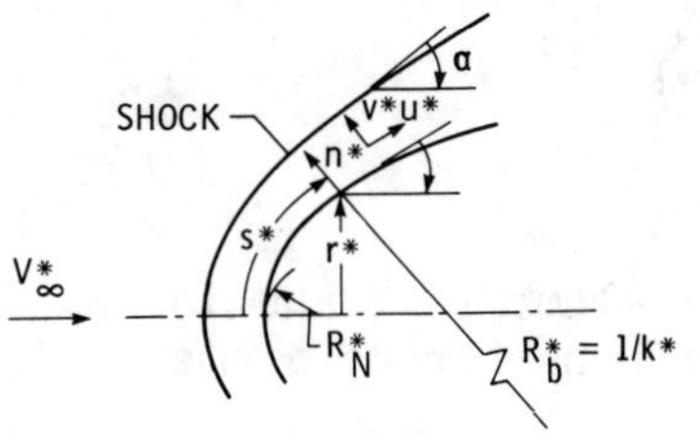

Fig. 1b Coordinate system.

present case can be expressed as[10,11]

Continuity:

$$(\partial/\partial s)(\zeta\rho u) + (\partial/\partial n)(\Gamma\zeta\rho v) = 0 \quad (1)$$

s momentum:

$$\rho[u/\Gamma)(\partial u/\partial s)+n(\partial u/\partial n)+uv\kappa/\Gamma]+\Gamma^{-1}(\partial p/\partial s)$$
$$= \varepsilon^2[(\partial/\partial n)\ (\mu\psi)+\mu(2\kappa/\Gamma+\cos\ \theta\ /\zeta)\psi] \quad (2)$$

n momentum:

$$\rho[(u/\Gamma)(\partial v/\partial s)+\partial v/\partial n-u^2\kappa/\Gamma+\partial p/\partial n = 0 \quad (3)$$

Energy:

$$\rho[(u/\Gamma)(\partial H/\partial s)+v(\partial H/\partial n)]-v(\partial p/\partial n)+\rho\kappa u^2 v/\Gamma$$
$$= \varepsilon^2[\partial\phi/\partial n+(\kappa/\Gamma+\cos\ \Theta/\zeta)\phi]-\text{div}\ q_r \quad (4)$$

Elemental continuity:

$$\rho[u/\Gamma)(\partial\tilde{C}_\ell/\partial s) + v(\partial\tilde{C}_\ell/\ n)]$$
$$= \{(\partial/\partial n)[(\Gamma\zeta\mu Le/Pr)(\partial\tilde{C}_\ell/\partial n)]\} \quad (5)$$

State:

$$p = \rho TR^*/\bar{M}^*\ C^*_{p,\infty} \quad (6)$$

where

$$\Gamma = 1 + n\kappa,\ \zeta = r + n\cos\theta \quad (7a)$$

$$\varepsilon = [\mu^*_{ref}/(\rho^*_\infty\ V^*_\infty\ R^*_N)]^{1/2} \quad (7b)$$

$$\psi = \partial u/\partial n-u\kappa/\Gamma \quad (7c)$$

$$\phi = (\mu/Pr)[\partial H/\partial n + (Le-1)\sum_{i=1}^{N} h_i(\partial C_i/\partial n) + (pr-1)\ u\ (\partial u/\partial n) - Pr\ \kappa\ u^2/\Gamma] \quad (7d)$$

$$H = h + u^2/2,\ \tilde{C}_\ell = \sum_{i=1}^{N} \delta_{i\ell}(M^*_\ell/M^*_i)C_i \quad (7e)$$

In Eq. (5), $\tilde{C}_\ell$ is the mass fraction of element ℓ; and in the definition of $\tilde{C}_\ell$, $\delta_{i\ell}$ represents the number of atoms of the ℓth element in species i.

The set of governing equations presented has a hyperbolic-parabolic nature. The hyperbolic nature enters through the normal momentum equation. If the shock layer is assumed thin, then the normal momentum equation can be expressed as

$$\rho u^2 \kappa/(1 + n\kappa) = (\partial p/\partial n) \tag{8}$$

If Eq. (3) is replaced with Eq. (8), then the resulting set of equations is parabolic. These equations can therefore be solved by using numerical procedures similar to those used in solving boundary-layer problems.[8,12]

In order to solve the preceding set of governing equations, it is essential to specify appropriate boundary conditions at the body surface and at the shock. At the body surface (wall), no-slip and no temperature jump conditions are used. Consequently, $u_W = v_W = 0$ and the wall temperature is either specified or calculated. The conditions in form of the shock are obtained from the freestream entry condition. The conditions immediately behind the shock are obtained by using the Rankin-Hugoniot relations.[8,9]

The heat transferred to the wall due to conduction and diffusion is referred to here as the convective heat flux and is given by the relation

$$q_{c,w} = -\varepsilon^2 \left[k \left(\frac{\partial T}{\partial n}\right) + \left(\frac{\mu Le}{Pr}\right) \sum_{i=1}^{N} \frac{\partial C_i}{\partial n} h_i \right]_w \tag{9}$$

The convective heat transfer is often described by a dimensionaless parameter called Stanton number. For the viscous shock-layer flow, the Stanton number is given by

$$St = q_{c,w}/(H_\infty - H_w) \tag{10}$$

The governing equations and boundary conditions presented here essentially describe the flowfield in the shock layer. It now remains to specify the equilibrium gas composition and obtain appropriate relations for the radiative flux and thermodynamic and transport properties.

Radiation Transport Model

The tangent slab approximation for radiative transfer is used in this study. This implies that the radiative energy transfer along the body is negligible in comparison to that transferred in the direction normal to the body. It should be noted that the tangent slab approximation is used only for the radiative transport and not for other flow variables. For nonscattering boundary surfaces, a one-dimensional expression for the net radiation flux in the shock layer is given by[9,13]

$$q_r = q_r^+ - q_r^- \tag{11}$$

where q_r^+ represents the energy transfer toward the shock and q_r^- the energy transfer toward the body. The expressions q_r^+ and q_r^- are available in the cited references.

The radiative flux, q_r, is calculated with the radiative transport code RAD[14] which accounts for detailed nongray radiation absorption and emission processes. The chemical species considered for determining the radiative transport are N, N_2, N^+, N^-, N_2, H, H_2, H^-, e^-, C, C^-, C^+, and CN.

Physical Conditions and Data Source

As pointed out earlier, the entry body considered for this study is a 45 deg sphere cone at a zero angle of attack (Fig. 1). The body temperature is assumed to be 2000 K and, for most cases, the body nose radius is taken to be 0.2 m.

Entry Trajectory and Freestream Conditions

As discussed in Ref. 4, a nominal aerocapture mission has two critical phases: approach navigation and atmospheric flight. Approach trajectory correction maneuvers outside the atmosphere can be controlled by commands computed on Earth. However, once the vehicle enters the atmosphere, trajectory maneuvers must be computed and commanded on board. For the Titan aerocapture mission, entry trajectories have been generated by JPL. The altitude history for an aerocapture vehicle is illustrated in Fig. 2 for two different (shallow and steep) entry angles. The entry trajectories and freestream conditions used in this study are given in Tables 1-5.

Chemical Composition of Gas Mixture

Analyses of chemically reacting flows are usually simplified by assuming the chemical equilibrium behavior

Table 1 Altitude and freestream conditions: trajectory I
(L/D = 1.2, γ = -45 deg, β = 800 kg/m^2, U_E = 10 km/s)

Time, s	Altitude, km	ρ_∞, g/cm^3	p_∞, mb	T_∞, K	V_∞, km/s	Mach No.
40	230.965	0.1265E-6	0.10355	159.10	9.929	29.13
50	169.824	0.3473E-6	0.24562	139.44	9.803	30.72
60	114.238	0.1022E-5	0.64627	119.17	9.431	31.90
70	70.409	0.3197E-5	1.5836	103.16	8.448	30.78
78	50.922	0.5157E-5	2.5901	96.28	7.185	27.10
90	48.539	0.5512E-5	2.7449	95.56	5.502	20.83
100	60.654	0.3944E-5	1.9738	99.26	4.727	17.56
150	95.613	0.1604E-5	0.89555	112.46	3.406	11.89
220	173.831	0.3728E-5	0.23159	140.84	3.015	9.40

Table 2 Altitude and freestream conditions: trajectory II
(L/D = 1.2, γ = -25 deg, β = 800 kg/m^2, U_E = 10 km/s)

Time, s	Altitude, km	ρ_∞, g/cm^3	p_∞, mb	T_∞, K	V_∞, km/s	Mach No.
50	321.264	0.3952E-7	0.3249E-1	160.00	9.953	29.11
100	213.217	0.1654E-6	0.1335E+0	155.60	9.689	28.74
110	198.471	0.2063E-6	0.1638E+0	150.30	9.583	28.91
120	185.306	0.2694E-6	0.1967E+0	145.10	9.451	29.03
130	173.617	0.3288E-6	0.2323E+0	140.76	9.286	28.96
140	163.335	0.3812E-6	0.2683E+0	137.16	9.100	28.75
150	154.339	0.4687E-6	0.3139E+0	134.02	8.889	28.41
170	141.502	0.6712E-6	0.3909E+0	129.53	8.358	27.18
175	139.412	0.7052E-6	0.4053E+0	128.79	8.210	26.77
180	137.610	0.7213E-6	0.4215E+0	128.16	8.062	26.36
200	130.832	0.7855E-6	0.4875E+0	125.79	7.495	24.73

Table 3 Altitude and freestream conditions: trajectory III
(L/D = 1.2, γ = -45 deg, β = 800 kg/m^2, U_E = 6 km/s)

Time, s	Altitude, km	ρ_∞, g/cm^3	p_∞, mb	T_∞, K	V_∞, km/s	Mach No.
90	144.674	0.6142E-6	0.3719	130.00	5.847	18.93
100	111.891	0.1076E-5	0.6732	118.70	5.681	19.29
110	83.284	0.2349E-5	1.1359	108.10	5.379	19.14
120	61.655	0.3859E-5	1.9338	99.60	4.872	18.06
129	49.822	0.5318E-5	2.6615	95.90	4.327	16.35
140	45.041	0.6079E-5	2.9723	94.50	3.690	14.04
150	47.699	0.5643E-5	2.7995	95.30	3.245	12.30
160	54.496	0.4666E-5	2.3577	97.30	2.934	11.01
170	63.709	0.3693E-5	1.8516	100.50	2.724	10.05

Table 4 Altitude and freestream conditions: trajectory IV
(L/D = 1.2, γ = -45 deg, β = 800 kg/m^2, 8 km/s)

Time, s	Altitude, km	ρ_∞, g/cm^3	p_∞, mb	T_∞, K	V_∞, km/s	Mach No.
60	181.218	0.2927E-6	0.2070	143.48	7.883	24.35
70	134.830	0.7469E-6	0.4464	127.19	7.709	25.30
80	93.931	0.1689E-5	0.9283	111.87	7.329	25.63
90	63.296	0.3726E-5	1.868	100.32	6.515	24.07
97	50.882	0.5162E-5	2.592	96.26	5.783	21.81
110	46.913	0.5768E-5	2.850	95.07	4.554	17.28
120	54.841	0.4621E-5	2.335	97.17	3.961	14.84
130	67.354	0.3413E-5	1.706	101.94	3.603	13.20
150	98.351	0.1473E-5	0.8421	111.42	3.240	11.25

Table 5 Altitude and freestream conditions: trajectory V
(L/D = 1.2, γ = -45 deg, β = 800 kg/m^2, U_E = 13 km/s)

Time, s	Altitude, km	ρ_∞, g/cm^3	p_∞, mb	T_∞, K	V_∞, km/s	Mach No.
30	237.512	0.1143E-6	0.0937	159.75	12.905	37.77
40	158.551	0.4165E-6	0.2887	135.49	12.672	40.28
50	90.769	0.1863E-5	0.9899	110.77	11.777	41.40
60	51.374	0.5092E-5	2.560	96.41	9.361	35.28
61	49.557	0.5339E-5	2.671	95.90	9.084	34.33
70	49.557	0.5357E-5	2.678	95.86	7.047	26.64
80	67.533	0.3401E-5	1.698	102.01	5.912	21.66
100	101.803	0.1345E-5	0.7892	114.72	5.083	17.56
120	124.659	0.8488E-6	0.5381	123.63	4.762	15.84

of the gas mixture. The equilibrium chemical reactions considered for this study are confined to a system of carbon, hydrogen, and nitrogen. The aerotherm chemical equilibrium (ACE) computer program was used to determine various chemical species under different pressure and temperature conditions and for different freestream atmospheric compositions.

At the initiation of this study, the atmospheric conditions of Titan were not defined clearly. As such, different atmospheric compositions were assumed for a parametric study. It is now evident[15] that Titan, the largest moon in the solar system, is wrapped essentially in a dense atmosphere of nitrogen vapors (rather than methane, the best guess before Voyager 1). Thus, a realistic case would be to assume a very high concentration of nitrogen in the

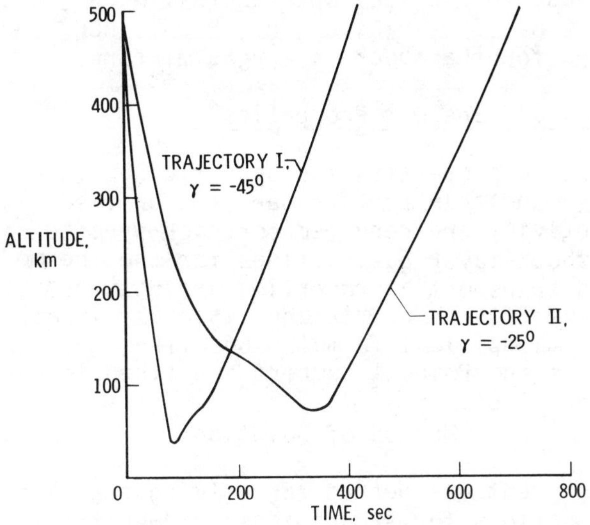

Fig. 2 The altitude history for an aerocapture vehicle.

freestream gas mixture. However, to study the effect of freestream gas composition on heating of the entry vehicle, different gas compositions were assumed.

The equilibrium composition is determined by a free energy minimization calculation as developed in Ref. 16. As mentioned earlier, the ACE computer program was used to determine various chemical species for different pressure, temperature, and freestream conditions. For initial study, 68 chemical species for carbon-hydrogen-nitrogen system were included in the matrix of calculations for a given freestream atmospheric composition. The matrix was

Pressure: 0.1, 0.5, 1.0, and 5 atm.
Temperature: 2000-10,000 K in 500 K increments.
Compositions: 99.5% N_2 + 0.5% CH_4, 90% N_2 + 10% CH_4, 75% N_2 + 25% CH_4, 50% N_2 + 50% CH_4, 25% N_2 + 75% CH_4, 10% N_2 + 90% CH_4.

For a freestream gas composition of 90% N_2 + 10% CH_4, the variation in mole fraction of different species, as a function of temperature, is illustrated in Fig. 3 for P=1 atm. For this case, there are 21 chemical species in the gas mixture. However, concentration of some species is less than 0.05% for the range of temperature considered. Similar trends were noted for other pressures and freestream gas compositions.[17] For this study, therefore, contributions of at least 18 chemical species (N_2, N, N^+, C_3, C_2, C, C^+, CH_4, C_3H, C_2H_2, C_2H, C_2N, CN, H_2, H, H^+, HCN, and e^-) were considered for the shock-layer gas mixture.

Thermodynamic and Transport Properties

Thermodynamic properties for specific heat, enthalpy, and free energy and transport properties for viscosity and thermal conductivity are required for each species considered in the shock-layer gas. Values for the thermodynamic[18,19] and transport[20] properties are obtained by using polynomial curve fits. The mixture viscosity is obtained by using the semiempirical formulas of Wilke.[7] In this study, the Lewis and Prandtl numbers are taken to be 1.1 and 0.64, respectively.[21]

Method of Solution

Davis[12] presented a method for solving the viscous shock-layer equations for stagnation and downstream flow. Mass[8] and Tiwari and Szema[9] applied this method of solution to reacting multicomponent mixtures. The entire solution procedure is discussed in detail in Refs. 8 and 9.

Results and Discussion

Results have been obtained for entry conditions given in Tables 1-5 to investigate the effects of freestream gas composition, entry velocity, and body nose radius on the stagnation point convective and radiative heating. Specific results were obtained to determine the extent of convective and radiative heating along the body for freestream gas composition of 90% N_2 + 10% CH_4 and 99.5% N_2 + 0.5% CH_4. Selected results for trajectories I and II are given in Tables 6 and 7 and extensive tabulations of all results are available in Ref. 17. Some specific results of the study are presented in this section. It should be noted that, for clarity, the asterisks on the dimensional quantities have been left out of the tables and figures.

The effects of freestream gas composition on the shock temperature are illustrated in Figs. 4 and 5. The results show that the shock temperature (as well as the temperature in the shock layer, Fig. 5) increases with increasing N_2 concentration. This is because N_2 provides less energy accommodation in comparison to CH_4. The stagnation shock temperatures are relatively higher for early entry times (Fig. 4); however, this would be expected because of relatively higher freestream velocities. The results of

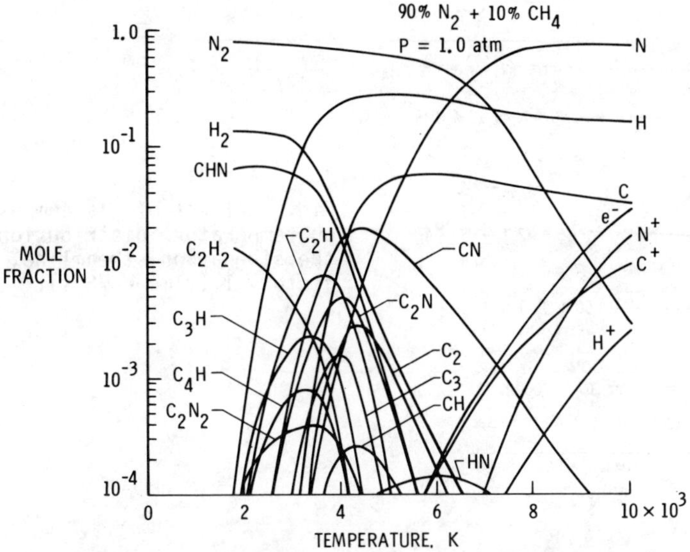

Fig. 3 Variation in mole fraction of different species for P = 1 atm and 90% N_2 + 10% CH_4.

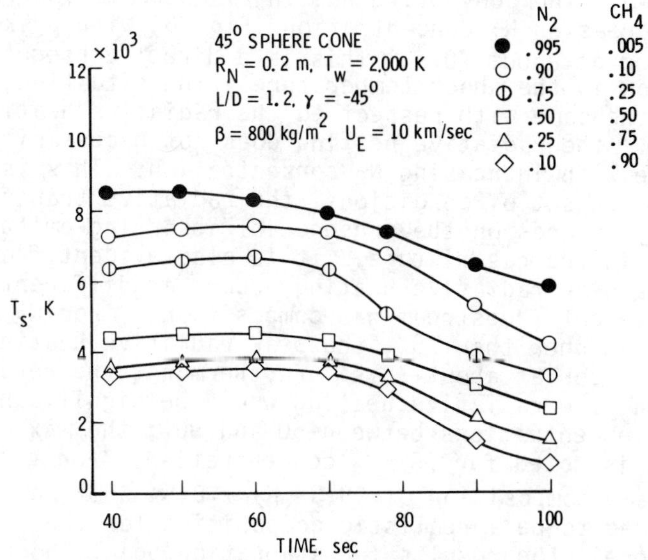

Fig. 4 Effect of gas composition on stagnation-point shock temperature, trajectory I.

Fig. 5 show that the temperature gradient in the shock layer is restricted essentially in the regions near the body surface for all freestream gas compositions.

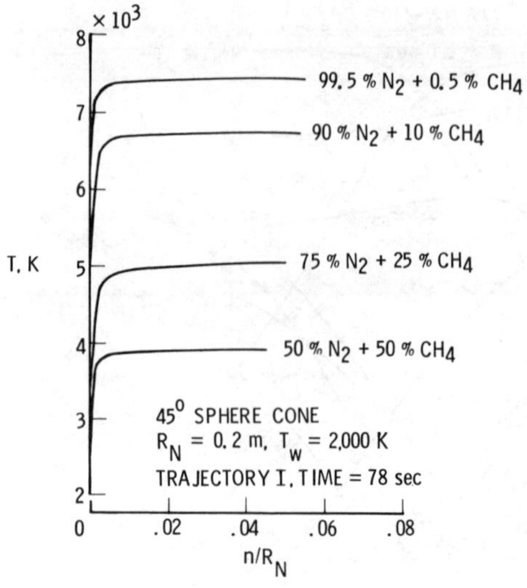

Fig. 5 Effect of gas composition on temperature distribution along the stagnation streamline, trajectory I (time = 78 s).

The effects of gas composition on the stagnation point convective and radiative heating are illustrated in Figs. 6-8. The convective heating is seen to increase with increasing N_2 concentration (Fig. 6) with peak heating occurring at about 70 s. This is a direct consequence of the variation in the shock temperature. The situation, however, is not the same with respect to the radiative heating (Fig. 7), i.e., the radiative heating does not necessarily increase with increasing N_2 concentration. This is because, for a given set of conditions, the radiative transfer strongly depends on the presence of absorbing-emitting species in the gas mixture. It is also evident from Fig. 7 that the peak radiative heating occurs at different times for different freestream gas compositions. For N_2 concentrations higher than 75%, the peak radiative heating is seen to occur at about 78 s. Furthermore, the results show that the radiative heating would be significant only for N_2 concentrations between 50 and 90%; the maximum heating is noted for 75% N_2 concentration. For the freestream composition of 99.5% N_2 + 0.5% CH_4 (which is considered to be a realistic composition for the Titan's atmosphere), the results for stagnation point shock temperature and convective and radiative heating are illustrated in Fig. 8 for trajectory I. It is noted that, for this case, the radiative heating is negligible as compared to the convective heating.

The effects of entry velocity on the stagnation point shock temperature and convective and radiative heating

Table 6 Stagnation results (sphere cone, R_N = 0.2 m, T_w = 2000 K): atmosphere of 99.5% N_2 + 0.5% CH_4, trajectory I

Time, s	ρ_∞, kg/m^3	p_s, atm	ρ_s/ρ_∞	T_s, K	n_s, cm	$q_{c,w}$, MW/m^2	$q_{r,w}$, MW/m^2
40	0.1265E-3	0.1163	18.13	8508	0.7963	5.783	0.0044
50	0.3473E-3	0.3112	17.93	8498	0.8241	8.558	0.0155
60	0.1022E-2	0.8473	17.76	8262	0.8381	12.265	0.0576
70	0.3197E-2	2.1194	16.75	7902	0.8927	14.818	0.2391
78	0.5157E-2	2.4560	15.08	7354	1.0011	11.188	0.3136
90	0.5521E-2	1.5121	12.10	6412	1.2341	4.688	0.2249
100	0.3944E-2	0.7888	10.48	5813	1.4109	2.290	0.1387

Table 7 Stagnation results (sphere cone, R_N = 0.2 m, T_w = 2000 K): atmosphere of 90% N_2 + 10% CH_4

Time, s	ρ_∞, kg/m^3	p_s, atm	ρ_s/ρ_∞	T_s, K	n_s, cm	$q_{c,w}$, MW/m^2	$q_{r,w}$, MW/m^2
			Trajectory I				
40	0.1265E-3	0.1165	18.56	7259	0.8029	5.247	0.0177
50	0.3473E-3	0.3112	17.95	7460	0.8440	8.126	0.0713
60	0.1022E-2	0.8455	17.15	7544	0.8880	11.637	0.3569
70	0.3197E-2	2.1108	15.74	7342	0.9619	13.498	2.0398
78	0.5157E-2	2.443	14.08	6733	1.0642	9.687	3.7697
90	0.5512E-2	1.512	12.10	5269	1.2120	3.543	3.8894
100	0.3944E-2	0.7999	12.10	4127	1.2208	1.683	1.5915
			Trajectory II				
140	0.3812E-3	0.2942	17.78	6953	0.8586	6.437	0.0862
150	0.4687E-3	0.3449	17.49	6895	0.8731	6.523	0.1190
170	0.6712E-3	0.4335	16.75	6729	0.9111	6.242	0.2044
175	0.7052E-3	0.4413	16.58	6665	0.9201	6.061	0.2206
180	0.7213E-3	0.4350	16.39	6590	0.9300	5.747	0.2287
200	0.7855E-3	0.4081	15.63	6330	0.9723	4.589	0.2728

rates are illustrated in Figs. 9-11. For the freestream gas composition of 90% N_2 + 10% CH_4, the results presented in Figs. 9, 10, and 11a show that the shock temperature and heating rates, in general, increase with increasing entry velocity for a fixed entry altitude (time). It is seen that the extent of convective heating is considerably higher than the radiative heating for all cases. As noted earlier, the convective and radiative heating peaks occur at different entry times (altitudes) for the same initial entry velocity. Similar trends in results are noted also for the

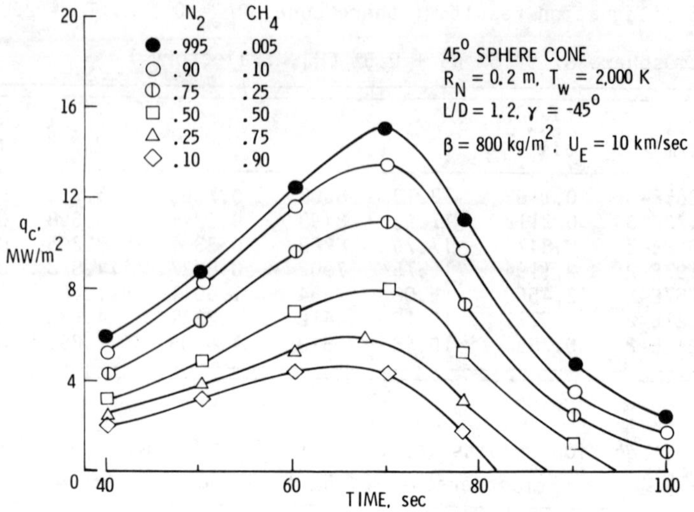

Fig. 6 Effect of gas composition on stagnation-point radiative heating, trajectory I.

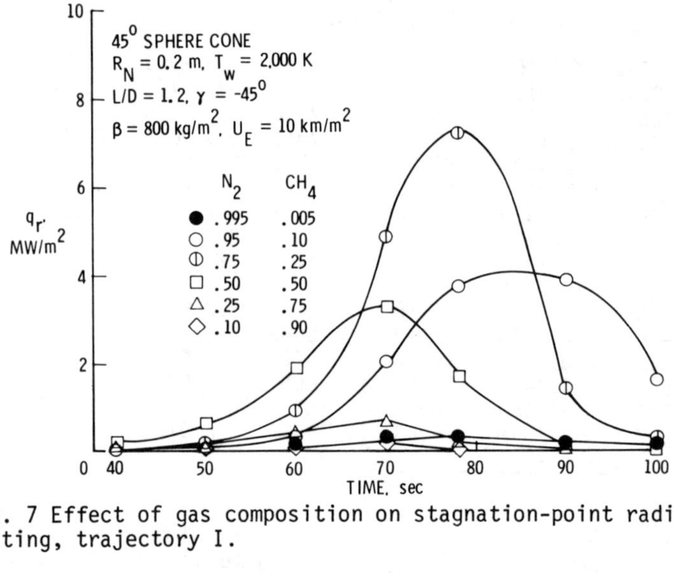

Fig. 7 Effect of gas composition on stagnation-point radiative heating, trajectory I.

gas composition of 99.5% N_2 + 0.5% CH_4[17]. For this composition, the radiative heating is negligible in comparison to the convective heating. One exception to this, however, is noted from the results presented in Fig. 11b. The radiative heating rate for entry velocity of 13 km/s is considerably higher than for other velocities. Thus, for the entry speed of 13 km/s (and for entry times between 30

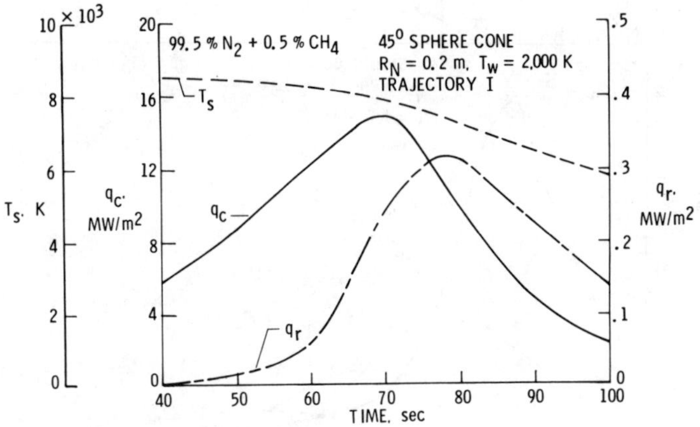

Fig. 8 Variation of stagnation-point shock temperature and convective and radiative heating for 99.5% N_2 + 0.5% CH_4.

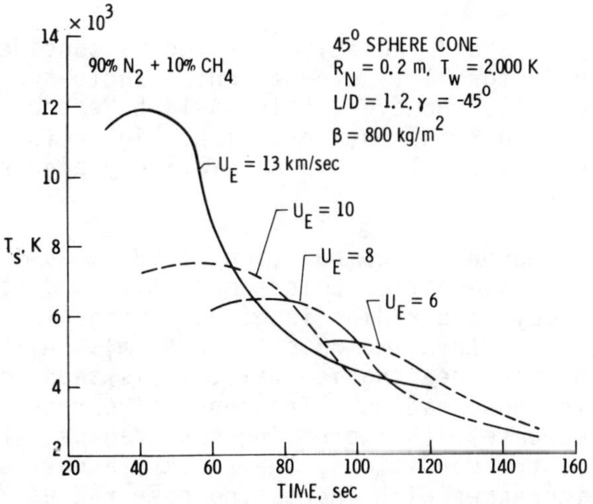

Fig. 9 Effect of entry velocity on stagnation shock temperature, 90% N_2 + 10% CH_4.

and 60 s), it is possible to have physical conditions in the shock layer to produce higher concentration of radiating species.

For the freestream atmospheric composition of 90% N_2 + 10% CH_4, the results for stagnation point convective and radiative heating are shown in Fig. 12 (and also in Table 7) for trajectories I and II. The results show that the extent

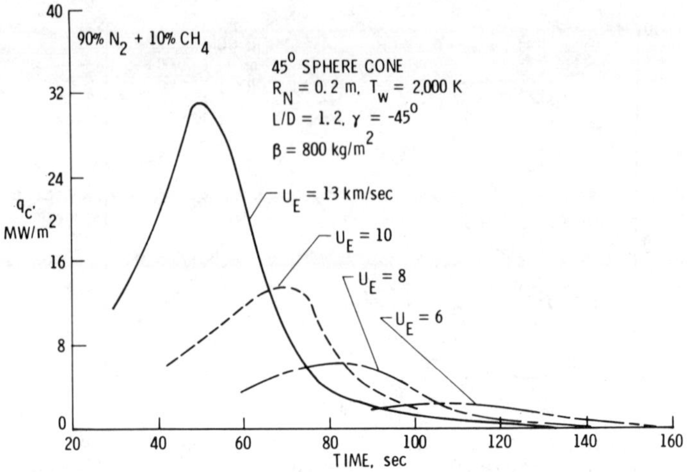

Fig. 10 Effect of entry velocity on stagnation-point convective heating 90% N_2 + 10% CH_4.

of both convective and radiative heating is considerably higher for trajectory I (a steeper entry angle trajectory) than trajectory II. However, this would be expected because the rate of viscous dissipation will be higher for the steeper trajectory resulting in a relatively higher shock temperature.

For the atmospheric composition of 99.5% N_2 + 0.5% CH_4, the variation of stagnation point convective and radiative heating with body nose radius is given in Fig. 13 for trajectory I and entry time 78 s (z=50.9 km). Although the extent of radiative heating is small, it is seen to increase with increasing nose radius. The convective rate, however, is seen to decrease with increasing nose radius. For a given set of entry conditions, the shock standoff distance, in general, increases with increasing nose radius.[9] This, in turn, results in different temperature, pressure, and species distribution in the shock layer. A combination of these changes influences the trend exhibited.

The results of heating rate along the body are illustrated in Figs. 14 and 15 for trajectory I and for the entry conditions at 78 s. Variations in shock temperature and heating rates are shown in Fig. 14 for the atmospheric composition of 99.5% N_2 + 0.5% CH_4. The results show that both convective and radiative heating essentially follow the trend of the shock temperature from the stagnation point to the tangency point (at about s=0.8). Beyond this

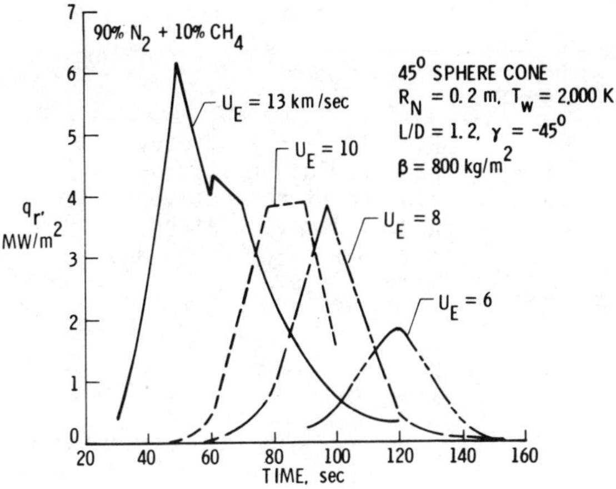

Fig. 11a Effect of entry velocity on stagnation-point radiative heating, 90% N_2 + 10% CH_4.

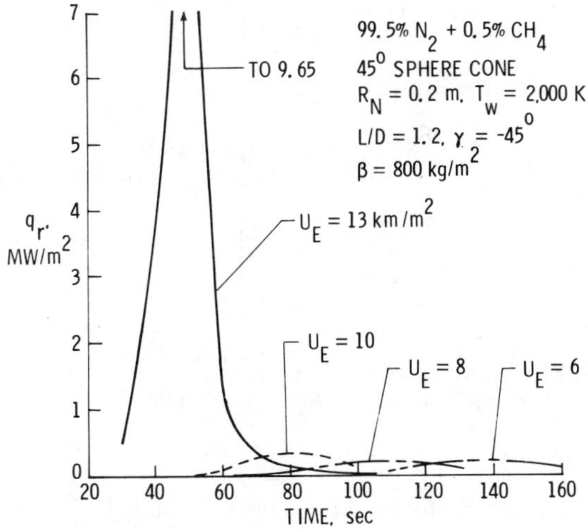

Fig. 11b Effect of entry velocity on stagnation-point radiative heating, 99.5% N_2 + 0.5% CH_4.

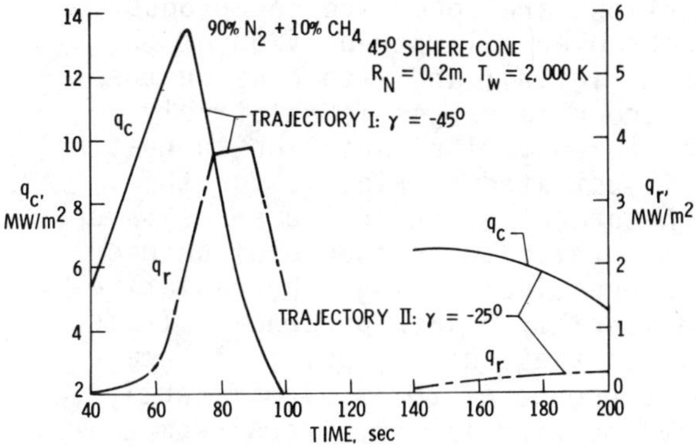

Fig. 12 Variation of stagnation-point convection and radiative heating for trajectories I and II, 90% N_2 + 10% CH_4.

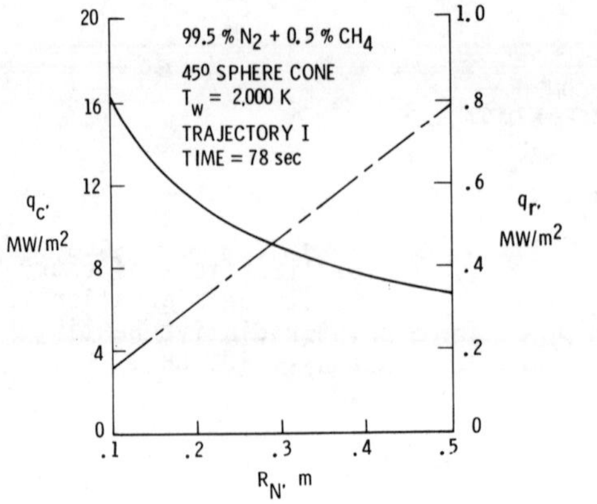

Fig. 13 Effect of body nose radius on stagnation-point convective and radiative heating for 99.5% N_2 + 0.5% CH_4, time = 78 s.

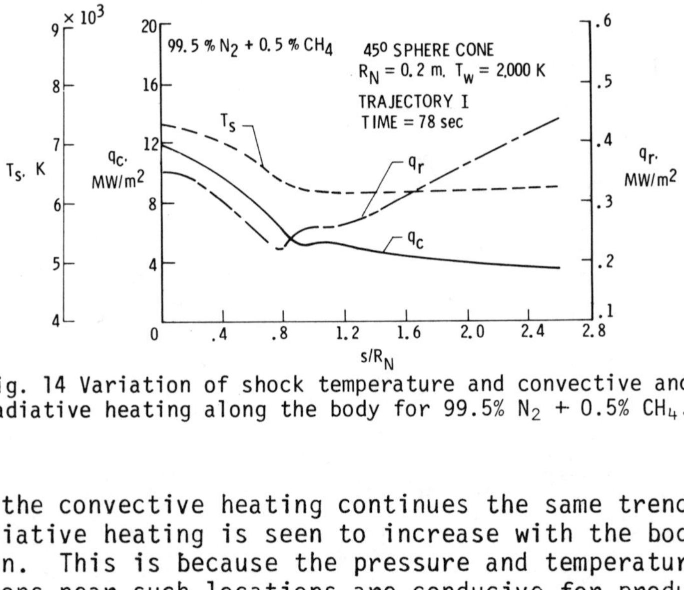

Fig. 14 Variation of shock temperature and convective and radiative heating along the body for 99.5% N_2 + 0.5% CH_4.

point, the convective heating continues the same trend but the radiative heating is seen to increase with the body location. This is because the pressure and temperature conditions near such locations are conducive for production of the radiating CN species over a larger portion of the shock-layer thickness (see Fig. 3); and also because the optical thickness of the shock-layer gas is relatively higher in the downstream regions. The variation in heating rates along the body is illustrated in Fig. 15 for the cases with and without CN concentration in the shock-layer gas. The results show that while the presence of CN has little influence on the convective heating, the radiative heating is increased considerably by its presence. It is important to note that after the tangency point, the rate of radiative heating in the presence of CN is significantly higher than the convective heating for the freestream com-

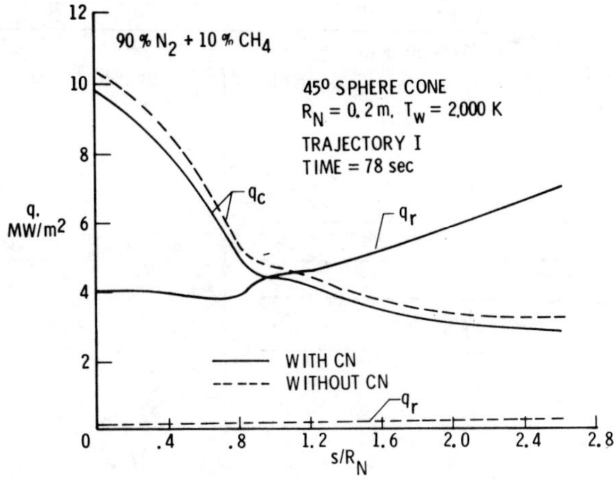

Fig. 15a Influence of CN on convective and radiative heating along the body for t = 78 s and 90% N_2 + 10% CH_4.

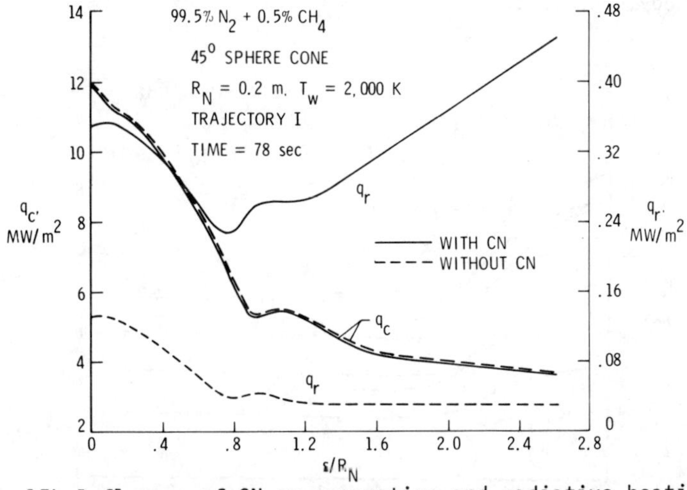

Fig. 15b Influence of CN on convective and radiative heating along the body for t = 78 s and 99.5% N_2 + 0.5% CH_4.

position of 90% N_2+ 10% CH_4 (Fig. 15a). The same trend is seen in Fig. 15b for the atmospheric composition of 99.5% N_2 + 0.5% CH_4, but the extent of radiative heating is considerably small.

For the freestream atmospheric composition of 99.5% N_2 + 0.5% CH_4, variations in important results with distance along the body surface, are illustrated in Figs. 16-18 for trajectory I and for critical entry times (altitudes). The results for shock density and shock standoff distance

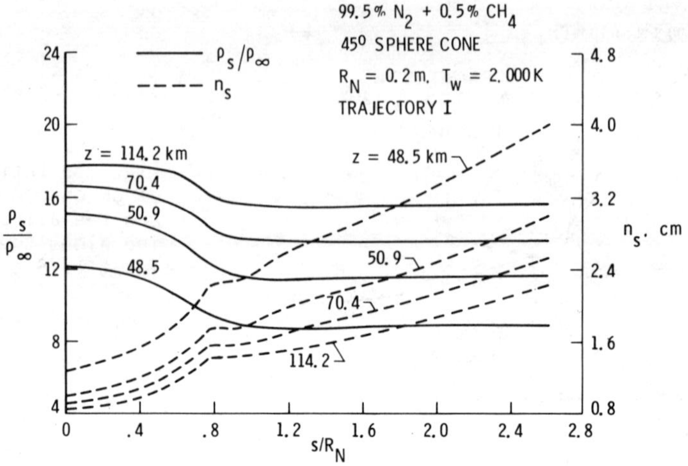

Fig. 16 Variation of shock density and shock-standoff distance with body coordinate for 99.5% N_2 + 0.5% CH_4.

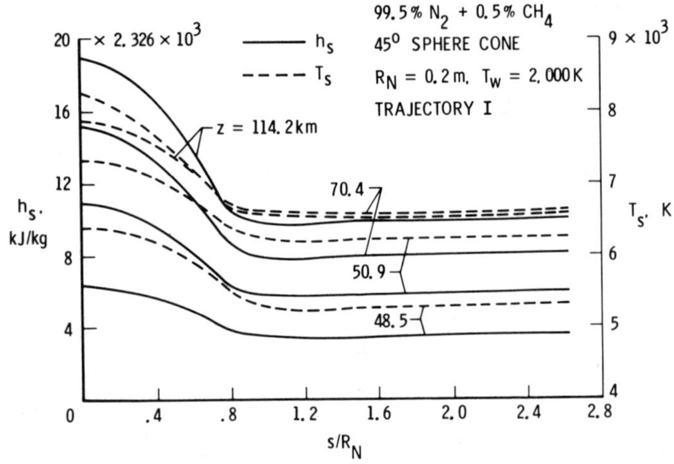

Fig. 17 Variation of shock temperature and enthalpy with body coordinate for 99.5% N_2 + 0.5% CH_4.

presented in Fig. 16 show that, for a given entry altitude, the shock standoff distance is seen to decrease with increasing altitude; this is because higher freestream velocities are associated with higher altitudes (see Table 1). The results for shock temperature and enthalpy presented in Fig. 17 show that both decrease along the body until the tangency point and that they essentially remain constant beyond that point. Because of higher freestream

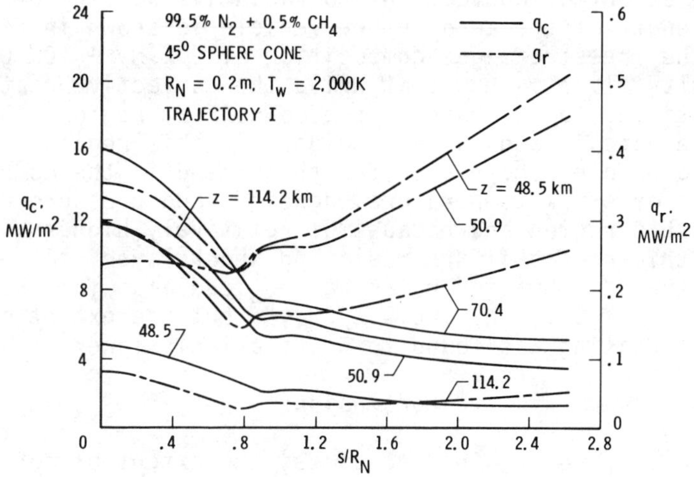

Fig. 18 Variation of convective and radiative heating along the body for 99.5% N_2 + 0.5% CH_4.

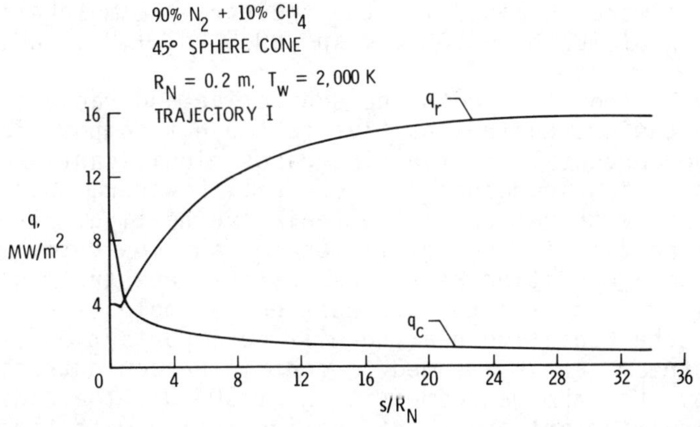

Fig. 19 Variation of convective and radiative heating along the body for 90% N_2 + 10% CH_4, t = 78 s.

velocities, the shock temperature and enthalpy are greater for higher altitudes. The variation in heating rates is shown in Fig. 18. As discussed earlier, the peak convective heating occurs for entry conditions at t=70 s (z=70.4 km) and peak radiative heating at t=78 s (z=50.9 km). These results clearly show that in the stagnation region the radiative heating is not important if the Titan's atmospheric composition is considered to be 99.5% N_2 + 0.5% CH_4.

The extent of convective and radiative heating over the entire length of the aerocapture vehicle is shown in Fig. 19 for the freestream gas composition of 90% N_2 + 10% CH_4. The results clearly show that while the convective heating rate continues to decrease in the downstream region, the radiative rate is considerably higher in this region. As discussed before, the reason for this trend is the combined influence of shock temperature, density, and pressure variation in this region and because of relatively higher optical thicknesses of the radiating shock layer. A similar trend in heating rates was noticed also for the gas composition of 99.5% N_2 + 0.5 % CH_4,[17] but the extent of radiative heating was found to be relatively lower.

Conclusions

Results were obtained to assess the extent of convective and radiative heating to an aerocapture vehicle for the Titan's aerocapture mission. Different compositions for the Titan's atmsophere were assumed and results were obtained for the entry trajectories specified by JPL. Specific results were obtained for the freestream atmospheric composition of 90% N_2 + 10% CH_4 and 99.5% N_2 + 0.5% CH_4.

Results show that both the convective and radiative heating rates are quite sensitive to the gas composition used. The convective heating increases significantly as the N_2 concentration increases. In general, however, this is not the case with regard to the radiative heating. The radiative heating is negligible for the shallow entry (γ = -25 deg) condition regardless of the freestream gas composition. But, for the steepest entry angle (γ = -45 deg), the radiative heating will be important only if the freestream gas is assumed to contain N_2 concentrations of 50-90%. For the gas composition of 90% N_2, the radiative heating is important in the stagnation region with the peak radiative heating rate being 30% of the corresponding convective heating rate (about 13 MW/m^2). For the freestream gas composition of 99.5% N_2 + 0.5% CH_4, the radiative heating in the stagnation region was found to be negligible (less than 3%) in comparison to the convective heating for all cases considered. For this gas composition, the peak convective heating was found to be about 15 MW/m^2. The amount of CN concentration in the shock-layer gas determines the extent of the radiative heating. For a given freestream gas composition, the radiative heating downstream of the stagnation region increases due to an increase in the CN concentration and the optical thickness of the shock layer.

Other results obtained in this study show that higher initial entry speeds produce higher shock temperatures which, in turn, result in higher heating rates. Results for the gas composition with 99.5% N_2 indicate that with the convective heating decreases, the radiative heating increases with increasing body nose radius.

Since the aerocapture vehicle is expected to traverse a part of its atmospheric flight in the transitional flow regime, it is suggested for further study to include both the shock and body slip conditions in the entire analysis. At this time, it might be advisable also to include the effects of nonequilibrium chemistry.

References

[1]Cruz, M. I., "Aerocapture Vehicle Mission Design Concept," AIAA Paper 79-0893, May 1979.

[2]Cruz, M. I., "Aerocapture Vehicle Mission Design Concepts for the Inner and Outer Planets," AIAA Paper 79-115, June 1979.

[3]Armento, R. F., "Mars Aerocapture Vehicle Definition Study Final Report," General Electric Co., Doc. 79SDR2258, Sept. 1979.

[4]French, J. R. and Cruz, M. I., "Aerobreaking and Aerocapture for Planetary Missions," Astronautics and Aeronautics, Vol. 18, No. 2, Feb. 1980, pp. 48-71.

[5]Cruz, M. I., "Technology Requirements for a Generic Aerocapture System," AIAA Paper 80-1493, July 1980.

[6]Florence, D. E., "Aerothermodynamic Design Feasibility of a Mars Aerocapture/Aeromaneuver Vehicle," AIAA Paper 81-0350, Jan. 1981.

[7]Bird, R. B., Stewart, W. E., and Lightfoot, E. N., Transport Phenomena, John Wiley & Sons, New York, 1960.

[8]Moss, J. N., "Reacting Viscous-Shock-Layer Solutions with Multi-component Diffusion and Mass Injection," NASA TR-411, June 1974.

[9]Tiwari, S. N. and Szema, K. Y., "Effects of Precursor Heating on Radiative and Chemically Reacting Viscous Flow Around a Jovian Entry Body," NASA CR-3186, Oct. 1979.

[10]Moss, J. N., "Stagnation and Downstream Viscous-Shock-Layer Solution with Radiation and Coupled Ablation Injection," AIAA Journal, Vol. 14, Sept. 1976, pp. 1311-1317.

[11]Tiwari, S. N. and Szema, K. Y., "Influence of Precursor Heating on Viscous Flow Around a Jovian Entry Body," Outer Planet Entry Heating and Thermal Protection, Progress in Astronautics and Aeronautics, Vol. 64, edited by R. Viskanta, AIAA, New York, 1979, pp. 80-107.

[12]Davis, R. T., "Numerical Solution of the Hypersonic Viscous-Shock-Layer Equations," AIAA Journal, Vol. 8, May 1970, pp. 843-851.

[13]Tiwari, S. N. and Subramanian, S. V., "Influence of Nonequilibrium Radiation and Shape Change on Aerothermal Environment of Jovian Entry Body," NASA CR-3432, Aug. 1980 (also Ph.D. Dissertation by S. V. Subramanian, Old Dominion University, Norfolk, Va., March 1980).

[14]Nicolet, W. E., "User's Manual for the Generalized Radiation Transfer Code (RAD/EQUIL)," NASA CR-116353, Oct. 1969 (also "User's Manual for RAD/EQUIL/1973, A General Purpose Radiation Transport Program," NASA CR-132470, Nov. 1973).

[15]Golden, F., "Visit to a Large Planet: A Thousand Rings Around Saturn, Icy Moons and Lakes of Liquid Nitrogen," Time Magazine, Vol. 116, No. 21, Nov. 24, 1980, pp. 32-41.

[16]Stround, C. W. and Brinkley, K. L., "Chemical Equilibrium of Ablation Materials Including Condensed Species," NASA TN D-5391, Aug. 1969.

[17]Tiwari, S. N. and Chow, H., "Aerothermodynamic Environment for a Titan Aerocapture Vehicle," NASA CR-NAG1-120, Jan. 1982 (also Master of Engineering Thesis by H. Chow, Old Dominion University, Norfolk, Va., Dec. 1981).

[18]Esch, D. D., Siripong, A., and Pike, R. W., "Thermodynamic Properties in Polynomial Form for Carbon, Hydrogen, Nitrogen, and Oxygen Systems From 300 to 1500 K," NASA CR-111989, Nov. 1970.

[19]McBride, B. J., Heimel, S., Ehlers, J. G., and Gordon, S., "Thermodynamic Properties to 6000 K for 210 Substances Involving the First 18 Elements," NASA SP-3001, 1963.

[20]Esch, D. D., Pike, R. W., Engel, C. K., Farmer, R. C., and Balhoff, J. F., "Stagnation Region Heating of a Phenolic-Nylon Ablator During Return From Planetary Missions," NASA CR-112026, Sept. 1971.

[21]Grier, N. T., "Calculation of Transport Properties and Heat-Transfer Parameters of Dissociating Hydrogen," NASA TN D-1406, Oct. 1962.

Optimization of Aerobraked Orbital Transfer Vehicles

D.G. Andrews* and V.A. Caluori†
Boeing Aerospace Company, Seattle, Wash.
and
F. Bloetscher‡
Goodyear Aerospace Corporation, Akron, Ohio

Abstract

A new method for utilizing the upper atmosphere to modify Earth orbital trajectories through use of a pressure-supported drag brake has been previously proposed. Analytical results and initial wind-tunnel test data indicate that there is considerable improvement available to the original aerobraking concept. Optimization of the drag brake with respect to size vs structural heating and the application of more advanced cooling techniques indicate that reductions in weight and volume of the aerobraking system by 50% should be possible. Additional analysis indicates that further improvements might be available through use of a alternate drag control mode that eliminates the need to throttle the vehicle main engine during the aerobraking maneuver.

Nomenclature

A	= Ballute area
C_D	= drag coefficient
P_i	= ballute internal pressure
P_{stag}	= freestream stagnation pressure
W	= vehicle weight at re-entry
ρ	= ambient density at time t
ρ_{design}	= atmospheric density in design trajectory at time t

---

Presented as Paper 81-1126 at the AIAA 16th Thermophysics Conference, Palo Alto, Calif., June 23-25, 1981,

*Senior Specialist Engineer.
+Upper Stages and Launch Vehicles Manager.
≠Senior Engineering Specialist.

I. Introduction

The recent and most successful Space Shuttle flight has marked the opening of a new era for frequent and low-cost manned access to near-Earth space. In the near future the Shuttle will serve as the platform from which many unmanned missions to the more distant and useful high orbits will be launched. Geosynchronous orbit, the most valuable of the distant space destinations, lies an additional 14,000 ft/s of velocity beyond the 25,000 ft/s attained by the Space Shuttle. As increasingly valuable assets (both in the sense of their purchased costs and their revenue production or ongoing strategic value) are deployed at these high altitudes, manned access to those destinations will become commensurately critical. The return to low-Earth orbit of this human payload adds another 14,000 ft/s velocity increment to an already difficult mission. This round trip is nearly equivalent in difficulty to our Apollo flights to the moon. Its technical solutions are further constrained by the throw weight and physical size limitations of the Shuttle as well as the practical limitations of vehicle structural fractions and propulsive efficiencies. Many complex and expensive solutions have been proposed which employ multiple Shuttle launches and "on-orbit" assembly of a "wagon train" of fuel tanks, propulsion systems, and payload. Although workable, these solutions require that each manned mission accomplish many events with high value in order to be cost effective. The long mission times associated with the stay at the destination altitude (to efficiently conduct the many and separated events), as well as in low Earth orbit while the "expedition is organized," reduce the flixibility of this approach, perhaps beyond the limits of practicality for certain applications. The early possible alternatives to this scenario require either a dramatic increase in the initial launch to low Earth orbit capability or use of the far reaches of the atmosphere to assist the vehicle in its return to low Earth orbit. This paper deals with the later alternative in which aerodynamic drag generated during a grazing pass through the upper atmosphere (Fig. 1) is substituted for a propulsive braking maneuver with a value of approximately 7000 ft/s.

Previous attempts to use the upper reaches of the atmosphere to enhance orbit transfer vehicle (OTV) performance were hampered by the dramatic impact on the vehicle's basic character, and therefore its "desirability" and efficiency, during the three-quarters of the mission when it operated in the space environment. In order to

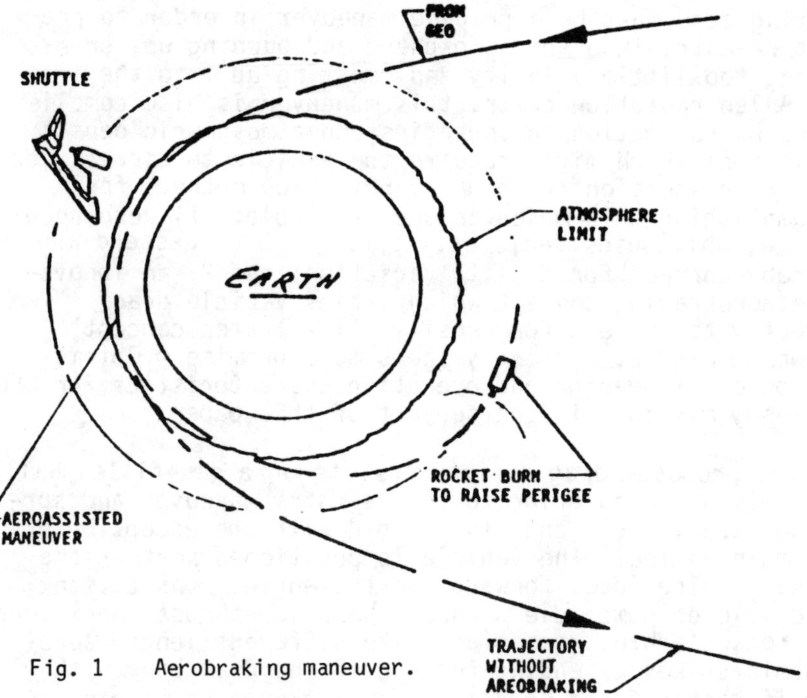

Fig. 1 Aerobraking maneuver.

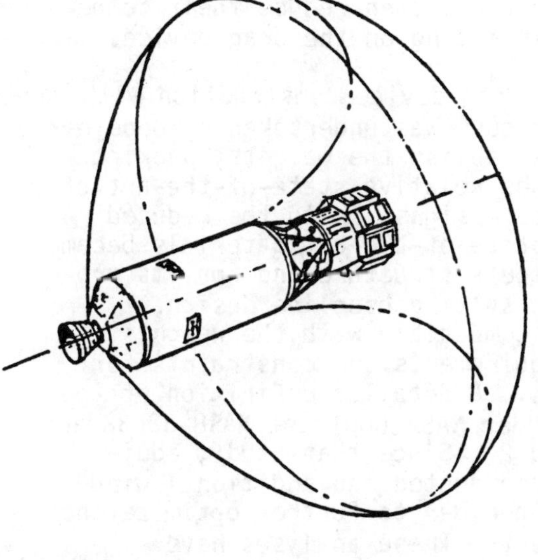

Fig. 2 Aerobraking orbital transfer vehicle.

successfully convert the 30% of total mission fuel consumed without an aerodynamic assist to useful payload, the means of accomplishing the aeroassist maneuver would require little impact to the overall vehicle weight. Since the grazing pass must be a precise maneuver in order to prevent re-entry into the atmosphere and burning up, or losing too little velocity and coasting up into the Van Allen radiation belts, this maneuver is also complicated by navigation inaccuracies and atmospheric density variations which might require the vehicle to correct for density dispersions as high as 50%. Two methods for accomplishing this maneuver are available: 1) aeromaneuvering, which uses vehicle lift to climb or descend and thereby correct for density variation; and 2) an innovative aerobraking concept which varies vehicle drag directly to correct for density. The latter concept, shown in Fig. 2, currently shows more promise. Optimization of its design and operating characteristics for GEO delivery missions is the subject of this paper.

The proposed drag device consists of a gas-filled Ballute‡ which is deployed prior to the re-entry maneuver and surrounds the vehicle and its payload with the exception of the main engine. The vehicle is positioned so that the rocket engine faces forward and the engine runs at tank head idle or pump idle. Under these low-thrust conditions the rocket exhaust gases act like different length aerodynamic spikes extending forward of the drag body with significant effects on drag. The rocket exhaust gases also mix with the aerodynamically heated atmospheric gases and, being of lower temperature, then reduce their temperature and the heating rates acting on the drag device.

Design analysis of the drag device construction with various heat protection techniques was undertaken by Goodyear Aerospace Corporation to establish the weights, packing volumes, interfaces, and the relative state-of-the-art of possible aerobraking device designs. With the reduced heating rates the use of state-of-the-art materials became possible for the drag device's structure and for its protective insulation. The resulting baseline design, although conservative, is compatible with the weights, volumes, and interface requirements and constraints for efficient aerobraking OTVs. A detailed definition of the aerobraked OTV was done under NASA Contract NAS8-33532 and is reported in Refs. 1 and 2. Since that study, additional analyses have been conducted and additional wind-tunnel testing scheduled in order to further optimize the proposed aerobraking concept. These analyses have

‡The term Ballute is the Goodyear Aerospace Corp. trademark for the ballonparachute.

examined the optimum sizing of the Ballute with respect to heating rates and vehicle stability, an alternate method of controlling vehicle drag, and the use of advanced materials or transpiration cooling for the Ballute. The wind-tunnel tests, which are scheduled to begin momentarily, will show the feasibility of the alternate method for varying drag and provide heat-transfer data for more accurate sizing of the thermal protection system. This paper will report on the results of the above analyses and testing. It will show the savings in weight and volume provided by resizing the Ballute and by using advanced materials or cooling techniques. It will also show the potential advantages of the alternate method of controlling vehicle drag.

II. Baseline Aerobraking Concept Definition

The requirements for the baseline aerobrake device include meeting the system nonoperating and operating requirements. Nonoperating requirements consist of surviving the transportation and storage environments prior to launch, the environment within the Space Shuttle payload bay from launch to low Earth orbit, and the space/OTV environment to geosynchronous orbit and back. The Ballute operating requirements consist of: 1) deploying completely prior to re-entry, 2) providing stability throughout re-entry, 3) providing the required drag including drag modulation throughout re-entry, 4) producing a surface large enough to protect the OTV and to reduce heating rates to levels acceptable for state-of-the-art materials, and 5) cleanly separating from the OTV and self-destructing after completion of the aerobrake maneuver.

The baseline Ballute structural design was based on a diameter of 80 ft to obtain a generous stability margin, an adequate drag margin, and heating rates which could be handled using available materials (Fig. 3). The Ballute

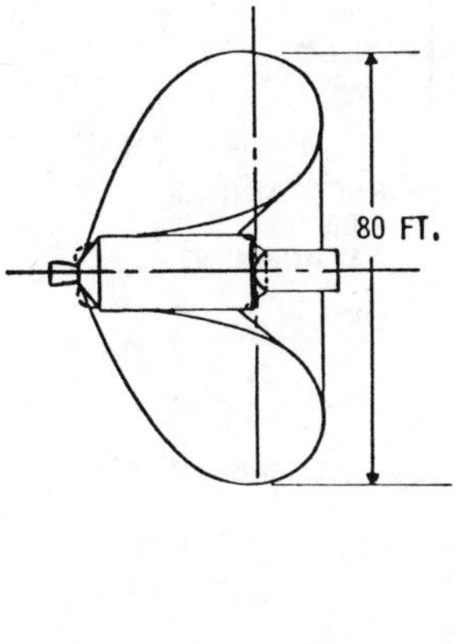

Fig. 3 Baseline ABOTV concept.

design shape is obtained when the tension loads in the Ballute's surface balance the internal and external loadings. The Ballute's surface consists of bias cloth gores and meridian straps that carry only tension loads. Goodyear Aerospace computer programs were used to establish the Ballute shapes and material tensions that balance the internal and external aerodynamic pressures. The design strength of the Ballute cloth and meridians were calculated based on the NASA AID shape[3] and the aerobraked OTV flight conditions at maximum drag. The calculated stresses in the minimum gage cloth currently available result in design safety factor of 21 at 600°F. The calculated meridian strap loads are also small and a design safety factor of six at 600°F can be obtained with lightweight webbing. The aramid (Kevlar) cloth used has approximately one-half of its room temperature strength at 600°F and starts to char at 800°F.

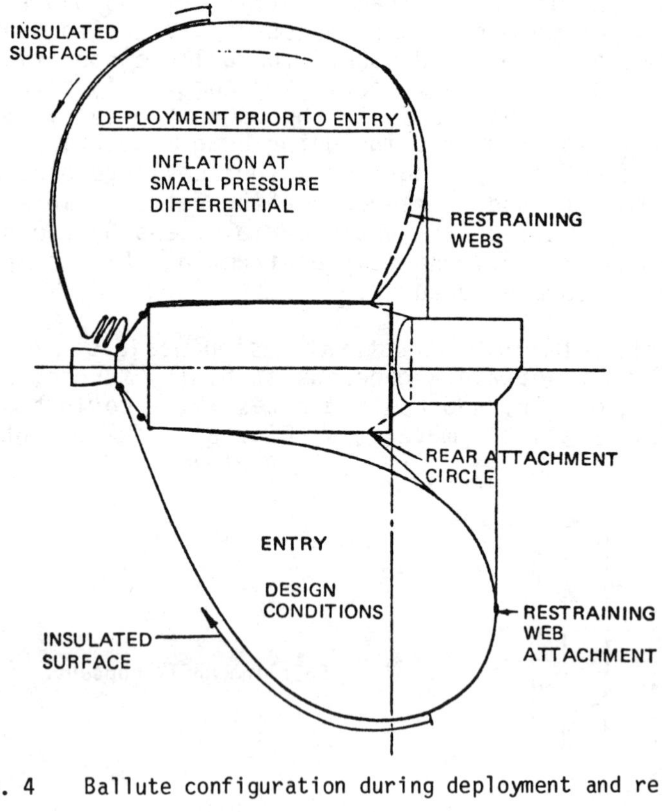

Fig. 4 Ballute configuration during deployment and reentry.

Schematic views of the Ballute being deployed prior to external loads and then later when reaching perigee conditions are presented in Fig. 4. The internal gas pressure unfolds the accordion pleats of the Ballute as it is expanded from its packing bag for deployment. The insulation over the forward region of the Ballute and the restraining webs that position the OTV within the Ballute are indicated. The schematic in the lower portion of the figure illustrates the nominal shape of the Ballute under its design operating condition. The Ballute maintains this design shape by controlling the internal pressure throughout the trajectory such that the ratio of internal pressure to stagnation pressure is always constant.

The baseline drag control mode used during the NASA-sponsored OTV study (NAS8-33532) was drag variation through use of engine thrust to vary the flowfield characteristics. Vehicle drag can be reduced by approximately an order of magnitude by switching from tank head idle to low pumped idle engine thrust levels. A detailed description of this operating mode is found in Ref. 2 and it will only be summarized here.

The basic operating scenario for a worst density trajectory using the baseline control mode is shown in Fig. 5. At t = 0 the LH_2 engine valves are opened to chill the lines. At t = 64 s the engine is started, running in the tank head idle mode. The engine stays at tank head idle until t = 100 s and during this period vehicle deceleration is measured using accelerometers. The measured deceleration is compared with decelerations associated with the design trajectory and the differences are input into an algorithm which determines when and for how long the engine thrust should be raised to pumped idle to correct for density variations. For a minimum density trajectory the engine will stay at tank head idle; for the maximum density trajectory shown, the engine will be at pumped idle for 40 s right through perigee.

After engine shutdown the Ballute is gradually allowed to cool and deflate until the vehicle has left the sensible atmosphere, at which time the forward restraint straps are released. This clears the auxiliary control system for use and enables the vehicle to orient itself for the apogee burn. The Ballute is discarded during the burn and re-enters on its next pass.

This completes the definition of the baseline aerobraking concept used throughout the NASA OTV Concept Definition

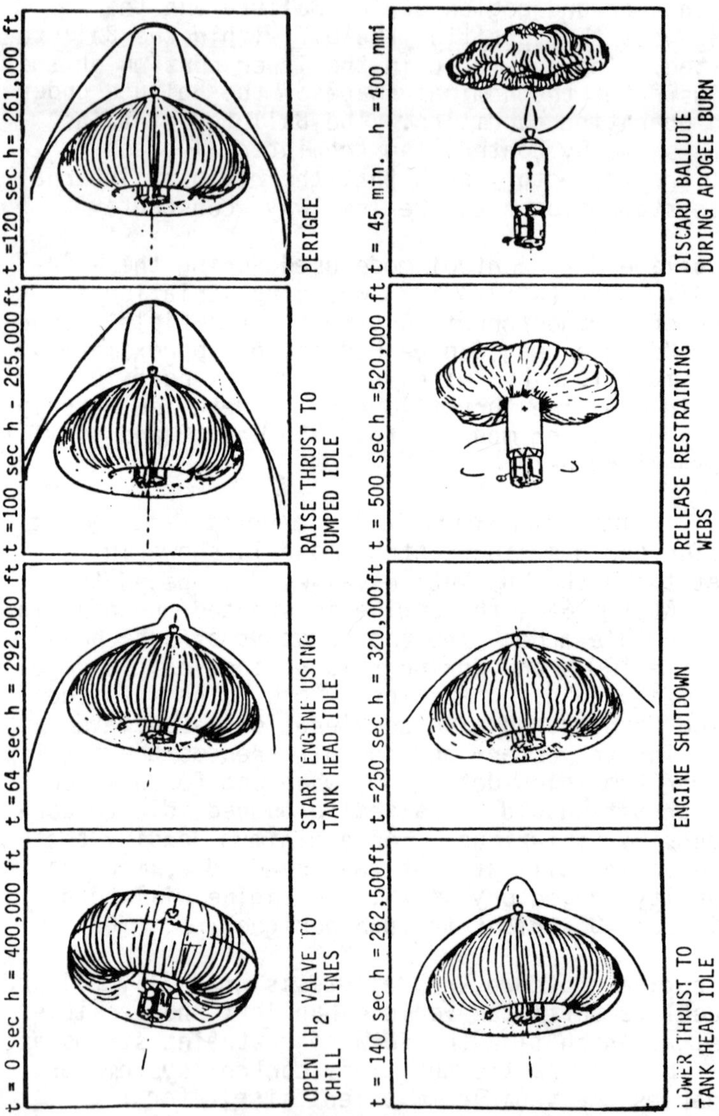

Fig. 5 Aerobraking operating scenario (high-density trajectory).

Study.[1] At the conclusion of that study some additional sizing trades were completed to optimize Ballute performance.

III. Ballute Sizing Trade

A parametric sizing trade has been completed to find the Ballute size which minimizes aerobraking system mass, protects the vehicle/payload combination from freestream air impingement, and meets all stability considerations. Four different Ballute diameters were investigated: 170, 140, 80, and 48 ft. At an assumed re-entry weight of 24,000 lb (roundtrip mission with manned capsule) and using the baseline Ballute shape these diameters correspond to ballistic coefficients (W/C_DA) of 1.0, 1.5, 4.5, and 12.5 lb/ft^2, respectively.

For each ballistic coefficient there is a corresponding ballistic trajectory such that the vehicle will lose exactly the right amount of velocity and leave the atmosphere in the proper phasing orbit. The air density at perigee varies linearly with ballistic coefficient for this condition. The altitude corresponding to this density is the maximum perigee altitude for which a vehicle with that ballistic coefficient can reach the phasing orbit in a single pass. The first four columns of Table 1 show the ranges of Ballute diameters, ballistic coefficients, design perigee altitudes, and perigee densities examined in this study.

Navigation inaccuracies and atmospheric uncertainties require the vehicle to be able to correct for density variations of $\pm 20\%$. In order to meet this design criterion, nominal trajectories are selected which have densities at perigee 20% greater than the densities at the design perigee altitudes. This allows the vehicle to reach the phasing orbit in a single pass even if the density is 20% less than nominal. At greater densities the

Table 1 Summary of aerobraking sizing trends

Ballute diam, ft	$\frac{W}{C_DA}$	Max H per ft	Min ρ max slug/ft^3 X10^{-7}	Max ρ max slug/ft^3 X10^{-7}	$\dot{H}$, Btu/ft/s	Insulation thickness, in.
170	1.0	300,000	0.048	0.075	1.7	0.25
140	1.5	292,500	0.075	0.117	2.1	0.31
80	4.5	273,000	0.225	0.352	3.7	0.50
48	12.5	252,000	0.625	0.977	6.2	0.70

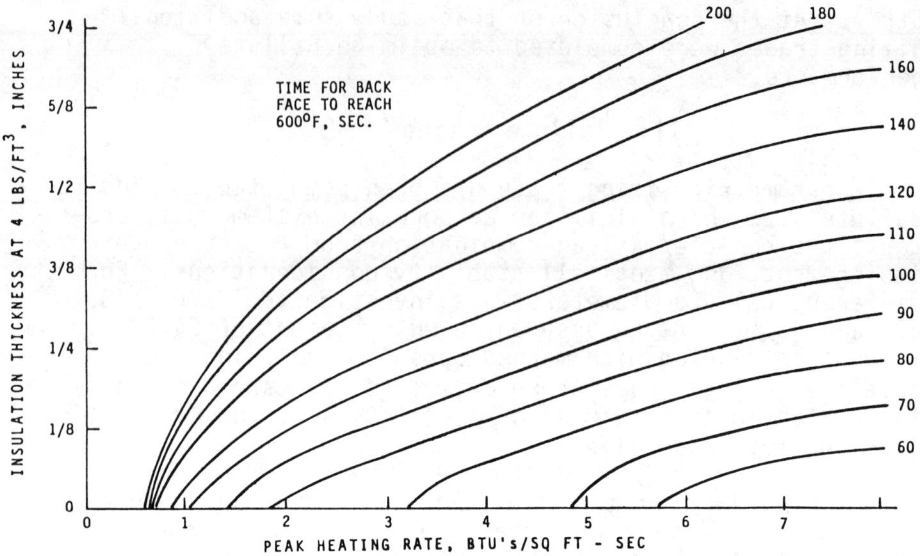

Fig. 6 Insulation thickness required vs peak heating.

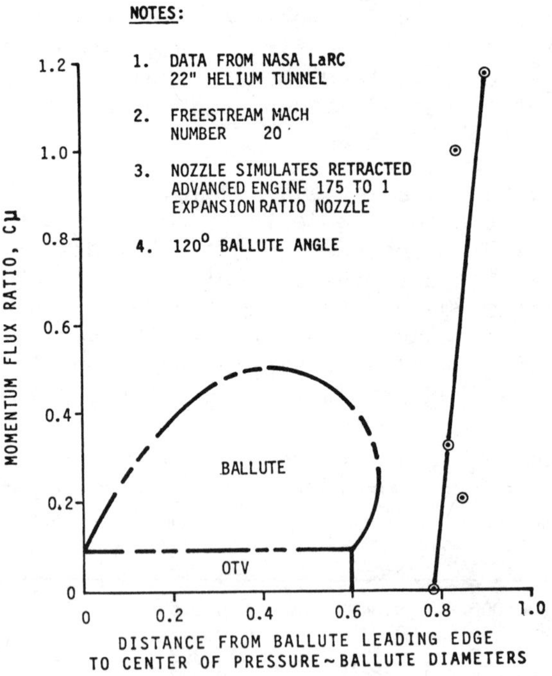

Fig. 7 Effect of jet flow on center of pressure.

Ballute drag is reduced via the method outlined in the last section in order to make the phasing orbit.

The aerodynamic heating rate is proportional $\rho^{0.5} V^3$ so the worst heating case is for the nominal plus 20% density case (50% greater than the design perigee altitude density). This density is shown in fourth column of Table 1 for each of the Ballute diameters, and the maximum heating rate corresponding to each of these maximum densities is shown in the fifth column. The insulation thickness required to protect the Kevlar Ballute structure for various heating rates is shown in Fig. 6. For the trajectories flown and the safety factor available, it has been estimated that the back face of the insulation should not reach 600°F until 200 s into the maneuver to insure Ballute survival. Note, that airloads on the Ballute peak at about 120 s and are reduced to approximately 10% of their peak values by 200 s.

Ballute weight estimates have been made using the insulation thickness required for each ballistic coefficient. Those estimates are shown in Table 2. As can be seen, increasing the Ballute diameter to reduce heating results in more, not less, insulation weight and therefore is a losing proposition. Accordingly, the optimum Ballute is the smallest Ballute which can meet vehicle stability requirements.

Vehicle stability data was obtained during the wind-tunnel test when the effect of jet flow on the center of pressure was investigated. The results of those tests are shown in Fig. 7. These data show that the Ballute is very stable and that the stability improves with increasing thrust, as

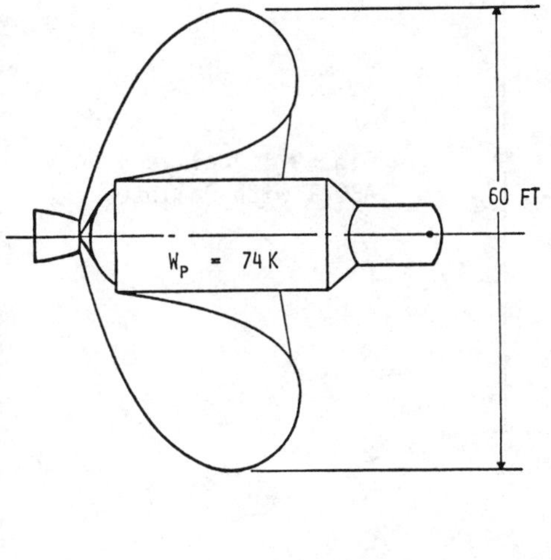

Fig. 8 Manned sortie ABOTV with new baseline Ballute.

was predicted prior to the testing. The location of the center of pressure at low thrust levels would allow the Ballute diameter to be reduced to approximately 50 ft and still be neutrally stable. However, to be on the safe side the baseline Ballute diameter was reduced only from 80 to 60 ft. The resulting vehicle configuration is shown in Fig. 8. For delivery-only missions the returning vehicle is much lighter and a smaller Ballute will provide an adequate ballistic coefficient. Since the center of gravity is well forward in this case, vehicle stability during the aerobraking maneuver is not a concern. A Ballute sized to provide the same ballistic coefficient as the new baseline manned vehicle (8.5 lb/ft^2) is shown in Fig. 9.

IV. Application of Advanced Technologies

In addition to resizing the Ballute, three alternate Ballute constructions were investigated to determine the potential for weight reduction through the application of advanced technology. The baseline Ballute construction and the three alternate constructions are summarized in Table 3. Of the three alternative Ballutes the first, transpiration cooling, offers large weight and volume reductions for additional operational complexity. The second, using Nextel 312 cloth, also offers large weight and volume reductions, but relies on a rather brittle material that does not like to be folded to tight radii, which could greatly aggravate the packing problem. The third option was not pursued further because no flexible gas-tight sealing material could be found with sufficient high-temperature capability.

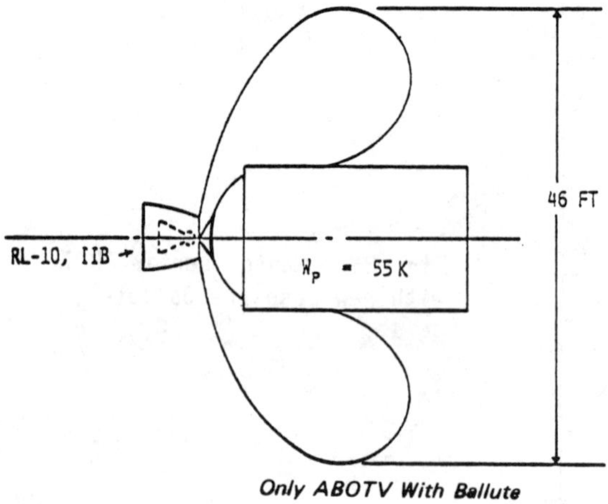

Fig. 9 Delivery only ABOVT with Ballute.

Table 2 Ballute weight estimates

	Diameter, ft			
	48	80	140	170
Kevlar cloth	109	302	925	1364
Kevlar webbing	23	64	196	289
Gas barrier coating	109	302	925	1364
Insulation	730	1450	2753	3274
Insulation cover cloth	49	136	417	614
Installation provisions, jettisonable	90	120	160	200
Margin	170	356	784	1065
(Total jettisonable)	(1280)	(2730)	(6160)	(8170)
Installation provisions, fixed	87	87	87	87
Margin	13	13	13	13
(Total fixed)	(100)	(100)	(100)	(100)
Total Ballute	1380	2830	6260	8270

Table 3 Ballute weight reduction studies

Ballute construction	Comments
Insulation to limit structure to < 600°F Fiberfrax felt blanket over silicone-coated Kevlar cloth structure	Baseline Insulations being investigated
Cooling to limit structure to < 600°F Transpiration cooling of porous Kevlar cloth structure using water or GH_2	Attractive option Water rate requirements compatible with on board supply
Ablation to limit structure to < 2600°F Silicone covered Nextel 312 cloth structure	Possible option Material and optical treatment suitability being investigated
Reradiation to limit structure to < 2600°F Treat the Nextel 312 cloth to provide the required surface optical properties	Possible option Material suitability being investigated

Investigation of transpiration cooling of fabric structures uncovered two different correlations for cooling effectiveness. The first one is the "classic" correlation shown in the Bartle and Leadon transpiration cooling study.[4] This correlation is summarized in Fig. 10 for different Mach numbers and different gases. The second correlation is based on test results from a hydrogen transpiration-cooled metal fabric nozzle extension fabricated and tested by the Goodyear Aerospace Corporation on a J-2 rocket engine.[5] As can be seen in Fig. 11, the Goodyear correlation provides markedly better cooling effectiveness for the same mass flow rates, but has not had as wide acceptance and use as the Bartle-Leadon correlation.

The coolant requirements to complete the aerobraking maneuver are shown in Fig. 12 as a function of Ballute diameter and the correlation equation assumed. For the 60 ft diam baseline Ballute, the coolant requirement ranges from a high of 1000 lb of water, using Bartle-Leadon correlation, to a low of 100 lb of hydrogen using the Goodyear correlation. It is obvious from the large spread in possible coolant requirements that more testing and analysis is required to define the transpiration-cooled Ballute system. For the purposes of preliminary design studies involving transpiration-cooled Ballutes, we have assumed that an aerobraked manned OTV using a 60 ft Ballute requires 500 lb of water and/or hydrogen coolant.

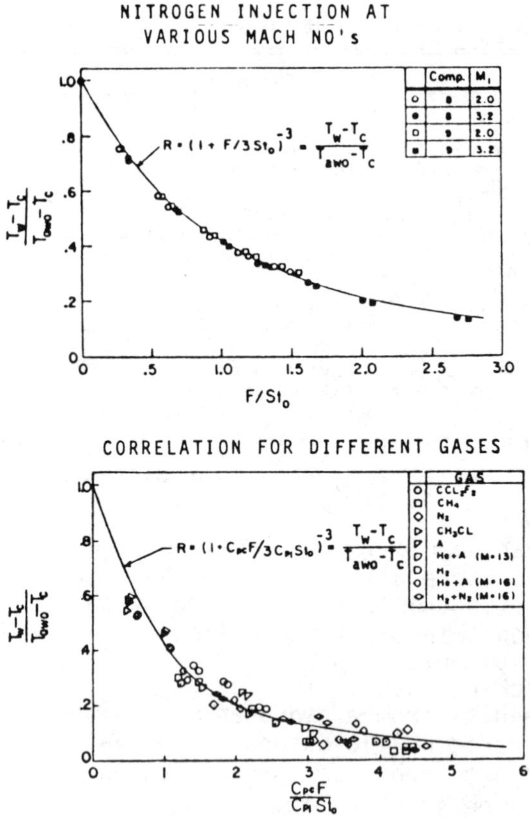

Nomenclature

- T = temperature
- C_p = specific heat
- F = injected flow to free stream flow ratio
- St = Stanton number

Subscripts

- c = coolant
- 1 = free stream
- w = wall
- aw = adiabatic wall
- o = no injection

Fig. 10 Bartle-Leadon transpiration cooling study.

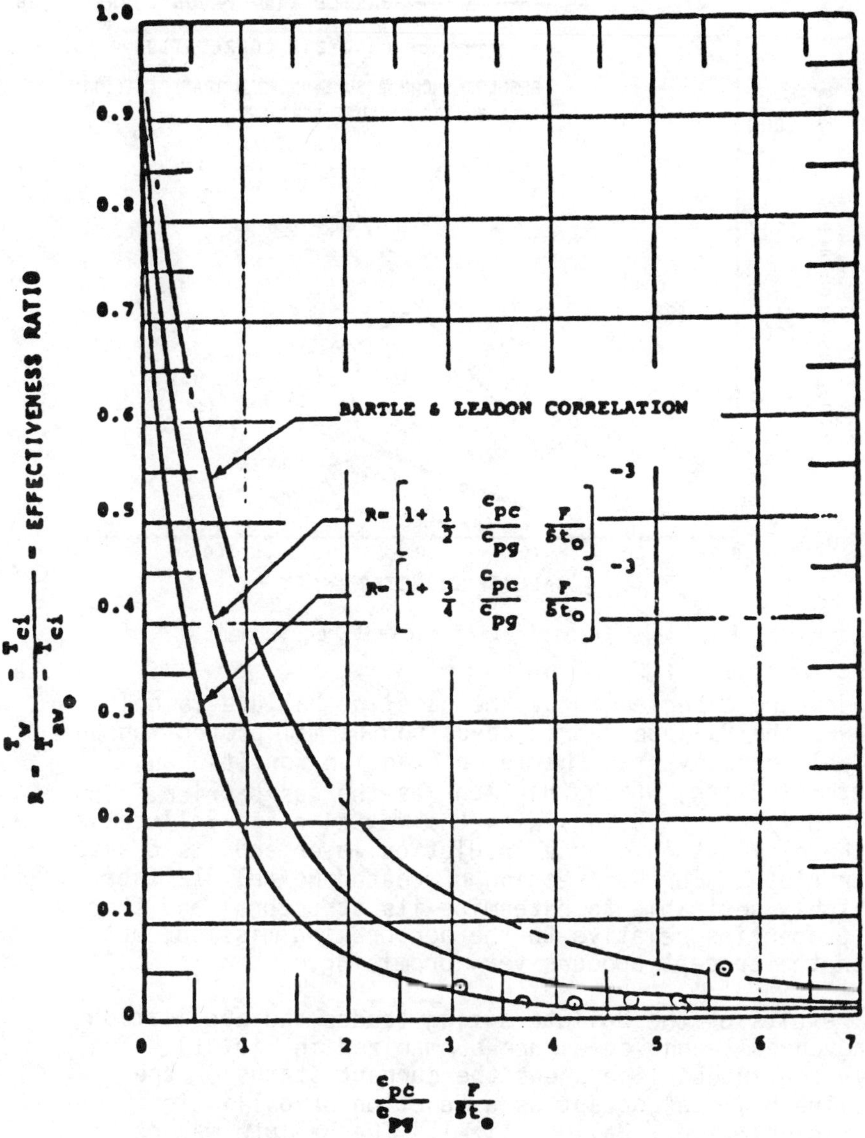

Fig. 11 Goodyear transpiration nozzle cooling study.

Investigation of a high-temperature ablator covered Ballute centered on the use of Nextel 312 ceramic fiber, a 3M product which has good structural properties to temperatures as high as 2600°F. The Nextel 312 fabric is coated with a silicone rubber barrier. Since the peak surface

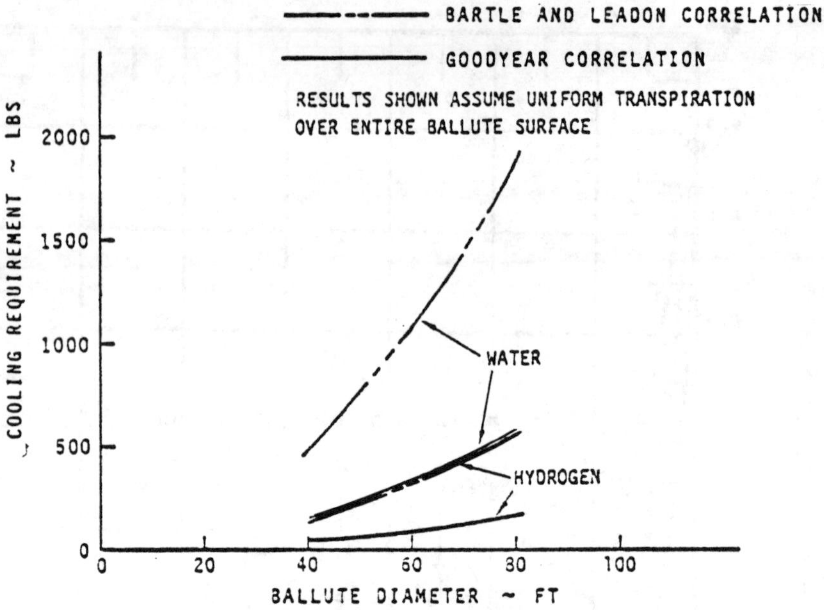

Fig. 12 Transpiration coolant requirement.

temperature experienced by the baseline Ballute is only 2000°F, the Ballute fabric needs no thermal protection and the ablator serves as thermal protection for its own innermost layer, which functions as the gas barrier. The principal weight savings with the Nextel-cloth Ballute is in the elimination of the insulation layer and its quartz cover cloth. Further testing of coated Nextel 312 fabric is highly desirable to determine its structural and thermal properties relative to the aerobraking mission, but on paper the concept appears very promising.

The results of the Ballute sizing trades and application of advanced technologies are summarized in Fig. 13. The heavy continuous line shows the current status of the baseline Ballute concept as a function of ballistic coefficient (i.e., Ballute size). The Ballute was resized to 60 ft diam based on the sizing trade and the wind-tunnel center-of-pressure data. This leaves an adequate stability margin to accept reasonable changes in the Ballute shape caused by pressure variations. Engineering estimates of the weights of the two advanced technology Ballute concepts are also shown. Note that when the weight of the transpiration coolant (500 lb) is added to the weight of the actively cooled Ballute, the total weight is very close to the weight of the Nextel 312

Ballute. Therefore, it appears that either of the advanced technology Ballute concepts will weigh approximately 1000 lb for a 60 ft diam Ballute.

This represents a considerable increase in mission payload relative to the baseline aerobrake system and both advanced Ballute concepts should be pursued in order to assess development technology readiness.

V. Alternate Drag Control Mode

The baseline drag control method developed during the OTV study involved maintaining a constant Ballute shape and using changes in engine thrust to vary the shape of the surrounding flowfield. This was the first drag control method investigated because it was the easiest method for which to obtain test data, since a solid Ballute model could be used. Once the test data were available, a workable drag control method became apparent and the alternate drag control modes were not pursued in order to conserve engineering effort. Recent in-house studies have examined the drag control issue again and decided to pursue the alternate drag control mode to determine its potential. The alternate control mode involves keeping the engine thrust constant and allowing the Ballute to change shape as the atmospheric density varies off the nominal. This method has the potential for automatically and continuously correcting for density dispersions, but is difficult to analyze and/or test.

A 4 in. diam flexible Ballute model has been designed and built by Goodyear Aerospace Corporation, and is scheduled to be tested in the NASA Langley Mach 20 helium tunnel shortly after this paper is submitted. The tunnel tests should produce definitive results with respect to Ballute shape as a function of internal to dynamic pressure ratio.

In the meantime, an analytical analysis of the variation of Ballute shape with changes in pressure ratio has been completed. The Ballute shape changes due to pressure ratio (internal pressure/stagnation point pressure) changes were calculated for a 120 deg included angle Ballute designed for the maximum entry conditions. The analysis preserved the length of the meridian webs and allowed the lightly loaded cloth between the 120 webs to change the amount it lobes between webs. The cloth yarns run at 45 deg to the meridian webs at design conditions.

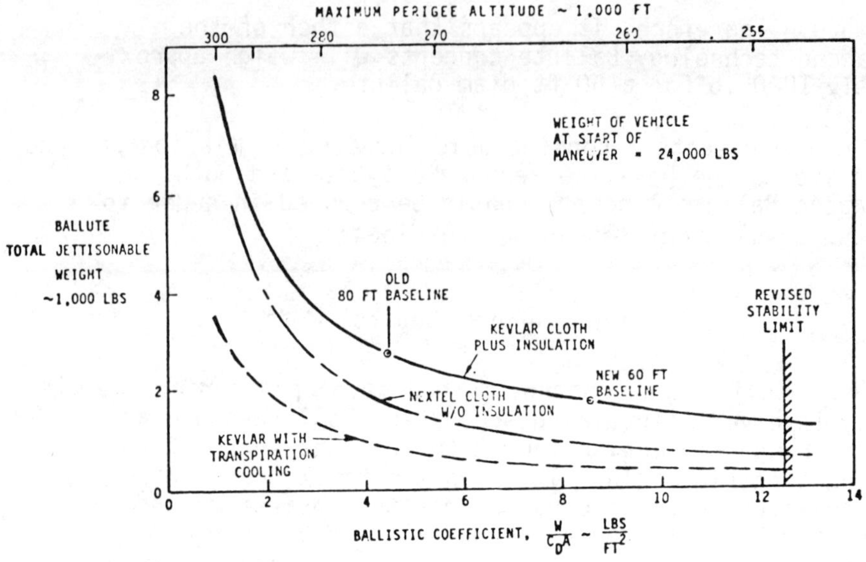

Fig. 13 Aerobraking sizing trade.

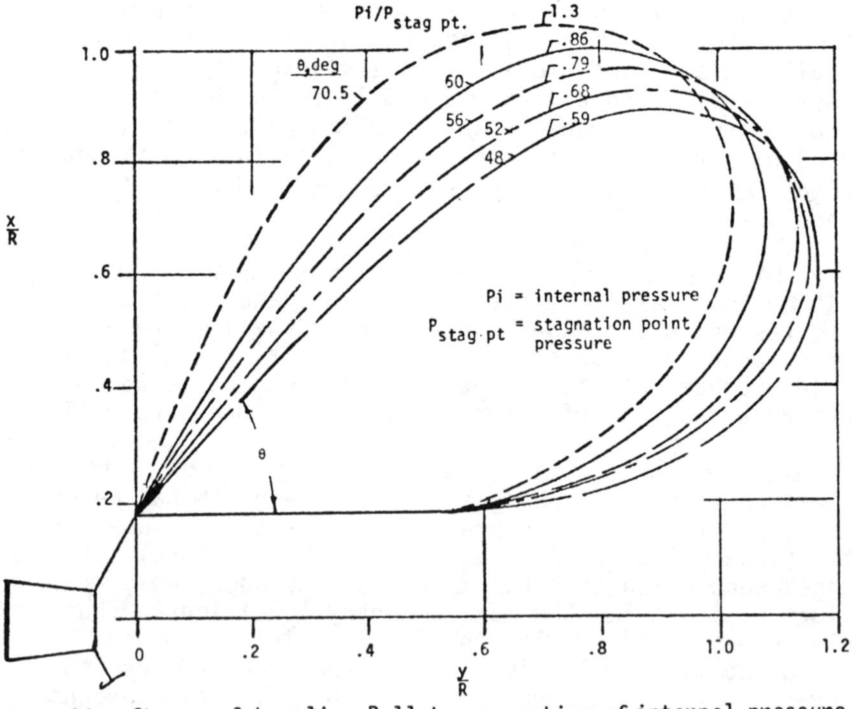

Fig. 14 Shapes of baseline Ballute vs. ration of internal pressure to stagnation point pressure.

At off-design conditions, the yarn angles change a few degrees, thus allowing the cloth in the gores to change length by changing the amount it lobes between webs.

The analysis consisted of an iterative process where a pressure ratio value is chosen for an initial angle and the shape of the Ballute is generated by segments. The process is repeated until all of the boundary conditions are met. The boundary conditions include the initial Ballute angle, the forward and rearward attachment point locations for the webbing, and the total length of the webbing. The method considered the varying aerodynamic pressure over the forward surface of the Ballute with its changing surface angle. The resulting variation in Ballute shape with varying pressure ratios is shown in Fig. 14.

The effective drag area (C_DA) for a series of these shapes has been calculated using modified Newtonian theory. The variation in effective drag area with the internal-to-stagnation pressure ratio is shown in Fig. 15. Area A is the cross-sectional area shown in Fig. 14 (A for the 60 deg design case is Pi). There is some scatter in the calcu-

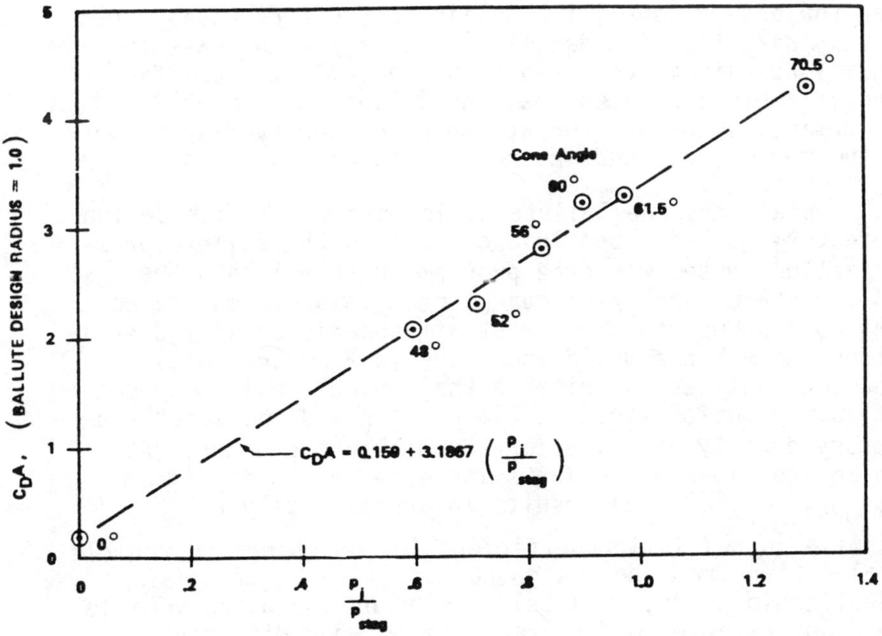

Fig. 15 Ballute effective drag area vs internal-to-stagnation pressure ratio.

lated values resulting from curvature approximations and measurement errors in the calculation process, but in general the effective drag area increases linearly with increases in the ratio of Ballute internal pressure to stagnation point pressure. There is an obvious upper limit on C_DA, since once increasing internal pressure drives the Ballute into a toroidal (tire) shape, no further increases in C_DA are possible. This limit for the Ballute shown in Fig. 14 would be around $C_DA = 6$.

To apply these calculated values toward drag modulation during a ballistic trajectory, certain assumptions must be made. First, assume that the Ballute system has a gas generator and bleed system that can be programmed to maintain an arbitrary Ballute internal pressure ratio of 0.86 as the Ballute traverses its design ballistic trajectory as discussed in Sec. III. This means the Ballute would maintain its 60 deg cone angle throughout its design trajectory and successfully reach phasing orbit. However, for off-design trajectories the densities encountered would vary, changing the stagnation pressures from design levels and hence changing the Ballute shape. To investigate this effect, the Ballute drag data have been replotted in Fig. 16 to show how the ballistic coefficient varies with changes in atmospheric density. For densities near the design level, the ballistic coefficient appears to vary directly with density (i.e., a 10% increase in density results in a 10% increase in ballistic coefficient). This indicates that the Ballute may be able to automatically correct for atmospheric density dispersions of the magnitude found in orbital operations.

In practice, the Ballute would be sized by its design trajectory as described in Sec. III and the Ballute pressurization system would be programmed to maintain the Ballute shape over this trajectory. This design trajectory corresponds to the lowest atmospheric density distribution the vehicle would encounter (3 σ error limit). Under current design criteria the nominal trajectory density distribution would be 25% greater and the worst trajectory density distribution would be 50% greater. As can be seen in Fig. 16 a 50% increase in density (from ρ/ρ_{DESIGN} = 1.0-1.5) results in approximately a 50% increase in ballistic coefficient for no change in vehicle deceleration from the design case. In fact, if this relationship holds, an accelerometer on board the vehicle would not be able to differentiate between different trajectories.

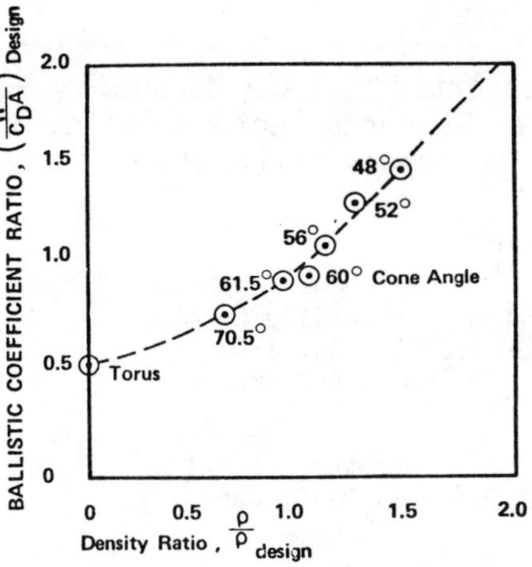

Fig. 16 Variation in Ballute ballistic coefficient with off-design atmospheric densities.

This result is based on preliminary analytical data since the flexible Ballute model has not yet been tested. Should the test results indicate this linear relationship between ballistic coefficient and atmospheric densities holds, then this mode of drag control should be given very serious consideration.

VI. Conclusions

The analytical results presented here indicate that there are many alternatives to the baseline aerobraking concept developed under NASA Contract NAS8-33532. There also appears to be considerable improvement available with respect to Ballute weight and bulk if some of the more advanced concepts are instituted. In fact, implementing transpiration cooling or a high-temperature fabric into the Ballute appears to reduce the weight and volume of the aerobraking system by about 50%. Further improvements might be available through use of an alternate drag control mode which eliminates the need to throttle the vehicle main engine and saves approximately 100 lb of propellant.

In summary, the more we investigate the aerobraking Orbital Transfer Vehicle, the more convinced we become that it has the most potential for space transportation system improvement outside of fundamental changes to the basic launch vehicle. We are convinced that aerobraking will work and the only remaining issues are what form the final system should take and what degree of performance will be attained.

Acknowledgment

The work reported here was supported by NASA Marshall Space Flight Center, NASA Langley Research Center, Boeing Aerospace Company, and Goodyear Aerospace Corporation.

References

[1]Caluori, V. A. et. al., "Orbital Transfer Vehicle Concept Definition Study," Boeing Doc. D180-26090, Vols. 1-6, Final Report, Contract NAS8-33532, Boeing Aerospace Co., 1980.

[2]Andrews, D. G. and Bloetscher, F., "Aerobraked Orbital Transfer Vehicle Definition," AIAA Paper 81-0279, Jan. 1981.

[3]Barton, R. R., "Development of Attached Inflatable Decelerators (AID's) for Supersonic Application," NAS-CR-66613, Report for Contract NAS1-7359, Goodyear Aerospace Corp., May 1968.

[4]Bartle, E. R. and Leadon, B. M., "The Effectiveness as a Universal Measure of Mass Transfer Cooling for a Turbulent Boundary Layer," Proceedings of the 1962 Heat Transfer and Fluid Mechanics Institute, June 13-15, 1962.

[5]Scott, C. N., Nordlie, R. W., and Sowa, W. W., "Investigation of Extendable Nozzle Concepts," Final Report GER 15240 for NASA Contract NAS8-21476, Goodyear Aerospace Corp., Nov. 1972.

Aerothermodynamic Design Feasibility of a Generic Planetary Aerocapture/Aeromaneuver Vehicle

Dwight E. Florence*
General Electric Company, Philadelphia, Pa.

Abstract

Lifting aerodynamic configurations have been screened and selected for a generic planetary aerocapture/aeromaneuver vehicle where a common configuration will be utilized for missions to any of several planets including Mars, Saturn and Uranus. Hypersonic convective and radiative heat transfer to these vehicles was estimated based on a combination of three-dimensional inviscid and viscous flowfield solutions and from engineering correlations developed for the numerous complete convective and radiative heat-transfer evaluations in CO_2 and H_2-He mixtures from the Pioneer Venus and Galileo programs. Thermal protection materials have been selected and sized. It is recommended that the thermal protection system be designed for each specific mission, thus maximizing the deliverable payload. Aerothermodynamic technology needs of this generic aerocapture concept are identified.

Nomenclature

A	=	reference (base) area, m^2
A/C	=	aerocapture
c.g., X_{CG}	=	center of gravity
CP	=	carbon phenolic
c.p., X_{CP}	=	center of pressure
C_A	=	axial force coefficient

Presented as Paper 81-1127 at the AIAA 16th Thermophysics Conference, Palo Alto, Calif., June 23-25, 1981.

*Staff Engineer, Aerothermal Performance Technology, Re-entry Systems Division.

C_D	=	drag coefficient
C_L	=	lift coefficient
C_N	=	normal force coefficient
C_m	=	coefficient of pitching moment, $M/q_\infty AL_A$
D	=	drag force
D_B	=	vehicle base diameter
G_E, g_E	=	Earth's gravitational acceleration, 9.8 m/s^2
GN&C	=	guidance, navigation, and control
h	=	heat-transfer coefficient, $\dot{q}/H_R - h_w$
h_w	=	enthalpy of gas evaluated at wall temperature, J/kg
H	=	altitude, km
H_R	=	recovery enthalpy, J/kg
L	=	lift force
L_A	=	apex length of vehicle, m
M	=	pitching moment
M	=	Mach number
MSR	=	Mars sample return
P	=	local pressure
p, q, r	=	roll, pitch, and yaw angular rates
q	=	dynamic pressure, kN/m^2
$\dot{q}$	=	local heat-transfer rate, W/cm^2
PN	=	phenolic nylon
P_S	=	stagnation pressure, atm
RCS	=	reaction control system
R_B	=	base radius, m
Re_θ	=	Reynolds number based on local properties and momentum thickness
Re_∞	=	freestream Reynolds number
R_I	=	intermediate radius of vehicle, m
R_N	=	nose radius
$\dot{S}$	=	material recession rate
SAM	=	silicone ablator material
T, t	=	time, s
TPS	=	thermal protection system
V_∞	=	freestream velocity, m/s
V_{EXIT}	=	exit velocity, m/s
W	=	weight of vehicle, kg
X	=	axial distance, m
Y	=	radial distance, m
α	=	angle of attack
α_T	=	trim angle of attack
δ_o	=	shock standoff distance
δ_N	=	nose bend angle, deg
δ_F	=	flap deflection angle, deg
β	=	angle of sideslip
γ	=	air relative flight path angle
ρ	=	material or gas density, kg/m^3
θ	=	cone angle

Subscripts

E	=	conditions at entry
e	=	local conditions
t	=	transition to turbulent flow
∞	=	freestream conditions
S	=	stagnation point

Introduction

Aerocapture is a technique that utilizes aerodynamic drag to provide the velocity depletion necessary to obtain a near circular orbit about a planet from a hyperbolic flyby trajectory. It is accomplished through an aerodynamically controlled atmospheric entry during which vehicle bank angle and thus altitude is varied to maintain a constant drag. When the required velocity depletion is obtained, the vehicle pulls full "lift up" and exits the atmosphere where a chemically fired kick stage produces a ΔV sufficient to raise periapsis out of the atmosphere.

It has long been recognized[1] that aerodynamic lift and drag can be employed at some of the planets to greatly increase the 10g (Earth) corridor width over that obtained with nonlifting vehicles for single-pass manned parabolic entry. It was also recognized[2] that use of aerodynamic lift and drag for a manned mission to Mars or Venus provides a large weight saving when comparing the thermal protection weight added to the weight of the chemical propellant that it replaced. The aerocapture concept has been analytically demonstrated[3-6] not only to more than double science payloads on some planetary spacecraft, but also to provide the potential for significantly decreasing interplanetary cruise time. The term generic aerocapture implies the utilization of a common aerodynamic configuration for missions to any of several planets.

A study was recently completed[7] where the primary objective was to develop a conceptual design of a generic aerocapture system with moderate-to-high L/D ratio to establish feasibility of aerocapture for the Mars Sample Return (MSR), Titan/Saturn Orbiter Dual Probe (SO2P), and Uranus Orbiter Probe (UOP) missions. The designs were selected to show mission feasibility based on state-of-the-art concepts. System optimization from an analytic or hardware standpoint was not performed. Emphasis was placed on selection of the aerodynamic configuration, definition of the aerodynamic loads, sizing of the aeroshell structure, selection and distribution of thermal protection materials on the structure, definition of control systems concepts, and aerocapture performance verification through flight control simulations.

The purpose of this paper is to summarize the design rationale, the aerodynamic configurations selected, the heat-transfer loads expected, and the thermal protection materials selected for the generic aerocapture vehicle.

Aerodynamic Configuration

Aerodynamic Configuration Screening Approach

The configuration selected for the generic aerocapture vehicle must be capable of packaging the required payload, fit within the length and width constraints of the Space Shuttle payload bay, and provide lift-to-drag ratio between 1 and 2 and a drag coefficient between 0.4 and 1.0,[3] see Table 1.

Several classes of configurations exist that meet the hypersonic performance requirements. These include axisymmetric and elliptical cross-section cones, biconics, and arbitrary cross-section bodies. Generally, the sphere cones are too long to meet the Space Shuttle payload bay length constraint and package the required payloads. The existing arbitrary bodies have been designed with other mission requirements in mind (e.g., subsonic landing

Table 1 System design and performance requirements

Envelope	
SO2P, Uranus	4.3 m diam, 6.3 m length
MSR	4.3 m diam, 9.6 m length
Launch mass	
SO2P	2660 kg (8 km/s), 2400 kg (13 km/s)
MSR	5000 kg
Uranus	2400 kg (10 year mission), 1800 kg (7 year mission)
Payload	
SO2P	1500 kg
MSR	4000 kg
Uranus	1250 kg
L/D	1.0-2.0
C_D	0.4-1.0
Launch load	6 g axial 3 g lateral
Control accuracy	± 25 m/s on exit velocity

capability or higher L/D) and, thus, they are geometrically more complex than necessary for the aerocapture/aeromaneuver vehicle. The class of configurations that lends itself to the packaging constraints most readily is an axisymmetric biconic with $\theta_{NOSE} > \theta_{FRUSTUM}$. These biconics can be trimmed at an angle of attack, simultaneously producing both lift and drag forces to provide maneuvering capability. Trimming can be accomplished by a variety of techniques, including c.g. offset, bent nose, aft frustum slice, or flaps.

Perhaps the least complex concept is that of a fixed bent nose (which provides a nearly fixed trim angle of attack) and bank to turn steering. This concept was thoroughly evaluated for the aerocapture mission and presents a feasible, well-characterized solution.

The GE 3D Flow Field (3DFF) capability[8,9] has been used to provide accurate and rapid estimates of the aerodynamic characteristics of candidate maneuvering aerocapture vehicle configurations. These techniques have been employed to generate the full set of aerodynamic coefficients for use in the mission/trajectory simulations for the selected configurations. The AFFDL/MDAC hypersonic arbitrary body computer program (HABP)[10] has been employed to determine the aerodynamic characteristics of the selected configuration at very high angles of attack ($>$20 deg). The HABP results compare quite well with the flowfield results in the intermediate angle-of-attack range where both techniques were used. Aerodynamic characteristics considered are the inviscid, skin friction, and induced pressure effects, although only the inviscid effects are of the first order.

Generic Aerocapture Vehicle Configuration Selection

Initially a goal was established to utilize a common aerodynamic configuration for the SO2P, MSR, and Uranus missions. A conical frustum extension could be added to the MSR aeromaneuvering configuration to accommodate the MSR aerocapture portion of the mission.

The configurations selected in the previous study for Mars aerocapture mission feasibility[6] were first considered as candidates for the generic aerocapture mission. Several additional configurations were evaluated in detail that offered such features as reduced surface area.

To maintain payload packaging flexibility of this generic aerocapture vehicle concept, the configuration that 1) maximizes the payload volume, 2) packages the larger spherical payload at the required aerocapture vehicle c.g. location, and 3) has the further aft c.g. location requirement is the more desirable. Configuration 7-6[6] was selected as representing the most attractive configuration, given the general missions requirements (Fig. 1). For the MSR aerocapture mission, a conic frustum extension was added to accommodate the additional payload.

Aerodynamic Characteristics of Selected Generic Aerocapture Vehicle. The force coefficients and moment coefficient derivatives of the generic aerocapture vehicle are summarized in Table 2. For the purpose of this feasibility study, the inviscid aerodynamic characteristics are assumed to be independent of gas composition, an assumption that will be evaluated in the next phase of the program. Turbulent flow is to be expected due to the magnitude of the freestream Reynolds number at pull up, 9 x 10^6 based on vehicle length. The skin friction contribution to CA_{TOTAL} is expected to vary with both gas composition and vehicle size, but due to the size of the inviscid term will have less than a 10% effect on the total axial force coefficient. These effects have been neglected in this feasibility study.

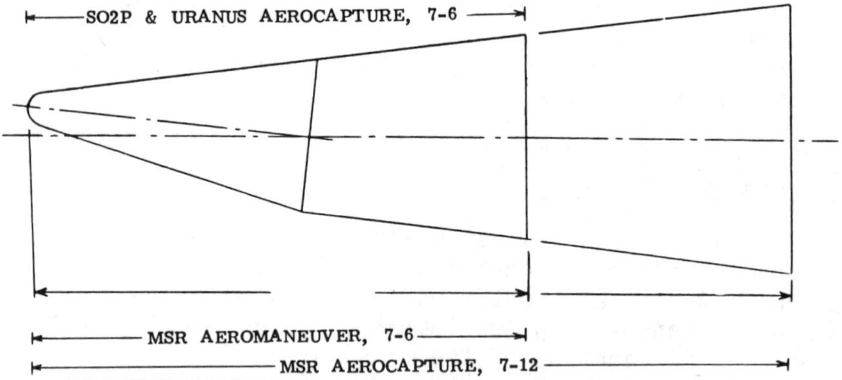

CONFIGURATION	L	θ_F	θ_N	R_N/R_I	R_I/R_B	R_B	L_A	R_I	R_N
7-6	6.3	7°	13°	0.2	.745	1.32	10.71	.98	.2
7-12	9.6	7°	13°	0.2	.570	1.72	14.01	.98	.2

Fig. 1 Generic aerocapture configurations selected.

Table 2 Generic aerocapture vehicle aerodynamics summary

α = 20 deg, M_∞ = 30, Re_∞ = 3.5 x 10^5/m

Force coefficients, CO_2

Flow state	C_A total	C_N total	Pitch X_{CP}/L_A	C_D	C_L	L/D
Laminar	0.262	0.690	0.736	0.482	0.559	1.16
Turbulent	0.308	0.682	0.736	0.522	0.536	1.03

Moment derivatives

C_{m_α}	C_{n_β}	C_{ℓ_β}	C_{Y_β}	$\partial C_{m_{CG}}/\partial C_N$
-2.58^{-3}	$+8.84^{-4}$	-2.43^{-4}	-3.20^{-2}	-8.4^{-2}

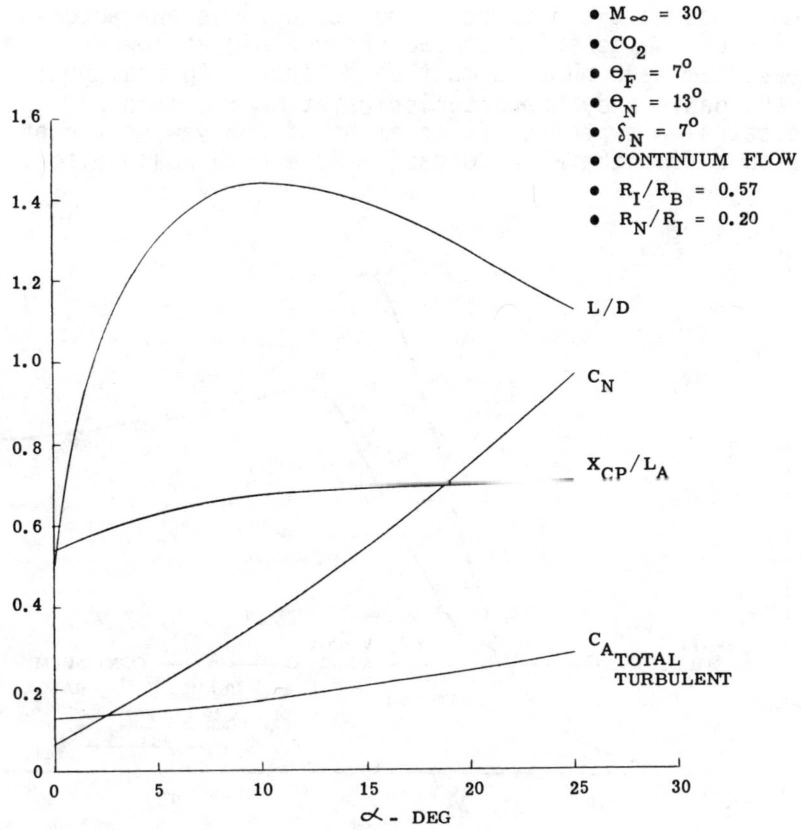

Fig. 2 Aerodynamic characteristics of MSR aerocapture vehicle.

Aerodynamic Characteristics of MSR Vehicle. Inviscid aerodynamic computations have been made for the MSR vehicle employing both the GE 3DFF and HABP codes in CO_2. Solutions were obtained on 3DFF for angles of attack of 10 and 15 deg and on HABP for angles of attack up to 25 deg. At 20 deg angle of attack, 3DFF would not produce a solution due to development of local subsonic flow regions. Axial and normal force coefficients, lift-to-drag ratio, and axial center-of-pressure location variation with angle of attack are illustrated in Fig. 2. The total axial force coefficient includes an estimate of the base drag and an increment due to turbulent skin friction and induced pressure in CO_2, scaled from results of the previous study.

Generic Aerocapture Configurations Applicability as Aeromaneuver Vehicle. The aeromaneuver portion of the MSR mission at Mars consists of a controlled lifting entry from close Mars orbit where the aerodynamic characteristics of the generic aerocapture vehicle at lower freestream Mach numbers must be defined. An examination of the basic body characteristics at $M_\infty = 4$ in air[6] indicates an expected aft movement of the yaw center of pressure from the $M_\infty = 20$ case. Base drag would also be

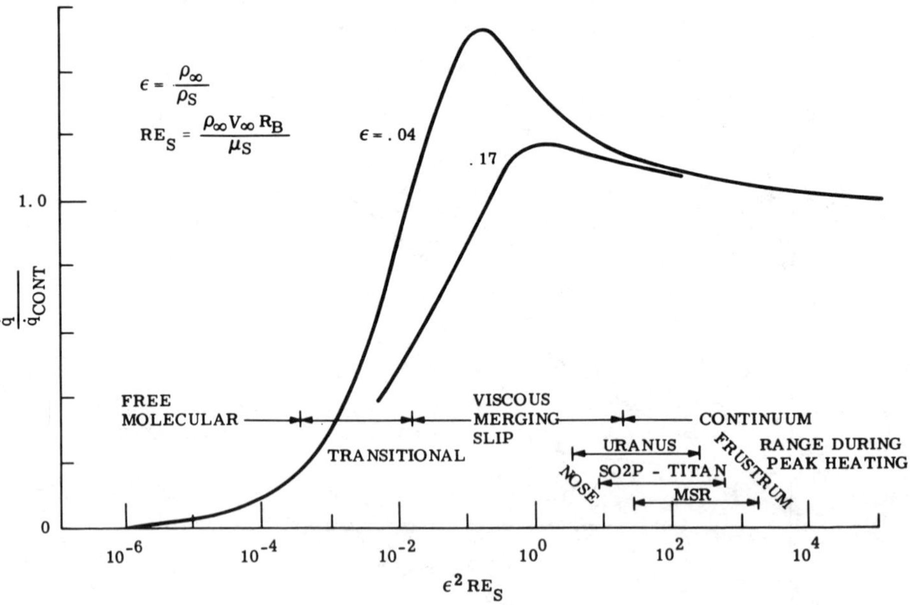

Fig. 3 Effect of flow regimes on heat transfer to blunt bodies.

increasing, thus contributing to a general reduction of L/D at the lower Mach numbers.

Predicted axial center-of-pressure location for the aeromaneuver vehicle is expected to vary somewhat with freestream Mach number. For a fixed center-of-gravity location, this results in a predictable small trim angle-of-attack variation. This effect is usually greater for the aeromaneuver portion than the aerocapture because of the larger freestream Mach number range. In addition, some uncertainty exists in the predicted vehicle aerodynamic characteristics. Typical flight experience has demonstrated a 3 σ aerodynamic coefficient axial and normal force prediction accuracy of $\pm 10\%$, given the angle-of-attack. The impact of these uncertainties should be addressed in future studies.

Dominant flow regimes for the aerocapture vehicles is continuum flow with turbulent flow present over most of the vehicle at pull-up.

Thermal Protection Subsystem

Aerothermodynamic Design Approach

State-of-the-art minimum mass thermal protection systems (TPS) that meet all atmospheric entry requirements have been selected for the generic aerocapture missions. Since the mass of the TPS and its supporting substructure is a major portion of the aerocapture vehicle, selection of a minimum mass TPS concept greatly enhances the potential science return benefits of the aerocapture mission concept.

Employing the predicted aerodynamic characteristics of the selected generic aerocapture configurations, entry trajectories were computed and control simulations conducted as a part of this study. A number of reference heating parameters were simultaneously computed employing engineering correlations for eventual use in sizing the thermal protection systems. Results from the selective calculations of the three-dimensional flowfield codes[8,9] were used to establish the shock shape, local pressure, local Reynolds number, and local laminar and turbulent heat-transfer rate distributions. The onset of local turbulent flow, which results in a significant increase in local heat-transfer rates, was based on available flight test transition data correlations analytically scaled to the aerocapture vehicle geometry. The thermal protection material thickness was sized to limit the maximum bondline/structure temperature to 600°F (590K), a contemporary design practice based on bond strength degradation. Typically, during the period of maximum entry loads, the bond and structure are at room temperature and reach their peaks during the long soak out after the heating period is over and the structural loads are minimal. TPS requirements were evaluated considering both the shallow and steep ends of the trajectory and a nominal initial termperature of 100°F (310K).

Convective Heat-Transfer Design Techniques

Several hypersonic flow regimes are traversed as the aerocapture vehicle enters the planetary atmosphere and executes the pull-up maneuver. Regimes encountered begin with free molecular flow and proceed through the transitional and viscous merging slip flows, finally entering full continuum flow during the constant "g" portion of the mission and then proceeding back through the rarefield flow regime to free molecular as the exit maneuver is performed. Free molecular and continuum drag and heat-transfer characteristics are well defined from both theory and ground and flight test data. Characteristics in the transition and slip flow regimes are less well defined and predictions are based on a shock Reynolds number semi-empirical bridging technique developed by Gilbert and Goldberg.[11] This bridging relationship represents a correlation of the numerical solutions of a theoretical model of the hypersonic viscous shock layer of air in chemical equilibrium over a range of shock Reynolds numbers of 50-10^4 and is illustrated in Fig. 3. The range of shock Reynolds numbers computed at peak heating for the various trajectories is illustrated in Fig. 3 and indicates that the vehicle frustum is operating in full continuum flow with the nose area operating in a more rarefield regime during this time. For purposes of this evaluation, only continuum heating will be considered.

The convective heating rate distributions have been calculated by means of the GE 3D Viscous Boundary Layer Code (3VFF).[9] This program calculates the Lees/Eckert reference enthalpy and Vaglio-Laurin laminar, transitional, and turbulent heating rate distributions, as well as skin friction and viscous drag over the biconic configurations at angle of attack. The code uses integral boundary-layer techniques, including entropy swallowing

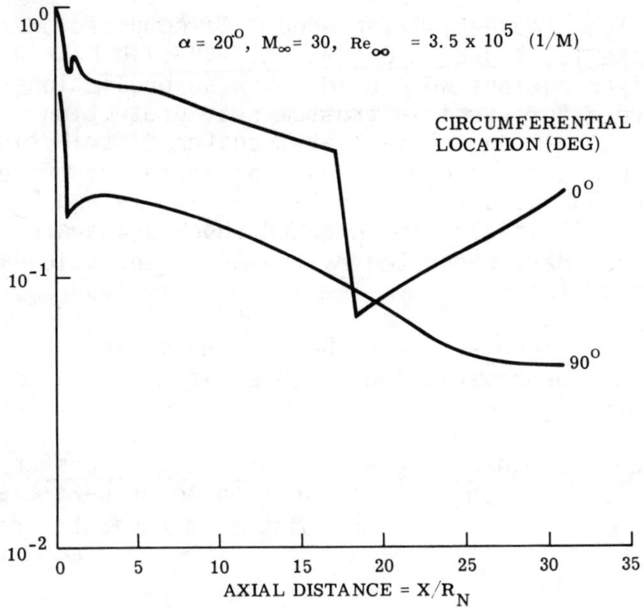

Fig. 4 Laminar heat-transfer distribution on generic aerocapture vehicle in CO_2.

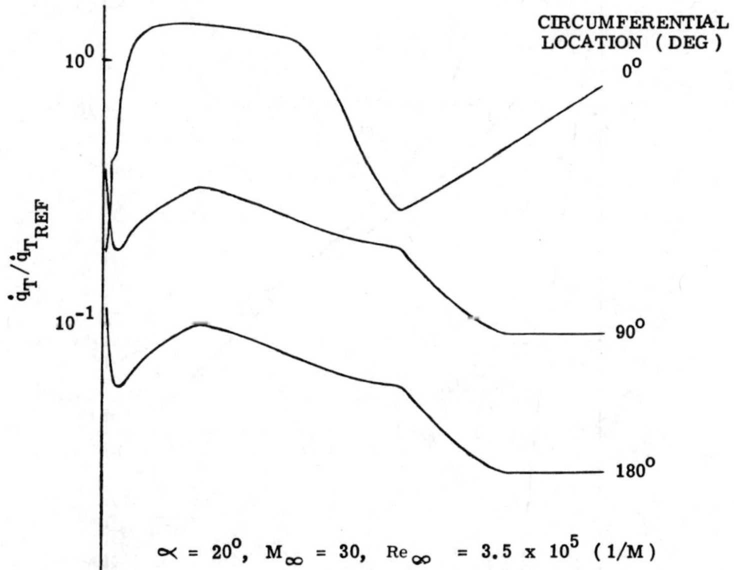

Fig. 5 Turbulent heat-transfer distribution on generic aerocapture vehicle in CO_2.

effects, along boundary-layer edge loci computed on the basis of the local pressure gradient. Vaglio-Laurin local heating distributions were used in these predictions and in all aerocapture vehicle frustum heat protection analyses. Local convective heat-transfer distributions have been computed at conditions representative of peak heating in CO_2 and are illustrated in Figs. 4 and 5. The inviscid flowfield pressure and shock shape are obtained from data generated by the GE 3D Inviscid Flow Field Program (3IFF).[8]

Atmosphere Models. Atmospheric composition, density, and speed of sound variations with altitude employed are presented in Figs. 6 and 7.

Reference Laminar Stagnation Point Heat Transfer. The reference laminar continuum stagnation point heat-transfer rate in CO_2 at the stagnation point of a hemispherical

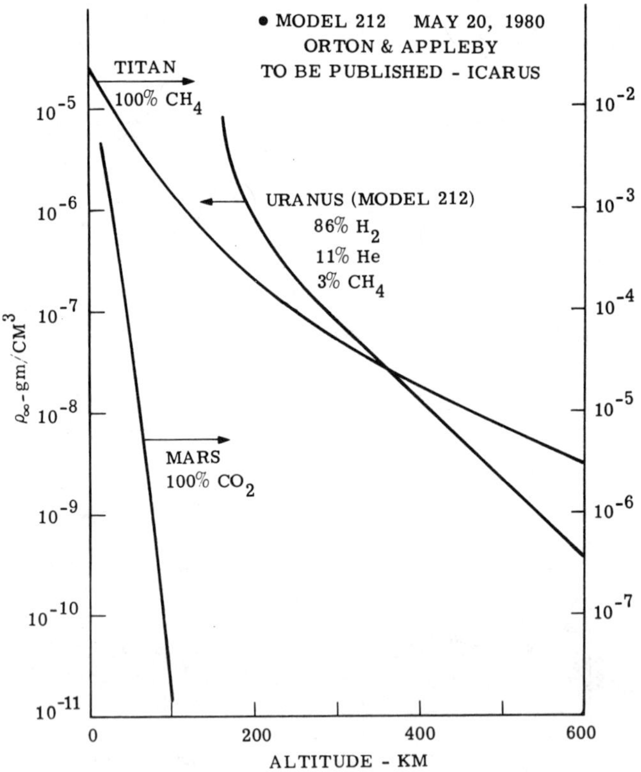

Fig. 6 Atmosphere models for generic aerocapture.

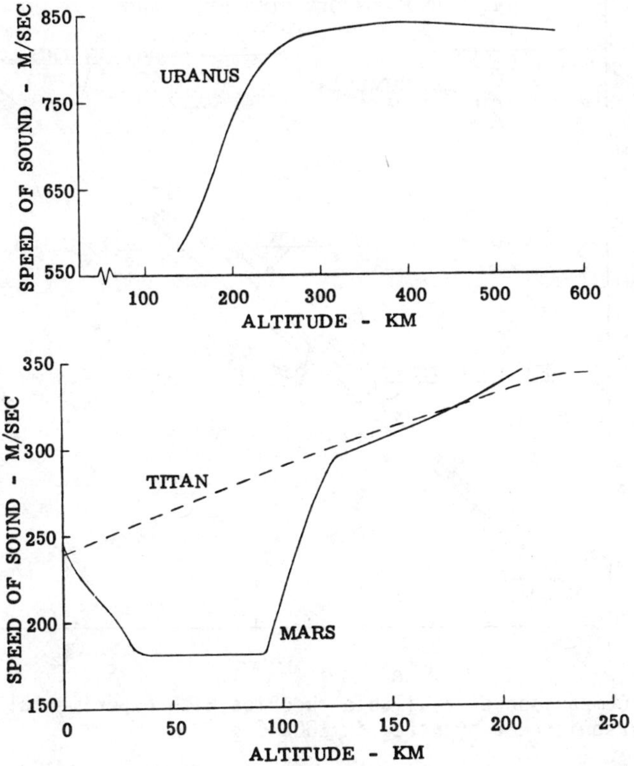

Fig. 7 Atmosphere models for generic aerocapture - speed of sound variations.

nose can be approximated by

$$\dot{q}_s = 1.156 \times 10^{-5} (H_R - h_w) \sqrt{P_S/R_N}, \quad W/cm^2$$

where P_S is in atm, R_N in m, and H_R, h_w in J/kg. This approach was selected after comparing the available hypersonic ground test data in CO_2 to theoretical and semiempirical approaches suggested by various investigators,[12] and to the three-dimensional viscous flowfield solutions in 100% CO_2 employing Lees' solution (Fig. 8). Estimates of the laminar stagnation point heat transfer in CH_4 were also computed with the above expression.

The reference laminar stagnation point heat-transfer

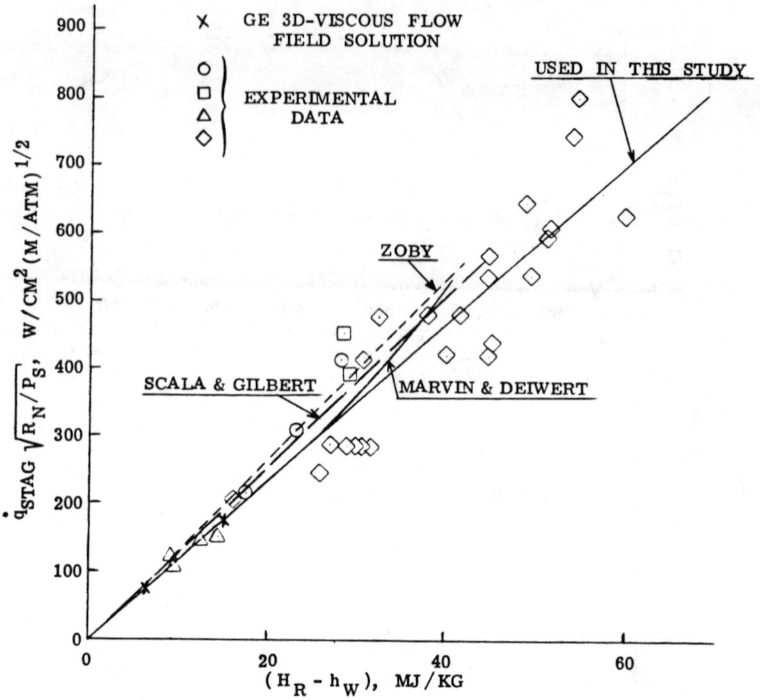

Fig. 8 Comparison of available theories with experimental data for hypersonic heat transfer in 100% CO_2.

rate in 89% H_2-11% He (by volume) can be approximated by

$$\dot{q}_s = 3.50 \times 10^{-6} (H_R - h_w) \sqrt{P_S/R_N}$$

This approach was selected for consistency with the CO_2 results and after examining results represented by Moss for Jovian entry,[13] see Fig. 9.

Reference Maximum Turbulent-Heat Transfer Rate. For the range of planetary entry conditions considered, transition to turbulent flow will occur prior to peak heating as far forward as $X/R_N = 3$, except for the shallow Titan entries where it occurs after peak laminar heating (see Fig. 19, below). Vehicle shape, angle of attack, and freestream density and velocity are the first-order parameters affecting magnitude of the turbulent heat-transfer rates. Gas composition is expected to be a second-order effect on q for CO_2 and CH_4.

The reference turbulent heat-transfer rate in CO_2 is taken as the maximum value of the turbulent heat-transfer rate occurring on a hemispherical nose at zero angle of attack. It can be shown that this maximum occurs at about 33 deg off the stagnation point, assuming a modified Newtonian pressure distribution and can be cast in terms of the freestream conditions,[14] as

$$\dot{q}_T R_N^{0.2} = 1.943 \times 10^{-7} \rho_\infty^{0.8} V_\infty^3$$

The constant has been increased for air by a factor of 1.24 to bring the estimate into line with air results from the 3VFF code. Estimates of the reference turbulent heat rate in CH_4 were also computed with the expression above.

The reference turbulent heat-transfer rate to a hemisphere in 89% H_2-11% He is approximated by

$$\dot{q}_T R_N\, 0.2 = 6.15 \times 10^{-8} \rho_\infty^{0.8} V_\infty^3$$

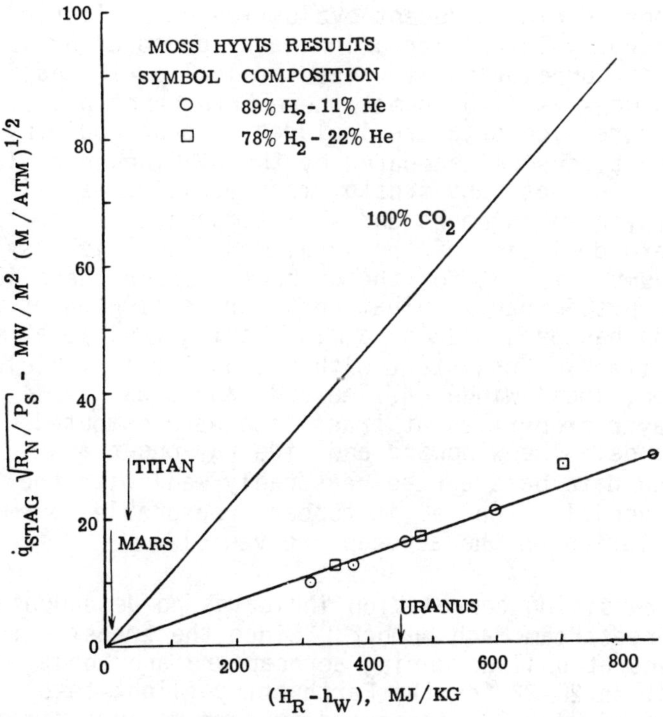

Fig. 9 Correlation of hypersonic heat-transfer results in H_2-He mixtures and comparison to CO_2.

The constant in this equation was selected after examination of the nonblowing turbulent heat-transfer results presented by Moss for Jovian entry.[15-17]

Turbulent Leeward. Turbulent leeward heating data in air for spherically blunted cones at angle of attack has been reviewed by Nestler.[18-22] He concluded that leeward turbulent heating levels are less than zero angle-of-attack values and are relatively independent of freestream Reynolds number, approaching a $h/h_{\alpha=0} = 0.3$ for α/θ_c values greater than about 2. Typical α/θ_c values of this study range about 3-4 for the aft cone. This value of $h/h_{\alpha=0} = 0.3$ was used throughout this study, independent of gas composition and angle of attack. This is considered a reasonable approach at this stage of evaluation, since the leeward heat protection weight is much less than the windward.

Boundary-Layer Transition. The boundary-layer transition criteria employed was a recently published Re_θ criteria.[23] This correlation, applicable to carbonaceous-nosed vehicles with carbon-phenolic frusta, was developed during a recent evaluation of ballistic vehicle boundary-layer transition data sponsored by AFFDL. This correlation is presented in Fig. 10 and employs an $Re_{\theta t}$ vs X/R_N format where both onset and forward progression data are included. Local boundary-layer properties were generated by the 3VFF boundary-layer code, including nose and frustum mass addition as well as boundary-layer entrainment and wall temperature effects. The standard deviation of the local momentum thickness Reynolds number is 22% for the graphite/carbon-phenolic subset. A preliminary evaluation of transition onset and progression has been made of a maneuvering vehicle at angle of attack. Consistent with the ballistic vehicle correlation, local windward, leeward, and side ray bounday-layer properties at transition were computed with the 3VFF code. The windward and side ray onset and progression data base agreed reasonably well with the ballistic vehicle. Hence, it appears reasonable to employ this correlation on the aerocapture vehicle.

This transition correlation indicates no dependence on local or freestream Mach number. Since the freestream Mach numbers at pull-up during aerocapture are about 30, in contrast to 20-22 for the Earth entry flight test correlation of Fig. 10, it seemed prudent to investigate the potential Mach number difference effect. One of the

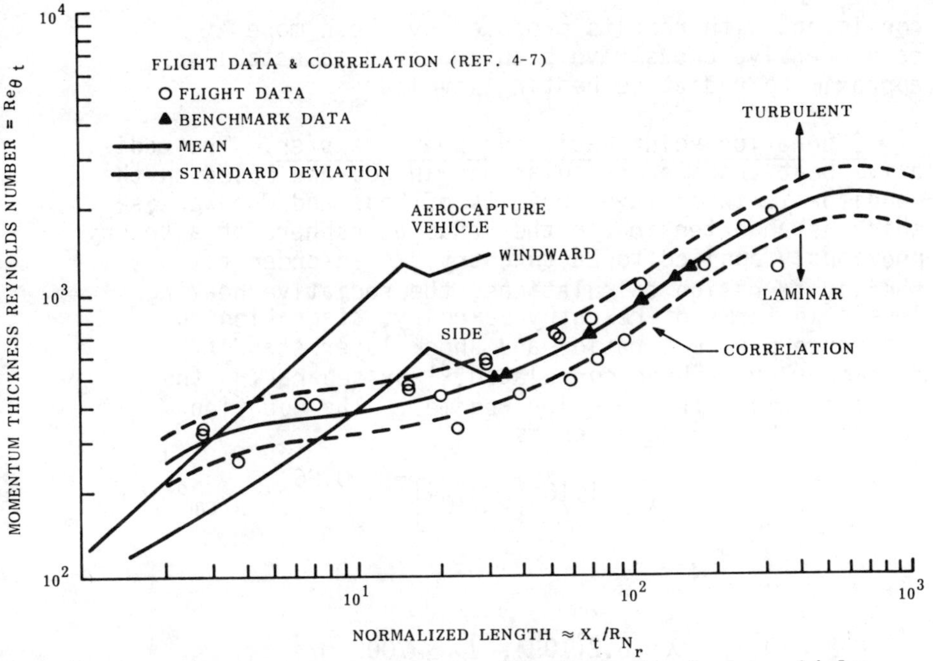

Fig. 10 Local Reynolds number at pull up, single-mission vehicles.

attempted correlations by Berkowitz[23] was an Re_θ vs M_e of carbonaceous-nosed phenolic frusta ballistic vehicles. Bounds of the flight data were compared to the predicted Re_θ-M_e variation of the single-mission aerocapture vehicle at pull-up. Employing this correlation, a probable turbulent windward meridan was predicted and a laminar side ray, whereas employing the correlation preferred by Berkowitz, a somewhat larger turbulent area is predicted. The correlation preferred by Berkowitz has been employed in this study for the preliminary feasibility assessment.

Radiative Heat-Transfer Design Techniques

The Pioneer Venus and Galileo programs have provided recent impetus for the development of full viscous shock-layer solutions for the radiative heat flux distribution on sphere cones,[13,24] as well as more approximate techniques.[25,26] Design studies have previously been completed using these techniques in both CO_2 and H_2-He mixtures.[16,27,28] For this initial assessment of radiative heat-transfer rates, engineering correlations consistent with results produced by these more comprehensive codes have been employed to establish approximate radiative heating levels.

Stagnation Point Radiative Heat Transfer. The radiative heat-transfer calculations in CO_2 are based on the nondiabatic shock-layer results of Page and Woodware,[29] which is an extension to the Venus atmosphere of a theory previously applied to Earth entry.[30] In order to facilitate design calculations, the radiative heating levels in terms of re-entry velocity, stagnation to reference pressure ratio, and shock-layer standoff distance[29] have been correlated,[31] by separating the results into three velocity regimes. The equation defining this correlation is

$$\dot{q}_{R_{NA_o}} = \lambda(p)^{1.16}\left[V_\infty/10^4\right]^\sigma \delta_o^{0.56} \quad W/cm^2$$

where:

$\sigma = 19.2, \quad \lambda = 6.2(10^5), \; V_\infty \leq 7000$ m/s
$\sigma = 4.1, \quad \lambda = 2(10^2), \; 7000 \leq V_\infty \leq 10{,}000$ m/s
$\sigma = 9.12, \quad \lambda = 2(10^2), \; 10{,}000 \leq V_\infty \leq 13{,}000$ m/s
p = local pressure, atm
δ_o = shock standoff distance, cm

Although the usually accepted value of the exponent of adiabatic shock standoff distance is 0.6, a value of 0.56 appeared to give a slightly better comparison with the results of Page and Woodward. Figure 11 is a plot of the radiative heat flux calculated using the correlation equation vs the original results.[29] The radiative rates of Page and Woodward were used since they are the most recent complete parametric results available.

Since the published Page and Woodward radiative heating rates are for a nonadiabatic shock layer, no correction factor was necessary to account for nonisothermal effects. This correlation of the Page and Woodward results has been applied to several Pioneer Venus design cases and compares favorably to the more complete calculations reported by Brewer[28] employing the computational technique developed by Falanga and Olstad.[25]

Also of concern is the effect of shock-layer cooling on the convective heating. The stagnation point convective heating results of Page and Woodward were

$$\dot{q}_R = \lambda \, (P/P_S)^{1.16} \, (V_\infty/10^4)^{\sigma} \, \delta_S^{.56}$$

$$0.0 < (V_\infty/10^4) < 0.7, \quad \sigma = 19.2, \quad \lambda = 4.28 \times 10^4$$

$$0.7 < (V_\infty/10^4) \leq 1.0, \quad \sigma = 4.1, \quad \lambda = 2 \times 10^2$$

$$1.0 < (V_\infty/10^4) \leq 1.3, \quad \sigma = 9.12, \quad \lambda = 2 \times 10^2$$

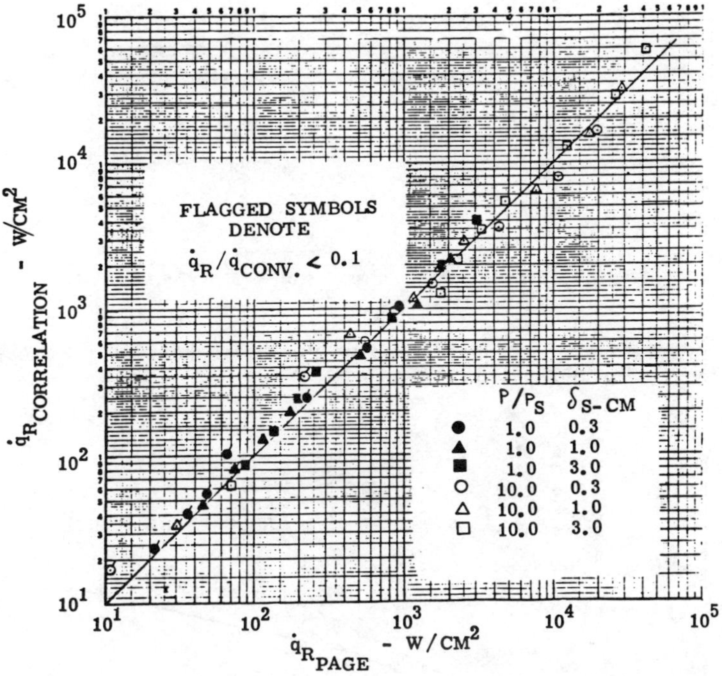

Fig. 11 Correlation of hypersonic radiative heat-transfer results in CO_2.

compared to the stagnation point rates predicted by the methods of Zoby[29] and Marvin and Deiwert.[34] No significant difference due to shock-layer cooling was indicated.

The radiative heat-transfer calculations in H_2-He mixtures are based on a correlation of the Project Galileo stagnation point results reported by Brewer, Brant, and Fogaroli.[26,27] The Galileo results were generated based on the viscous shock-layer solutions of Moss[13] for the radiative flux distribution along the surfaces of large-angle sphere cones and hyperboloids and the tabulated matrix of precalculated RAD/EQUAL code results of Nicolet.[35, 36] The correlations of the Project Galileo

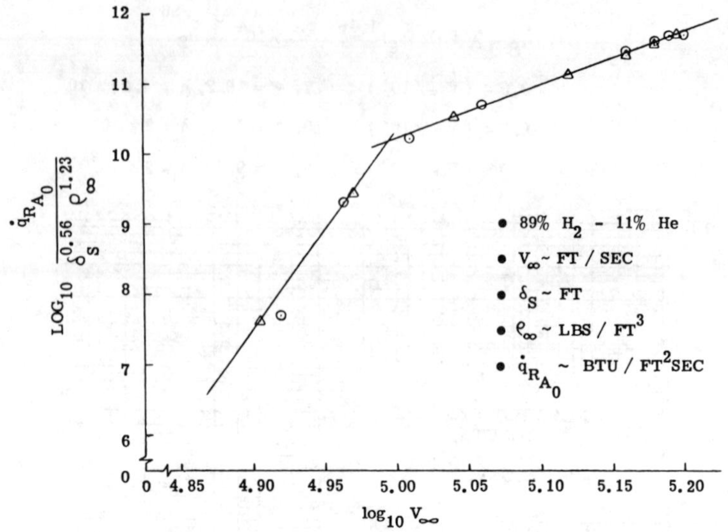

Fig. 12 Variation of stagnation point radiative heating parameter with velocity for 89%H_2 - 11% He.

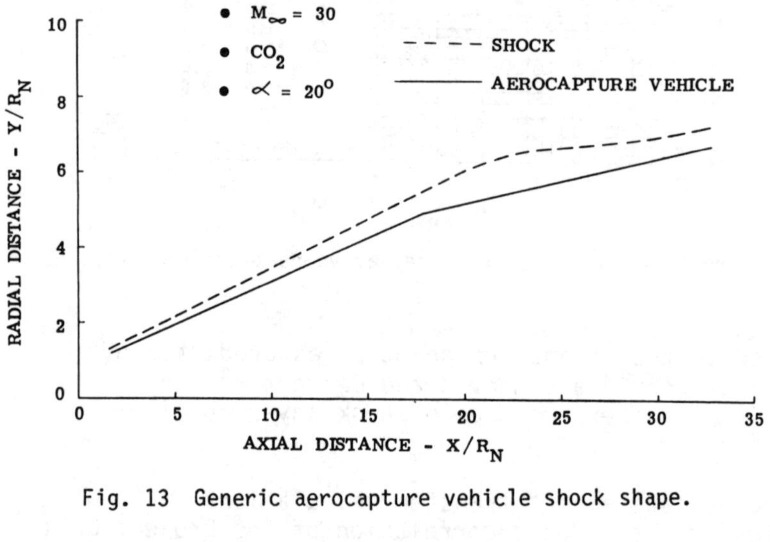

Fig. 13 Generic aerocapture vehicle shock shape.

results is modeled after that suggested by Stickford and Menard[37] for a 61% H_2-36% He-3% N_e mixture and takes the form (Fig. 12) of

$$\dot{q}_{R_{A_0}} = \lambda \rho_\infty{}^{1.23} \left[V_\infty/10^4\right]^\sigma \delta_0{}^{0.56}$$

Based on the results presented by Brewer et al.[26] the radiative heating can be expected to be a factor of two higher for the heavy atmosphere (78% H_2) and about half the nominal value for the light atmosphere (100% H_2).

Hypersonic radiative heat-transfer results in CH_4 are not yet available, and generation of them was beyond the scope of this study. For purposes of estimating the radiative heat transfer from radiating CH_4, the CO_2 correlation was employed. Preliminary results of computations in radiating CH_4-N_2 mixtures[38] indicate this to be a reasonably conservative approach.

Shock Standoff Distance. The shock standoff distance around the entry vehicle is obtained from the 3IFF results validated by Berman.[39] The full shock shape computed is shown in Fig. 13. The stagnation point adiabatic shock standoff distances in CO_2 and 89% H_2-11% He are compared in Fig. 14 to Serbin[40] and Li-Geiger[41] relationships for predicting the shock standoff distance in air. The flowfield result favors the Li-Geiger relationship. Ridyard and Storer[41] demonstrated that for values of $\rho 2/\rho 1 > 20$ the Li-Geiger relationship is preferred over Serbin. In addition, Serbin[40] and Storer[43] have recognized that the shock standoff distance can be correlated for various gases with a single parameter, $\rho 2/\rho_\infty$, the density ratio across the shock. Therefore, the agreement of the air and CO_2 results is not suprising.

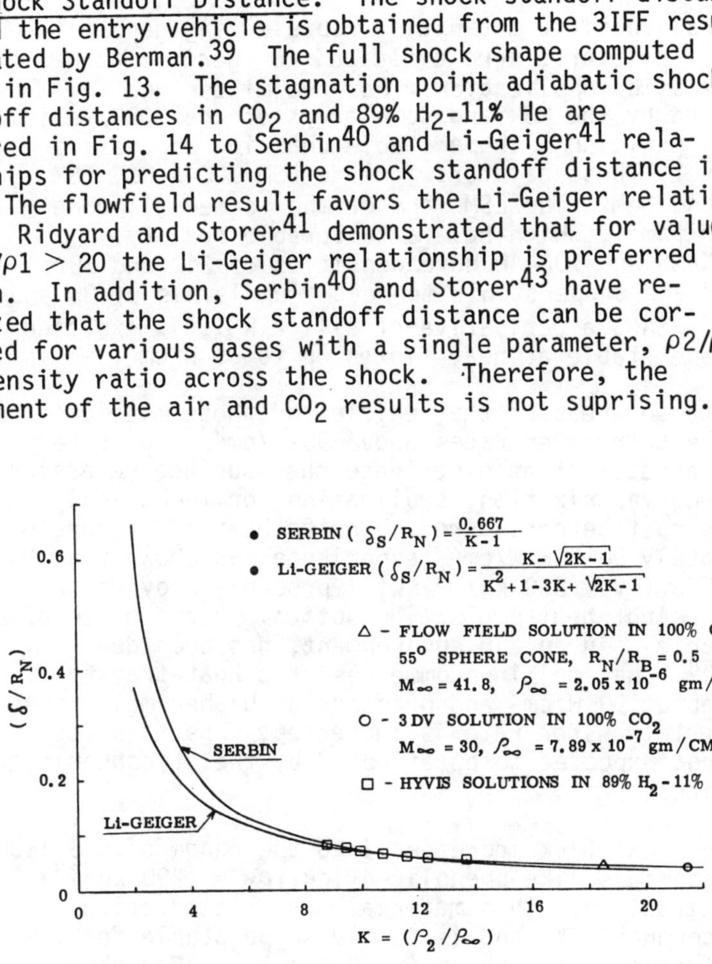

Fig. 14 Stagnation point adiabatic shock standoff distance.

This correlation was used to estimate the shock standoff distance in the CH_4 and H_2-He mixtures.

Radiative Heating over the Skirt. For purposes of this study, the radiative heating over the conical section of the vehicle has been estimated from a review of the sphere-cone distributions for H_2-He mixtures[26] and for CO_2.[28] Employing a tangent cone approach, the equivalent cone angle of our forward frustum is 40 deg at α = 20 deg. This results in the radiative heat flux dropping to about 20% of the stagnation point value at $X/R_N > 2$.

Ablative Mass Loss

Several specific materials, representing general classes of ablators, were evaluated for generic aerocapture mission application. For the milder heating experienced by the MSR and some areas of the SO2P vehicles, thermally efficient, shape stable charring ablators are attractive. These ablators include ESM 1004X-NS, ESM 1004AP, and SAM. The ESMs are of the foamed silicone elastomer family which have a tailorable density. The 1004X-NS and 1004AP have densities of 240 and 580 kg/m^3 and are shape stable to heat flux levels of 85 and 90 W/cm^2. SAM, a derivative of silica RSI, is expected to be shape stable at heat fluxes up to 85 W/cm^2.

For those areas of the vehicle and those missions that receive heat-transfer rates above 90 W/cm^2, higher performance ablators that experience char surface recession by melting, vaporization, sublimation, or mechanical scrubbing must be considered. In the heat flux range of approximately 90-565 W/cm^2, experience has shown that ESM 1004LPS (ρv = 880 kg/m^3) will probably provide the lightest weight heat protection option. Performance of this material, in an air environment, has been described.[44] Char melting commences at a heat-transfer rate of about 90 W/cm^2 and proceeds at higher melt rates as the heat-transfer rate is increased. The mass loss rate is not expected to be affected by the atmospheric gas composition.

As the heat flux increases into the range of 565-1130 W/cm^2, materials like phenolic-nylon (ρv = 1200 kg/m^3) may be attractive. This material has an attractive thermal conductivity and is nearly shape stable for low values of aerodynamic shear (≤ 48 N/m^2).[45] For shear levels above 48 N/m^2, the char fails mechanically and is swept away by the aerodynamic shear, thus forcing the use of a material like carbon-phenolic. For aerodynamic shear levels above 48 N/m^2, a correlation of char removal data[45] was made of the form $Q^* = \dot{q}/\dot{s}\,\rho_{surface} = f\,(H_R - h_w)$. For values greater than 27.8 MJ/kg (V_E = 7.5 km/s), $Q^* \longrightarrow 81.2$ MJ/kg. This value has been employed to generate material tradeoffs in this study.

For application in the higher heat flux areas, $\dot{q} >$ 1130 W/cm^2, materials of the carbon-phenolic family (ρv = 1450 kg/m^3) are usually attractive. The re-entry activity on the Pioneer Venus and Galileo programs and the many Earth entry scientific and strategic programs have provided a large bank of both analytical and test results. Ablation material surface reactions with the dissociated boundary layer have been modeled for several carbon-atmosphere combinations. Carbon ablation in CO_2 was evaluated for the Pioneer Venus program.[28] It was shown that the difference between performance in CO_2 and air is small in the sublimation region: in the diffusion-controlled region the mass loss is about 60% greater in CO_2 (Fig. 15). Carbon ablation in H_2-He mixtures was evaluated for the Galileo program.[26] In this case, a much greater difference was observed between the air and H_2-He results (Fig. 15). Sublimation is seen to commence earlier in the H_2-He mixture than in air or CO_2 and the wall gas enthalpy, the primary term that controls the carbon surface recession (higher h_w resulting in lower mass loss rates), is seen to be

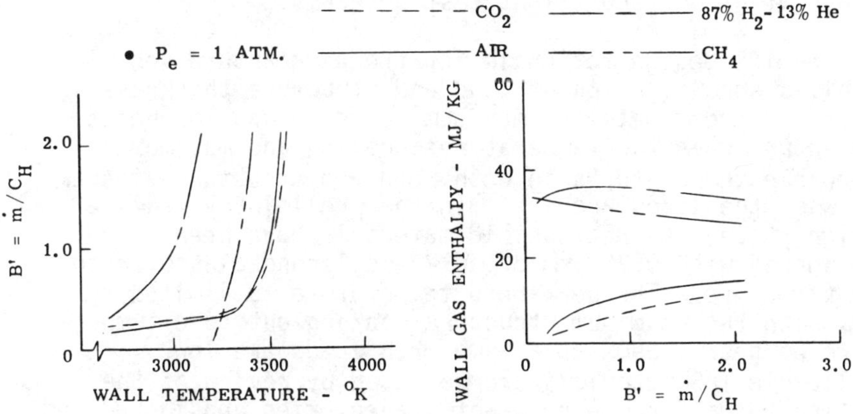

Fig. 15 Atmospheric composition has large effect on carbon ablative performance.

considerably higher for the H_2-He mixture. Carbon mass loss rates in dissociated CH_4 were unavailable at initiation of this study, but were generated as part of this study. These results are illustrated in Fig. 15 for a local pressure of 1 atm and have been employed in this study to generate carbon-phenolic thermal response and mass loss results.

Material Thermal Response

Having determined the heating rates, heating times, and time-integrated heating loads using the techniques discussed above, TPS trade studies have been conducted through use of design correlations derived from the results of the GE Reaction Kinetics Ablation Program (REKAP). The REKAP ablation analysis technique[46] was developed to provide predictions of the surface recession, degradation, and thermal penetration for a variety of materials over a wide range of flight conditions and environments. The program includes the decomposition of a plastic material and the flow of pyrolysis gases through a porous char, the effect of mass addition on aerodynamic heat transfer, surface recession caused by melting and/or vaporization, and the oxidation and vaporization of graphite-type materials. This code has made possible the analysis of numerous materials including the silicone and epoxy elastomers (e.g., ESM), carbon phenolic, the family of graphites and carbon-carbon composites, Teflon, silica phenolic, phenolic nylon, boron nitrides, quartz, and other silica-based glassy materials such as RSI and SAM. The REKAP computer code has been used successfully on a number of ground and flight test programs.

The TPS design rationale for the aerocapture vehicles involves specification of bond and structure thicknesses, sized to accommodate the maximum loads during IUS boost and entry, when the temperatures are low and maximum allowable temperatures in both bond and structure at soak out when the loads are low. In the preliminary vehicle design phase, the heat-shield materials have been sized and bonded with 0.254 mm of an RTV silicone elastomer to the structure. The peak bond temperature is limited to 590K with the aluminum structure soaking out to a temperature of nearly 590K at a time when loads are low. Applicable TPS materials are selected by review of the heating rates for the range of trajectories considered and usually sized by the larger heating loads of the shallow path angle trajectory.

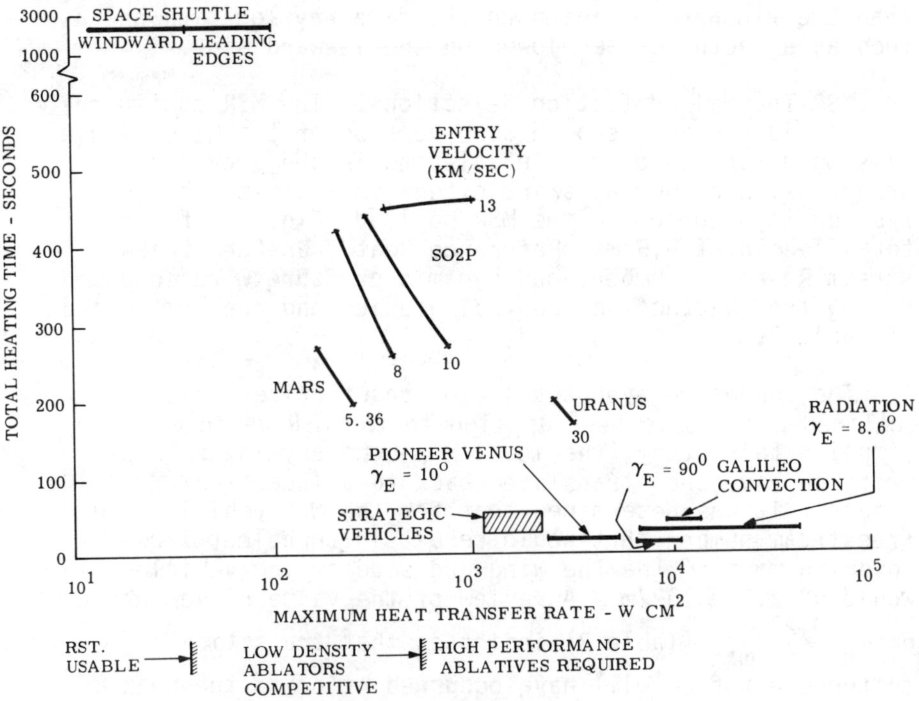

Fig. 16 Comparison of peak heating rates and heating times expected for generic aerocapture to current entry vehicles - windward forecone location.

Heat Protection Material Selection

General Considerations. Heat protection material selection is driven by the requirement to minimize the total TPS weight while choosing from contemporary materials or their derivatives where possible. It has been found that a meaningful parameter to indicate potential range of a materials application is the peak heat-transfer rate expected. The variation of the peak heat-transfer rate on the aerocapture vehicle forecone on the windward meridan across the entry corridors for the MSR, SO2P, and Uranus entry vehicles is illustrated in Fig. 16. It is seen that the peak heating rates are much greater than those experienced by the Space Shuttle and approach those experienced by strategic entry vehicles and the Pioneer Venus probes. The peak heat-transfer rate also varies significantly over the aerocapture vehicle, as shown in Fig. 17, being as much as a factor of five lower than the windward meridian on the side ray location and as much as a factor of ten lower on the leeward side.

MSR Thermal Protection Selections. The MSR configuration evaluated is a scaled down version of the Mars single-mission aerocapture vehicle selected in the previous study,[6] with a conical skirt extension added to the aft frustum to accommodate the MSR payload (Fig. 1), for a total length of 9.6 m. Reference heat transfer, free-stream Reynolds number, and dynamic pressure were computed during the guidance and control studies and are summarized in Table 3.

The boundary-layer transition onset criteria logic outlined above have been applied to the MSR vehicle. To simplify this study, the local Re_θ boundary-layer trip criteria have been translated back to a freestream Reynolds number. It was determined that for the MSR vehicle, the freestream number that would result in turbulent flow covering most of the the windward side of the vehicle would be 2.2×10^5/m. A review of the value of Re_θ at $\left[\dot{q}_T R_N^{0.2}\right]_{max}$ (Table 3) indicates that transition to turbulent flow will have occurred prior to the peak of $\dot{q}_T R_N^{0.2}$. Employing techniques previously described, heat-transfer rate time histories for a representative forward cone body point on the windward meridian (X/R_N = 12.5) have been generated for the range of entry path angles and are illustrated in Fig. 18. Heat-transfer rate distribution over the vehicle at the time of peak heating is illustrated for the steep angle in Fig. 17. The general heat protection material selection criteria have been superimposed on this figure. The TPS materials selected from consideration of the peak heat flux values of the steep trajectory (Fig. 17) have been sized for the time integrated heat flux and heating times experienced during the shallow trajectory. These heat protection material requirements are summarized in Table 4.

The MSR vehicle requires a thermal protection material on the aeromaneuver vehicle that can function in a reusable mode. It must function during the aerocapture maneuver and then subsequently during the entry maneuver. Low-density ablators such as ESM and SAM are expected to be reusable with some small loss of thermal efficiency because of the charring experienced during the aerocapture

Table 3 Summary of reference heating parameters

Planet	m/C_DA, kg/m^2	m/C_LA, kg/m^2	V_{∞_E}, km/s	γ_E, deg	D_{REF}, ge	$q_{\infty_{MAX}}$, kN/m^2	Re_∞ at $\dot{q}_{T_{MAX}}$, 1/m	$\dot{q}\sqrt{R_N}_{MAX}$, W/cm^2	$\int \dot{q}\sqrt{R_N}dt$, $W\ s/cm^2$	$\dot{q}_T R_N^2{}_{MAX}$, W/cm^2	$\int \dot{q}_T R_N^2 dt$, $W\ s/cm^2$	$1/2\rho_\infty V_\infty^3{}_{MAX}$, W/cm^2	$\int 1/2\rho_\infty V_\infty^3 dt$, $W\ s/cm^2$	θ_{TURB}, s
Mars	977	771	5.36	14.5	1.5	14.5	8E5	76	10300	106	13000	6200	7.2E5	260
				15.5		15.2	8.5E5	82.5	9900	115	12700	6700	7.2E5	250
				16.5		22.0	10.9E5	99	9000	155	12000	9700	7.2E5	210
Titan	513	546	8.0	23	3.0	15.0	3.6E5	110	28000	125	27000	6800	13.255	420
				28		15.4	3.1E5	145	24200	165	26000	8550	13.3E5	310
				33		22.5	4.7E5	176	22000	228	24700	13000	13.2E5	240
			10.0	23	4.0	20.0	1.8E5	200	47000	180	44500	9300	21.5E5	460
				28		20.5	3.6E5	265	40000	285	42000	14300	21.2E5	310
				33		34.5	6.5E5	340	35000	440	39800	24700	21.3E5	210
			13.0	23.5		19.9	2.0E5	425	88000	360	80000	15800	97.3E6	420
				26.0		20.5	2.6E5	480	80000	440	76000	20000	77.1E6	350
				29.0		29.0	3.9E5	585	76500	600	76000	29200	63.8E6	290
Uranus	596	527	30.0	8	15.0	89	5.0E5	1610	1.7E5	1390	1.24E5	1.9E5	15.8E6	200
				12	15.0	103	6.4E5	1880	1.5E5	1706	1.18E5	2.4E5	16.0E6	165
				8	20.0	133	8.6E5	1776	1.5E5	1785	1.17E5	2.7E5	15.5E6	175
				12	20.0	137	8.4E5	2022	1.37E5	2054	1.14E5	3.1E5	15.5E6	145

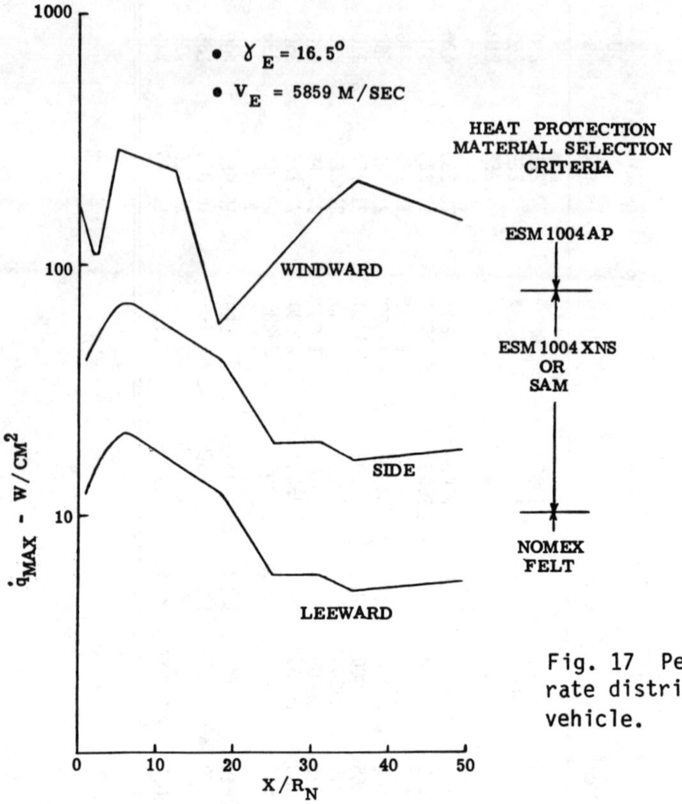

Fig. 17 Peak heat-transfer rate distribution for MSR vehicle.

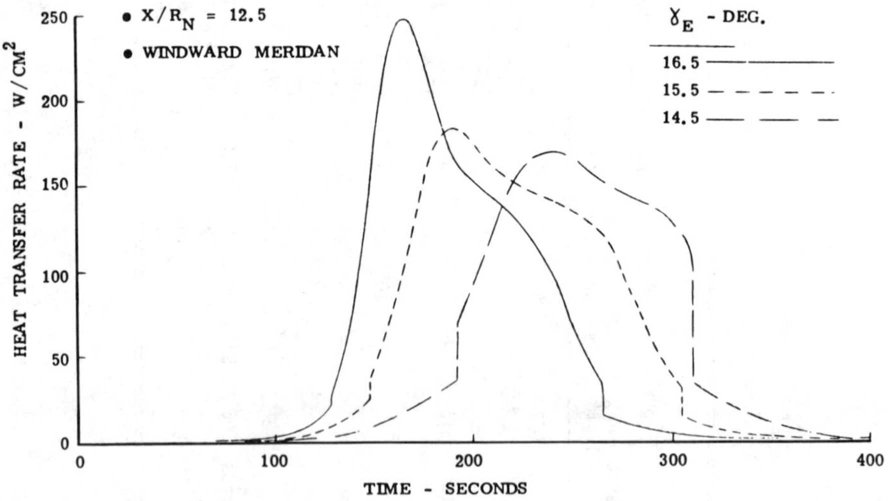

Fig. 18 Heat-transfer rate history for MSR vehicle.

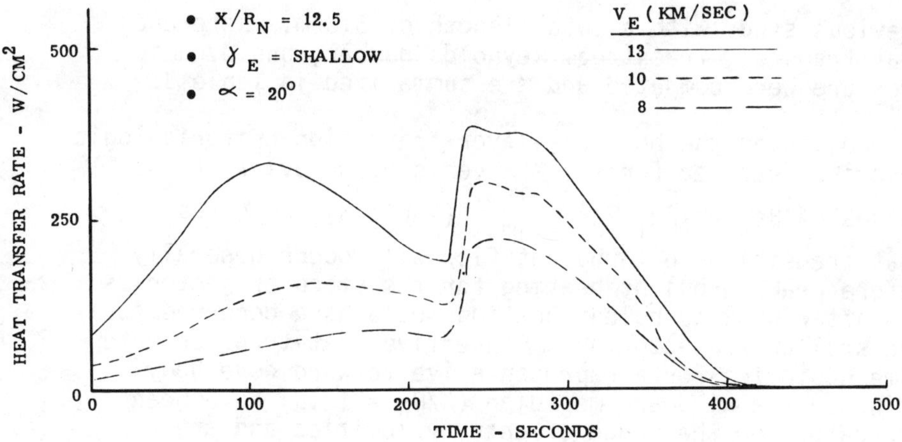

Fig. 19 Convective heat-transfer rate history for SO2P vehicle Titan entry.

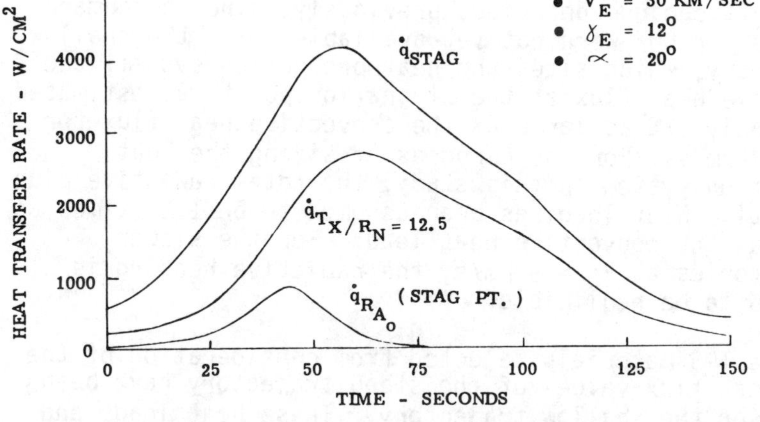

Fig. 20 Heat-transfer rate history for Uranus vehicle.

manuever. It was found in the previous study that due to the milder entry conditions of the out-of-orbit entry, the heat load was a factor of five less and the heating time a factor of six less than the aerocapture maneuver. Thus, it is expected that use of the precharred albator material for aeromaneuver will result in a peak bondline temperature of much less than 590K.

<u>SO2P Thermal Protection Selection.</u> The SO2P configuration evaluated is a scaled-down version of the Mars single-mission aerocapture vehicle selected in the

previous study with a total length of 6.3 m. Reference heat transfer, freestream Reynolds number, and dynamic pressure were computed and are summarized in Table 3.

Following the boundary-layer transition criteria logic described earlier for the MSR vehicle, a review of the values of Re_θ at $\left[\dot{q}_T R_N{}^{0.2}\right]_{max}$ (Table 3) indicates that transition to turbulent flow will occur generally before peak turbulent heating for the steep trajectories and after peak turbulent heating would have occurred for the shallow trajectories. Convective heat-transfer rate time histories for a representative forward cone body point on the windward meridian ($X/R_N = 12.5$) have been generated for the range of entry velocities and are illustrated in Fig. 19.

Radiative heat transfer to the stagnation point area in the methane environment has been estimated, employing the CO_2 technique described previously, since benchmark results for CH_4 were not yet available. For the shallow trajectory, which sizes the heat protection system, the radiative heat flux at the stagnation point was estimated to be only 17% as large as the convective heat flux for $V_E = 13$ km/s. For the purposes of sizing the heat protection system in this study, the total radiative plus convective heat load has been assumed to be 1.25 times as large as the convective heat load. For the slower trajectories at $V_E = 8$ km/s, the radiative heating is assumed to be negligible.

The TPS materials selected from consideration of the peak heat flux values of the steep trajectory have been sized for the shallow trajectory. These heat loads and heat protection material requirements are summarized in Table 4.

The materials selected for the high-speed entry have also been sized for the low-speed ($V_E = 8$ km/s) entry. However, consideration of the peak heat-transfer rates experienced during the slow-speed entry indicates that low-density charring ablators could be used over the entire SO2P vehicle for the low-speed entry and offer a lighter weight heat protection system (Fig. 20). The materials selected as providing the lightest weight locally while meeting the constraint of performing adequately for the peak heat fluxes expected for the steep trajectories are identified with their thickness requirements in Table 4.

Table 4 Comparison of candidate thermal protection materials thermophysical properties

		MSR V_E=5859 m/s		V_E = 8 km/s		SO2P = Titan V_E = 13 km/s		Uranus V_E = 30 km/s	
Vehicle station, X/R_N	Peripheral location, deg	Material selected	Thickness, cm	Material selected	Thickness, cm	Material selected	Thickness, cm	Material Selected	Thickness, cm
5	0	ESM1004AP	1.2	ESM1004LPS	2.8	Carbon phenolic	4.1	Carbon phenolic	3.3
31				ESM1004AP	1.5	Carbon phenolic	2.9	Carbon phenolic	2.5
35.6		ESM1004AP	1.2						
5	90	SAM	1.9	ESM1004X	2.8	Carbon phenolic	2.5	Carbon phenolic	1.9
31				ESM1004X	2.0	Carbon phenolic	1.5	Carbon phenolic	1.2
35.6		SAM	1.8						
5	180	SAM	1.8	ESM1004X	2.1	ESM1004AP	1.4	Carbon phenolic	1.2
25				SAM	2.0	ESM1004X	2.0	Carbon phenolic	1.0
35.6		Nomex felt	1.9						

Several tradeoff studies were conducted in arriving at these final selections. Employing the char mass loss models described previously, recession histories were generated for ESM, phenolic nylon, and carbon phenolic for the windward meridian location at $X/R_N = 12.5$ (Fig. 21). It is seen that the ESM experiences nearly 6.8 cm of recession, phenolic nylon about 3 cm, and carbon phenolic a negligible amount in the methane environment. The shape stability of the carbon phenolic is highly desirable, although as currently constructed it is not as good an insulator as ESM or PN.

The thermal response of carbon phenolic, when exposed to a typical methane heating environment on the windward meridian of the forward cone, has been predicted with REKAP for both 8 and 13 km/s entries. The predicted surface temperature reached a maximum value of a little over 2200K. This was not sufficient to initiate sublimation (Fig. 15). The carbon-phenolic thickness vs maximum bondline temperature occurring at soak out are illustrated in Fig. 22. These data were factored into our carbon-phenolic thickness requirement data bank for air, and a design correlation for use in heated methane generated for this study.

Carbon phenolic, although attractive as a heat protection material for short periods of intense heating, is not a very weight-efficient insulator. This can be confirmed from a review of the thermal properties in Table 5 and a comparison of the carbon phenolic and low-density ablator weights summarized for the 8 km/s entry of SO2P in Fig. 23. An experimental development material, PD 218,[47] is a low-density carbon foam with excellent thermal properties and adequate structural properties for this application. Thermal response calculations were performed on a carbon-phenolic/PD 218 composite. The carbon-phenolic outer layer is employed to help suppress the surface temperature response and minimize the char mass loss. The PD 218 is employed as a lightweight insulator to meet the bondline temperature requirements. The computations indicate that a combination of carbon phenolic plus PD 218 can reduce the weight by a factor of over three from the all carbon-phenolic design. This concept provides the potential for significant weight reduction for the V_E = 13 km/s SO2P Titan entries.

Uranus Thermal Protection Selection. The Uranus aerocapture configuration evaluated is the scaled-down version

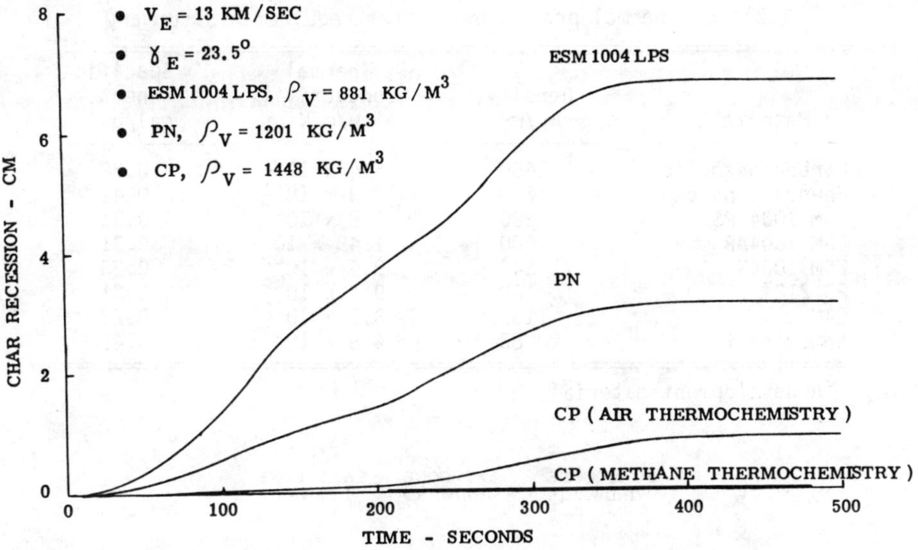

Fig. 21 Ablator char recession for SO2P entry.

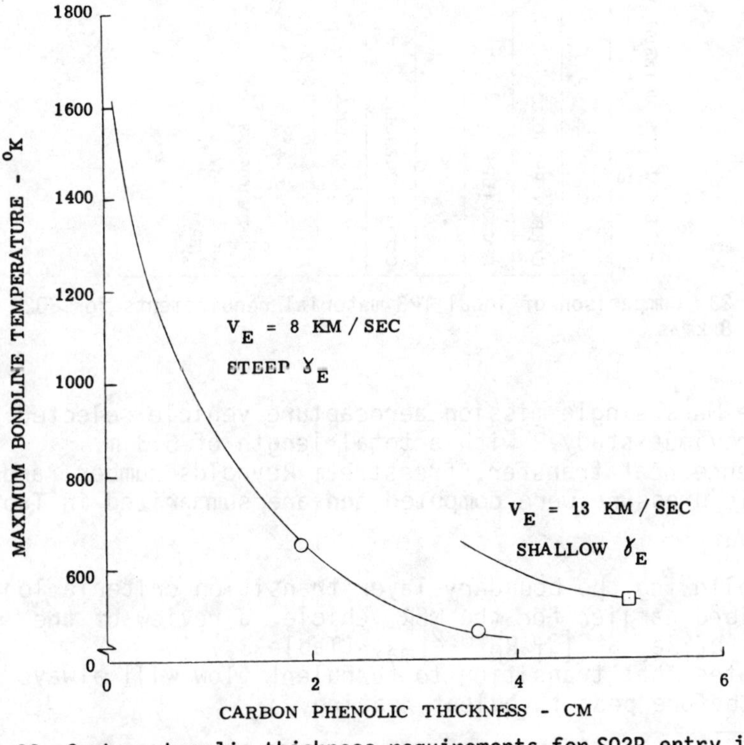

Fig. 22 Carbon-phenolic thickness requirements for SO2P entry into a methane atmosphere.

Table 5 Thermal protection system requirements summary

Material	Density, kg/m^3	Thermal conductivity, W/m·K	Specific heat, cal/g·K
Carbon phenolic	1450	6.4×10^{-1}	0.28
Phenolic nylon	1200	2.3×10^{-1}	0.41
ESM 1004LPS	880	1.8×10^{-1}	0.31
ESM 1004AP	580	1.48×10^{-1}	0.31
ESM 1004X	240	9.6×10^{-2}	0.36
PD 218[a]	160	9.3×10^{-2}	0.21
SAM	130	3.1×10^{-2}	0.23
Nomex felt	83	4.8×10^{-2}	0.49

[a]A development material.

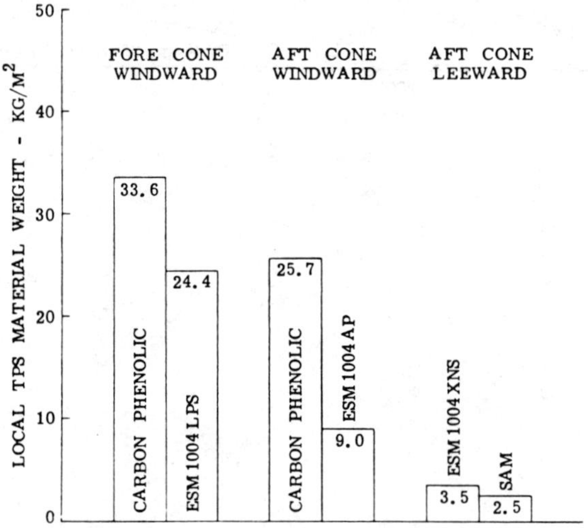

Fig. 23 Comparison of local TPS material requirements for SO2P entry, V_E = 8 km/s.

of the Mars single mission aerocapture vehicle selected in the previous study,[6] with a total length of 6.3 m. Reference heat transfer, freestream Reynolds number, and dynamic pressure were computed and are summarized in Table 3.

Following the boundary-layer transition criteria logic described earlier for the MSR vehicle, a review of the values of Re_∞ at $[\dot{q}_T R_N^{0.2}]_{max}$ (Table 3,) indicates that transition to turbulent flow will always occur before peak turbulent heating.

Convective heat-transfer rate time histories for the stagnation point and a representative forward cone windward location where turbulent flow predominates have been generated for the 12 deg entry path angle case for the nominal atmosphere (Fig. 20).

Radiative heat transfer to the stagnation point area in the 89% H_2-11% He mixture has been estimated employing an engineering correlation of Galileo results described previously. The convective heat-transfer rate dominates the TPS response (Fig. 20). The stagnation point convective heat-transfer rate peaks at 3700 Btu/ft^2·s (4200 W/cm^2), the adiabatic nonblowing radiative heat-transfer rate at the stagnation point peaks at about 780 Btu/ft^2·s (880 W/cm^2), the turbulent heat-transfer rate at X/R_N = 12.5 peaks at 2400 Btu/ft^2·s (2700 W/cm^2) with the radiative decreasing to about 20% of the stagnation point values at that station. Based on these comparisons, the radiative heating has been considered negligible for this study, although in subsequent design studies it must be included.

With peak heat-transfer rates over most of the vehicle being high enough that carbon mass loss will be significant, the TPS evaluated was all carbon phenolic over all sides of the vehicle. The required nominal ablator thicknesses are summarized in Table 4. Some weight reduction might be possible by use of a carbon-phenolic/PD 218 composite, where the carbon-phenolic thickness is sufficient to cover the expected char mass loss with margin and the PD 218 is used to provide a lighter weight insulation layer.

Control Surface Heat Protection. Trailing windward flaps are being considered as one of the control options for trimming the c.g. offset effects. The basic vehicle configuration is modified to accommodate the flaps by addition of a body slice on the aft frustum. Typically, these flaps introduce significant flowfield perturbations and experience heating rates well above the local approach conditions. The flaps are expected to be deflected at about 5 deg with a differential deflection from that point of ± 5 deg. This creates an effective total impact angle to the flow of 27-37 deg. The forecone has an effective impact angle of 40 deg. For the purposes of this study, the heat protection requirements for the flaps is assumed to be equivalent to the requirement on the forecone. For SO2P, if the heating to the flaps is a factor of

Table 6 Generic aerocapture mass comparisons

Aerothermodynamic
Gasdynamics
Thermodynamic and transport properties
CH_4, N_2
Turbulent heat-transfer validation
Gas composition
Configuration
Radiative heat-transfer evaluation
CH_4, N_2
Configuration
Leeside and base heat transfer
Gas composition
Configuration
Aerothermochemistry
Surface gas phase
Carbon, silica
CH_4, N_2
Thermal penetration
Carbon phenolic, ESM
H_2-He, CH_4, N_2
Thermal protection materials
Evaluation of contemporary materials and reuse capability
CH_4, N_2, H_2-He
Development of new composites
Evaluation of new composites
Evaluation of ablator gap shape change at α
Aeromechanic
Configuration refinement
Nose bend and aft slice trades
Accommodate flaps
Reduce $\dot{q}$ to forecone
Flap geometry and sizing
Minimize $\dot{q}$
Enhancement of yaw stability
Six degree-of-freedom aerodynamic characteristics
Force and moment coefficients and derivatives
Asymmetric vehicle effects
Ablative mass addition phase lag effects
Flap shape change
Skewed laminar/turbulent transition front effect
Ablator roughness effects
Basic body
Flap
Gas composition effects
Six degree-of-freedom trajectory and control simulations

two higher, the flap thermal protection requirements will increase by 30%.

Technology Needs

Based on the results of this study, it is strongly recommended that existing aerothermodynamic design technology be expanded to meet generic aerocapture needs. Emphasis should be placed on thermal protection systems and associated vehicle interactions.

A summary listing is presented in Table 6 of the primary technology needs identified. These include the gasdynamics, aerothermochemistry, and thermal protection material needs summarized under the aerothermodynamics category and the vehicle configuration refinement, six degree-of-freedom aerodynamic characteristics, and six degree-of-freedom trajectory and control simulation needs summarized under the aeromechanics category.

Implementation of a program addressing these needs will insure a high degree of generic aerocapture technology readiness.

Summary

Potential maneuvering vehicle basic body configuration candidates have been screened for the generic aerocapture missions considering the packaging constraints and the desired aerodynamic performance characteristics. A single biconic configuration was selected for the SO2P and Uranus aerocapture, and MSR aeromaneuver vehicle applications with a conical frustum added to perform the MSR aerocapture application (Fig. 1). These were selected based on meeting the aerodynamic characteristics requirements, ease of packaging the given payloads in the vehicles and the vehicles in the Space Shuttle payload bay, minimizing the nose bend required to trim the vehicle, minimizing the vehicle surface area where possible, and providing some degree of passive longitudinal and directional static stability control.

Dominant flow regimes for the aerocapture vehicles is continuum flow with turbulent flow present over most of the vehicle at pull up. Aerodynamic characteristics considered the inviscid, skin friction, and induced pressure effects, although only the inviscid effects are of first order and are summarized in Fig. 2 and Table 2.

Thermal protection materials have been selected and sized for a generic aerocapture vehicle for entry at Mars, Titan, and Uranus. Hypersonic convective and radiative heat transfer to these vehicles was estimated based on a combination of three-dimensional inviscid and viscous flowfield solutions generated for the aerocapture vehicle and engineering correlations developed from the numerous complete convective and radiative heat-transfer evaluations conducted in CO_2 and H_2-He mixtures for the Pioneer Venus and Galileo Programs.

A summary is presented in Table 3 of the peak and time integrated convective and radiative heat-transfer rates, the peak dynamic pressure, the freestream Reynolds number at peak turbulent heating, and the heating times. It was concluded that generally transition to turbulent flow is expected to occur prior to peak heating except for the shallow path angle entries into Titan.

The thermal protection material selection and sizing is driven by the peak and time-integrated heat-transfer rates experienced and the heating times summarized in Fig. 16. It is seen that appropriate contemporary materials are available ranging from low-density ablators for MSR and low-speed Titan entry to high-performance ablators for Uranus and high-speed Titan entry. The long heating times associated with use of the high-performance ablators, however, fall outside the demonstrated capability of current entry vehicle applications and thus need to be experimentally demonstrated. Thermochemical response of graphitic heatshields in CO_2 and H_2-He mixtures was available from the previous studies; response in a CH_4 mixture was not available and thus was generated during this study.

Sizing of the thermal protection system was accomplished by employing design correlations derived from the results of GE REKAP. Comparisons between the thermal protection material weight required of representative windward meridian locations are presented in Fig. 24. For the low-speed Mars and Titan entries, low-density, near shape-stable ablators offer attractive solutions. Low-density ablators were selected over a metallic reradiative type or an RSI approach due to the limited reuse required, simplicity of design, the fabrication and applications data base existing, and cost. For the 13 km/s Titan entry in CH_4, carbon phenolic is predicted to be shape-stable since there is no oxidation experienced and the peak surface temperature expected is well below the sublimation

Table 7 Generic aerocapture technology needs

Planet	V_E, km/s	Launch capability, kg	Payload, kg	Allowable aerocapture system mass, kg	Estimated aerocapture system mass, kg	TPS mass, kg
Mars	5.36	5000	4000	1000	600	247
						ESM CP
Titan	8	2260 (SEGA)	1500	1160	570	214 463
	13	2400	1500	900	1020	743
Uranus	30	2400	1250	1150	1090	795

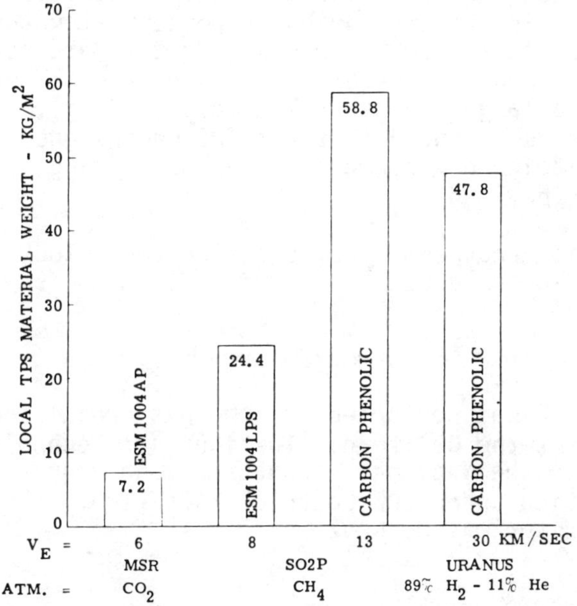

Fig. 24 Comparison of local windward TPS material requirements on the forecone.

temperature. A developmental material, PD 218 (a low-density carbon foam), used as an insulative layer is shown to reduce the local TPS weight requirement by a factor of three. For Uranus entry, a high-performance ablator such as carbon phenolic is required and experiences some mass loss due to sublimation.

Generic aerocapture configurations and thermal protection system design concepts have been developed to allow further evaluation of the mission design concepts. A set of allowable launch masses for a specific mission concept is presented in Table 7. As indicated the SO2P (at 8 km/s), MSR, and Uranus missions show positive mass margins. The SO2P (at 13 km/s) has a negative margin of 120 kg with the current payload. Further evaluations are required to trade the increased entry velocity with the required payload capability. In addition, utilization of new heat-shield materials, such as PD 218, could significantly decrease the aeroshell mass.

In general, it is concluded that a generic aerocapture configuration can be utilized for the missions evaluated. The variation in dynamic pressure over the range of missions does not require a large change in structural mass. As a result, identical aluminum honeycomb structures appear feasible for SO2P, Uranus, and the Mars aeromaneuver missions. However, the range of peak heating does cause a large change in TPS mass (670 kg). Therefore, it is recommended that the TPS be designed for the specific mission environments, thus maximizing the deliverable payload.

Specific aerothermodynamic design technology needs are identified in Table 6.

Acknowledgment

This work was performed for the Jet Propulsion Laboratory of the California Institute of Technology, sponsored by the National Aeronautics and Space Administration under JPL Contract NAS 7-100. The JPL Project Monitor was M. Cruz.

References

[1]Chapman, D.R., "On the Corridor and Associated Trajectory Accuracy for Entry of Manned Spacecraft into Planetary Atmospheres," Xth International Astronautical Congress, edited by F. Hecht, Springer-Verlag, Vienna, 1960, pp. 254-267.

[2]Repic, E.M., Bovlar, M.G. and Chapel, F.G., "Aerobraking as a Potential Planetary Capture Mode," Journal of Spacecraft and Rockets, Vol. 5, Aug. 1968, pp. 921-926.

[3]Cruz, M.I., "The Aerocapture Vehicle Mission Design Concept," AIAA Paper 79-0893, May 1979.

[4]Cruz, M.I., Armento, R.F., and Giles, W.H., "Aerocapture - A System Design for Planetary Exploration," Paper 79-160 presented at the XXXth IAE Congress, Munich, FRG, Sept. 1979.

[5]Hassett, R.J., "Design Integration of Aerocapture and Aeromaneuver Vehicles for a Mars Sample Return Mission," AIAA Paper 80-0728, May 1980.

[6]Florence, D.E., "Aerothermodynamic Design Feasibility of a Mars Aerocapture/Aeromaneuver Vehicle," AIAA Paper 81-0350, Jan. 1981.

[7]"Generic Aerocapture Atmosphere Entry Study Final Report," GE Rept. 80SDR2226, Oct. 24, 1980.

[8]Daywitt, J., Brant, D., and Bosworth, F., "Computational Technique for Three Dimensional Inviscid Flow Fields About Re-entry Vehicles," SAMSO-TR-79-5, April 1978.

[9]Hecht, A.M. and Nestler, D.E., "A Three Dimensional Boundary Layer Computer Program for Sphere-Cone Type Re-entry Vehicles," AFFDL-TR-78-67, July 1978.

[10]Gentry, A.E., Smith, D.N., and Oliver, W.R., "The Mark IV Supersonic-Hypersonic Abirtrary--Body Program," Vols. I, II, and III, AFFDL-TR-73-159, Nov. 1973.

[11]Gilbert, L.M. and Goldberg, L., "A Reynolds Number Scaling Theory for Hypersonic Ablation," AIAA Paper 67-155, 1969.

[12]Brewer, R.A., "A Comparison of Experimental Data and Theoretical Predictions for Laminar Stagnation Point Convective Heat Transfer in the Martian Entry Environment," GE-TDM-8151-107.

[13]Moss, J.N., Anderson, E.C., and Beiz, C.W., Jr., "Viscous-Shock Layer Solutions with Radiation and Ablation Injection for Jovian Entry," AIAA Paper 75-671, May 1975.

[14]Haviland, R., Handbook of Satellites and Space Vehicles, D. Van Nostrand Co., Inc. New York, 1965, pp. 354-374.

[15]Moss, J.N., Simmons, A.L., and Anderson, E.C., "The Impact of Turbulence on a Radiating Shock Layer with Coupled Ablation Injection," AIAA Paper 78-1186, July 1978

[16]Moss, J.N., "A Study of the Aerothermal Entry Environment for the Galileo Probe, "AIAA Paper 79-1081, July 1979.

[17]Moss, J.N., "Flow Field Analysis for Future Planetary Entry," AIAA Paper 79-0918, May 1979.

[18]Nestler, D.E., Personal communication, June 1980.

[19]Feldhuhn, R.H. and Pasiuk, L., "An Experimental Investigation of the Aerodynamic Characteristics of Slender Hypersonic Vehicles at High Angles of Attack," NOL Tech. Rept. 68-52, May 1968.

[20]Moyer, R.E., "Skybolt Heat Transfer Data Analysis Angle of Attack," GE MSD TDM-8151-004, Nov. 1962.

[21]Widhopf, G.F., "Turbulent Heat Transfer Measurements on a Blunt Cone at Angle of Attack," Aerospace Rept. TR-0059 (S6816-66)-1, Feb. 22, 1971.

[22]Widhopf, G.F., "Transitional and Turbulent Heat Transfer Measurements on a Yawed Blunt Conical Nosetip," AIAA Paper 72-212, 1972.

[23]Berkowitz, A.M., Kyriss, C.L., and Martellucci, A., "Boundary Layer Transition Flight Test Observations," AIAA Paper 77-125, Jan. 1977.

[24]Sutton, K., "Fully Coupled Nongray Radiating Gas Flows with Ablation Product Effects About Planetary Entry Bodies," AIAA Paper 73-672, July 1973.

[25]Falanga, R.A. and Olstad, W.B., "An Approximate Inviscid Radiating Flow Field Analysis for Sphere-Cone Venusian Entry Vehicles," AIAA Paper 74-75, July 1974.

[26]Brewer, R.A., Brant, D.N., and Fogaroli, R.P., "Development of a Steady-State Shape Change Ablation Code for the Design of Outer Planet Probes," AIAA Paper 77-95, Jan. 1977.

[27]Brewer, R.A. and Brant, D.N., "Thermal Protection System for the Galileo Mission Atmospheric Entry Probe," AIAA Paper 80-0358, Jan. 1980.

[28]Brewer, R.A., "Selection of a Heat Protection System for Venusian Entry," AIAA Paper 75-731, May 1975.

[29]Page, W.A. and Woodward, H.T., "Radiative and Convective Heating During Venus Entry," AIAA Journal, Vol. 10, Oct. 1972, pp. 1379-1381.

[30]Page, W.A., Compton, D.L., Borucki, W.J., Ciffone, D.L., and Cooper, D.M., "Radiative Transport in Inviscid Nonadiabatic Stagnation Region Shock Layers," AIAA Paper 68-784, 1968.

[31]Florence, D.A., Hecht, A.M., and Vogel, E., "Hypersonic Entry Heat Transfer," Pioneer Venus, GE-RSD, 1972.

[32]Wolf, F. and Spiegel, J.M., "Status of Basic Shock-Layer Radiation Information for Inner-Planet Atmospheric Entry," Journal of Spacecraft and Rockets, Vol. 4, Sept. 1967, pp. 1166-1173.

[33]Zoby, E.V., "Empirical Stagnation-Point Heat Transfer Relaxation in Several Gas Mixtures at High Enthalpy Levels," NASA TND-4799, Oct. 1968.

[34]Marvin, J.G. and Deiwert, C.S., "Convective Heat Transfer in Planetary Gases," NASA TR R-224, July 1965.

[35]Nicolet, W.E., Morse, H.L., Neuner, G.J., and Gomez, A.V., "Outer Planet Probe Entry Thermal Protection," Aerotherm Final Rept. 74-92, Feb. 10, 1975.

[36]Nicolet, W.E., Private communication, Nov. 18, 1975.

[37]Stickford, G.H. and Menard, W.A., "Bow Shock Composition and Radiation Intensity Calculations for a Ballistic Entry into the Jovian Atmosphere," AIAA Paper 68-787, 1968.

[38]Moss, J.M., Personal communication, Sept. 1980.

[39]Berman, R.J., "Bow Shock Shape About a Spherical Nose," AIAA Journal, Vol. 3, April 1965, pp. 779-780,

[40]Serbin, H., "Supersonic Flow Around Blunt Bodies," Journal of Aerospace Sciences, Vol. 25, 1958, pp. 58-59.

[41]Li, T.-Y. and Geiger, R.E., "Stagnation Point of a Blunt Body in Hypersonic Flow," Journal of Aerospace Sciences, Vol. 24, 1957, pp. 25-32.

[42]Ridyard, H.W. and Storer, E.M., "Stagnation Point Shock Detachment of Blunt Bodies in Supersonic Flow," Journal of Aerospace Sciences, Vol. 29, 1962, pp. 751-752.

[43]Storer, E.M., "A Simple Method for Calculating the Stagnation Point Shock Detachment of Spherical Bodies in Supersonic Flow," GE-RESD Aerodynamics Fundamentals Memo 64-2, May 6, 1964.

[44]Florence, D.E.,Thibault H., and Hiltz, A., "Selection, Development, Characterization and Flight Test of a Thermal Protection System for an Earth Entry Satellite Vehicle," AIAA Paper 78-861, May 1978.

[45]Kottock, S., "The Development of a Phenolic Nylon Model for a Reaction Kinetics Ablation Program," GE-TIS-65SD320, Sept. 1965.

[46]Cline, P.B. and Schultz, F.L., "Investigation of the Effect of Material Properties on Composite Ablative Material Behavior," NASA CR-72142, April 1967.

[47]Juneau, P.W., Jr., Brazel, J., Metzger, J.W., Fehl, C., and Fenton, R.M., "Ablative Materials for High Heat Loads - Part III - Insulative Layers for Carbon/Phenolic Heatshields," AFML-M-79-95, III, June 1973.

Author Index for Volume 82